Concepts of Development

CONCEPTS OF

DEVELOPMENT

Edited by Jay Lash
University of Pennsylvania

and J. R. Whittaker
*Wistar Institute of Anatomy and Biology
and University of Pennsylvania*

SINAUER ASSOCIATES, INC. · PUBLISHERS
Stamford, Connecticut

CONCEPTS OF DEVELOPMENT
First Edition

Library of Congress Catalog Card Number: 73-94061

ISBN 0-87893-450-2

Contents

Preface

Developmental biology has become so diverse that it is no longer possible for any one author to cover the field in depth or detail. As a result, many university courses in the subject have become rather superficial. We have attempted to solve this problem by inviting a team of experts to explore in depth a number of selected areas of animal development. Because many introductory biology courses now provide a rather sophisticated introduction to developmental biology, we think there is a need for a textbook for upper division courses that begin at a more advanced level. In addition, we feel that our book should prove to be a valuable supplement to other textbooks and lecture courses.

Developmental biologists commonly differ in their interpretations of experimental results. We have intentionally emphasized the spirit of inquiry and debate as well as the research that underlies some of the prevalent concepts of development.

It is difficult to impose a general theme on a multiauthored book; in fact, a completely integrated treatment might actually be contrary to the purposes that we should like this book to serve. The contributors have their own specific themes and concepts, which are based on raw data. We have not attempted to impose our own views, but we have attempted to organize the chapters into a logical framework that should give the reader a broad view of the problems of animal development.

The book is organized into four parts. Part I deals with current concepts about the onset of development: gametogenesis, fertilization and cleavage. Part II emphasizes molecular and cellular phenomena; a discussion of molecular problems is followed by chapters on erythropoiesis, pigmentogenesis and myogenesis. Part III covers cell and tissue interactions, morphogenesis and the development of the immune system. Part IV deals with developmental aspects of genetics, endocrinology and enzymology in the intact organism. This section also includes chapters on regeneration, congenital malformations and aging—topics we believe to be too important to neglect.

Even in a book as up-to-the-minute as this one, certain topics must be omitted because some fields are developing so rapidly that the reader would be better advised to consult the current literature. We hope that the topics we have selected will be a satisfying sample of what is truly an abundance of riches.

Some of the contributions in this book were based on a series of lectures given in an advanced developmental biology course at the Institute for Medical Research in Camden, N.J.; these were revised by the contributors to bring them up to date for inclusion here. Other contributions were solicited from investigators who had not originally participated in the course but whom we felt could add things originally lacking.

We are grateful to Dr. Lewis Coriell, Director of the Institute, for creating the circumstances that eventually led to this book. We should also like to acknowledge the valuable assistance of Natalie Lash and Noel de Terra Whittaker.

JAY LASH
J. R. WHITTAKER
Philadelphia
January, 1974

Contributors

EVERETT ANDERSON, The Laboratory of Human Reproduction and Reproductive Biology, Harvard Medical School, Boston

ROBERT AUERBACH, Department of Zoology, University of Wisconsin, Madison

C. R. AUSTIN, Physiological Laboratory, Cambridge University, Cambridge, England

R. E. BILLINGHAM, Department of Cell Biology, University of Texas Southwestern Medical School, Dallas

PATRICIA A. BUCKLEY, Department of Biology, University of Virginia, Charlottesville

VINCENT J. CRISTOFALO, Wistar Institute of Anatomy and Biology, Philadelphia

C. R. FILBURN, Department of Biology, Yale University, New Haven

ELIZABETH D. HAY, Harvard Medical School, Boston

LAIRD JACKSON, Division of Medical Genetics, Jefferson Medical College of Thomas Jefferson University, Philadelphia

KURT E. JOHNSON, Department of Anatomy, Duke University Medical Center, Durham

IRWIN R. KONIGSBERG, Department of Biology, University of Virginia, Charlottesville

JAY LASH, Department of Anatomy, School of Medicine, University of Pennsylvania, Philadelphia, and Marine Biological Laboratory, Woods Hole

FRANK J. LONGO, Department of Anatomy, University of Tennessee, Knoxville

MARTIN NEMER, The Institute for Cancer Research, Fox Chase Center for Cancer and Medical Sciences, Philadelphia

IVAN T. OLIVER, Department of Biochemistry, University of Western Australia, Nedlands, Western Australia

R. RAPPAPORT, Department of Biological Sciences, Union College, Schenectady, and The Mount Desert Island Biological Laboratory, Salisbury Cove, Maine

RICHARD A. RIFKIND, Department of Medicine and Department of Human Genetics and Development, College of Physicians and Surgeons, Columbia University, New York City

BRIAN S. SPOONER, Division of Biology, Kansas State University, Manhattan

DAVID T. SUZUKI, Department of Zoology, The University of British Columbia, Vancouver, Canada

J. R. WHITTAKER, Wistar Institute of Anatomy and Biology, Philadelphia, and Department of Anatomy, School of Medicine, University of Pennsylvania

CHARLES E. WILDE, JR., School of Dental Medicine, University of Pennsylvania, Philadelphia, and The Mount Desert Island Biological Laboratory, Salisbury Cove, Maine

G. R. WYATT, Department of Biology, Queen's University, Kingston, Ontario, Canada

I

THE ONSET
OF DEVELOPMENT

CHAPTER ONE

Gametogenesis

FRANK J. LONGO AND
EVERETT ANDERSON

Gametogenesis may be defined as all the highly specialized nuclear and cytoplasmic events that lead to the formation of the male (spermatozoon) and female (egg) gametes. In the female organism the process of gametogenesis is commonly referred to as *oogenesis*. Sometimes, however, this term specifically denotes the meiotic events in the female gamete, to the exclusion of concurrent changes in the cytoplasmic components. In the male, the process of gametogenesis is commonly referred to as *spermatogenesis*, although this term, too, has a specific meaning: it is frequently used to denote the process of chromosome reduction or meiosis. Structural changes in the cytoplasm and the nucleus are subsumed under the term *spermiogenesis*—that is, the metamorphosis of an immature sperm cell (spermatid) into a spermatozoon.

There are four major aspects to gametogenesis: (1) the origin of the primordial germ cells, (2) the proliferation of the primordial germ cells by mitosis, (3) the reduction of the chromosomal number by meiosis, and (4) those events responsible for chemical and morphological changes in the cytoplasm and nucleoplasm which lead to the development of mature (fertilizable) gametes. As demonstrated by early cytologists, these events are highly variable from one group of animals to another; however, the end result is invariably the same —the production of specialized cells for fertilization. Prior to the turn of the century, biologists recognized that gametogenesis is in fact the prelude to fertilization and development; that is, embryogenesis begins in gametogenesis. Consequently, for a clear understanding of the features that enable the egg and sperm to unite so as to produce a new individual, a firm knowledge is needed of the origin and the formation of the gametes.

Although there is a considerable body of descriptive and experimental information concerning the aforementioned processes, we will only present the salient features as they assist in our understanding of the phenomenon of gametogenesis. Moreover, it is difficult, if not impossible, to mention every aspect of major

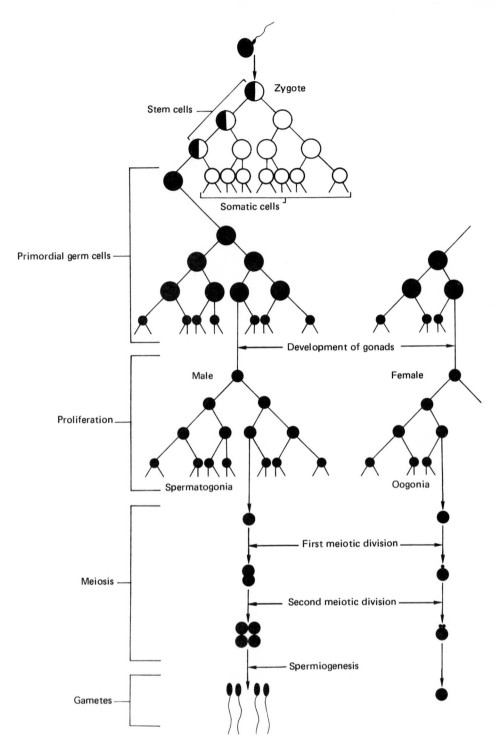

FIGURE I. *Origin and fate of the male (left) and the female (right) gametes. The number of mitotic divisions is larger than depicted. (Adapted from E. B. Wilson, "The Cell in Development and Heredity," Macmillan, New York, 1925.)*

and minor interest in such a brief discussion. What follows then illustrates the general nature of this information and some of the problems which have captured the interest of early and contemporary biologists.

Origin of the Germ Cells

According to Wilson (1925), the gametes of animals are originally derived from stem cells which are capable of producing by mitosis both germ cells and somatic cells (Fig. 1). From the stem cells arise the primordial germ cells which after a certain number of divisions may enter a quiescent period. They are then converted into gonia; cells which aggregate to form a portion of the gonads and give rise to the mature gametes. The primordial germ cells initially have the same general characteristics in both sexes, but later assume structural features of the male or female gamete.

The germ cell lineage in a wide variety of animals is traceable to pregastrula stages and in some organisms as early as the first few cleavages of the embryo. This is particularly true of invertebrates such as insects, crustaceans, nematodes and chaetognaths (Wilson, 1925). The identification of the primordial germ cells is based on certain distinguishing features of the nucleus and/or cytoplasm. For example, in the chicken the primordial germ cells, which may be found between the ectoderm and the endoderm, are recognized by their large size and prominent nuclei (Willier, 1937). In some cases—for example, in the copepod *Cyclops*—the cytoplasm of the primordial germ cell is marked by distinctive pigment granules (see Mahowald, 1971 for other examples).

One of the more outstanding investigations of the origin of primordial germ cells is that by Boveri (1899), who performed his experiments with the embryos of the nematode *Ascaris megalocephala*. In *Ascaris* the germ line may be traced to a stem cell that is distinguishable in the two-cell embryo from the second cell, the somatic cell. Boveri, investigating centrifuged and polyspermic eggs, was able to establish that the origin of the germ cells in *Ascaris* is induced by cytoplasmic-derived substances. Investigators examining the origin of germ cells in other organisms have also been able to demonstrate the basis for this dichotomy: that is, the differentiation of somatic and germ cell lines depends on the presence of cellular components (germ cell determinants) in one cell, or group of cells, and not in the other. Thus far, the causative agent (or agents) and its mode of action have not been identified. Nevertheless, the differentiation of cells that form the germ line represents one of biology's more dramatic examples of cytoplasmic localization of a substance responsible for cellular morphogenesis.

In many vertebrates the primordial germ cells migrate to a specific site within the embryo, where in conjunction with other cells they form the gonad. In these cases the origin of the primordial germ cells is at some distance from the site of the definitive gonad—either they come from another region of the embryo, or they may be extraembryonic.

The Chromosomal Events of Gametogenesis: Mitosis and Meiosis

MITOSIS

In many animals a definitive gonad with nests of germinal cells is established during the larval stage. From this period to sexual maturity—that is, the time at which the gametes are made ready for fertilization and then shed—much time may elapse. During this period, and in many cases at specific intervals during the organism's life cycle, the germ cells replicate by mitosis. This constitutes the proliferative phase of gametogenesis (Fig. 1). The replication of germ cells is not unlike that found during the renewal of somatic cells. In most mammals, the replication of female germ cells by mitosis ceases in the fetus prior to birth, whereas in the male the germ cells are able to proliferate mitotically and meiotically following birth. Figure 2 illustrates some of the important features of mitosis.

In addition to a proliferation of germ cells

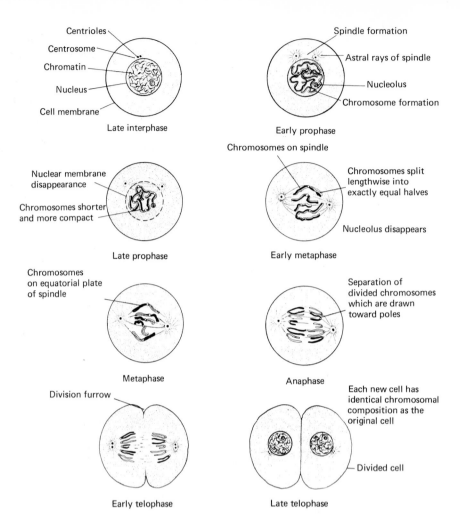

FIGURE 2. *Mitosis. (From R. Rugh and L. B. Shettles, "From Conception to Birth," Harper & Row, New York, 1971.)*

via mitosis, there may also be a degeneration of the potential gametes at specific stages in the life history of an organism. For example, in mammals, degeneration (*atresia*) of germ cells occurs during the formation of the ovary and at certain stages in the female's reproductive cycle.

MEIOSIS

The particular stage in the life history of the developing gamete in which meiosis takes place is known as the *meiotic phase*. In the male it takes place in the testes during the formation

of the spermatozoa. In the female, however, portions of meiosis may be delayed until after the egg has attained its maximum growth— that is, following ovulation, or even after the sperm has entered the egg during fertilization.

During meiosis the chromosome number characteristic of the species is reduced by one half; that is, the double or diploid chromosome number is reduced to the haploid state, such that each gamete contains one member of each pair of chromosomes. Generally, the symbol n is used to signify the haploid chromosome number; n also symbolizes the number of pairs of homologous chromosomes in diploid cells.

Essentially, meiosis involves only one DNA

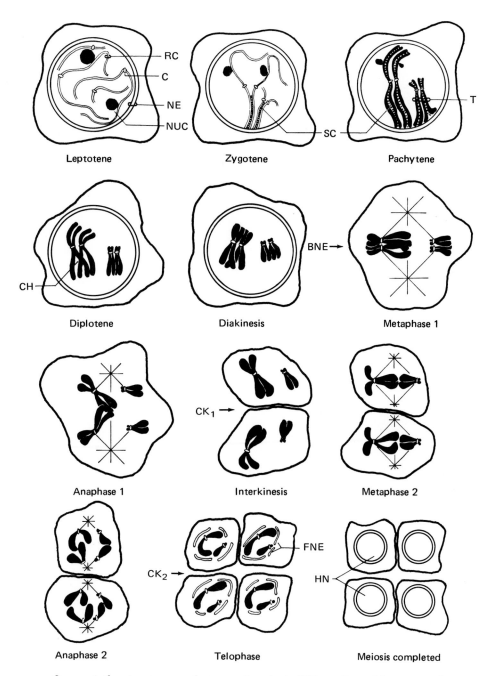

FIGURE 3. *Some of the structures and stages of meiosis. RC, replicated leptotene chromosome; C, centromere; NE, nuclear envelope; NUC, nucleolus; SC, synaptinemal complex; T, tetrad; CH, chiasmata; BNE, breakdown of nuclear envelope; CK_1 and CK_2, cytokinesis; FNE, formation of nuclear envelope; HN, haploid nuclei.*

synthetic period, which is followed by two mitoses, called the *meiotic* or *maturation divisions*. During the meiotic divisions, the sorting of the chromosomes into haploid groups is completed. The prophase of the first meiotic division is characteristically prolonged and is distinguished by five stages based on the morphogenesis of the chromosomes. The first meiotic prophase is customarily divided into *leptotene, zygotene, pachytene, diplotene,* and *diakinesis*. Figure 3 illustrates some of the features of these stages. Leptotene is characterized by long thread-like chromosomes that are initially distributed at random throughout the nucleus. However, it is important to note that although the chromosomes may appear as single structures, they have in fact undergone DNA replication and actually consist of two unresolved chromatids.

Zygotene is that phase of prophase in which the homologous pairs of chromosomes become intimately apposed to one another, a condition known as *synapsis*. Synapsis involves the lateral association of genetically homologous regions of the chromosomes that usually start at a sin-gle region and continue along the rest of the chromosome. The resultant paired structure is referred to as a *bivalent*. During this stage the *synaptinemal complex* is first observed (Figs. 3 and 4). The synaptinemal complex is a tripartite structure located within the region where the homologous chromosomes are juxtaposed (Moses, 1968). It consists of one central and two lateral elements or structures that are composed primarily of protein. Comings and Okada (1971) indicate that the lateral elements consist of two individual structures, presumably one from each chromatid. Protein fibers emerge as a series of loops from the lateral elements and fuse to form the central structure, which is devoid of DNA. The synaptinemal complex apparently pulls the homologous chromosomes into tight apposition; this is accomplished by the sequential synthesis of the lateral loops, which fuse to form the central element (Comings and Okada, 1971). Pairing of the homologues is followed by molecular pairing, which involves the recognition and complementary pairing of short stretches of chromosomal DNA in the region around the synap-

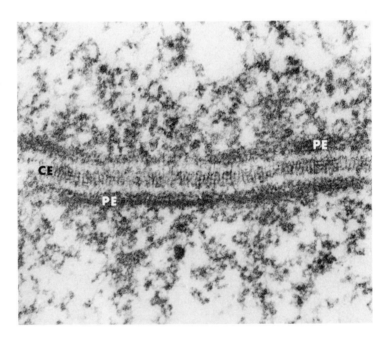

FIGURE 4. *Electron micrograph of a section through a synaptinemal complex of a* Bombyx *spermatocyte. PE, peripheral elements; CE, central element.* × *82,500. (From R. C. King and H. Akai,* J. Morphol. 134, *181, 1971.)*

tinemal complex. This may then lead to chiasmata formation, or what is commonly recognized as genetic crossing-over. It is generally held that the synaptinemal complex must be present for genetic recombination but that the latter does not necessarily follow synaptinemal complex formation (Moses, 1968).

Pachytene follows the synapsis of the homologous chromosomes: during this stage the bivalents become thicker, apparently a result of coiling of the chromosomes. During the subsequent period (diplotene) chiasmata become evident (it is assumed, however, that they are formed during pachytene). A *chiasma* results from a physical exchange between two of the four chromatids—that is, one chromatid of each homologous pair breaks and the ends subsequently rejoin each other in a reciprocal fashion. Chiasmata are regarded as physical counterparts of genetic crossing-over; however, rigorous proof of this is lacking and there is not a one-to-one correspondence between the two.

Stern and Hotta (1967) have examined the biochemical events of meiotic prophase in the plants *Lilium* and *Trillium* and have shown that during this period DNA synthesis occurs to the extent of 0.3 to 0.4 percent of the total DNA complement. It has been shown that this DNA is not special to meiosis but is present also in somatic cells. The relation of this synthesis to the chromosome is unknown. Inhibition of DNA synthesis by adenosinedeoxyriboside during early zygotene results in the absence of synaptinemal complexes (Roth and Ito, 1967). The application of adenosinedeoxyriboside during late zygotene or pachytene, when the synaptinemal complexes are present, has no effect on these structures. Therefore, some of the DNA synthesis observed during prophase may be involved with chromosomal pairing. The relation of the synthesis of DNA that occurs during pachytene to crossing-over is not clearly understood, and investigators have found no differences between the pattern of synthesis in chiasmata and achiasmata cells (Stern and Hotta, 1969).

At diplotene the bivalents separate, except at those regions where they are associated by chiasmata. During this period the chromatids may become visible. Each bivalent consists of four chromatids—a structure called a *tetrad;*

the number of chromatids is therefore tetraploid or double the diploid number. During diplotene the chromatids of most species uncoil slightly—a phenomenon that, in developing eggs, is correlated with intense RNA synthesis and cell growth. A more detailed consideration of the events associated with the uncoiling of the chromatids is presented in the sections discussing vitellogenesis and lampbrush chromosomes.

At diakinesis the chromatids of the tetrad condense and become thicker and shorter. By this time the chiasmata have moved to the ends of the bivalents; this movement of the chiasmata is referred to as *terminalization*. The nucleolus then disappears, the nuclear envelope is dismantled, and the tetrads become aligned on the metaphase plate of the first meiotic spindle (metaphase I).

At anaphase I, the tetrads split lengthwise, whereas the centromeres remain intact (this differs significantly from the anaphase of mitosis, where chromosome separation also involves division of the centromere). The chromatids of one homologue thus separate from the other. The homologue that consists of two chromatids constitutes a *dyad;* it should be borne in mind that if recombination occurred, the chromatids constituting a dyad might not be genetically equivalent. During or following this operation cytokinesis occurs, so that two cells are formed, each receiving one group of dyads representing the (*n*) number of pairs of homologous chromosomes characteristic of the species.

The period following the first meiotic division and preceding the second is called *interkinesis*. Generally, it is rather short in duration and quite variable. Subsequent to this period the dyad chromosomes become reorganized on a second meiotic spindle in preparation for the second meiotic division (metaphase II). Anaphase II is similar to anaphase I; however, the centromere joining the chromatids that comprise each dyad divides. Consequently, the chromatids of the dyads are separated, and each moves to its respective pole of the meiotic spindle. Telophase ensues, following the migration of the chromatids to the poles of the spindle. Prior to or during this period the cell divides, so that each daughter cell contains the haploid number of chromosomes. Meiosis of one spermatogonium results in the formation of four spermatozoa, whereas an oogonium only gives rise to one gamete and to two or three smaller cells—the polar bodies (Fig. 1).

Diplotene is, in almost all cases, the longest period of oogenesis and the stage during which the deposition and accumulation of yolk begins. It and, to some extent, pachytene serve as stages for the storage of oocytes during the long pause that intervenes between the continuation of meiotic prophase and sexual maturity (Davidson, 1968). In many animals, the diplotene chromatids may have an unusual configuration that is reminiscent of lamp brushes used to clean the soot from chimneys of oil-burning lamps; these are therefore referred to as *lampbrush chromosomes*. The study of lampbrush chromosomes has contributed greatly to our understanding of chromosome structure and function, and since they play such a vital role during vitellogenesis, they will be discussed in some detail in connection with the development of the female gamete.

POLAR BODY FORMATION AND FUNCTION

The function of the polar bodies was recognized by Weismann (1887, cited in Voeller, 1968), who correctly inferred that there must be a diminution of hereditary determinants in the egg before the association of the male and female pronuclei at fertilization, and that this must be brought about by a reduction in the number of chromosomes. Polar body formation consists of two stages: (1) the protrusion of a cytoplasmic mass from one pole of the egg or zygote, and (2) the subsequent development of a cleavage furrow along the base of this protrusion (Fig. 5). This series of events occurs subsequent to anaphases I and II, prior to or following insemination, depending upon the egg in question (Longo and Anderson, 1969b, 1970). The result of this operation is the formation of a small mass of cytoplasm containing maternally derived chromatin, the polar body. In many animals the site of polar

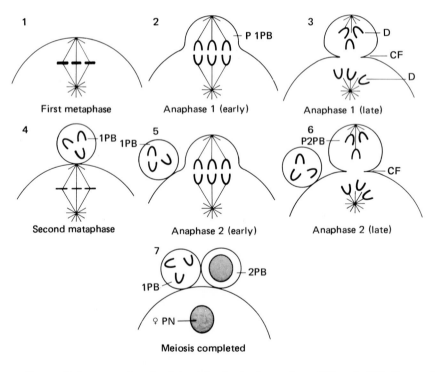

FIGURE 5. *Some of the events involved in polar body formation. 1PB and 2PB, first and second polar bodies; P1PB and P2PB, presumptive first and second polar bodies; ♀ PN, female pronucleus; CF, cleavage furrow; D, dyads.*

body formation marks the animal pole of the egg or zygote. Structurally, the polar body is much smaller than the egg or zygote from which it is derived, but it contains many of the same organelles. The chromatin that remains in the egg or zygote following polar body formation—that is, subsequent to the completion of the meiotic divisions of the maternal chromatin—becomes organized into a nucleus referred to as the *female pronucleus*. This structure contains half the diploid number of chromosomes (one genomic set) and is therefore haploid.

Development of the Female Gamete

ACCESSORY CELLS

Associated with the developing eggs of most organisms are accessory cells. These are frequently responsible for various aspects of gamete differentiation, such as the accumulation of yolk materials and the formation of investing layers found surrounding the eggs of a number of animals (egg coats). There are two types of accessory cells: (1) the nurse cell, and (2) the follicle cell. Nurse cells are joined to the oocyte by intercellular bridges, which are the remnants of the nurse cells' formative divisions produced by incomplete cytokinesis of the primordial germ cell (Fig. 6). The result of incomplete cytokinesis is that each nurse cell is in cytoplasmic continuity with another nurse cell and/or egg. In many animals the nucleus of the nurse cell is polyploid; this increase in genomic content is a reflection of the cell's high metabolic activity (Davidson, 1968).

Follicle cells form a layer surrounding the developing egg that is designated as the follicular epithelium. In most cases, the follicle cells are not associated with the egg by intercellular bridges. Nevertheless, the egg and follicle cells may be in close apposition with little intervening extracellular space. From the surface of both the egg and the follicle cells, cytoplasmic extensions of microvilli may project and interdigitate with one another. The functional role

of these structures has not been elaborated in all cases; however, their association is suggestive of some type of transport phenomenon from the follicle cell to the egg (Anderson and Telfer, 1969; Huebner and Anderson, 1970).

The accessory cell–egg relationship is well developed in insects. Generally, there are two modes of associations, exemplified by (1) the panoistic ovariole, and (2) the meroistic ovariole (Fig. 7). In the panoistic ovariole, which is found in such insects as the cockroach *Periplaneta*, true nurse cells are absent, and the egg is surrounded by a layer of follicle cells. In the meroistic ovariole (in which follicle cells are also present), the egg is connected to nurse cells by intercellular bridges. Based on the manner in which the nurse cells are associated with the egg via intercellular bridges, the meroistic type of ovariole may be subdivided into the polytrophic and telotrophic types. In the polytrophic ovariole the morphological relation of the egg and the nurse cells is relatively simplistic, with individual groups of nurse cells associated with each oocyte. The nurse cells are near to the egg; hence the cytoplasmic bridges connecting the various cells with one another are relatively short (Fig. 7). This system, then, is actually a syncytium, where one nucleus and surrounding cytoplasm differentiate into what will become the egg, while the remaining nuclei and cytoplasm take on the morphological characteristics and functions of nurse cells. How this dichotomy is established has not been determined; however, in most cases, morphological structures are probably involved. For example, in the fruit fly *Drosophila*, the oocyte and 15 associated nurse cells are the mitotic descendants of a single primordial germ cell. Three-dimensional reconstruction made from low-power electron micrographs show that the egg–nurse cell system (germarium) form a complex branching chain of cells (Koch et al., 1967). To generate such a pattern, the cleavage of the primordial germ cell and of its descendants through four mitotic divisions must be oriented in a specific manner. Theoretically, the nuclei of these 16 cells should be equivalent; yet two specific cells undergo a pattern of morphogenesis different from that of the remaining 14. The two cells enter into meiotic prophase and initially have synaptinemal complexes. These two cells are invariably those that possess four intercellular bridges. Later the synaptinemal complexes in one of the two cells disappear, whereas they persist in the other cell, which differentiates into an egg. These facts suggest that the cells that form the egg–nurse cell system in *Drosophila* are programmed to develop as nurse cells, unless they have four intercellular bridges. It follows, then, that if this system were to develop without a single cell that had four inter-

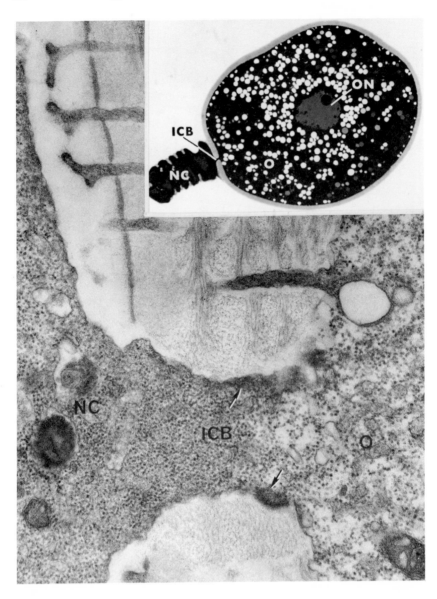

FIGURE 6. *Sections showing an intercellular bridge (ICB) between a nurse cell (NC) and an oocyte (O) of* Diopatra cuprea. *Note the local thickening on the inner aspect of the plasma membrane limiting the intercellular bridge (arrows). ON, oocyte nucleus. Electron micrograph × 45,000; photomicrograph × 300. (From E. Anderson and E. Huebner,* J. Morphol. 126, *163, 1968.)*

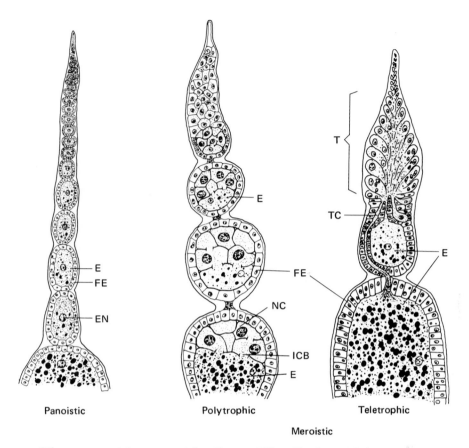

FIGURE 7. *Three types of insect ovarioles. E, egg; FE, follicular epithelium; EN, egg nucleus; NC, nurse cell; ICB, intercellular bridge; T, tropharium; TC, trophic cords.*

cellular bridges, only nurse cells would be produced. The results of investigations by Koch et al. (1967) with the homozygous *fes* (female sterile) mutant of *Drosophila* indicate that whereas 14 interconnected cells may be produced, none possesses four intercellular bridges and thus none differentiates into an oocyte.

Unlike the polytrophic ovariole the telotrophic has nurse cells confined to an anterior region referred to as the *trophic chamber*, posterior to which is the *vitellarium*, that contains the eggs and accompanying follicle cells (Fig. 7). Each egg within the vitellarium is in cytoplasmic continuity via a trophic or nutrient cord with groups of nurse cells. Thus, in this type of ovariole, the eggs and nurse cells are far removed from one another (Huebner and Anderson, 1972a,b,c).

YOLK FORMATION (VITELLOGENESIS)

The nutritive material of the embryo formed during egg development is commonly referred to as yolk or deutoplasm. However, it should not be taken for granted that the yolk of an egg serves no other purpose than that of food supply. Various investigators have proposed that yolk may also play a role in controlling development—either by supplying limiting amounts of essential substrates or by releasing specific inducing substances at specific developmental stages (Flickinger, 1957). The types and amounts of yolk within eggs vary greatly within the animal kingdom. In mammals, for example, yolk is produced in small quantities, whereas in birds it is produced in large amounts. In addition to the variation in the amount of yolk stored, its distribution may vary a great deal in different groups. On the basis of yolk distribution, early investigators classified eggs into (1) homolecithal (yolk evenly distributed throughout egg), (2) telolecithal (yolk localized at one pole of the egg), and (3) centrolecithal (yolk located in the center of the egg).

The exact meaning of the term "yolk" may be unclear if not properly defined. "Yolk" may refer either to reserve materials in general or only to special inclusion bodies present in the egg and embryo. Moreover, investigators have shown that in some organisms (e.g., amphibians) the formed yolk bodies have a different role from such other nutrients as glycogen and lipid reserves (Williams, 1967). The formed bodies comprising the yolk of a wide variety of eggs are composed of protein, lipid, and carbohydrate—the structure and composition of which varies for the organism in question. In addition, there may also be different types of yolk bodies in a given egg (Williams, 1967).

The yolk bodies most frequently investigated are those whose chemical composition is protein–carbohydrate. The morphology of yolk formation varies markedly from species to species, and there is considerable confusion about this process in the early literature. However, recent comparative studies indicate that precursors of protein–carbohydrate yolk may be derived from several sources: from the oocyte itself (autosynthetically), from an extra oocyte source (heterosynthetically), or both (Schjeide et al., 1970; Williams, 1967; Schechtman, 1955). *In general, synthesis both within and without the egg seems to be necessary for yolk production.*

Role of Accessory Cells

It has been recognized that one of the functions of accessory cells is the formation of deutoplasmic materials. How the accessory cells go about this task has been elaborated for only a few groups of animals. The role of nurse cells in the polytrophic ovary has been demonstrated by the labeling experiments of Bier (1963, 1965) and in the telotrophic ovary by Mays (1972). Bier (1963, 1965) determined

that much of the RNA in the egg is produced by the nurse cells and transported to the egg via intercellular bridges. Concomitantly, little or no RNA synthesis is undertaken by the egg during its growth phase. These data indicate that nurse cells provide the growing egg with preformed materials (macromolecules) by way of the intercellular bridges. It is suggested that if macromolecules such as RNA are transported to the egg, other substances necessary for yolk production may be also. In addition to the transport of macromolecules from the nurse cells to the developing egg, the movement of various organelles has been demonstrated (Wilson, 1925; Steinert and Urbani, 1969; Huebner and Anderson, 1970).

In the meroistic ovary, oogenesis is completed very rapidly—for example, 8 to 9 days in the fruit fly. In the panoistic type of ovary, which lacks nurse cells, meiotic prophase is usually longer: in the cricket, for example, it is 3 to 6 months in duration. This temporal difference may be due to the cooperative function of the nurse cells in preparing the egg for fertilization and embryogenesis. Hence the cellular associations observed in the meroistic ovary may enable the organism to dispense with a prolonged meiotic prophase and period of vitellogenesis (Davidson, 1968).

In amphibians, birds, and insects some yolk precursors are synthesized outside the ovary and are transported to the egg via the general circulation. In most cases the yolk precursor (or precursors) then passes through the follicular epithelium to reach the egg for incorporation. The high rate of RNA and protein synthesis in follicle cells suggests that the latter function in ways other than transport, although unequivocal proof of this supposition is lacking for all animals. Autoradiographic studies by Anderson (1971) and Anderson and Telfer (1969) show that a component of the yolk body in the cecropia moth is produced by the follicle cells. The state of permeability of the follicle epithelium has also been shown to be a determinant in the egg's acquisition of yolk precursors in the mosquito *Aedes* (Roth and Porter, 1964; see also Anderson and Spielman, 1971). Several hours after a blood meal the cells of the follicular epithelial layer separate, fully exposing the egg to the hemolymph. The egg is then able to pick up various components for yolk production. Similar results observed in other organisms indicate that the follicle epithelium may selectively regulate the amounts of specific yolk precursors that reach the oocyte. In some cases (e.g., cecropia), the follicle cells may induce the egg to take up yolk substances. In connection with these data, studies of the cecropia moth and mosquito ovarioles show also that the basement lamina lining the distal surface of the follicle epithelium acts much like a mechanical barrier, for it is capable of limiting the size of materials that gain access to the egg (Telfer, 1961, 1965; Anderson, 1964; Anderson and Spielmann, 1971; Roth and Porter, 1964).

In addition to the role of follicle cells in yolk production and in the formation of egg coats (see the following section), Raven (1970) has indicated that six cortical areas of the snail (*Limnaea*) egg are related to the location of the six follicle cells that surround the egg during its development. In this case the follicle cells are thought to function in determining the specific topology of cytoplasmic areas for future deposition of gene products. Results of experiments with mammals and amphibians indicate that follicle cells may also regulate oocyte maturation and supply the egg with metabolites that are needed for its energy requirements (Biggers et al., 1967; Peters et al., 1965; Schuetz, 1967).

Role of Micropinocytosis

As previously indicated, yolk protein may be manufactured outside the ovary (for example, in the liver of birds, mammals, and some amphibians and in the fat body of insects) and carried to the ovary (Telfer, 1960, 1965; Knight and Schechtman, 1954; Flickinger, 1960; Schjeide and and Urist, 1956; Wallace and Dumont, 1968). This material then becomes sequestered within the ooplasm by the process of micropinocytosis. Once within the cytoplasm the material may become organized into yolk bodies.

Deposition of protein yolk in those organisms that obtain precursors heterosynthetically is usually correlated with an increase in micropinocytotic activity, which is recognizable by the accumulation of pits and vesicles along the oolemma (the egg's plasma membrane). The cytoplasmic surface of the pits is distinguished by a bristle coat (Fig. 8), and the resulting vesicles have been referred to as *coated vesicles*. In addition to the layer rich in polysaccharide (glycocalyx), the inner surface of the coated vesicle displays a rather dense material. A number of investigators are of the opinion that this dense material accumulates on the glycocalyx by selective adsorption from the perivitelline space. Anderson and Spielman (1971), using tracers of various molecular weights, have demonstrated ultrastructurally that exogenous proteins and polysaccharides are taken into the egg more rapidly than inorganic materials and that the foreign substances are not separated but are incorporated and concentrated within the same pinocytotic pits. The pits pinch off and become coated vesicles, which carry the absorbed material into the cytoplasm of the egg.

The extent of the contribution made by micropinocytosis and the ultimate fate of the incorporated material is often not easily determined, particularly when the egg may possess several different morphologically distinguishable yolk bodies. Investigators, utilizing such tracers as horseradish peroxidase or ferritin, have been able to follow, with the aid of the electron microscope, the fate of the substances incorporated by micropinocytosis and have determined to some extent the contribution of this process to overall yolk production (Roth and Porter, 1964; Anderson, 1964, 1969; Dumont 1969). In the many organisms studied thus far (e.g., the mosquito, the common white worm *Enchytraeus*, and the cockroach), the micropinocytotic vesicles lose their coat and fuse together to form small yolk bodies, which in turn coalesce to form the large proteid yolk bodies characteristic of the mature egg (Fig. 8). Thus the bounding membrane of the yolk body is derived from the membrane that limited the pinocytotic pit—that is, a modified portion of the oolemma.

The contribution to yolk provided by micropinocytosis in eggs ranges from almost total in such forms as the cockroach, mosquito, moth, and cricket; through moderate in the guppy, the white worm *Enchytraeus*, and amphibia; to apparently negligible in the crayfish, freshwater mussel, and in some tunicates and coelenterates.

Role of the Endoplasmic Reticulum and the Golgi Apparatus

The ooplasm of those forms in which micropinocytosis plays a major role in yolk sequestration is frequently distinguished by the absence of well-developed endoplasmic reticulum and Golgi systems. On the other hand, many eggs, in which micropinocytosis plays a seemingly minor role in yolk production, possess both a well-developed endoplasmic reticulum and Golgi complexes, which appear to be involved in the formation of yolk bodies. In the eggs of the crayfish and lobster the rough endoplasmic reticulum is extensively developed and is involved in the synthesis (autosynthetic) of yolk protein (Beams and Kessel, 1962, 1963). During vitellogenesis, material is produced in the cisternae of the rough endoplasmic reticulum that develops into small granules (Ganion and Kessel, 1972) (Fig. 9). The granules within the cisternae are then apparently transported to connected elements of the smooth endoplasmic reticulum, where they aggregate and are transformed into yolk bodies.

Ultrastructural studies by a number of investigators have shown that a portion of the yolk body in many animals (e.g., the common whiteworm *Enchytraeus*, the pipefish *Syngnathus* and killifish *Fundulus*, the dragonfly *Aeschna*, the horseshoe crab and the hydrozoan jellyfish) is formed in part by the Golgi com-

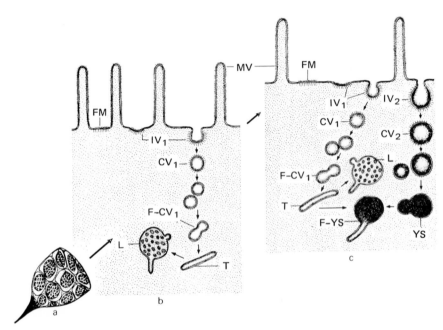

FIGURE 8. *Maturation of oogonia (a), to previtellogenic (b) and vitellogenic oocytes (c). During development the oolemma produces microvilli (MV), areas adorned with a matted substance (FM) and pits or invaginations (IV$_1$ and IV$_2$). In the previtellogenic oocyte (b), only the small coated vesicles (CV$_1$) are produced. These fuse with each other (F-CV$_1$) to form tubular elements (T), which become confluent with lysosomes (L). In vitellogenic oocytes (c), both small (CV$_1$) and large (CV$_2$) coated vesicles are produced. The small ones fuse (F-CV$_1$) to form tubular elements; the larger ones fuse to form densely cored yolk bodies (YS). Some of the tubular elements may fuse with lysosomes (L), whereas others fuse with yolk bodies (F-YS). In addition to its contribution to the developing yolk bodies, the content of a portion of the smaller vesicles may be a source of metabolites during oogenesis. (From E. Anderson, J. Microsc. 8, 721, 1969.)*

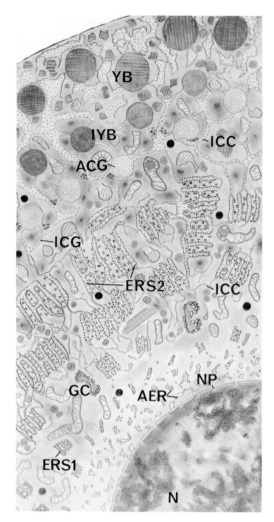

FIGURE 9. *Differentiation and growth in the crayfish oocyte. The stages in the differentiation and activity of the endoplasmic reticulum and its role in the formation of proteinaceous yolk are emphasized. The structures shown are nucleus (N), nuclear pores (NP), Golgi complex (GC), agranular endoplasmic reticulum (AER), forming cisternal stacks of rough-surfaced endoplasmic reticulum (ERS1), differentiated stacks of rough-surfaced endoplasmic reticulum (ERS2), intercommunicating smooth-surfaced cisternae (ICC), intracisternal granules (ICG), aggregates of intracisternal granules (ACG), immature yolk bodies (IYB), and mature proteinaceous yolk bodies (YB). (From H. W. Beams and R. G. Kessel, J. Cell Biol. 18, 621, 1963.)*

plex (Kessel, 1968c; Beams and Kessel, 1969; Anderson, 1968b; Dumont, 1969; Dumont and Anderson, 1967). During vitellogenesis the Golgi complexes enlarge and frequently migrate from a perinuclear position to the cortical region of the egg. A close association develops between the endoplasmic reticulum and the Golgi complex, which is reflected by what appears to be the transfer of material via coated vesicles from the former structure to the latter (Fig. 10). The yolk bodies, which appear in association with the Golgi region, are a collection of small vesicles that gradually acquire a core of dense material (Fig. 11). Exactly how the vesicles increase in diameter has not been elucidated in all cases; however, it is possible that they fuse with each other, thereby giving rise to the large yolk bodies of the egg.

The specific contributions of materials to the yolk body by the endoplasmic reticulum and/or the Golgi complex are not entirely clear in all animals. The transfer of proteinaceous material from the endoplasmic reticulum to the Golgi complex is well documented by Caro and Palade (1964) and Jamieson and Palade (1967a,b) in cytological and biochemical studies of the exocrine pancreas. The possibility should not be excluded that, in the developing egg, substances synthesized by the endoplasmic reticulum are transferred to the Golgi complex for further biochemical modification and/or packaging.

Recent investigations have demonstrated that a function of the Golgi apparatus in various cells is the synthesis of complex carbohydrates (Neutra and Leblond, 1966a,b) and the sequestration and packaging of hydrolytic enzymes (Friend and Farquhar, 1967; Novikoff et al., 1964). Possibly the Golgi complex may be involved in similar functions during vitellogenesis. Although the fact that carbohydrates may be found in yolk bodies cannot be taken as conclusive evidence for their origin in the Golgi complex, their presence coincides with our present knowledge regarding the functions of this organelle. Moreover, it is also known that hydrolytic and other enzymes are associated with the yolk bodies in some eggs (Brachet, 1960; Dalcq, 1963). The function of these enzymes in the utilization of yolk during embryogenesis has not been ascertained. Nevertheless, their presence in the deutoplasm of

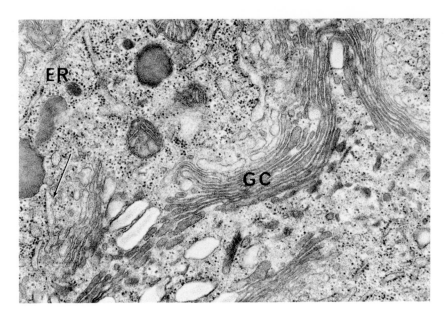

FIGURE 10. *Section through the ooplasm of an oocyte of the amphineuran (Mollusca)* Mopalia mucosa *illustrating the saccules of the Golgi complex (GC) and cisternae of the endoplasmic reticulum (ER) of the rough variety that are involved in vitellogenesis. Note an evagination of the reticulum at the arrow.* × 45,000 *(From E. Anderson,* J. Morphol. 129, *89, 1969.)*

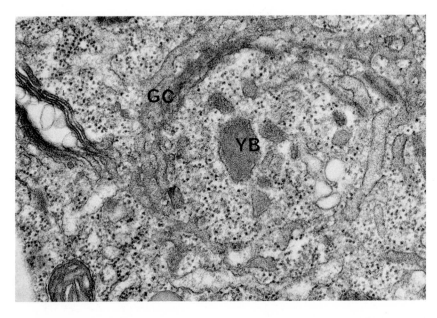

FIGURE 11. *Section of a young oocyte of* Diopatra cuprea *showing "immature" yolk bodies (YB) in association with a Golgi complex (GC).* × 45,000. *(From E. Anderson and E. Huebner,* J. Morphol. 126, *163, 1968.)*

some animals and the role of the Golgi complex in yolk production and in the packaging of hydrolytic enzymes in somatic cells suggest that this organelle may contribute enzymes as well as carbohydrates to the developing yolk body.

Role of Mitochondria

In addition to micropinocytosis, yolk production in amphibians occurs also by the formation of yolk crystals within mitochondria (Ward, 1962; Massover, 1971) (Fig. 12). This is not to imply, however, that intramitochondrial yolk formation is confined exclusively to the anurans, for it has also been observed in other organisms such as the mollusk *Planorbis* (Favard and Carasso, 1958). The yolk crystals

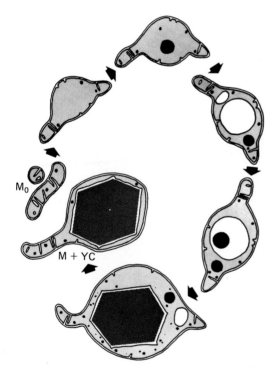

FIGURE 12. *Proposed sequence of organelle modifications and structural additions during the formation of a yolk-crystal inclusion within a mitochondrion (M_0) in bullfrog oocytes. Typical mitochondrion having a yolk-crystal inclusion (M + Yc). (From W. H. Massover, J. Cell Biol. 48, 266, 1971.)*

observed in mitochondria have been widely assumed to be a variant form of the crystals found within the yolk bodies, but whether or not they are of nutrient value to the embryo has not been ascertained.

Massover (1971) has demonstrated the morphological heterogeneity of mitochondria in the developing bullfrog's egg and has shown that a portion of the population of mitochondria undergo a morphogenesis involving the formation of a yolk crystal within their matrix, while the remainder appear unchanged (Fig. 12). In addition to describing intermediate stages of intramitochondrial yolk crystal formation, these studies vividly demonstrate that several morphologically distinct classes of mitochondria exist in a common cytoplasm, thereby suggesting that some portion of the genetic specificity for their function and morphogenesis may reside in the organelle itself.

When vitellogenesis is completed, the yolk bodies of most organisms are architecturally complex; in many cases they are composed of a dense medulla enringed by a more dispersed cortex (Figs. 13 through 15). In sea urchins, particularly *Arbacia* and *Strongylocentrotus*, the yolk is vesicular. Amphibian protein yolk bodies consist of a crystalline structure surrounded by a superficial granular layer. Wartenberg (1964) has shown that the crystalline portion of the frog yolk platelet is derived from material absorbed to the concave surface of forming coated vesicles during micropinocytosis. The substance(s) comprising the superficial layer of the yolk body probably represents "extraneous material" that was part of the fluid surrounding the egg and taken up in coated vesicles along with the yolk precursors. During crystallization of the yolk this material is excluded from the interior of the yolk body.

THE NUCLEUS

One of the distinguishing features of the egg nucleus prior to, and during, meiotic prophase is its relatively large size. During this period the egg nucleus is often referred to as the *germinal vesicle* (Wilson, 1925). Generally the germinal vesicle of young previtellogenic eggs is much smaller than the nucleus of vitellogenic eggs. Moreover, the size of the germinal vesicle is related to the presence of nurse cells —that is, eggs associated with nurse cells usually have a smaller germinal vesicle than those without. In general, nuclear as well as nucleolar activities appear to diminish toward the end of oogenesis.

The bilaminar nuclear envelope of the ger-

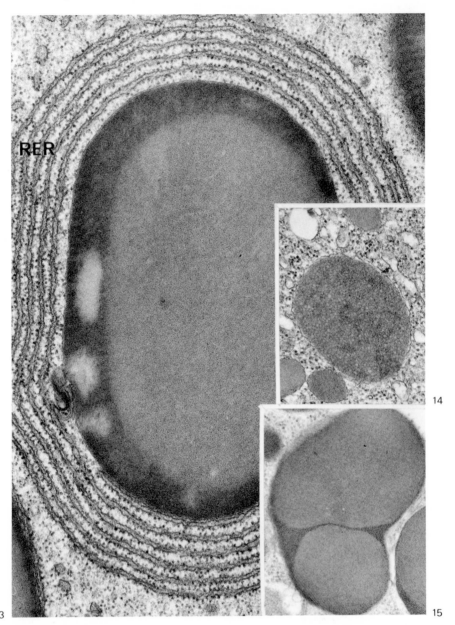

FIGURES 13–15. *Yolk bodies of the oocytes of an amphineuran (Mollusca)* Mopalia *(Fig. 13), the sea urchin* Arbacia *(Fig. 14), and the frog* Xenopus *(Fig. 15). RER, cisternae of rough endoplasmic reticulum circumscribing the yolk body. Fig. 13,* × *45,000 (From E. Anderson, J. Morphol. 129, 89, 1969); Fig. 14,* × *30,000; Fig. 15,* × *36,000 (Fig. 15 courtesy of James N. Dumont.)*

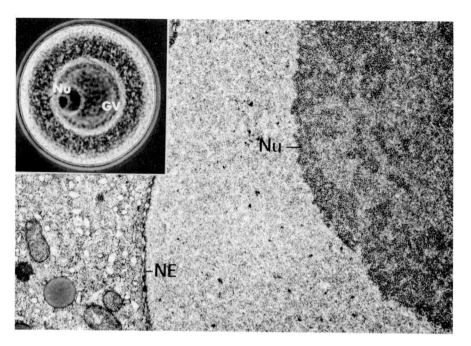

FIGURE 16. *Electron micrograph of a portion of the germinal vesicle and nucleolus (Nu) of a* Spisula *oocyte. NE, nuclear envelope Inset: phase contrast photomicrograph demonstrating the germinal vesicle (GV) and bipartite nucleolus (Nu). Fig. 16 × 20,000; inset, × 660. (From F. J. Longo and E. Anderson,* J. Ultrastruct. Res. *33, 495, 1969.)*

minal vesicle is perforated at intervals by pores or annuli and limits a granulofibrillar nucleoplasm in which are suspended chromosomes and a nucleolus (or nucleoli) (Fig. 16). The germinal vesicle of many organisms such as amphibians, echinoderms, spiders, insects, crayfish, tunicates, and fish is often encompassed by clumps of electron-dense material of a fibrous or granular texture that have projections extending into the interior of the nucleus —through the nuclear envelope via nuclear pores. The aggregates of dense material present in the perinuclear zone of many eggs have been interpreted as nucleoprotein "in passage" from the nucleus (or nucleolus) to the cytoplasm (Stevens, 1967; Anderson and Beams, 1956; Miller, 1962; Franke and Scheer, 1970). Results of cytochemical and radioautographic studies at the ultrastructural level of investigation by Eddy and Ito (1971) *do not support* this view in its entirety. Their results show that although significant amounts of nucleic acids are not present in the dense perinuclear cytoplasmic bodies of amphibian eggs, the lat-

ter do have an appreciable protein content. The function of the dense perinuclear material is unknown.

LAMPBRUSH CHROMOSOMES

Lampbrush chromosomes have a widespread distribution but are not found in the developing eggs of every organism. This general distribution suggests that they play a fundamental role during oogenesis (Davidson, 1968). In the organisms that seem to lack them it appears either that their eggs have a short diplotene period or that the chromosomes remain clumped.

The structure of the lampbrush chromosome is suggestive of intensive and widespread gene activity (Fig. 17): the linear alternating loop-chromosome arrangement may represent the orderly arrangement of genetic sites of the genome (Callan, 1967). The structure of the lampbrush chromosome is not static, for the loops increase and decrease in size apparently in concert with the activity of the egg. Al-

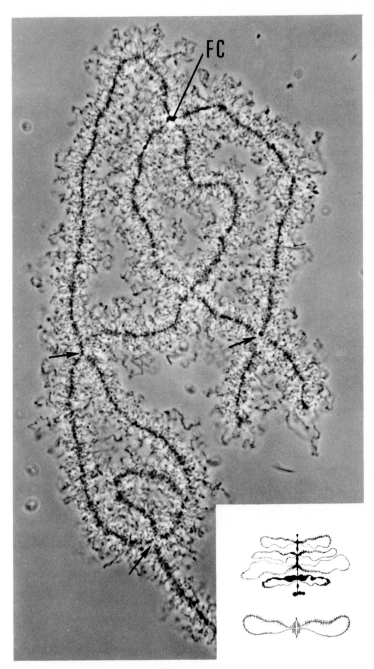

FIGURE 17. *Phase contrast photomicrograph of an isolated lampbrush chromosome from an oocyte nucleus of the newt* Triturus viridescens. *Multiple, paired loops extend from the axes of the homologues, which are loosely held in a tetrad by chiasmata (arrows). FC, fused centromere regions of the paired chromosomes.* × *400. Inset: (top) diagrammatic representation of a small portion of a lampbrush chromosome showing the characteristic paired loops extending laterally from the main axis; (bottom) interpretation of the chromosome structure in terms of two continuous chromatids.* × *520. (Micrographs courtesy of Joseph G. Gall.)*

though the loops have a wide variety of forms, they invariably have one feature in common: one end of the loop which is thin gradually grades into a broader region (Gall, 1963) (Fig. 17). It has been estimated that about 5 percent of the total calculated length of the lampbrush chromosomes is confined to the loop regions (Gall, 1955; Callan, 1963). Electron microscopic observations and DNAase digestion have shown that each loop is equal to one DNA helix, while the main axis contains two (Callan, 1963; Gall, 1963).

Associated with the lampbrush chromosomes of the newt *Triturus* are a total of approximately 20,000 loops; hence there are 5×10^3 loops per haploid set of chromosomes (Callan, 1963). If each loop were equal to one gene, this would mean that there are 5×10^3 genes per haploid set—or seemingly far too few to carry out all the functions required of the cells and tissues comprising the animal (Davidson, 1968). Furthermore, calculations based on the DNA content per loop indicate that the one-loop-equals-one-gene hypothesis is very unlikely, particularly with our current knowledge of gene action. At present there is no reliable evidence as to the number of gene products a loop produces.

In many cases the lampbrush chromosomes contain four times as much DNA as expected. In light of the structure of these chromosomes, the nature of this excess DNA is obscure. In addition to DNA the lampbrush chromosomes have also been shown by staining methods and enzyme digestion to contain RNA and protein. The ratios of RNA to DNA and of protein to DNA in the lampbrush chromosomes are higher than those normally obtained from somatic cells. These facts suggest that there is an accumulation of gene products on the chromosomes.

When the lampbrush chromosomes are exposed to labeled precursors of RNA, the main sites of incorporation are found to be the loops (Gall, 1958; Gall and Callan, 1962). In most loops, the incorporation is uniform; however, in some, the uptake is polarized (Callan, 1963) (Fig. 18). The latter situation occurs when ^{3}H-uridine is injected into the amphibian's coelomic cavity and the uptake of the labeled precursors is followed for a period of days. The loops show the incorporation of ^{3}H-uridine

after 1 day; the radioactivity is limited to the thin region of the loop. During the next 14 days the area of radioactivity gradually extends to the thicker region until the whole loop is labeled. Since the label does not appear at intermediate points, it is doubtful if a wave of synthesis is passing over the loop. Hence it is suggested that there is an actual movement of materials from one side to the other and that RNA synthesis is initiated at the thin region of the loops that show the polarized label (Gall and Callan, 1962). The kinetics of this heterogenous labeling pattern are difficult to interpret, primarily because it is impossible to dilute out

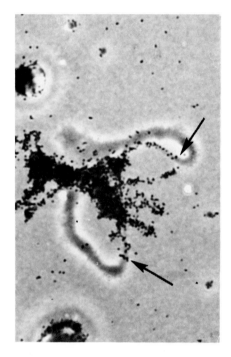

FIGURE 18. *Autoradiograph of a pair of loops of the lampbrush chromosomes of* Triturus viridescens. *Unlike most loops that are uniformly labeled, this is a pair of sequentially labeling loops. The radioactivity extends only a short distance up the thin arm of each loop (arrow). The bulk of the loops is not radioactive 5 hours after administration of the isotope. However, over a period of days, radioactivity would extend farther toward the thick end of the loops. At about 10 days the loops would be uniformly labeled. × 1250. (Micrograph courtesy of Joseph G. Gall.)*

the precursor at a suitable time and thus to observe the change in the pattern of incorporation. Consequently, the movement of material from one region of the loop to another has been interpreted as either (1) the movement of the RNA gene product itself along the loop while the DNA remains stationary, or (2) the movement of the RNA gene product and the corresponding DNA template. Gall and Callan (1962) and Gall (1963) have considered these alternatives and suggest that the DNA present as a chromomere of the lampbrush chromosome is spun out into a loop and makes its way to the next chromomere.

During the lampbrush chromosome stage considerable amounts of ribosomal RNA are formed, as well as a heterogeneously sedimenting class of RNA referred to as *informational RNA* (Davidson, 1968). DNA–RNA hybridization methods have indicated that 3 percent of the total genome of the African clawed frog *Xenopus* is involved in the production of informational RNA (Davidson et al., 1966; Crippa et al., 1967). The value "3 percent" may be looked upon as an index of the approximate amount of DNA involved in the coding of various proteins employed in the functioning of the cell. Moreover, it is similar to the number of available genetic sites estimated on the basis of the total length of the lampbrush chromosomes of *Triturus* that are confined to the loops (5 percent). This relation, however, may be merely fortuitous.

Following the lampbrush stage, RNA synthesis drops to a low level. Nevertheless, large amounts of template active RNA are present in the mature egg. Experiments by Davidson et al. (1966) show that approximately 65 percent of the species of RNA synthesized at midlampbrush stage are present in the mature egg. The time span from midlampbrush stage to a fully matured egg is on the order of 6 to 12 months. How these templates are stored has not been ascertained; however, Callan (1969) has suggested that they may be associated with protein, whose function is to inactivate them until a later period.

The function of the informational RNA that is stored during vitellogenesis has not been clearly established. It has been suggested that this RNA might be required for the maintenance of the high metabolic activity and cell division characteristic of the cleaving embryo (Davidson, 1968). Hence the informational RNA would be translated to produce various proteins necessary for early embryogenesis—for example, mitotic spindle proteins and glycolytic enzymes. However, experiments have shown that the major portion of the stored informational RNA is not employed for such functions (Davidson, 1968).

NUCLEOLI AND RIBOSOMES

The egg nucleolus is variable in structure, showing a spectrum of morphological forms that depend upon the source and the activity of the organelle itself. For example, the large yolk-filled egg of arthropods and vertebrates has a large number of apparently equivalent nucleoli measuring 5 to 10 μm in diameter and composed of granular and fibrillar matrices (Fig. 19, inset). In many invertebrates the nucleolus is a large spheroidal structure consisting of a granular and fibrillar portion. Generally, eggs that possess this type of nucleolus are characterized by relatively less yolk than those having multinucleoli. Some invertebrate eggs have nucleoli that are a combination of the two types previously cited; that is, they contain a large nucleolus and smaller accessory nucleoli. [For a recent review of nucleolar function and morphology in developing cells, the reader is directed to Hay (1968) and Miller and Beatty (1969).]

The multinucleoli or extrachromosomal nucleoli of amphibian eggs contain a high concentration of ribosomal RNA and a considerable amount of protein (Brown, 1966, 1967). The multinucleoli can be isolated and dispersed for electron microscopy. The dispersed core of an isolated nucleolus can be shown to be formed of a long coarsely coated filament, the continuity of which is maintained by a single DNA molecule irregularly coated with ribonuclear protein (Miller, 1966; Miller and Beatty, 1969). When the filament is stretched and examined with the electron microscope, it is seen to possess a periodic pattern of matrix-free and matrix-covered regions (Fig. 19). The matrix-covered regions consist of fibers that are ribonuclear protein. RNA polymerase is believed to be situated at the axis of each fiber (Miller and Beatty, 1969). These observations suggest that segments of DNA coding for ribosomal RNA are separated by regions of DNA that are not used in transcription.

The connection between the extrachromosomal nucleoli and nucleolar organizing region —that is, the region of the chromosome that codes for ribosomal RNA—has been demonstrated in amphibian oocytes (Miller and Beatty, 1969; Birnstiel et al., 1966). In amphibian eggs, extrachromosomal nucleoli have been observed originating from a specific region on one of the lampbrush chromosomes. Because of the presence of ribosomal RNA and its origin, the multinucleoli are regarded as being homologous with the nucleoli of somatic cells.

Recent evidence such as the cessation of ribosomal RNA synthesis following the administration of actinomycin D, the presence of DNAase labile material and the results of DNA-RNA hybridization experiments demonstrate that the multinucleoli contain DNA that codes for ribosomal RNA. Therefore, at some point in oogenesis there is a replication of the ribosomal RNA genome resulting in the differential multiplication of the ribosomal RNA cistrons. These cistrons, which contain the genetic code for ribosomal RNA, are then liberated from the nucleolar organizing region and assume a peripheral location within the germinal vesicle. Since the egg is arrested in meiotic prophase at this time and is tetraploid, the number of nucleoli that the germinal vesicle would normally be expected to contain is four. However, replication of the nucleolar organizing region persists until approximately 600 to 1000 multi- or extrachromosomal nucleoli are formed (Brown and Dawid, 1968). Whether other cistrons undergo endoreplication during the course of oogenesis remains to be demonstrated.

The multinucleoli have been shown to be the sites of ribosomal RNA synthesis and are most active at the lampbrush stage when the loops are maximally distended (Brown and Dawid, 1968). Very little of the ribosomal RNA formed by the multinucleoli is turned over, and almost all of it is conserved throughout oogenesis. The ribosomes are stored in the egg cytoplasm and apparently remain "inert" until after fertilization.

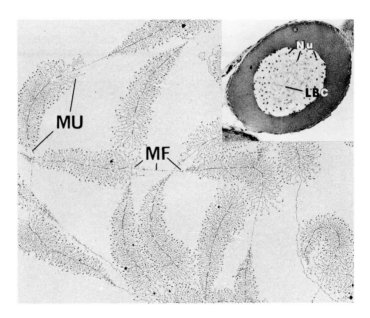

FIGURE 19. *Electron micrograph of a portion of dispersed nucleolar core matrix units (MU) and matrix-free axis segments (MF). (From O. L. Miller and B. R. Beatty,* Science 164, 955, *1969.) Inset: photomicrograph showing the peripherally placed extrachromosomal nucleoli (Nu) within the germinal vesicle of a* Xenopus *oocyte. LBC, lampbrush chromosomes. (Micrograph courtesy of James N. Dumont.)*

CORTICAL GRANULES

During oogenesis a population of bodies that stain positive for mucopolysaccharides and that are of varied sizes and internal configurations appears within the ooplasm. Initially, these structures are randomly distributed but later come to lie within the egg's cortex. Because of their location in the peripheral aspect of the egg at the time of its insemination, they have been called cortical granules in both invertebrates and vertebrates. The function of the cortical granules in organisms such as the surf clam *Spisula* remains unexplained. However, the cortical granules of eggs of a wide variety of animals are involved in the initial aspects of fertilization.

Ultrastructural observations of developing fish, amphibians, and echinoderm eggs indicate that the Golgi complex and endoplasmic reticulum play a major role in the production of cortical granules (Kessel, 1968b; Ward and Ward, 1968; Anderson, 1968a,b; Szollosi, 1967). The protein component of the cortical granule is thought to be manufactured in the cisternae of the rough endoplasmic reticulum; it is then transferred to the Golgi complex, where the polysaccharide component (or components) is added. Direct proof for this contention is lacking; however, the formation of protein–carbohydrate complexes in other cells suggests that this is a likely mode of formation. Vesicular components that are believed to contain the precursors to the cortical granules seemingly pinch off the tips of the Golgi saccules (Fig. 20). These vesicles subsequently increase in diameter by coalescing with others, thereby producing larger membrane-bounded granules.

The mature cortical granules of many animals are morphologically complex and consist of several components surrounded by a unit membrane (Fig. 21). In the pipefish the cortical granules consist of a filamentous component embedded within a reticular matrix. The cortical granules of the sea urchin *Arbacia* are composed of a granular stellate central portion surrounded by a filamentous substance. In relation to the cortical granules of the organisms just cited, those of most mammals are smaller and relatively simple architecturally (Szollosi, 1967; Zamboni, 1971).

ANNULATE LAMELLAE AND HEAVY BODIES

During the morphogenesis of the egg, a membrane system may be produced known as

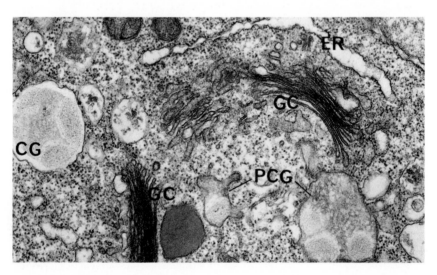

FIGURE 20. *Section through a cortical granule forming region of the ooplasm of a young oocyte of the sea urchin* Arbacia punctulata. *ER, cisterna of the endoplasmic reticulum; GC, Golgi complexes; PCG, presumptive cortical granules; CG, a miniature cortical granule.* × *30,000. (From E. Anderson, J. Cell Biol. 37, 514, 1968.)*

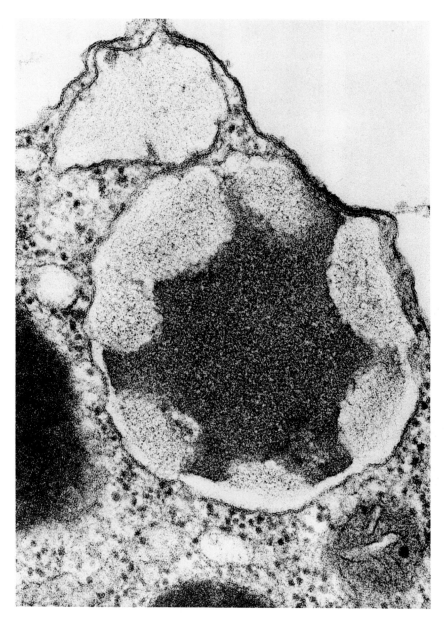

FIGURE 21. *Section of a cortical granule of the egg of the sea urchin* Arbacia punctulata. *(From E. Anderson,* J. Cell Biol. 37, *514, 1968.)*

annulate lamellae (Figs. 22 and 23). Annulate lamellae develop by a blebbing of the nuclear envelope (Figs. 22 and 23). The cisternae that are produced coalesce, thereby forming a membrane system structurally similar to the nuclear envelope, in that it consists of parallel, double membranes that are interrupted at intervals by pores (Fig. 23). Occasionally this process takes place within the nucleus and gives rise to intranuclear annulate lamellae (Kessel, 1968a). Because they originate in the nuclear envelope, it has been suggested that annulate lamellae may carry to various areas of the cytoplasm nuclear-derived information that would provide

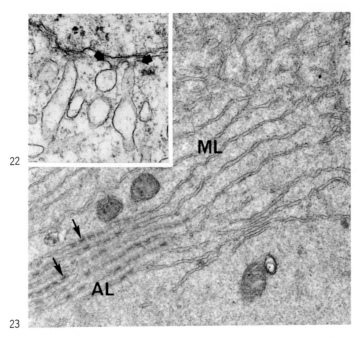

FIGURES 22 and 23. *Formation of annulate lamellae in a frog oocyte. Fig. 22: production of perinuclear cisternae by blebbing of the outer lamina of the nuclear envelope (arrows). Fig. 23: some of the membranous lamellae (ML) have developed pores (arrows) and hence constitute annulate lamellae (AL). Fig. 22, × 37,500; Fig. 23, × 22,500. (From R. G. Kessel, Z. Zellforsch. 94, 454, 1969.)*

a means for establishing control over remote areas of the ooplasm. Proof for such a contention, however, is lacking. Several investigators suggest that this membrane system may be involved in various processes of embryogenesis (Tilney and Marsland, 1969; Longo, 1972; Harris, 1969).

In the eggs of some organisms, such as sea urchins, annulate lamellae are associated with aggregations of ribosome-like granules, and together they constitute the heavy body. The term "heavy body" originates from the fact that in some species of sea urchins, but not all, these structures are displaced centrifugally upon centrifugation of the egg. The heavy bodies are scattered randomly throughout the cytoplasm, and they are often found associated with the female pronucleus. Some investigators contend that this association of the heavy bodies with the female pronucleus is indicative of their origin (Harris, 1969). Histochemical staining of the heavy bodies for light microscopy indicates that they contain RNA. Harris

(1969) has speculated that a portion of this RNA may be "masked" or inactive messenger RNA that is activated (translated) following fertilization (Monroy, 1965).

OTHER ATTRIBUTES OF THE OOPLASM

Many of the ooplasmic organelles become localized in specific areas of the egg during its ontogeny. In *Diopatra* a major portion of the mitochondria moves to one pole of the egg (Anderson and Huebner, 1968). This stratification has permitted investigators to refer to the non-yolk-filled, mitochondrial-rich area adjacent to the site of polar body formation or the nucleus as the *animal pole*. The yolk-filled area, which is relatively devoid of mitochondria, is called the *vegetal pole*. The eggs of many organisms possess such a stratification and specific localization of organelles, particularly those with copious amounts of yolk—for example, ascidians and amphibians.

In addition to a host of organelles found in the differentiating egg, considerable amounts of cytoplasmic DNA are also present. In many cases the amount of DNA is greater than that detected in the diplotene nucleus. The base composition of the cytoplasmic DNA is different from that extracted from somatic cell nuclear DNA, for only 0.1 to 0.5 percent is complementary. There is good evidence to show that most of the cytoplasmic DNA is derived from mitochondria (Pikó et al., 1967; Dawid, 1966).

EGG COATS

During the early stages of oogenesis the oolemma may lack specialized surface structures; however, in many cases, as differentiation ensues, the surface of the egg becomes morphologically specialized by the production of microvilli. Associated with the extracellular surface of the microvilli is a substance usually rich in acid mucopolysaccharide. This material is referred to as the *vitelline envelope.*

Despite the numerous papers written on the coverings of the egg in various groups of organisms, information on the origin of these investments is somewhat confused and incomplete. In addition, there is no general agreement with respect to nomenclature of the noncellular materials that are assembled on the egg's surface during development. Accordingly, we have adopted the terminology of Ludwig (1874), who states that if the vitelline envelope is produced by the egg, it is classified as a *primary envelope.* If the coat is produced by the encompassing follicle cells, it is designated as a *secondary envelope,* whereas if it is produced by the oviduct or other maternal structures not immediately connected with the egg, it is termed *tertiary.*

The layers investing the egg may be homogeneous in composition; however, in many cases they are heterogeneous and morphologically complex (Figs. 24 and 25). The material comprising the primary coat may be separated from the surface of the microvilli by a space; the coat in this case would be extraneous or an unattached glycocalyx. On the other hand, the material may be in the form of a fuzzy coat similar to that shown for the enteric layer of the intestinal microvilli (Ito, 1969). Follicle cells are involved in the production of the secondary coat in, for example, the insects *Drosophila* (fruit fly), *Rhodnius* (reduvid bug), and *Aeschna* (dragonfly). In these cases the follicle cells enveloping the oocyte exhibit phases of secretory activity that involve the hypertrophy of the Golgi complex and the rough endoplasmic reticulum and the release of an electron-dense component into the perivitelline space. The released material becomes organized into the morphological pattern that is typical for the secondary coat of the species (Quattropani and Anderson, 1969; Huebner and Anderson, 1972b; Beams and Kessel, 1969).

The tertiary coat is formed by the oviduct or other maternal structures (e.g., the shell gland) (Anderson et al., 1970). Such a covering would include the gelatinous material found on the eggs of the frog, the complex egg case of certain elasmobranchs, the shell of the egg of the brine shrimp *Artemia*, and the albumin and shell of the chicken. The tertiary envelope may be formed before fertilization (e.g., the frog) or after fertilization (e.g., the brine shrimp).

Development of the Male Gamete

As previously indicated, the term "spermatogenesis" is used primarily to refer to the chromosomal events of meiosis. During this period there is some alteration of the spermatocyte's cytoplasmic components that are part of the many processes involved in sculpturing the cell into a mature spermatozoon. By and large, however, the major morphological and biochemical changes in the nucleoplasm and cytoplasm occur after meiosis, at the spermatid stage. The processes of spermatid differentiation (spermiogenesis) are in a sense entirely directed toward the transference of the male chromosomal complement to the ovum. Without exception, the cellular elements of the spermatid undergo profound chemical and/or structural modifications during the course of this process. Cytoplasmic organelles that are

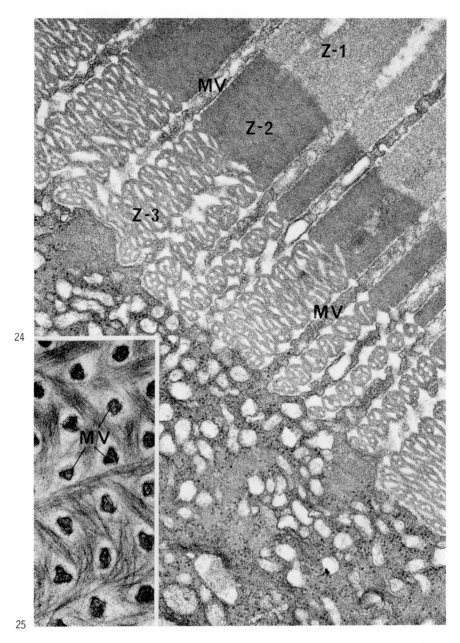

FIGURE 24. *Tangential section through the primary coat of an oocyte of the sea horse* (Hippocampus erectus) *depicting the regions that make up this structure:* (Z-1), (Z-2), *and* (Z-3). *Note the longitudinal profiles of microvilli of the oocyte* (MV). × 45,000. *(From E. Anderson, J. Cell Biol. 35, 193, 1967.)*

FIGURE 25. *Tangential section of a relatively small part of the primary coat of the oocyte of* Diopatra *showing its filamentous component organized around the microvilli* (MV) *of the egg shown in cross section.* × 30,000. *(From E. Anderson and E. Huebner, J. Morphol. 126, 163, 1968.)*

not cast off during spermatid differentiation become characteristic structures of the mature spermatozoon, whose function is primarily concerned with locomotion and penetration into the egg in order to accomplish fertilization.

The sperm of different species of animals exhibit remarkable differences of size, form, and internal structure. Underlying this great structural diversity, there exists a fundamental and common plan of organization. In general, the animal spermatozoon is a motile, flagellate cell; however, nonflagellate forms occur in several groups of invertebrates, such as the nematodes and the crustaceans. In some cases, nonflagellate sperm may perform slow movements, either by amoeboid changes or by the operation of spine-like processes. The structure of the nonflagellate forms is variable among different species.

Owing to the great diversity of form and structure in different species, investigators have experienced considerable terminological difficulty with respect to sperm morphology and morphogenesis. Consequently, the following discussion employs the nomenclature as defined by Fawcett (1965); it is based on the morphology of mammalian sperm. In its more typical form, the flagellate sperm is a greatly elongate cell that commonly consists of two parts (Fig. 26): (1) a head consisting of the acrosome and nucleus, and (2) the tail, which may be further subdivided (from anterior to posterior) into the neck, the middle piece, the principal piece, and the end piece. The architectural modifications that these components demonstrate are generally correlated with the nature of the environment in which fertilization takes place rather than, for example, with phylogeny. The so-called "primitive type" of sperm are recognized as those sperm that belong to marine and freshwater invertebrates and are discharged into the water, where fertilization may be accomplished (Franzen, 1956). These sperm typically have a spheroidal nucleus, a short middle piece consisting of a relatively small number—generally four—of spheroidal mitochondria situated around the proximal and distal centrioles, and a single flagellum. In the primitive type of spermatozoon the acrosome shows considerable variation from one animal to another; the nucleus may be structurally variable, but the mitochondria and flagellum are similar from species to species.

Animal sperm that are transferred directly to the female during copulation (internal fertilization) typically have a slender, elongate form. Generally, the nucleus is relatively long, and the tail is modified considerably (Fig. 26). These specializations are believed to be adaptations to the specific condition under which fertilization takes place; in addition, these variations may be due to the length of time they are stored in the male as well as in the female reproductive tract. However, only a few variations have been shown to be related to a particular feature of fertilization (Fawcett, 1970).

INTERCELLULAR BRIDGES

In thin sections of the testis, groups of spermatids at the same stage of differentiation are found to be connected by intercellular bridges, 0.5 to 1.0 μm in diameter (Fig. 27). These bridges have been described in numerous species of vertebrates and invertebrates and therefore appear to be of general occurrence. They are formed in the same manner as those observed with the egg and associated nurse cells. That is, during their mitotic and meiotic divisions, the cleavage furrows are arrested, and the daughter cells are left in cytoplasmic continuity with one another via intercellular bridges. Thus this interconnected group of cells is actually a large syncytium—a multinucleated mass of cytoplasm, from which a number of spermatozoa are derived. The identity, however, of the spermatogonia, the spermatocytes, and the spermatids has been sufficiently documented for them to be referred to as cells. The number of interconnected spermatogonia is variable. In some mammals as many as 74 or more spermatocytes may be in cytoplasmic continuity (Moens and Go, 1972). In insects the number of spermatids per cyst (sperm packet) is approximately equal to the integral power of 2. The number of spermatozoa of a given cyst, which presumably arise from synchronous mitotic and meiotic divisions, may range from 2^5 to 2^8, depending upon the species (Phillips, 1970a).

The presence of intercellular bridges provides the precise synchrony of differentiation as well as the simultaneous mitotic and meiotic divisions characteristic of the spermatids that are interconnected (Fawcett, 1961). In connection with this, the extensive and complex system of bridges that develops in some organisms (e.g., the annelids) may be involved in the nutrition of the developing sperm cells (Anderson et al., 1967). Hence it is possible that the anucleate mass of cytoplasm that interconnects the developing spermatids may func-

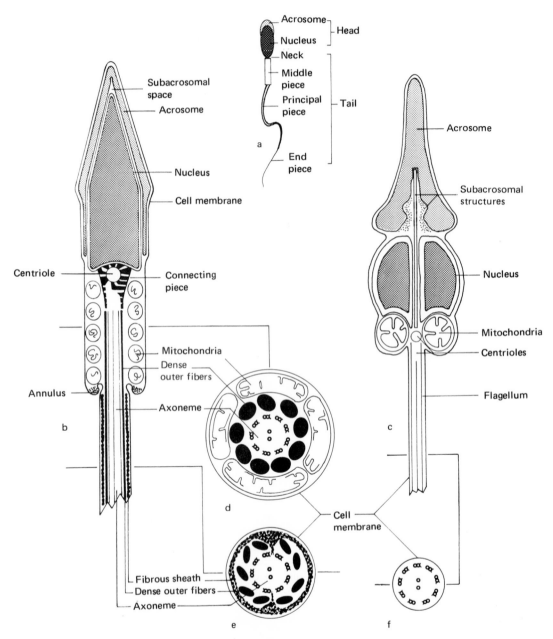

FIGURE 26. *Some structural features of animal spermatozoa. (a) Anatomical components of a a representative spermatozoon. (b) Mammalian spermatozoon. (c) Primitive spermatozoon. (d, e, and f) Cross-sections of sperm tails at the levels indicated in b and c.*

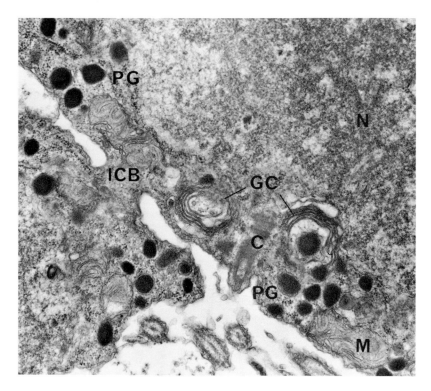

FIGURE 27. *Section of two spermatids of the surf clam* Spisula *associated by an intercellular bridge (ICB). One of the spermatids contains two Golgi complexes (GC) associated with pro-acrosomal granules (PG).* C, *centrioles;* M, *mitochondria;* N, *nucleus.* × 32,000. *(From F. J. Longo and E. Anderson,* J. Ultrastruct. Res. 27, 435, 1969.)

(Potswald, 1967).

tion in a manner similar to that of a nurse cell

NUCLEAR MORPHOGENESIS

During spermiogenesis, the chromatin of the spermatid nucleus undergoes extensive reorganization and condensation. In addition to the locomotive advantage achieved with the reduction of nuclear volume by chromatin condensation, this process may make the paternal genome resistant to physical and/or chemical injury or mutation during its storage and transit to the site of fertilization.

Generally three classes of spermatic nuclear forms are recognized: (1) nuclei in the form of elongate fibers, (2) spheroidal nuclei, and (3) flattened ovoid nuclei. Within these broad categories a number of variations may be found. Biochemically, sperm nuclei fit into at least

three groups with respect to whether their DNA is complexed with protamines (e.g., salmon, trout, snail, squid), histones (e.g., carp, sea urchins), or with nonhistone–nonprotamine proteins (e.g., mammals) (DuPraw, 1968). How these macromolecules are organized within the sperm nucleus is based largely on studies of a few invertebrate species by X-ray diffraction, polarization optics, and radioautography. Inoué and Sato (1966) have demonstrated that the DNA molecules are arranged in parallel fashion and are loosely coiled into a secondary helix in the sperm nucleus of the cave cricket (*Ceuthophilus*). The helical arrangement of chromatin filaments has been demonstrated in ultrastructural studies of insect sperm by Fawcett et al. (1971). In mammalian spermatozoa sectioned for electron microscopy the condensed nucleoplasm usually appears homogeneous or coarsely granular. However, Koehler (1970) has demonstrated by polarization optics and freeze fracture replication a high degree of order in the form of lamellae arranged parallel to the flattened surfaces of rabbit sperm nuclei.

Inoué and Sato (1966) have also demonstrated that the haploid number of chromosomes lie in a single line along the length of the sperm nucleus. This conclusion is supported by the radioautographic investigations of Taylor (1964) with grasshopper (*Romalea*) sperm. Using tritiated thymidine, Taylor (1964) was able to show that the heterochromatic X chromosome replicates asynchronously with respect to the autosomes. There-

fore, spermatocytes were found in which the label was located over the X chromosome exclusively, whereas in other cells it was found only over the autosomes. Examination of mature spermatozoa revealed that some had nuclei with only one short segment labeled, whereas in others only one short segment was unlabeled. Based on this evidence, Taylor (1964) concluded that the chromosomes retain their individuality during spermiogenesis and lie in tandem within the confines of the sperm nucleus. The exact position of the X chromosome, however, is not rigid and varies along the length of the nucleus from one sperm to

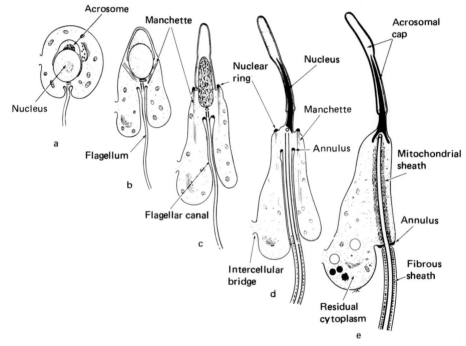

FIGURE 28. *Successive stages of spermiogenesis in the guinea pig. (a) The nucleus and acrosome of the spermatid are centrally located. (b) With the completion of the acrosome, the Golgi migrates into the caudal cytoplasm, microtubules appear tangential to the caudal half of the nucleus, and spermatid elongation begins. (c) Caudal displacement of cytoplasm brings the plasma membrane into close apposition to the acrosome. Nuclear condensation begins; the microtubules become arranged in a cylinder (manchette), originating in a dense nuclear ring at the caudal margin of the acrosome. (d) Nuclear condensation and spermatid elongation continue. Nuclear ring moves back to the level of the base of the flagellum; flagellar canal lengthens, and the fibrous sheath begins to form around the axoneme of the free portion of the flagellum. (e) The nuclear ring and manchette disappear; annulus migrates caudally to the anterior margin of the fibrous sheath; mitochondria gather around the base of the flagellum anterior to the annulus and form the mitochondrial helix of the middle piece. (From D. W. Fawcett, W. A. Anderson, and D. M. Phillips, Develop. Biol. 26, 220, 1971.)*

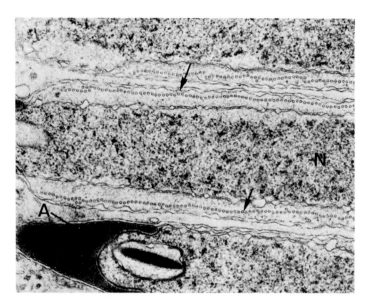

FIGURE 29. *Longitudinal section of a pigeon spermatid nucleus at an early stage in its elongation and condensation. Cross sections of a set of circumferentially oriented microtubules (arrows) are located along the periphery of the nucleus (N). A, acrosome. (Micrograph courtesy of J. R. McIntosh.)*

another. Similar results have been obtained by Bianchi and de Bianchi (1969), whose investigations indicate that the chromosomes are associated end to end in the sperm nucleus, and that the Y chromosome and perhaps the autosomes as well occupy a constant position in rat sperm.

The structural transformation of chromatin during its condensation does not follow the same course in all organisms. The chromatin of early spermatids is generally diffuse and gradually becomes electron-dense; it frequently lacks any discernible substructure in the mature spermatozoon. The intermediate stages between these two extremes differ widely from one animal to another. For a more complete discussion of nuclear differentiation in cricket spermatids that correlates the transformation of the fine structural elements and chemical components, the reader is referred to the works by Kaye and McMaster-Kaye (1966) and Kaye (1969).

During its differentiation, the spermatid nucleus gradually assumes the shape characteristic of the species. Concomitantly, the chromatin becomes condensed, which suggests that modeling of the nucleus is related to this proc-

ess. Fawcett et al. (1971) suggest that the shape of the sperm nucleus is largely determined by a specific genetically controlled pattern of aggregation of the molecular subunits of DNA and protein during condensation of the chromatin. They dismiss the possibility that the form of the sperm nucleus is a consequence of external modeling by means of pressure applied to the condensing spermatid nucleus by perinuclear microtubules or by filaments in the ectoplasm of juxtaposed supporting cells. On the other hand, a number of investigators have argued that cytoplasmic microtubules—whether they be organized in a sleeve-like organelle (the manchette) encircling the nucleus, which appears prior to the onset of chromatin condensation in some vertebrates (mammals), or as a set of helically wound microtubules disposed circumferentially with respect to the long axis of the condensing nucleus (birds)— act to bring about the elongation of the spermatid nucleus (Figs. 28 and 29) (Clark, 1967; McIntosh and Porter, 1967; Kessel, 1970). At present there is no direct evidence to show that microtubules are the causal agents of elongation.

Reduction in nuclear volume during sper-

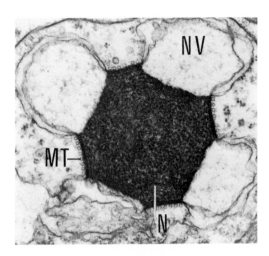

FIGURE 30. *Section through a spermatid of the annelid* Spirorbis *showing blebbing of the nuclear envelope (redundant nuclear envelope). Eventually there is a progressive pinching off of the four vesicles (NV). N nucleus; MT, microtubules.* × *39,000. (From H. E. Potswald,* Z. Zellforsch. 83, *231, 1967.)*

miogenesis is accomplished by an alteration in the nuclear envelope to accommodate this transformation. In many animals the nuclear envelope that is in excess (redundant nuclear envelope) may remain with the mature spermatozoon and take various forms. It may be reflected into the neck region to form two sacs (the guinea pig), or it may develop into a pair of scrolls (the bat). In many animals the redundant nuclear envelope is not retained, and the excess is pinched off into the perinuclear cytoplasm in the form of lamellae, which are eventually sloughed with the residual cytoplasm (teleosts and annelids) (Fig. 30).

Associated with the spermatocyte and spermatid nucleus of mammals, birds, reptiles, crustaceans, and insects is a cytoplasmic component known as the *chromatoid body.* This structure consists of a granular matrix, which some investigators have shown to be positive for basic proteins and RNA at the light microscope level of observation (Sud, 1961). However, recent studies employing cytochemical techniques specific for demonstration of nucleic acid at the electron microscope level of observation indicate that RNA is not present in the chromatoid body of the rat and guinea

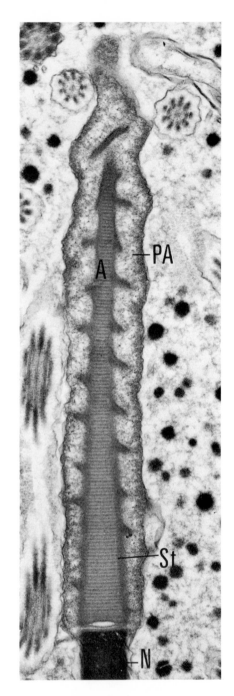

FIGURE 31. *Longitudinal section through the cork-screw-shaped acrosome (A) of an octopus spermatozoon. PA, periacrosomal material; St, striations of the acrosome; N, nucleus.* × *22,500. (From F. J. Longo and E. Anderson,* J. Ultrastruct. Res. 32, *94, 1970.)*

pig (Eddy, 1970). Although the chromatoid body may become closely associated with the spermatid nucleus, the functional interaction that may exist between these two structures is conjectural. The function and the fate of the chromatoid body are unknown; there is no clear morphological evidence indicating that it contributes directly to the formation of any definitive structure of the spermatozoon.

ACROSOME FORMATION

The acrosome exhibits a great diversity of form, varying from a small membrane-bounded vesicle (the sea urchin) to a large conical or spine-like process (the cricket) that in extreme cases (the urodele) is provided with a prominent barb similar to that of a fish hook. In many animals the content of the acrosome may be amorphous, or regional differences in densities may exist. The contents of the acrosome in a number of species (e.g., the octopus) demon-strate a complex organization (Fig. 31). Some animals, however, such as many of the higher teleosts, lack an acrosome. The acrosomes of various mammals have been shown to contain acid proteases and hyaluronidase. These findings have fostered the idea that this structure is a special type of lysosome.

Early and contemporary literature show that the acrosome forms in association with the Golgi complex. In many of the species that have been studied to date, one large Golgi complex develops in the spermatid, and within the confines of this organelle the acrosomal vesicle is produced (Fig. 32). Initially, the vesicle is small (sometimes referred to as a proacrosomal granule); gradually it enlarges by the fusion of smaller vesicles along its parameter; and it is then deposited adjacent to the spermatid nucleus. However, in some animals (the lamellibranchs *Mytilus* and *Spisula*), a number of Golgi complexes of the spermatocyte produce a host of small vesicles filled with an electron-dense substance (proacrosomal gran-

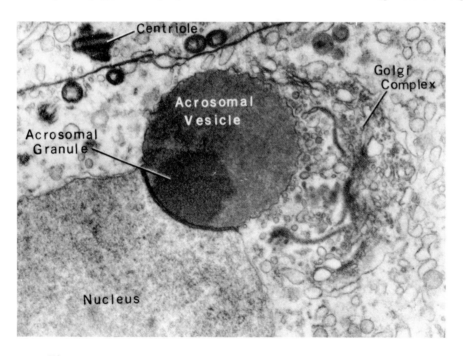

FIGURE 32. *Electron micrograph of a portion of a guinea pig spermatid. After fusion of the proacrosomal granules to form a single large acrosomal granule, the Golgi complex remains closely associated with its surface. Numerous small vesicles derived from the Golgi complex continue to coalesce with the developing acrosome, adding their contents to it. × 12,000. (From D. W. Fawcett, and R. D. Hollenberg, Z. Zellforsch. 60, 276, 1693.)*

ules) (Fig. 27). These vesicles do not remain associated with the Golgi complexes and are distributed throughout the cytoplasm. During the spermatid stage the proacrosomal granules fuse together and eventually form one large vesicle, which migrates to the nucleus. Then the spherical acrosome undergoes a series of morphogenic events that involve an alteration in its shape and simultaneous changes in the density of its contents. These transformations are quite variable and are characteristic of the species.

Within the intracellular space between the acrosome and the nucleus of the sperm of many invertebrates, amphibians, and birds is an aggregation of material that may take the form of a rod (the chicken and the mussel *Mytilus*) (Fig. 26). In some invertebrates where this material is well developed, there is good evidence to show that it participates in the fusion of the egg and sperm at fertilization. In mammals, the subacrosomal material is amorphous in appearance, and its volume is relatively small; there is no evidence to show that it functions during fertilization.

THE MIDDLE PIECE

The mitochondria of the middle piece assume various configurations in different animals and during spermiogenesis undergo extreme alterations in shape and substructure (Fig. 33) (Favard and André, 1970).

The middle piece of the primitive spermatozoon is relatively simple architecturally and is represented by a single ringlet or a cluster of four or five mitochondria. This arrangement suggests that the primitive spermatozoon may not be heavily dependent upon the mitochondria for providing energy of locomotion (Fawcett, 1970). During spermiogenesis there is a progressive loss of mitochondria and a concomitant increase in the size of those that remain. In the late spermatid, a number of mitochondria are found along the caudal region of the nucleus that surrounds the developing flagellum. How these events are accomplished has not been well documented. However, it is generally assumed that the simultaneous increase in size and decrease in number of mitochondria is brought about by their fusion.

The mitochondria in early spermatids of

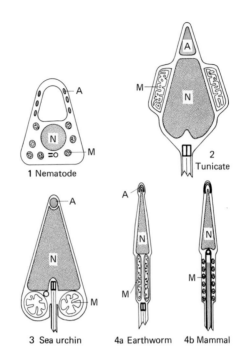

FIGURE 33. *Location of mitochondria in the sperm of most animals. (Adapted from P. Favard and J. Andre, in "Comparative Spermatology" (B. Baccetti, ed.), p. 415, Academic Press, New York, 1970.)*

many insects, as well as of a number of other invertebrates, aggregate and subsequently fuse together to form two large interwoven mitochondria collectively referred to as the *nebenkern* (Fig. 34) (Pratt, 1968; Phillips, 1970a). Later the nebenkern "divides" into two bodies of equal size (nebenkern derivatives or mitochondrial derivatives), which take up positions on either side of the base of the developing flagellum, where they undergo internal reorganization. Differentiation of the nebenkern derivatives involves their elongation and the precise alignment of the mitochondrial cristae. The cristae become arranged in evenly spaced parallel lamellae that are oriented perpendicular to the long axis of the organelle. Subsequently, small areas of structured material may appear in the mitochondrial matrix. This material, which has a paracrystalline structure, may accumulate within the nebenkern derivatives until it replaces the mitochondrial matrix; in some insects it may comprise most of the mitochondrial volume.

In mammals, the mitochondria are elongate structures arranged end to end in a helix around a portion of the flagellum. During spermiogenesis the mitochondria of the spermatid migrate to the proximal region of the flagellum, where they elongate, decrease in volume, and ultimately become arranged to form a helix. During this operation the mitochondria may undergo extensive internal reorganization (Fawcett, 1970).

DEVELOPMENT OF THE NECK REGION AND THE SPERM FLAGELLUM

In general, the neck region of the primitive type of sperm is reduced or almost nonexistent (Fig. 26). Structurally, it is quite simple in comparison to the more complex arrangement of the elongate sperm of mammals. In many organisms the flagellum is formed early during sperm development. For example, in the sea urchin the flagellum is produced in the spermatocyte stage (Longo and Anderson, 1969a).

The spermatids of insects have two centrioles, one of which serves as the basal body for the growing flagellum, while the other is oriented perpendicular to the first (Phillips, 1970). Phillips (1970) has found that the centrioles disappear during insect spermiogenesis and cannot be found at the base of the flagellum in the mature spermatozoon.

Figure 35 diagrammatically summarizes some of the stages in the morphogenesis of the connecting piece, a complex cross-striated organelle found in the neck region of mammalian sperm (Fawcett and Phillips, 1970). At the periphery of the early spermatid one of the centrioles, destined to become the distal, gives rise to the flagellum. At a later stage of development the centrioles and the base of the flagellum move inward and become fixed to the caudal pole of the nucleus. At this location dense strands of material emerge from the interspaces between the triplet tubules of the centrioles and fuse to form the complex struc-

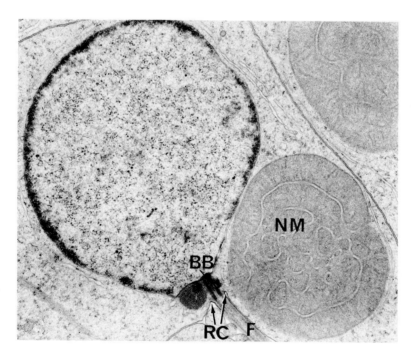

FIGURE 34. *The nebenkern (NM) of a young spermatid of the stinkbug* Euschistus. *The nebenkern at the stage shown is composed of fusing interlocked mitochondria. RC, ring centriole; BB, basal body; F, flagellum.* × *12,750. (From D. M. Phillips, J. Cell Biol. 44, 243, 1970.)*

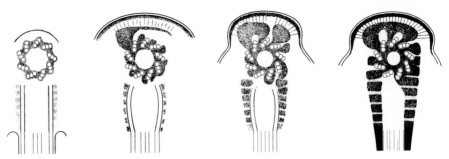

FIGURE 35. *Four successive stages in the differentiation of the connecting piece. (From D. W. Fawcett and D. M. Phillips, in "Comparative Spermatology" (B. Baccetti, ed.), p. 2, Academic Press, New York, 1970.)*

ture of the connecting piece. While the connecting piece is taking form, the distal centriole disintegrates, although the second (proximal) persists and is present in the mature spermatozoon.

In comparison to other regions of the differentiating spermatid, the formation of the flagellum has received relatively little attention. In some animals (insects) where sperm tail development has been studied, the flagellum is initiated deep within the cytoplasm (Phillips, 1970a) (Fig. 36). A vesicle becomes associated with the end of one of the centrioles (basal body) from which the flagellar tubules arise. As the flagellum develops, the vesicle becomes cup-shaped to cover the tips of the growing doublet tubules. The cup-shaped vesicle then elongates as the flagellar tubules lengthen and eventually comes in contact and fuses with the cell membrane. The nine doublet tubules of the axonemal complex of the flagellum appear to be initiated by the accumulation and polymerization of microtubule protein at the ends of the triplet tubules in the wall of the basal body. The microtubules then elongate by accretion to their distal ends until the flagellum has reached its definitive length.

In most flagellate sperm, the axoneme consists of nine doublets and two central singlet tubules (9 + 2) (Fig. 26). There are variations in this pattern, however, particularly among the insects (Phillips, 1970b). In addition to the typical axonemal complex, the mammalian sperm tail also possesses nine outer dense fibers and a fibrous sheath. The outer fibers circumscribe the axoneme, such that one fiber is positioned just lateral to one doublet tubule (Fig. 26). During spermiogenesis the dense fibers

arise in association with a double tubule. The fibrous sheath that begins at the annulus lies just beneath the plasma membrane and circumscribes the outer dense fibers (Fig. 26). Insect sperm flagella have a pattern similar to that of mammals; however, the outer nine ele-

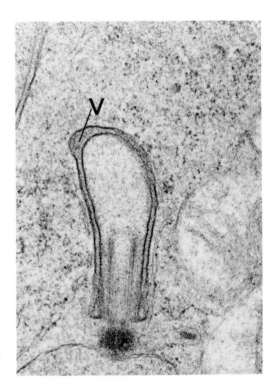

FIGURE 36. *Longitudinal section of a developing flagellum in a spermatid of the mosquito Culex. The flagellum extends into a flattened, cup-like vesicle (V). × 37,000. (From D. M. Phillips, J. Cell Biol. 44, 243, 1970.)*

ments are tubules rather than dense fibers, and there is no fibrous sheath.

LOSS OF RESIDUAL CYTOPLASM

One of the final events of spermiogenesis is the loss of excess cytoplasmic organelles and components that become localized within one region of the spermatid (the forming residual body). This elimination appears to occur by sloughing of residual materials rather than by the intracellular degradation of these structures by lysosomes—which is the process common to most somatic cells. The cytoplasm that is discarded is surrounded by a plasma membrane and may contain ribosomes, membranous elements, and a Golgi complex. The removal of this body results in the formation of a spermatozoon that lacks many of the cytoplasmic organelles present in young spermatids.

SUSTENTACULAR CELLS

During spermatogenesis and spermiogenesis in mammals and in many nonmammalian forms, the differentiating sperm cells are associated with supportive or sustentacular cells. In mammals, they are referred to as *Sertoli cells* (Dym and Fawcett, 1970). Cells that may be functionally similar to the Sertoli cells of mammals have been identified in such invertebrates as the sea urchin (Longo and Anderson, 1969a).

The Sertoli cell of mammals extends from the basement lamina into the lumen of the seminiferous tubule (Fig. 37). Spermatocytes and early spermatids are arranged in rows around and project into the cytoplasm along its length. As the early germ cells differentiate, they slowly move from a basal–lateral position to the apical region of the Sertoli cell. Along the apical region are located clusters of spermatids in various stages of morphogenesis (Fig. 37).

Sertoli cells are believed to provide mechanical support and protection for the developing sperm cells. Although it has not been clearly established, it is possible that these cells also participate in the nutrition of the germ cells. In this respect, the sustentacular cells of

the testes may function in a manner similar to that demonstrated for the accessory cells of the developing egg. Sertoli cells have also been implicated in the coordination of sperm cell development and in the release of spermatozoa that are differentiated from the spermatids that project into their apical surface. Because of the way that the seminiferous tubule is constructed, the Sertoli cells "isolate" the spermatogonia within a basal compartment separate from other differentiating germ cells (Fig. 37).

In many animals, notably mammals, following the differentiation of the spermatid, the spermatozoon leaves the seminiferous epithelium but does not attain its definitive structure until a later period. For example, Fawcett and Hollenberg (1963), Fawcett and Phillips (1969) and Bedford and Nicander (1971) have shown that the acrosome of a number of mammalian spermatozoa do not achieve their final form until they are in segments of the epididymal ducts. The sperm of animals other than mammals also demonstrate morphological changes following their release from the testis (Phillips, 1966a, b). In this connection, it is pertinent to recall the physiological transformations that mammalian sperm must undergo in order to achieve insemination. This process, known as capacitation, occurs in the female reproductive tract (Austin, 1961).

References

ANDERSON, E. (1964). Oocyte differentiation and vitellogenesis in the roach *Periplaneta americana*. *J. Cell Biol.* 20, 131–155.

ANDERSON, E. (1968a). Oocyte differentiation in the sea urchin *Arbacia punctulata*, with particular reference to the origin of cortical granules and their participation in the cortical reaction. *J. Cell Biol.* 37, 514–539.

ANDERSON, E. (1968b). Cortical alveoli formation and vitellogenesis during oocyte differentiation in the pipefish *Syngnathus fuscus* and the killifish *Fundulus heteroclitus*. *J. Morphol.* 125, 53–60.

ANDERSON, E. (1969). Oogenesis in the cockroach *Periplaneta americana*, with special reference to the specialization of the oolemma and the fate of coated vesicles, *J. Microsc. 8,* 721–738.

ANDERSON, E., AND BEAMS, H. W. (1956). Evidence from electron micrographs for the passage of material through the pores of the nuclear envelope. *J. Biophys. Biochem. Cytol.* 2, Suppl. 4, 439–445.

ANDERSON, E., AND HUEBNER, E. (1968). Development of the oocyte and its accessory cells of the

polychaete *Diopatra cuprea* (Bosc). *J. Morphol.* 126, 163–197.

ANDERSON, E., LOCHHEAD, J. H., LOCHHEAD, M. S., AND HUEBNER, E. (1970). The origin and structure of

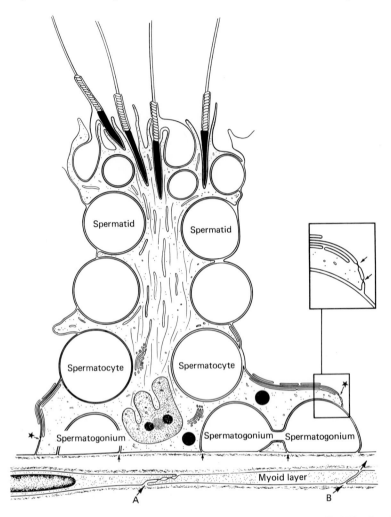

FIGURE 37. *Relation between the germ cells and a columnar Sertoli cell. The localization of the blood-testis barrier and the compartmentalization of the germinal epithelium by tight junctions between adjacent Sertoli cells is depicted also (inset). Material gaining access to the base of the epithelium by passing through open junctions (B) in the myoid layer is free to enter the intercellular gap between the spermatogonia and the Sertoli cells. Deeper penetration is prevented by occluding junctions (stars) on the Sertoli–Sertoli boundaries. In effect, the Sertoli cells and their tight junctions delimit a basal compartment in the germinal epithelium, containing the spermatogonia and early preleptotene spermatocytes, and an adluminal compartment, containing the spermatocytes and spermatids. Substances traversing open junctions in the myoid cell layer have direct access to cells in the basal compartment, but to reach the cells in the adluminal compartment, substances must pass through the Sertoli cells. (A) Tight apposition of cell membranes in the myoid layer. (From M. Dym and D. W. Fawcett, Biol. Reprod. 3, 308, 1970.)*

the tertiary envelope in thick-shelled eggs of the brine shrimp *Artemia*. *J. Ultrastruct. Res. 32*, 497–525.

ANDERSON, L. M. (1971). Protein synthesis and uptake by isolated cecropia oocytes. *J. Cell Sci. 8*, 735–750.

ANDERSON, L., AND TELFER, W. H. (1969). A follicle cell contribution to the yolk spheres of moth oocytes. *Tissue and Cell, 1*, 633–644.

ANDERSON, W. A., AND SPIELMAN, A. (1971). Permeability of the ovarian follicle of *Aedes aegypti* mosquitoes. *J. Cell Biol. 50*, 201–221.

ANDERSON, W. A., WEISSMAN, A., AND ELLIS, R. A. (1967). Cytodifferentiation during spermiogenesis in *Lumbricus terrestris*. *J. Cell Biol. 32*, 11–26.

AUSTIN, C. R. (1961). "The Mammalian Egg." Charles C. Thomas, Springfield, Illinois.

BEAMS, H. W., AND KESSEL, R. G. (1962). Intracisternal granules of the endoplasmic reticulum in the crayfish oocyte. *J. Cell Biol. 13*, 158–162.

BEAMS, H. W., AND KESSEL, R. G. (1963). Electron microscope studies on developing crayfish oocytes with special reference to the origin of yolk. *J. Cell Biol. 18*, 621–649.

BEAMS, H. W., AND KESSEL, R. G. (1969). Synthesis and deposition of oocyte envelopes (vitelline membrane, chorion) and the uptake of yolk in the dragonfly (Odonata: Aeschnidae). *J. Cell Sci. 4*, 241–264.

BEDFORD, J. M., AND NICANDER, L. (1971). Ultrastructural changes in the acrosome and sperm membranes during maturation of spermatozoa in the testis and epididymis of the rabbit and monkey. *J. Anat. 108*, 527–543.

BIANCHI, N. O., AND DE BIANCHI, M. S. (1969). Y chromosome replication and chromosome arrangement in germ line cells and sperm of the rat. *Chromosoma 28*, 370–378.

BIER, K. (1963). Synthese, interzellulärer Transport, und Abbau von Ribonukeinsäure im Ovar der Stubenfliege *Musca domestica*. *J. Cell Biol. 16*, 436–440.

BIER, K. (1965). Zur Funktion der Nährzellen im meroistischen Insektenovar unter besonderer Berücksichtiguing der Oogenese adephager Coleopteren. *Zool. Jahrb. Physiol. 71*, 371.

BIGGERS, J. D., WITTINGHAM, D. S., AND DONAHUE, R. P. (1967). The pattern of energy metabolism in the mouse oocyte and zygote. *Proc. Nat. Acad. Sci. U.S.A. 58*, 560–567.

BIRNSTIEL, M., WALLACE, L. H., SIRLIN, J. L., AND FISCHBERG, M. (1966). Localization of the ribosomal DNA complements in the nucleolar or-ganizer region of *Xenopus laevis*. *Nat. Cancer Inst. Monogr. 23*, 431–447.

BOVERI, T. (1899). Die Entwicklung von *Ascaris megalocephala* mit besonderer Rucksicht auf die Kernverhaltnisse. *Festschr. für C. von Kupffer*, 10a (XIII).

BRACHET, J. (1960). "The Biochemistry of Development." Permagon Press, Elmsford, New York.

BROWN, D. D. (1966). The nucleolus and synthesis of ribosomal RNA during oogenesis and embryogenesis of *Xenopus laevis*. *Nat. Cancer Inst. Monogr. 23*, 297–309.

BROWN, D. D. (1967). The genes for ribosomal RNA and their transcription during amphibian development. *Curr. Top. Develop. Biol. 2*, 48–73.

BROWN, D. D., AND DAWID, I. B. (1968). Specific gene amplification in oocytes. *Science 160*, 272–280.

CALLAN, H. G. (1963). The nature of lampbrush chromosomes. *Int. Rev. Cytol. 15*, 1–34.

CALLAN, H. G. (1967). The organization of genetic units in chromosomes. *J. Cell Sci. 2*, 1–7.

CALLAN, H. G. (1969). Biochemical activities of chromosomes during the prophase of meiosis. *In* "Handbook of Molecular Cytology" (A. Lima-de-Faria, ed.), pp. 540–552. North-Holland, Amsterdam.

CALLAN, H. G., AND LLOYD, L. (1960). Lampbrush chromosomes of crested newts *Triturus cristatus* (Laurenti). *Phil. Trans. Roy. Soc. B 243*, 135–219.

CARO, L. G., AND PALADE, G. E. (1964). Protein synthesis, storage, and discharge in the pancreatic exocrine cell. An autoradiographic study. *J. Cell Biol. 20*, 473–495.

CLARK, A. W. (1967). Some aspects of spermiogenesis in a lizard. *Amer. J. Anat. 121*, 369–400.

COMINGS, D. E., AND OKADA, T. A. (1971). Fine structure of the synaptonemal complex. *Exp. Cell Res. 65*, 104–116.

CRIPPA, M., DAVIDSON, E. H., AND MIRSKY, A. E. (1967). Persistence in early amphibian embryos of informational RNA's from the lampbrush chromosome stage of oogenesis. *Proc. Nat. Acad. Sci. U.S.A., 57*, 885–892.

DALCQ, A. M. (1963). "Lysosomes." Ciba Foundation Symposium. (A. V. S. de Reuch and M. P. Cameron, eds.), pp. 226–281. Little, Brown, Boston.

DAVIDSON, E. H. (1968). "Gene Activity in Early Development." Academic Press, New York.

DAVIDSON, E. H., CRIPPA, M., KRAMER, F. R., AND MIRSKY, A. E. (1966). Genomic function during the lampbrush chromosome stage of amphibian oogenesis. *Proc. Nat. Acad. Sci. U.S.A., 56*, 856–863.

DAWID, I. B. (1966). Evidence for the mitochondrial origin of frog egg cytoplasmic DNA. *Proc. Nat. Acad. Sci. U.S.A. 56*, 269–276.

DUMONT, J. N. (1969). Oogenesis in the annelid *Enchytraeus albidus* with special reference to the origin and cytochemistry of yolk. *J. Morphol. 129*, 317–344.

DUMONT, J. N., AND ANDERSON, E. (1967). Vitellogenesis in the horseshoe crab *Limulus polyphemus. J. Microsc. 6*, 791–806.

DUPRAW, E. J. (1968). "Cell and Molecular Biology." Academic Press, New York.

DYM, M., AND FAWCETT, D. W. (1970). The blood-testis barrier in the rat and the physiological compartmentation of the seminiferous epithelium. *Biol. Reprod. 3*, 308–326.

EDDY, E. M. (1970). Cytochemical observations on the chromatoid body of the male germ cells. *Biol. Reprod. 2*, 114–128.

EDDY, E. M., AND ITO, S. (1971). Fine structural and radioautographic observations on dense perinuclear cytoplasmic material in tadpole oocytes. *J. Cell Biol. 49*, 90–108.

FAVARD, P., AND ANDRÉ, J. (1970). The mitochondria of spermatozoa. *In* "Comparative Spermatology" (B. Baccetti, ed.), pp. 415–429. Academic Press, New York.

FAVARD, P., AND CARASSO, N. (1958). Origine et ultrastructure des plaquettes vitellines de la planorbe. *Arch. d'Anat. Microsc. 47*, 211–234.

FAWCETT, D. W. (1961). Intercellular bridges. *Exp. Cell Res. 22*, Suppl. 8, 174–187.

FAWCETT, D. W. (1965). The anatomy of the mammalian spermatozoon with particular reference to the guinea pig. *Z. Zellforsch. 67*, 279–296.

FAWCETT, D. W. (1970). A comparative view of sperm ultrastructure. *Biol. Reprod.*, Suppl. 2, 90–127.

FAWCETT, D. W., ANDERSON, W. A., AND PHILLIPS, D. M. (1971). Morphogenetic factors influencing the shape of the sperm head. *Develop. Biol. 26*, 220–251.

FAWCETT, D. W., AND HOLLENBERG, R. D. (1963). Changes in the acrosome of guinea pig spermatozoa during passage through the epididymis. *Z. Zellforsch. 60*, 276–292.

FAWCETT, D. W., AND PHILLIPS, D. M. (1969). Observations on the release of spermatozoa and on changes in the head during passage through the epididymis. *J. Reprod. Fert.*, Suppl. 6, 405–418.

FAWCETT, D. W., AND PHILLIPS, D. M. (1970). Recent observations on the ultrastructure and development of the mammalian spermatozoon. *In* "Comparative Spermatology" (B. Baccetti, ed.), pp. 2–28. Academic Press, New York.

FLICKINGER, R. A. (1957). The relation between yolk utilization and differentiation in the frog embryo. *Amer. Natur. 91*, 373–379.

FLICKINGER, R. A. (1960). Formation, biochemical composition and utilization of amphibian egg yolk. *In* "Symposium on Germ Cells and Development," p. 29. Institut International d'Embryologie and Fundazione A. Baselli, Milan.

FRANKE, W. W., AND SCHEER, U. (1970). The ultrastructure of the nuclear envelope of amphibian oocytes: a reinvestigation. II. The immature oocyte and dynamic aspects. *J. Ultrastruct. Res. 30*, 317–327.

FRANZEN, A. (1956). On spermiogenesis, morphology of the spermatozoon, and biology of fertilization among invertebrates. *Zool. Bidr. Uppsala 31*, 355–482.

FRIEND, D. S., AND FARQUHAR, M. G. (1967). Functions of coated vesicles during protein absorption in the rat vas deferens. *J. Cell Biol. 35*, 357–376.

GALL, J. G. (1955). Problems of structure and function in the amphibian oocyte nucleus. *Symp. Soc. Exp. Biol. 9*, 358–370.

GALL, J. G. (1958). Chromosomal differentiation. *In* "The Chemical Basis of Development" (W. D. McElroy and B. Glass, eds.), pp. 103–135. Johns Hopkins Press, Baltimore.

GALL, J. G. (1963). Kinetics of deoxynibonuclease action on chromosomes. *Nature* (London) *198*, 36–38.

GALL, J. G., AND CALLAN, H. G. (1962). ^{3}H-uridine incorporation in lampbrush chromosomes. *Proc. Nat. Acad. Sci. U.S.A. 48*, 562–570.

GANION, L. R., AND KESSEL, R. G. Intracellular synthesis, transport, and packaging of proteinaceous yolk in oocytes of *Orconectes immunis. J. Cell Biol. 52*, 420–437.

HARRIS, P. (1969). Relation of fine structure to biochemical changes in developing sea urchin eggs and zygotes. *In* "The Cell Cycle: Gene-Enzyme Interactions" (E. M. Padilla, G. L. Whitson, and I. L. Cameron, eds.), pp. 315–340. Academic Press, New York.

HAY, E. D. (1968). Structure and function of the nucleolus in developing cells. *In* "The Nucleus" (A. J. Dalton and F. Haguenau, eds.), pp. 1–79. Academic Press, New York.

HUEBNER, E., AND ANDERSON, E. (1970). The effects of vinblastine sulfate on the microtubular organization of the ovary of *Rhodnius prolixus*. *J. Cell Biol. 46*, 191–198.

HUEBNER, E., AND ANDERSON, E. (1972a). A cytological study of the ovary of *Rhodnius prolixus*. I. The ontogeny of follicular epithelium. *J. Morphol. 136*, 459–494.

HUEBNER, E., AND ANDERSON, E. (1972b). A cytological study of the ovary of *Rhodnius prolixus*. II. Oocyte differentiation. *J. Morphol. 137*, 385–416.

HUEBNER, E., AND ANDERSON, E. (1972c). A cytological study of the ovary of *Rhodnius prolixus*. III. Cytoarchitecture and development of the trophic chamber. *J. Morphol. 138*, 1–40.

INOUÉ, S., AND SATO, H. (1966). Desoxyribonucleic acid arrangement in living sperm. *In* "Molecular Architecture in Cell Physiology" (T. Hayashi and A. G. Szent-Györgyi, eds.), pp. 209–248. Prentice-Hall, Englewood Cliffs, New Jersey.

ITO, S. (1969). Structure and function of the glycocalyx. *Fed. Proc. 28*, 12–25.

JAMIESON, J. D., AND PALADE, G. E. (1967a). Intracellular transport of secretory proteins in the pancreatic exocrine cell. I. Role of the peripheral elements of the Golgi complex. *J. Cell Biol. 34*, 577–596.

JAMIESON, J. D., AND PALADE, G. E. (1967b). Intracellular transport of secretory proteins in the pancreatic exocrine cell. II. Transport to condensing vacuoles and zymogen granules. *J. Cell Biol. 34*, 597–615.

KAYE, J. S. (1969). The ultrastructure of chromatin in nuclei of interphase cells and in spermatids. *In* "Handbook of Molecular Cytology" (A. Lima-de-Faria, ed.), pp. 361–380. North-Holland, Amsterdam.

KAYE, J. S., AND MCMASTER-KAYE, R. (1966). The fine structure and chemical composition of nuclei during spermiogenesis in the house cricket. I. Initial stages of differentiation and the loss of nonhistone protein. *J. Cell Biol. 31*, 159–179.

KESSEL, R. G. (1968a). Annulate lamellae. *J. Ultrastruct. Res., Suppl. 10*, 5–82.

KESSEL, R. G. (1968b). An electron microscope study of differentiation and growth in oocytes of *Ophioderma panamensis*. *J. Ultrastruct. Res. 22*, 63–89.

KESSEL, R. G. (1968c). Electron microscope studies on developing oocytes of a coelenterate medusa with special reference to vitellogenesis. *J. Morphol. 126*, 211–248.

KESSEL, R. G. (1970). Spermiogenesis in the dragonfly with special reference to a consideration of the mechanisms involved in the development of

cellular asymmetry. *In* "Comparative Spermatology" (B. Bacetti, ed.), pp. 531–552. Academic Press, New York.

KNIGHT, P. F., AND SCHECHTMAN, A. M. (1954). The passage of heterologous serum proteins from the circulation into the ovum of the fowl. *J. Exp. Zool. 127*, 271–304.

KOCH, E. A., SMITH, P. A., AND KING, R. C. (1967). The division and differentiation of *Drosophila* cystocytes. *J. Morphol. 121*, 55–70.

KOEHLER, J. K. (1970). A freeze-etching study of rabbit spermatozoa with particular reference to head structures. *J. Ultrastruct. Res. 33*, 598–614.

LONGO, F. J. (1972). An ultrastructural analysis of mitosis and cytokinesis in the zygote of the sea urchin *Arbacia punctulata*. *J. Morphol. 138*, 207–238.

LONGO, F. J., AND ANDERSON, E. (1969a). Sperm differentiation in the sea urchins *Arbacia punctulata* and *Strongylocentrotus purpuratus*. *J. Ultrastruct. Res. 27*, 486–509.

LONGO, F. J., AND ANDERSON, E. (1969b). Cytological aspects of fertilization in the lamellibranch *Mytilus edulis*. I. Polar body formation and development of the female pronucleus. *J. Exp. Zool. 172*, 69–96.

LONGO, F. J., AND ANDERSON, E. (1970). An ultrastructural analysis of fertilization in the surf clam *Spisula solidissima*. I. Polar body formation and development of the female pronucleus. *J. Ultrastruct. Res. 33*, 495–514.

LUDWIG, H. (1874). Uber die Eibildung im Thierreiche. *Arb. Physiol. Lab. Wirtzburg 1*, 287.

MAHOWALD, A. P. (1971). Origin and continuity of polar granules. *In* "Origin and Continuity of Cell Organelles" (J. Reinert and H. Ursprung, eds.), pp. 158–169. Springer-Verlag, New York.

MASSOVER, W. H. (1971). Intramitochondrial yolkcrystals of frog oocytes. I. Formation of yolkcrystal inclusions by mitochondria during bullfrog oogenesis. *J. Cell Biol. 48*, 266–279.

MAYS, U. (1972). Stofftransport in Ovar von *Pyrrhocoris apterus* L. *Z. Zellforsch. 123*, 395–410.

MCINTOSH, J. R., AND PORTER, K. R. (1967). Microtubules in the spermatids of the domestic fowl. *J. Cell Biol. 35*, 153–173.

MILLER, O. L. (1962). Studies on the ultrastructure and metabolism of amphibian oocytes. *Proc. 5th Int. Congr. Electron Microsc. 2*, NN-8.

MILLER, O. L. (1966). Structure and composition of peripheral nucleoli of salamander oocytes. *Nat. Cancer Inst. Monogr. 23*, 53–66.

MILLER, O. L., AND BEATTY, B. R. (1969). Nucleolar structure and function. *In* "Handbook of Molecular Cytology" (A. Lima-de-Faria, ed.), pp. 605–619. North-Holland, Amsterdam.

MOENS, P. B., AND GO, V. L. W. (1972). Intercellular bridges and division patterns of rat spermatogonia. *Z. Zellforsch.* 127, 201–208.

MONROY, A. (1965). "Chemistry and Physiology of Fertilization." Holt, New York.

MOSES, M. J. (1968). Synaptinemal complex. *Ann. Rev. Genet.* 2, 363–412.

NEUTRA, M., AND LEBLOND, C. P. (1966a). Synthesis of the complex carbohydrate of mucus in the Golgi complex as shown by electron microscope radioautography of goblet cells from rats injected with glucose-H^3. *J. Cell Biol.* 30, 119–136.

NEUTRA, M., AND LEBLOND, C. P. (1966b). Radioautographic comparison of the uptake of galactose-H^3 and glucose-H^3 in the Golgi region of various cells secreting glycoproteins or mucopolysaccharides. *J. Cell Biol.* 30, 137–150.

NOVIKOFF, A. B., ESSNER, E., AND QUINTANA, N. (1964). Golgi apparatus and lysosomes. *Fed. Proc.* 23, 1010–1022.

PETERS, H., LEVY, E., AND CRONE, M. (1965). Oogenesis in rabbits. *J. Exp. Zool.* 158, 169–180.

PHILLIPS, D. M. (1966a). Observations on spermiogenesis in the fungus gnat *Schiara coprophila*. *J. Cell Biol.* 30, 477–497.

PHILLIPS, D. M. (1966b). Fine structure of *Sciara coprophila* sperm. *J. Cell Biol.* 30, 499–517.

PHILLIPS, D. M. (1970a). Insect sperm: their structure and morphogenesis. *J. Cell Biol.* 44, 243–277.

PHILLIPS, D. M. (1970b). Insect flagellar tubule patterns: theme and variations. *In* "Comparative Spermatology" (B. Baccetti, ed.), pp. 263–273. Academic Press, New York.

PIKÓ, L., TYLER, A., AND VINOGRAD, J. (1967). Amount, location, priming capacity, circularity and other properties of cytoplasmic DNA in sea urchin eggs. *Biol. Bull.* 132, 68–90.

POTSWALD, H. E. (1967). An electron microscope study of spermiogenesis in *Spirorbis* (*Laeospira*) *morchi* Levinsen (Polychaeta). *Z. Zellforsch.* 83, 231–248.

PRATT, S. A. (1968). An electron microscope study of nebenkern formation and differentiation in spermatids of *Murgantia histrionica* (Hemiptera, Pentatomidae). *J. Morphol.* 126, 31–66.

QUATTROPANI, S. L., AND ANDERSON, E. (1969). The origin and structure of the secondary coat of the egg of *Drosophila melanogaster*. *Z. Zellforsch.* 95, 495–510.

RAVEN, C. P. (1970). The cortical and subcortex of cytoplasm of the *Lymnaea* egg. *Int. Rev. Cytol.* 28, 1–44.

ROTH, T. F., AND ITO, M. (1967). DNA-dependent formation of the synaptinemal complex at meiotic prophase. *J. Cell Biol.* 35, 247–255.

ROTH, T. F., AND PORTER, K. R. (1964). Yolk protein uptake in the oocyte of the mosquito *Aedes aegypti* L. *J. Cell Biol.* 20, 313–332.

SCHECHTMAN, A. M. (1955). Ontogeny of the blood and related antigens and their significance for the theory of differentiation. *In* "Biological Specificity and Growth" (E. G. Butler, ed.), pp. 3–31. Princeton University Press, Princeton, New Jersey.

SCHUETZ, A. W. (1967). Action of hormones on germinal vesicle breakdown in frog (*Rana pipiens*) oocyte. *J. Exp. Zool.* 166, 347–354.

SCHJEIDE, O. A., GALEY, F., GRELLERT, E. A., I-SAN LIN, R., DE VELLIS, J., AND MEAD, J. F. (1970). Macromolecules in oocyte maturation. *Biol. Reprod.*, Suppl. 2, 14–43.

SCHJEIDE, O. A., AND URIST, M. R. (1956). Proteins and calcium in serums of estrogen-treated roosters. *Science* 124, 1242–1244.

STEINERT, G., AND URBANI, E. (1969). Communications intercellulaires dans les ovarioles de *Dytiscus marginalis* L. *J. Embryol. Exp. Morphol.* 22, 45–54.

STERN, H., AND HOTTA, Y. (1967). Chromosome behavior during development of meiotic tissue. *In* "The Control of Nuclear Activity" (L. Goldstein, ed.), pp. 47–76. Prentice Hall, Englewood Cliffs, New Jersey.

STERN, H., AND HOTTA, Y. (1969). Biochemistry of meiosis. *In* "Handbook of Molecular Cytology" (A. Lima-de-Faria, ed.), pp. 520–539. North-Holland, Amsterdam.

STEVENS, A. R. (1967). Machinery for exchange across the nuclear membrane. *In* "The Control of Nuclear Activity" (L. Goldstein, ed.), pp. 189–271. Prentice-Hall, Englewood Cliffs, New Jersey.

SUD, B. N. (1961). Morphological and histochemical studies of the chromatoid body and related elements in the spermatogenesis of the rat. *Quart. J. Microsc. Sci.* 102, 495–505.

SZOLLOSI, D. (1967). Development of cortical granules and the cortical reaction in rat and hamster eggs. *Anat. Rec.* 159, 431–446.

TAYLOR, J. H. (1964). The arrangement of chromosomes in the mature sperm of the grasshopper. *J. Cell Biol.* 21, 286–289.

TELFER, W. H. (1960). The selective accumulation of blood proteins by the oocytes of saturniid moths. *Biol. Bull.* 118, 338–351.

TELFER, W. H. (1961). The route of entry and localization of blood proteins in the oocytes of saturniid moths. *J. Biophys. Biochem. Cytol. 9*, 747–759.

TELFER, W. H. (1965). The mechanism and control of yolk formation. *Ann. Rev. Entomol. 10*, 161–184.

TILNEY, L., AND MARSLAND, D. (1969). A fine structural analysis of cleavage induction and furrowing in the eggs of *Arbacia punctulata*. *J. Cell Biol. 42*, 170–184.

VOELLER, B. R. (1968). "The Chromosome Theory of Inheritance: Classic Papers in Development and Heredity." Appleton-Century-Crofts, New York.

WALLACE, R. A., AND DUMONT, J. N. (1968). The induced synthesis and transport of yolk proteins and their accumulation by the oocyte in *Xenopus laevis*. *J. Cell Physiol. 72*, Suppl. 1, 73–102.

WARD, R. T. (1962). The origin of protein and fatty yolk in *Rana pipiens*, II: Electron microscopical and cytochemical observations of young and mature oocytes. *J. Cell Biol. 14*, 309–341.

WARD, R. T., AND WARD, E. (1968). The origin and growth of cortical granules in the oocytes of *Rana pipiens*. *J. Microsc. 7*, 1021–1030.

WARTENBERG, H. (1964). Experimentelle Untersuchungen über die Stoffaufnahme durch Pinocytose während der Vitellogenese des Amphibienoocyten. *Z. Zellforsch. 63*, 1004–1019.

WILLIAMS, J. (1967). Chemical constitution and metabolic activities of animal eggs. *In* "The Biochemistry of Animal Development" (R. Weber, ed.), Vol. 1, pp. 13–71. Academic Press, New York.

WILLIER, B. H. (1937). Experimentally produced sterile gonads and the problem of the origin of germ cells in the chick embryo. *Anat. Rec. 70*, 89–112.

WILSON, E. B. (1925). "The Cell in Development and Heredity." Macmillan, New York.

ZAMBONI, L. (1971). "Fine Morphology of Mammalian Fertilization." Harper & Row, New York.

CHAPTER TWO

Fertilization

C. R. AUSTIN

The process of fertilization commonly involves the union of specialized male and female gametes and the combination of genetic material from paternal and maternal sources. As such, fertilization is known to exist throughout the animal and vegetable kingdoms—except in the bacteria, where variable portions of a single chromosome are transferred, and in certain Protista, where whole organisms temporarily combine (conjugate) in order to exchange modified nuclei. Fertilization is complemented in its function by meiosis, which entails the reassortment of genetic factors; the two together thus provide the basis for biparental inheritance and adaptive evolution. These genetic functions have become associated with processes of multiplication to constitute sexual reproduction.

An account of fertilization necessarily includes treatment of certain ancillary matters, such as (1) the transport of gametes to the site of fertilization (a complex series of events in higher animals), (2) the cytological, physiological, and biochemical features of gamete

union, and (3) the nature of the regulatory mechanisms that determine the precise outcome of the process. In addition, this chapter deals with the investigative approach to fertilization, and with some of the anomalies that appear spontaneously or following experimental intervention.

Assembly of Gametes

SPERM MOTILITY

The spermatozoa of most members of the Metazoa are capable of movement; generally this takes the form of free-swimming activity. Swimming movements are to be attributed to the possession of a tail or flagellum, or of an undulating membrane, the beating of which propels the cell through the medium. In some instances the spermatozoon has two flagella (as in the fish *Opsanus*), or two indulating membranes (as in the amphibian *Pseudobranchus*). Motility may also be produced through flowing movements of the cytoplasm, as in the amoeboid spermatozoa of the mesozoan *Dicyema* and the round worm *Ascaris* or the elongated strap-like spermatozoa of the liver fluke *Fasciola*. Another form of activity is seen in the bulbous spermatozoa of the tick *Argas:* fine, wave-like elevations passing along the cell surface are responsible for moving the bulky cell forward. Finally, there are immotile spermatozoa, the best known probably being those of the decapod crustaceans—rigid stellate or pronged spermatozoa that depend on purely passive transport. (Fuller details and additional references regarding sperm structure may be found in the reviews by Bishop and Austin, 1957, and Fawcett, 1970.)

The flagellum in most flagellate spermatozoa contains an axial filament consisting of an array of microtubules similar to that found in cilia—a ring of nine doublets with two single microtubules running through the center. Mammalian spermatozoa possess additionally an outer ring of nine large "coarse" fibrils, each located peripherally to a doublet of the axial filament. Coarse fibrils of smaller girth are seen in the spermatozoa of some nonmammals (snake, sparrow, snail, grasshopper, honeybee, and fruitfly).

These fibrillar and filamentous structures are considered to be the active elements in sperm motility (see Nelson, 1967). Flagellar movement consists of more or less sinusoidal bending waves that pass distally and that are two-dimensional in most animal spermatozoa and three-dimensional in mammalian spermatozoa. The energy for this movement apparently depends upon the presence of ATP associated chiefly with a characteristic myosin-like protein known as *spermosin*, which is thought to bind with an actin-like protein ("flactin") in a manner analogous to the actin–myosin interaction in muscle. The energy-rich phosphorus compounds are regenerated by endogenous or exogenous catabolic processes—only the former process occurs in spermatozoa such as those of the sea urchin, whereas both processes occur in those of mammals. Endogenous metabolism consists of the oxidation of intracellular phospholipid. The phospholipid in question has been shown to be plasmalogen—and not lecithin, as proposed by Lardy and Phillips; its metabolism involves hydrolytic cleavage of the fatty acids, which, on oxidation, can provide fully for energy requirements (see Mann, 1964).

A wide range of extracellular substrates, including sugars, organic acids, and alcohols, can be metabolized by mammalian spermatozoa, either by oxidative (aerobic) or glycolytic (anaerobic) pathways; the two systems are likely to be involved to different degrees at various times in the life of the spermatozoon. In the female tract the oxygen partial pressure is probably adequate to maintain oxidative metabolism, except when the spermatozoa are first deposited and exist together in very large numbers; under these circumstances glycolytic processes may well predominate. The principal glycolysable carbohydrate in the semen of most mammals is fructose. (For further details on sperm metabolism, see Mann, 1964).

CHEMOTAXIS

Substances that attract the spermatozoa to the egg cell are known to exist in a variety of primitive plants, such as the algae, ferns, mosses, horsetails, and liverworts. In many instances the chemical nature of the attractant, commonly a simple compound, has been identified—for example, sucrose in *Funaria* and L-malic acid in *Pteridium*. For many years it was thought that the chemical attraction of spermatozoa was restricted to the plant kingdom, and early reports of apparent chemotaxis in animal gametes were explained on the basis of trap action. Yanagimachi's (1957) observations on *Clupea*, however, clearly pointed to the existence of a chemotactic mechanism: the spermatozoa are immotile until they are brought into close relation with the egg chorion, especially in the vicinity of the micropyle, whereupon they begin swimming with great vigor in the direction of the micropylar canal, which they then pass through (Fig. 1). More recently, evidence for the operation of chemotaxis has been presented for certain hydroids, such as *Campanularia* (Miller, 1966; Fig. 2).

The suggestion has been made that chemotaxis may also play a role in mammals—in the meeting of egg and spermatozoon at the site of

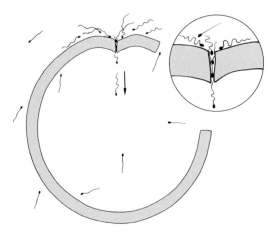

FIGURE 1. *Behavior of spermatozoa near the chorion of the herring* (Clupea) *egg, after removal of the vitellus. The spermatozoa are motionless until they drift onto the surface of the chorion near the micropyle. They then show vigorous motility and soon swim into and through the micropyle. (Redrawn from R. Yanagimachi, Anat. Zool. Japan 30, 114, 1957. From C. R. Austin, "Fertilization," Prentice-Hall, Englewood Cliffs, New Jersey, 1965.)*

fertilization: in mated rats and mice it has often been observed that spermatozoa can be found in all the eggs (nearly always one per egg), and yet there appear to be no free spermatozoa in the vicinity. But the idea still lacks actual experimental evidence, and the observations could well be explicable by the difficulty of finding spermatozoa among the follicle cells and the efficiency with which the "zona reaction" (discussed later in the chapter) excludes supernumerary spermatozoa from eggs. (Various aspects of chemotaxis are discussed by Trinkaus, 1969.)

TRANSPORT OF SPERMATOZOA

In all species there are mechanisms that help to bring together the egg and spermatozoon; this is true both for motile and immotile spermatozoa but is naturally more important for the immotile. Transport begins at the site of sperm production and leads to the exterior, after which there may be a phase of free-swimming activity. Probably the simplest situation

exists in marine organisms that reproduce by external fertilization: a minimum amount of transport is involved in the shedding of the male gamete, and thereafter it propels itself through the medium, in which eddies and currents may well assist in bringing it to the female gamete.

Where fertilization is internal, a remarkable range of different devices and mechanisms is involved in transporting spermatozoa into the vicinity of eggs. The male spider constructs a specialized "sperm" web, on which a drop of semen is deposited. The fluid is taken up in the palps, which are equipped with a duct and receptacle for holding the semen, and by means of which the male deposits the material in the spermathecae of the female. In the scorpion *Opisthophthalmus* the spermatozoa are embedded in a rod-like spermatophore that adheres to the ground; when the female, moving over the spermatophore, brushes against it, the structure "snaps" in the middle, projecting the spermatozoa into her genital pore (see Savory, 1964). In the onychophoran *Peripatopsis*, the bug *Cimex*, and the leech *Clepsine*, spermatophores are attached by the male to the external body surface of the female. The body wall immediately beneath the spermatophore becomes eroded, and spermatozoa escape into the tissues

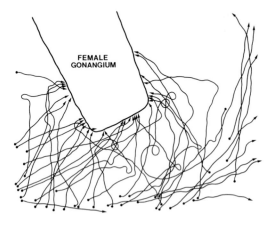

FIGURE 2. *Tracks of spermatozoa swimming in the region of the female gonangium of* Campanularia flexuosa. *Except for a few spermatozoa on the right, almost all move in an oriented manner toward the gonangium. (Reproduced with small modifications from R. L. Miller, J. Exp. Zool. 162, 23, 1966.)*

of the female, through which they migrate to reach the eggs (see Davey, 1965). Among the cephalopods one "arm" is modified to form a so-called "hectocotyl" (see Morton, 1958). Spermatozoa are released from the male duct packaged within specialized spermatophores, a bundle of which is grasped by the hectocotyl arm and transferred to the mantle cavity of the female. Contact with the seawater initiates an explosive reaction in the spermatophores, liberating the spermatozoa, which then find their way into the seminal receptacle of the female (see Austin et al., 1964, for *Loligo*, and Mann et al., 1970, for *Octopus*).

Most is known about sperm transport in mammals. The spermatozoa leave the testis through the efferent ducts in a stream of fluid actuated by secretion pressure and the beating of cilia in the rete testis. Thus they reach a single, much-coiled tube known as the *epididymis*, in which the greater part of the suspending fluid is absorbed, so that the sperm suspension becomes dense and viscous. Passing through this organ, the spermatozoa undergo both structural and functional changes, which will be discussed later. Leaving the epididymis, spermatozoa reach the thick-walled vas deferens and from here are conducted into the male urethra. At coitus the spermatozoa become suspended in the secretions from several glands associated with the vas and the urethra. The mixture, which composes the semen, is then conveyed to the female tract by contractions of smooth muscle layers surrounding the urethra.

The principal male sex glands are the ampullary glands, the prostate, the seminal vesicles, the coagulating glands, the bulbourethral (Cowper's) glands, and the urethral (Littré's) glands; their relative development and importance differ widely in different species. The activity of the glands is under hormonal control and is especially sensitive to blood levels of testosterone (see Mann, 1964).

The site of semen deposition in mammals is the vagina, or the cervix or lumen of the uterus, depending on species; from this point, sperm transport is attributable to muscular activity in the walls of the uterus and oviducts, and to currents produced by cilial beat in the oviducts. The motility of the spermatozoa seems to play little effective role in transport,

its main function probably being that of maintaining the spermatozoa in suspension; but it may have some significance in the passage through the cervical mucus in the human uterus, in the traversing of the uterotubal junction, and in the penetration of the spermatozoa through the egg investments. Other factors also seem likely to influence passage through the cervix—at least in the rabbit; treatment of semen with univalent nonagglutinating antibody, which had no noticeable effect upon motility, was found to reduce conception rate dramatically (Metz and Anika, 1970). The effect was thought possibly to be due to inhibition of a sperm enzyme involved in penetration through the cervical mucus.

During the course of transport through the female tract, the sperm suspension undergoes considerable dilution, so that the originally vast numbers are reduced to a very few (Fig. 3). The rate of transport is rapid in most species—in several instances only about 5 or 10 minutes are required for the spermatozoa to achieve the site of fertilization from the time of deposition (Table 1). The rate of transport

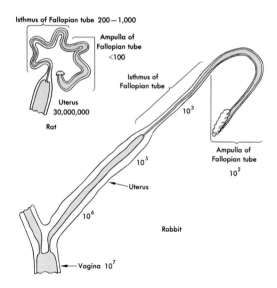

FIGURE 3. *Changes in lumen diameter in the genital tracts of the rat and the rabbit, and the numbers of spermatozoa reaching different regions. Ejaculation in the rat is into the uterus, and in the rabbit into the vagina. (From C. R. Austin, "Fertilization," Prentice-Hall, Englewood Cliffs, New Jersey, 1965.)*

TABLE 1

INTERVAL BETWEEN EJACULATION AND APPEARANCE OF SPERMATOZOA IN OVIDUCT[a]

Species	Time and Area
Mouse	15 min: upper region of oviduct
Rat	15–30 min: ampullae
Hamster	2 min: ampullae
	60 min: ampullae
Rabbit	3 hr: in oviduct
	5–6 hr: sufficient spermatozoa in oviduct to fertilize all eggs
Guinea pig	15 min: middle of oviduct few seconds; in oviduct
Dog	2 min: isthmus
	Few hours: in oviduct
Sow	15 min: in ampullae
Cow	2 min: in ampullae
	13 min: in ampullae
Ewe	8–30 min: in ampullae
	6 min: in ampullae
	2½ hr: in ampullae
	5 hr: in ampullae
Man	68 min: in oviduct in extirpated tracts *in vitro*
	30 min: in oviduct removed at surgery

[a]From R.J. Blandau, Gamete transport—comparative aspects, *in* "The Mammalian Oviduct" (E. S. E. Hafez and R. J. Blandau, eds.), University of Chicago Press, Chicago, 1969. Copyright © 1969, The University of Chicago Press.

depends a good deal upon the hormonal status of the female: it is most rapid at estrus when estrogen influence is predominent and may depend upon the stimulating action of oxytocin from the posterior pituitary upon the musculature of uterus and oviduct. Prostaglandins in the semen could also play a role by increasing muscular activity in parts of the female tract.

TRANSPORT OF EGGS TO SITE OF FERTILIZATION

Where fertilization is external, the eggs are simply released through body pores or ducts leading from the ovary to the exterior, but eggs often receive one or more enclosing coats during passage through the ducts; presumably these investments, together with those deposited in the ovary, have a protective function and may also tend to prevent entry by heterologous spermatozoa. In animals with internal fertilization, sperm entry may take place while the oocyte is still in the ovary, as in certain insects (see Davey, 1965, p. 56) and some insectivores (members of the Tenrecidae—see Strauss, 1950). This event can also occur in man—but very rarely (see Ashley, 1959). Otherwise, the site of internal fertilization is represented by a specialized chamber such as a seminal receptacle or, in birds and mammals, by the ovarian region of the oviduct.

In mammals, the eggs are released onto the surface of the ovary and soon come under the influence of fluid currents set up by the beating of cilia lining the inner surface of the oviduct infundibulum. The cilia may also exert traction directly on the granulosa-cell mass surrounding each egg—the main function of this investment could be to provide such a "handle." The infundibulum is often expanded into a funnel-like structure and may have a fringe (fimbria) of finger-like processes or be deeply fissured, so that the ciliated surface is very extensive (Fig. 4). The eggs are thus carried into the ampulla of the oviduct, pausing usually in a somewhat distended region where the greater part of the fertilization process occurs.

In some mammals (*Dasyurus, Didelphis, Setonix,* and *Echidna;* see Austin, 1961, for references), the eggs traverse the oviduct and

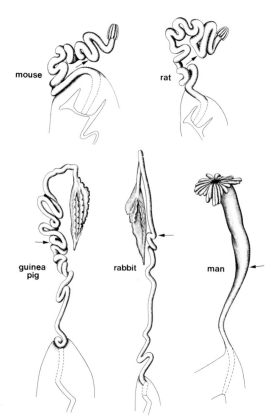

FIGURE 4. *Oviducts of five mammals. Arrow indicates junction between ampulla and isthmus. (From O. Nilsson and S. Reinino, in "The Mammalian Oviduct" (E. S. E. Hafez and R. J. Blandau, eds.), University of Chicago Press, Chicago, 1969. (Copyright © The University of Chicago Press.)*

enter the uterus during fertilization; in the majority this happens during the eight-cell-to-morula stage of cleavage. Movements of eggs within the oviduct are attributable also to the contractile activity of the walls of this organ. (The subject of egg transport in mammals has recently been treated in detail by Blandau, 1969.)

CHANCE ENCOUNTER BETWEEN EGG AND SPERMATAZOON

Under all circumstances chance appears to play some part in the meeting of eggs and spermatozoa. In the animal kingdom, probably the least is left to chance in the insects. The honey-bee, for example, lays approximately a quarter of a million eggs in her fertile lifetime, most of which are fertilized; since the queen is known to be inseminated only on her nuptial flight, the spermatozoa stored in her spermathecae must be made to suffice, and it has been calculated that roughly 1 in every 30 spermatozoa succeeds in fertilizing an egg. By contrast, some hundreds of millions of spermatozoa are normally deposited in the vagina for the fertilization of a single human egg. The actual numbers determining the chances of sperm–egg meeting, however, are of a much lower order: at the site of fertilization in mammals there may be only a few dozen or at most a few hundred spermatozoa in striking distance of the egg. The prospects of effective sperm–egg encounter may be improved by the granulosa-cell mass, which would provide a larger target than the naked egg, and by the radial arrangement of the granulosa cells, which might well direct the spermatozoa toward the egg.

Chance clearly plays a major role in the meeting of gametes in the marine invertebrates, where both eggs and spermatozoa are released into the sea; but even here there are mechanisms that promote the meeting. For one thing, gamete shedding is rarely at random: it is often related to specific times such as dawn or dusk, and it appears to be stimulated by close approximation of male and female organisms. Furthermore, substances shed with the gametes appear to stimulate shedding in nearby individuals of the same species.

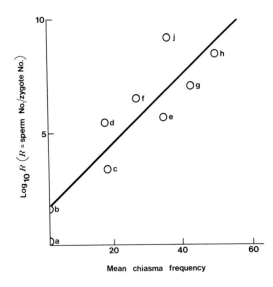

FIGURE 5. *Sperm number/zygote number and mean chiasma frequency for species for which both figures have been published. (a) Drosophila; (b) Apis; (c) Schistocerca; (c) Chinese hamster; (e) mouse; (f) golden hamster; (g) pig-tailed macaque; (h) man; (j) bull. (From J. Cohen, Nature 215, 826, 1967.)*

Where wastage of male gametes is high, as in mammals, there would seem to be circumstances appropriate for gamete selection. Pursuing this idea, Cohen (1971) noted that the chiasma frequency was broadly correlated with sperm number over a wide range of species (Fig. 5). He surmised that the risks of error in spermatogenesis are related to the chiasma frequency and, consequently, that animals with high frequencies must produce many defective spermatozoa, an effect to be compensated for by large sperm numbers.

SPERM–EGG ATTACHMENT

When spermatozoa and eggs have finally come together, attachment between the two precedes fusion (see Metz, 1967). The probable underlying mechanism of this attachment in the sea urchin was first studied in detail by Lillie (1919), who identified the fertilizin–antifertilizin system. Fertilizin is a major component of the egg jelly coat, and it gradually passes into solution in the surrounding sea-

water. When in solution, it has the property of agglutinating spermatozoa, the surface of which carries the complementary substance antifertilizin. The effect has several features in common with an antigen–antibody reaction. Under normal circumstances, fertilizin is unlikely to accumulate to agglutinating concentrations in the surrounding medium, and the fertilizin–antifertilizin interaction would have the function of attaching spermatozoa to the egg surface.

Similar conditions have been demonstrated in many different species, vertebrate as well as invertebrate. The apparent lack of such a mechanism in some instances is possibly attributable to inappropriate experimental conditions, and, therefore, a not unreasonable inference is that the fertilizin–antifertilizin mechanism, or something like it, exists in all species that show sperm–egg attachment. The reaction has species specificity; thus, in addition to ensuring a high probability of fertilization, it may well have an important function in limiting hybridization. With invertebrate spermatozoa, fertilizin may also be the substance responsible for initiating the acrosome reaction, the nature of which is discussed later.

Acquisition and Loss of Fertility in Gametes

MATURATION AFTER RELEASE FROM GONAD

Sometimes gametes recovered from the gonads can be shown to be capable of taking part immediately in fertilization, but more commonly some degree of maturation or ripening is necessary before fertilization can occur (similarities and differences show little or no link with taxonomic relationships).

Eggs

In the sea urchin, the eggs are released after they have completed the two polar divisions, and they are then in a state ready for fertilization. In the starfish, however, the eggs are shed as primary oocytes with intact germinal vesicles. Maturation apparently depends on chemical substances released from the ovary at the time of shedding. For *Asterias*, the substance has been identified as 1-methyladenine (Kanatani et al., 1969); it can be used under experimental conditions to induce maturation of oocytes recovered from dissected ovaries; it will also induce shedding. Only after such maturation are the eggs capable of taking part in fertilization, though precocious sperm entry may occur. In the cone worm *Pectinaria*, the eggs begin to undergo maturation soon after shedding, but if sperm penetration occurs, maturation stops and the male pronucleus fails to form (Austin, 1963). The actual stage at which fertilizability is achieved differs in different species—in some before meiosis, in others after the formation of the first polar body, and in others again after the formation of the second polar body (Table 2).

Spermatozoa

The occurrence of some form of terminal maturation in mammalian spermatozoa has already been noted as taking place during their passage down the epididymis. Ripening of spermatozoa in this manner after release from the gonad is rather a rare phenomenon among animal species other than mammals. As they leave the testis in mammals, the spermatozoa superficially resemble those of the semen but do in fact differ in both morphology and functional ability. Young (1931), who briefly reviewed the earlier literature, noted that spermatozoa from the cauda epididymidis were more motile than those from the caput and corpus, and, following artificial insemination, showed a higher fertility. However, they also gave rise to embryonic development marred by a high incidence of absorption and resorption. The motility and fertility changes were attributed to progress in functional development, and the

TABLE 2

STAGE OF EGG MATURATION AT WHICH SPERM PENETRATION OCCURS
IN DIFFERENT ANIMALS[a]

Young primary oocyte	Fully grown primary oocyte	First metaphase	Second metaphase	Female pronucleus
The annulate worm *Dinophilus*	The round worm *Ascaris*	The nemertine worm *Cerebratulus*	The lancelet *Amphioxus*	Coelenterates, e.g., anemones
The polychaete worm *Histriobdella*	The mesozoan *Dicyema*	The polychaete worm *Chaetopterus*	The amphibian *Siredon*	Echinoids, e.g., sea urchins
The flatworm *Otomesostoma*	The sponge *Grantia*	The mollusk *Dentallium*	Most mammals	and starfish
The onychophoran *Peripatopsis*	The polychaete worm *Myzostoma*	The cone worm *Pectinaria*		
The annulate worm *Saccocirrus*	The clam worm *Nereis*	Many insects		
	The clam *Spisula*			
	The echiuroid worm *Thalassema*			
	Dog and fox			

[a]From C. R. Austin, "Fertilization," Prentice-Hall, Englewood Cliffs, New Jersey, 1965.

embryonic losses to the consequences of aging of the spermatozoa.

Later Lagerlöf (1934) observed in bull spermatozoa a change in position and loss of the *kinoplastic* (or *cytoplasmic*) *droplet*, associated with passage through the epididymis; this relation has subsequently often been confirmed. In addition, other workers have reported changes in the shape and composition of the acrosome (Bedford, 1965; Fawcett and Hollenberg, 1963), increased resistance to cold shock, and alterations in metabolism and membrane permeability (see Salisbury and Lodge, 1962).

More recently, Bedford (1966) and Orgebin-Crist (1967), working with rabbits, showed that though fertilizing ability was demonstrable in spermatozoa from the corpus epididymidis, it reached normal levels (more than 90 percent of eggs fertilized) only in the cauda epididymidis. In part, at least, the low fertilizing ability seemed attributable to defective attachment of spermatozoa to eggs, reflecting a late development of appropriate cell membrane properties in the spermatozoa, and possibly a longer time requirement for capacitation (see below). With spermatozoa from the lower corpus or the cauda epididymidis about 10 percent of eggs had three or four pronuclei. Additionally, Orgebin-Crist reported that the initiation of cleavage was delayed and that the pre- and postimplantation losses were well above normal. In an additional communication (see Orgebin-Crist, 1969), these effects were attributed to slow passage of epididymal spermatozoa to the site of fertilization and to contact with eggs. The eggs were therefore fertilized in an aged state, which is well known to result in an increased incidence of polyploidy (discussed in the last part of this chapter).

Capacitation of Spermatozoa

Even with spermatozoa recovered from the distal epididymis, the vas deferens, or the ejaculated semen, the story is not yet complete, for the spermatozoa are still incapable of taking part *immediately* in fertilization: a change known as *capacitation* is first required. The need for capacitation has been demonstrated in several mammals, and estimates of the time in-

volved are available for some: about 5 hours in the rabbit, 3 to 6 hours in the pig, 2 to 3 hours in the rat, 2 to 4 hours in the hamster, $1\frac{1}{2}$ hours in the sheep, $3\frac{1}{2}$ to $11\frac{1}{2}$ hours in the ferret, and an indeterminately brief time in the mouse (for references, see Austin, 1970b). Data are consistent with a period of 5 or 6 hours in the rhesus monkey (Marston and Kelly, 1968) and man (Edwards et al., 1969). Capacitation has been induced *in vitro* with the spermatozoa of the golden hamster (Barros and Austin, 1967), Chinese hamster (Pickworth and Chang, 1969), man (Edwards et al., 1969), mouse (Iwamatsu and Chang, 1969), and guinea pig (Yanagimachi, 1972).

When considering possible mechanisms of capacitation, preference is commonly given to the idea that some sort of stabilizing surface layer is thereby removed from the spermatozoon; but despite prolonged investigation, evidence for the actual existence on the sperm surface of an extraneous coat that could exert a protective or stabilizing influence, and be removed under the conditions for capacitation, remains fragmentary and largely circumstantial. Extraneous coats, usually mucopolysaccharide in nature, have been demonstrated on several other types of cells by enzyme treatment or by light and electron microscopy, but these methods have not as yet yielded any significant results with spermatozoa.

Ejaculated rabbit spermatozoa are readily stainable supravitally with the fluorofor tetracycline hydrochloride, and the removal of the stain has been proposed as an index of capacitation, since the fluorescence disappeared after incubation in the estrous uterus (see Ericsson, 1967). The idea that removal of fluorescence can be equated with capacitation is denied by Vaidya et al. (1969), who find that staining disappears in both uterus and oviduct in about one fifth the time required for demonstrable capacitation; the loss of staining may nevertheless represent direct evidence that removal of the surface coat is involved in the capacitation process.

Some evidence on the surface-coat problem may be forthcoming from observations on the "decapacitation" of rabbit spermatozoa—the apparent reversal of capacitation brought about by submitting the spermatozoa to resuspension in seminal plasma (Chang, 1957; Bedford and

Chang, 1962); but arguments along these lines are rather speculative. Information on the hormonal control of capacitation in the rabbit and on the role of genital secretions and cellular elements in the process has recently been summarized by Soupart (1971).

Better prospects for an improvement in understanding of the nature of capacitation seem to be offered by experiments with the gametes of the golden hamster, because with this species capacitation can be obtained *in vitro* (Yanagimachi and Chang, 1964). Barros and Austin (1967) showed that eggs recovered from immediately preovulatory follicles could be fertilized at high incidence with epididymal spermatozoa—after an incubation period of a little less than 4 hours; thus an essential role in fertilization could not be attributed to any component of the female tract secretions in this species. It was inferred that the agent inducing capacitation probably originated in the ovarian follicle. This idea was developed by Yanagimachi (1969), who interpreted his data as showing that there were in fact two essential factors and that these existed in the follicular fluid. Fractionation of follicular fluid by dialysis and salt precipitation led him to the conclusion that one factor stimulated motility and the other induced the acrosome reaction. The former was dialysable and heat stable, and the latter nondialysable and heat labile.

A somewhat different point of view was initiated by Bavister (1969, 1971), who stressed the importance of culture conditions, notably the ionic composition of the medium, the use of CO_2–air mixtures, and the maintenance of an appropriate pH in the medium. He was in fact able to define a medium—essentially Tyrode's solution with added sorbitol and polyvinylpyrrolidone, and a pH of 7.8—in which hamster spermatozoa often survived well, showed increased motility, and apparently achieved the state of capacitation after 3 to 4 hours incubation. On being transferred to a preparation containing eggs and follicular fluid, the spermatozoa displayed acrosome reactions within 15 minutes and began to penetrate eggs about $\frac{1}{2}$ hour later. From this work it is inferred that the requirements for capacitation of hamster spermatozoa can be met by a defined medium and that only the acrosome reaction depends upon a specific factor, which is present in follicular fluid (see also Austin et al., 1972). The nature of the factor remains unknown.

Capacitation evidently makes it possible for the sperm acrosome to undergo its characteristic reaction in which enzymes are released. More will be said about the acrosome reaction later.

Something resembling capacitation has been observed in the anurans (Shivers and James, 1971; Wolf and Hedrick, 1971). Frog and toad eggs have been shown to be fertilizable only if surrounded by a jelly coat through which the spermatozoa must pass; denuded eggs can be fertilized provided the spermatozoa have previously been in contact with the jelly coat. The time relations in the anurans are much shorter than in mammals—minutes rather than hours. A change corresponding to capacitation seems also to be a necessary preliminary to fertilization by spermatozoa of the hydrozoan *Campanularia* (O'Rand, 1971).

LOSS OF FERTILITY IN GAMETES

Neither spermatozoa nor eggs are notable for prolonged vitality—with a few striking exceptions. The queen honeybee's store of spermatozoa may retain fertility for 2 to 3 years (Snelgrove, 1946; Davey, 1965), and intervals of the same order or even longer have been reported for sperm fertility in certain reptiles (*Malaclemmys centrata, Terrapene carolina, Drymarchon corais, Leptodeira annulata polysticta;* see Parkes, 1960). In the domestic hen, fertile eggs continue to be laid up to 3 weeks after a single mating or artificial insemination (Polge, 1955).

Among mammals, the record for sperm longevity is held by certain bats, notably *Myotis* and *Eptesicus* spp., in which early embryonic development was demonstrated 138 and 156 days, respectively, after coitus and isolation (hibernation) (see Parkes, 1960). Of the domestic mammals, the horse probably has the spermatozoa with the longest fertile life—about 5 or 6 days, though the estimate for the dog is not much shorter (Table 3). Fertile life in human spermatozoa has been difficult to assess owing to the problem of detecting ovulation; one expert committee recently gave an esti-

TABLE 3

DURATION OF SPERM FERTILITY AND MOTILITY IN MAMMALIAN OVIDUCT

Animal	Fertility (hr)	Motility (hr)
Mouse	6	13
Rat	14	17
Guinea pig	21–22	41
Rabbit	28–32; 43–50	—
Ferret	36–48; 126	—
Dog	134 (est.)	268
Sheep	24–48	48
Cow	24–48	96
Pig	24–48	—
Horse	144	144
Bat	138–156 (days)	149–156 (days)
Man	24–48	48–60

mate of about 2 days (WHO Technical Report Series, No. 333, 1966), and another remarked that "figures suggest that the fertile period has a maximum duration of 4 days, but the average is probably much less" (WHO Technical Report Series, No. 360, 1967).

In mammals there is some evidence that fertilization by spermatozoa near the end of their fertile life (aging spermatozoa) may lead to developmental anomalies (see Bishop, 1970; Salisbury and Hart, 1970).

The fertile life of the egg is subject to much

TABLE 4

ESTIMATED FERTILIZABLE LIFE OF SOME MAMMALIAN EGGS

Animal	Length of fertilizable life
Rabbit	6–8 hours
Guinea pig	Not more than 20 hours
Rat	12–14 hours
Mouse	8–12 hours
Hamster	5 hours, 12 hours
Ferret	Up to about 36 hours
Pig	About 20 hours
Sheep	15–24 hours
Cow	22–24 hours
Mare	About 24 hours
Rhesus monkey	Probably less than 24 hours
Man	Not more than 24 hours

TABLE 5

PROPORTIONS OF NORMAL AND ABNORMAL PREGNANCIES
AND LITTER SIZES AFTER INSEMINATION AT VARIOUS TIMES IN
THE GUINEA PIG[a]

	Pregnancies (%)	Litter size	Abnormal pregnancies (%)
Controls	83	2.6	12
8 hr after ovulation	67	1.7	34
14 hr after ovulation	56	1.6	73
20 hr after ovulation	31	1.3	90
26 hr after ovulation	7	0	100

[a]Data from R. J. Blandau and W. C. Young, *Amer. J. Anat. 64*, 303, 1939.

less variation than that of the spermatozoon (for mammals, see Table 4). It has long been recognized that the eggs of marine invertebrates are very limited in this respect, and, further, that fertilization of "stale" eggs is highly prone to result in embryonic abnormality (see Costello et al., 1957). The effect has been studied experimentally in anurans, notably by Witschi and Laguens (1963), who reported a variety of anomalies associated with chromosomal aberrations. In mammals information was obtained by performing artificial insemination or by allowing mating at increasing times after ovulation, and noting the effects on development (Tables 5 and 6). The mechanisms involved seem chiefly to concern the behavior of chromosomes in the second meiotic division and the functioning of the blocks to polyspermy (see reviews by Witschi, 1969; Austin, 1970a). The contribution that aging of the eggs appears to make toward prenatal loss in man is illustrated diagrammatically in Fig. 6.

TABLE 6

PROPORTIONS OF NORMAL AND ABNORMAL PREGNANCIES
AND LITTER SIZES AFTER INSEMINATION AT VARIOUS TIMES
IN THE RAT[a]

	Pregnancies (%)	Litter size	Abnormal pregnancies (%)
10–12 hr before ovulation	50	3.5	0
Near time of ovulation	83	6.7	11
6 hr after ovulation	47	4.6	48
10 hr after ovulation	22	1.8	79
12 hr after ovulation	4	0	100

[a]Data from R. J. Blandau and E. S. Jordan, *Amer. J. Anat. 68*, 275, 1941, and A. L. Soderwell and R. J. Blandau, *J. Exp. Zool. 88*, 55, 1941.

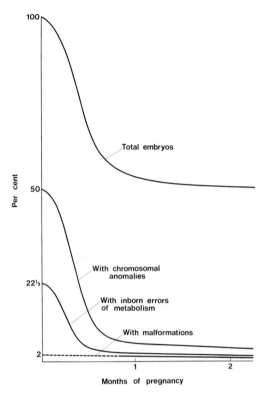

FIGURE 6. *Major losses of embryos occur during human pregnancy, principally within the first few weeks, and about half these losses are of embryos with chromosomal anomalies. Approximately 20 per cent of the chromosomal anomalies take the form of triploidy which has often been shown in experimental animals to result from the fertilization of ageing eggs.*

Passage Through Egg Investments

TYPES OF INVESTMENTS

In all species the egg proper is a cytoplasmic body limited by a plasma membrane. In addition, there are generally one or more coatings, sometimes of considerable thickness, composed of mucoprotein or mucopolysaccharide, though an outer hull of keratinous material may also be found (see Costello et al., 1957; Austin, 1961). Where keratinous or similar tough hulls exist, as in most if not all insects, cephalopods, and fish (other than lampreys), they regularly display one or more micropiles

through which the spermatozoa find access to the egg. Mucoprotein coats must be penetrated by the spermatozoon with the aid of enzymes. The thickness of these coats varies greatly—from the broad jelly coat of the sea urchin egg to the relatively narrow chorion of starfish, mussel, and clam eggs, and the zona pellucida of mammalian eggs (especially thin in the monotreme egg). In some instances, as in the rabbit among mammals and in *Nereis*, a jelly coat is deposited or extruded after sperm penetration; these coats are impermeable to spermatozoa.

ROLE OF ENZYMES

Sperm enzymes, capable of exerting a lytic action on egg investments, are clearly of significance in sperm penetration. In some instances these have been well characterized (see next section). In mammals the sperm enzymes exert a rapid lytic action upon the jelly-like mucoprotein matrix of the granulosa-cell mass, which invests the freshly ovulated egg. (The effect can be inhibited by antibodies that also prevent sperm penetration—Metz et al., 1972.) Similarly, lytic agents (of a character as yet unidentified) are believed to be implicated in the passage of the spermatozoon through the chorion surrounding the eggs of the toad *Discoglossus*, the keyhole limpet *Megathura*, the mussel *Mytilus*, the serpulid worm *Hydroides*, the clam worm *Nereis*, and the echiurid worm *Urechis* (see Dan, 1967, for references).

An anomaly apparently exists, however, in the case of the zona pellucida of the mammalian egg. Ultrastructural studies have shown that, by the time the spermatozoon has reached the surface of the zona pellucida, the acrosome reaction has gone to completion, and the enzymatic contents of this organelle have been dissipated. Nevertheless, some kind of enzymatic reaction would seem to be necessary for the passage of the spermatozoon through the zona pellucida. At this stage the anterior part of the sperm head is lined by what was previously the inner acrosome membrane, which is now continuous posteriorly with the cell membrane of the spermatozoon; it is thus surmised that molecules having a lytic action must be rather firmly attached to the inner acrosome membrane, perhaps in a manner similar to the association of proteases with cell membranes. Thus disposed, the lytic agent would not diffuse from the sperm head, and we should have an explanation for the fact that the spermatozoon passing through the zona pellucida leaves only a narrow slit behind.

RELEASE OF ENZYMES: ACROSOME REACTION

Spermatozoa of the majority of animal species possess acrosomes. These generally take the form of a vesicular structure lying at the anterior tip of the sperm head between the nucleus and the cell membrane. Sometimes the nucleus is indented to accommodate the acrosome; in other cases the acrosome is molded over the anterior part of the nucleus. The spermatozoa of *Ascaris* and other nematodes are amoeboid in character and carry a number of acrosome-like bodies in part of the peripheral cytoplasm (Foor, 1970).

The contents of the acrosome are inferred to be enzymatic, and in a few instances this has been demonstrated experimentally. Observations on the acrosomal contents of mammalian spermatozoa—obtained by treating the spermatozoa with surface-active agents, followed by differential centrifugation, and then extraction of the isolated acrosomes—indicate that the contents of these bodies is a lipoglycoprotein complex. The material is probably responsible for the PAS reactivity of the acrosome (Onuma and Nishikawa, 1963). The complex is made up of galactose, mannose, fucose, glucosamine, galactosamine, and sialic acid, as well as several enzymes—including one that has a lytic effect on the zona pellucida of the egg (Srivastava et al., 1965). Most abundant are hyaluronidase and a trypsin-like enzyme; also present are catalase, carbonic anhydrase, and lactic dehydrogenase (Stambaugh and Buckley, 1969, 1970). Because the acrosome contains this array of enzymes, together with acid phosphatase, aryl sulphatase, β-N-acetylo-glucosaminidase, and phospholipase A, Allison and Hartree (1970) maintained that it is lysosomal in nature. Consistently, both sperm acrosomes and lysosomes are vesicular structures that arise in association with the Golgi apparatus. Both fluoresce bright red when the cells are treated with acridine orange and irradiated in the near ultraviolet (Bishop and Smiles, 1957; Allison and Young, 1969).

PATTERNS OF ACROSOME REACTIONS

The acrosome reaction, whereby the contents are released from the organelle, is similar in principle throughout all the species investigated so far but differs greatly in detail (see Austin, 1968; Bedford, 1970; Franklin, 1970). Essentially, the reaction consists in the fusion of the outer wall of the acrosome membrane with the overlying cell membrane. The acrosome membrane thus becomes continuous with the cell membrane, with the result that an aperture is formed into the cavity of the acrosome. In the type of acrosome reaction typified by spermatozoa of marine invertebrates, the aperture formed into the acrosome widens rapidly until the acrosome as a vesicular structure ceases to exist. These events occur in very much the same way as is seen in the approach of secretory and excretory vesicles to the cell membrane of other kinds of cells, their fusion with this membrane, and evacuation to the exterior.

In spermatozoa of many marine invertebrates and certain other species (e.g., the lamprey *Lampetra*—Kille, 1960), one or more protrusions develop from the posterior region of the acrosome as the aperture enlarges. These are the acrosome filaments which, as has been known for some time (see Dan, 1967), will make first contact with the surface of the egg cell. The mammalian acrosome reaction differs from this in two particulars: fusion between acrosome and cell membranes occurs at numerous points leading to the formation of many apertures, and an acrosome filament is not developed, the spermatozoon making contact with the egg surface with a lateral region of its head. The multiple fusions of the mammalian acrosome reaction lead to vesiculation of the two membranes involved and their complete removal from the sperm head.

The acrosome reaction in mammals seems to be triggered by contact with an egg investment, or by the action of some soluble agent emanating from the egg. In mammals, and perhaps the anurans, the acrosome reaction must be preceded by the process of capacitation, which has already been discussed.

Gamete Union

ADHESION

As noted earlier, encounter between the egg and the fertilizing spermatozoon is followed by attachment between the two cells. This involves, first, a close approximation of the two cell membranes, which necessarily means that electrostatic forces of repulsion must be overcome. Some years ago Pethica (1961) pointed out that repulsion between cells is reduced and close approximation facilitated if the radius of curvature of the apposing surfaces is approximately 0.1 μm or less. It is therefore surely significant that the tip of the acrosome filament is of approximately this dimension. For the mammal, Pikó and Tyler (1964) and Piko (1969) have stressed that the initial contact of spermatozoon and egg involves the microvilli that project from the egg surface. These structures, too, have a radius of curvature of the order indicated. Close approach of the membranes apparently leads in some way to their destabilization—perhaps through the action of associated proteases—and fusion occurs between them (see Trinkaus, 1969).

MEMBRANE FUSION

With the gametes of the marine invertebrates, fusion is between the former inner acrosome membrane and the plasma membrane of the egg, whereas in mammals the sperm membrane involved apears to be the cell membrane overlying the posterior half of the sperm head (the "postnuclear cap" region). In either instance continuity is established between the two gamete membranes, so that effectively a single cell is formed at this juncture. Egg cytoplasm begins to flow around the sperm organelles and gradually the spermatozoon is drawn into the confines of the egg. The details in-

volved in sperm–egg fusion, with particular reference to the changes undergone by the sperm cell membrane and acrosomal membrane, have been the subject of many elegant electron microscope studies (see Austin, 1968; Barros and Franklin, 1968; Stefanini et al., 1969; Yanagimachi and Noda, 1970a, b; Bedford, 1972).

In the early stages of sperm penetration, the sperm membrane tends to pass into the egg, so that a deep fold of composite egg-and-sperm membrane is produced around the site of sperm entry. Some of the membrane fold undergoes vesiculation in the egg cytoplasm, and indeed, according to certain investigators, essentially the whole of the sperm cell membrane will undergo this process (see Tyler, 1965). Other workers, however, maintain that the bulk of the sperm cell membrane is left on the surface of the egg as a mosaic patch (Colwin and Colwin, 1964). At present, experimental evidence is inadequate to resolve this difference of opinion.

CYTOPLASMIC CONTRIBUTION OF MALE GAMETE

Sperm cytoplasmic components are necessarily incorporated in the egg (zygote) cytoplasm—these are: the material lying within the acrosome filament (when present) and in the subacrosomal region, and, in mammals, the specialization known as the *rod* or *perforatorium* and the material surrounding the nucleus in the postnuclear cap region of the sperm head. In addition, it appears to be the general rule that the sperm centrioles are carried into the egg cytoplasm along with the mitochondria and other components of the tail; however, they may suffer a different fate, according to the species. In some mammals the sperm tail is regularly drawn into the egg, while in others it is equally regularly left outside; in still others either event may happen.

Among mammals the relatively small tails of rabbit, rat, mouse, golden hamster, and human spermatozoa appear regularly to pass into the egg cytoplasm, whereas the relatively large tail of the Chinese hamster spermatozoon is left outside. The obvious inference from these observations is that components of the spermatozoon other than the nucleus and perhaps the centrioles have no further relevance for embryonic development; observations by electron microscopy tend to support this conclusion. In the rat the sperm mitochondria soon show degenerative changes, and the coarse fibres of the tail together with the axial filament fade out and disappear (Szollosi, 1965).

The precise role of the sperm centrioles is still a largely unresolved problem. In some animals a sperm centriole appears to be involved in the establishment of one pole of the first cleavage spindle, while in others both poles of the cleavage spindle seem to arise from the division of the egg centriole. The fact that cleavage can follow parthenogenetic activation of eggs certainly supports the idea that development of the cleavage spindle is not dependent upon a sperm centriole.

Pronuclear Development and Syngamy

DISAGGREGATION OF SPERM CHROMATIN

As the egg cytoplasm passes around the sperm nucleus, the nuclear envelope begins to separate into a number of small vesicles that become dispersed in the cytoplasm. This change does not necessarily involve the whole expanse of nuclear envelope, which may retain its integrity at the anterior and posterior poles of the nucleus. The peripheral regions of nuclear material that do become exposed begin to expand, and fine irregular strands move out into the egg cytoplasm (Fig. 7). Progressively, the whole body of nuclear chromatin becomes converted into a much less dense, but intricate, matrix of fine fibrils and granules. The changes somewhat resemble those produced in rabbit spermatozoa *in vitro* by treatment with the reagents sodium dodecyl sulphate and dithiothreitol, which have the property of disrupting the disulphide bonds in protein (Fig. 8) (Calvin and Bedford, 1971). The implication is that changes of a similar chemical nature may take place in the egg in the first stages of fertiliza-

tion. At such a time the sperm DNA is known to become associated with histone that is less arginine-rich, and it is inferred that the chromosomes resume their fully expanded condition.

PRONUCLEAR STRUCTURE

Soon after the major expansion of sperm chromatin, the new nuclear envelope develops by the union of a series of vesicular bodies that have accumulated about the expanded nucleus. One or several spherical nucleoli make their appearance within the pronucleus. The female pronucleus forms in much the same manner, and both increase greatly in size and move together to meet in the center of the egg. In many species fusion does not take place; when the two pronuclei have come together, the nuclear envelopes disintegrate, and the chromosomes recondense within the areas of the pronuclei and become arranged as the metaphase group on the first cleavage spindle. This pattern of events is shown in the rat (Austin, 1951), the rabbit, the mussel, *Mytilus*, and the clam *Spisula* (Longo and Anderson, 1969a, b, 1970). In the sea urchin, on the other hand, presumably because the female pronucleus is already formed at the time of sperm entry, the male pronucleus remains small and never achieves the distinctive vesicular form. After contact in the middle of the egg, pronuclear fusion occurs to form a single zygote nucleus (Longo and Anderson, 1968). (Ultrastructural aspects of fertilization in plants and animals have been reviewed by Austin, 1968, and in mammals by Pikó, 1969, and Zamboni, 1971).

Response of Egg to Gamete Union

METABOLIC ACTIVITY

The biochemical changes occurring in the egg at fertilization have been most extensively studied in certain marine invertebrates, especially the sea urchin (see Monroy, 1965; Epel et al., 1969; Vacquier et al., 1972). Initial changes seem to involve the permeability char-

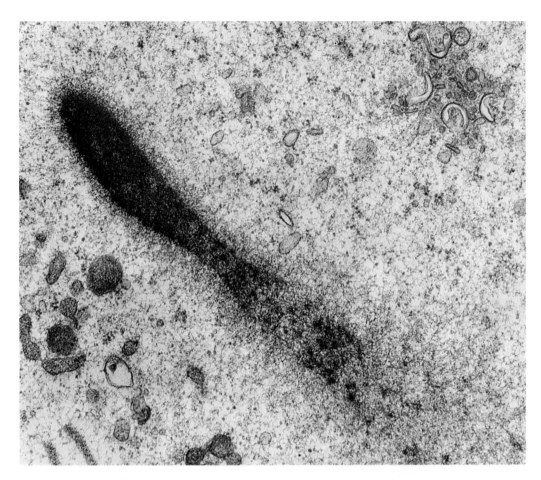

FIGURE 7. *Electron micrograph of part of a hamster egg showing changes undergone by the sperm nucleus during an early phase of its expansion.*

acteristics of the egg cortex and the movement of ions. These then lead to changes in respiration and the synthesis of macromolecules (Figs. 9 and 10).

Respiratory Metabolism

Very different measurements have been obtained for the change in oxygen consumption of sea urchin eggs following sperm penetration —a slow steady rise, or a sharp rise leading to a slow fall, or a sudden peak succeeded by a steady rise, or a peak succeeded by a constant uptake at a high level. The differences can be ascribed to variations in species, habitat, experimental conditions, and so forth, but more

work is needed to clarify the situation. Where increase in oxygen uptake occurs, the effect seems likely to be attributable to activation of the hexosemonophosphate shunt, leading to synthesis of TPNH and augmentation of the oxidative utilisation of carbohydrates. There is also evidence of a disinhibition of certain enzymes associated with the pentosephosphate cycle, such as G-6-P dehydrogenase, which apparently had been inhibited through combination with structural protein.

Synthesis of Macromolecules

In the unfertilized sea urchin egg, all the machinery for protein synthesis appears to be

present, and most of the enzymes can be demonstrated as functional under appropriate conditions *in vitro*. The system, however, is
blocked in some way, and the nature of this
block has long been a problem. In recent years,
some detailed information has come to light.
For one thing there seems to be a barrier of a
protein nature (removable by protease) that
prevents association between messenger RNA
(mRNA) and ribosomes. Release of protease
can in fact be demonstrated as one of the
earliest events of activation, and recently the
breakdown of lysosomes has been implicated
in this process (Allison and Hartree, 1970).
Lysosomal breakdown is evidently preceded by
changes in the ion permeability of the egg

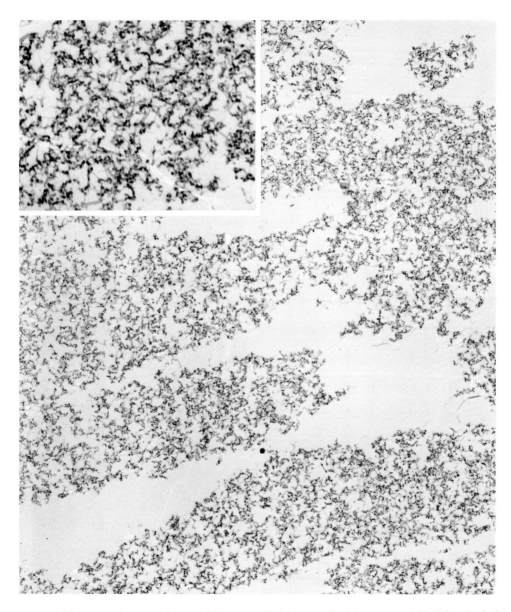

FIGURE 8. *Electron micrographs of rabbit sperm heads treated with $2 \times 10^{-3}M$ dithiothreitol in 1 percent sodium dodecyl sulfate. (Photographs courtesy of J. M. Bedford.)*

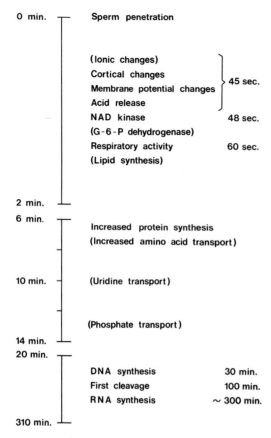

0 min.	Sperm penetration	
	(Ionic changes)	
	Cortical changes	
	Membrane potential changes	45 sec.
	Acid release	
	NAD kinase	48 sec.
	(G-6-P dehydrogenase)	
	Respiratory activity	60 sec.
	(Lipid synthesis)	
2 min.		
6 min.	Increased protein synthesis	
	(Increased amino acid transport)	
10 min.	(Uridine transport)	
	(Phosphate transport)	
14 min.		
20 min.		
	DNA synthesis	30 min.
	First cleavage	100 min.
	RNA synthesis	~ 300 min.
310 min.		

FIGURE 9. *Succession of structural and functional changes from the start of fertilization in the egg of the sea urchin* Strongylocentrotus purpuratus, *maintained at 17°C. Precise times can be given for several of the events; events for which only approximate times are known are shown in parentheses.*
[*From the data of D. Epel, B. C. Pressman, S. Elsaesser, and A. M. Weaver, in* "The Cell Cycle"(*G. M. Padilla, I. L. Cameron, and G. L. Whitson, eds.), chap. 13, Academic Press, New York, 1969.*]

cytoplasm, but just how the two processes are related has yet to be explained.

As a result of the establishment of normal functional relationship between mRNA and ribosomes, protein synthesis is initiated and continues throughout the early stages of cleavage. In addition, renewed RNA synthesis is also detectable. DNA replication apparently occurs during the pronuclear phase in the mouse egg, as was first demonstrated by Alfert

(1950). In the sea urchin *Strongylocentrotus*, replication takes place rather late in fertilization, after pronuclear fusion (Hinegardner et al., 1964).

BARS TO POLYSPERMY

In most animals and plants the participation of more than one spermatozoon in the process of fertilization leads to anomalous development; several devices have been evolved, however, that prevent this occurrence (Table 7). Animal eggs are said to show either physiological or pathological polyspermy. In physiological polyspermy several spermatozoa may enter and commonly do; they form male pronuclei, but the supernumerary pronuclei become relegated to the periphery of the egg cell, and only one combines with the female pronucleus. Physiological polyspermy is characteristic of animals with large eggs, especially the birds, the reptiles, the urodeles, and the elasmobranch fishes. For the majority of animal eggs, however, polyspermy is pathological; they depend upon changes in the egg investments or in the surface properties of the egg cytoplasm to prevent entry of more than one spermatozoon.

In the sea urchin egg, primary protection is vested in a change in the egg surface of an unknown nature; a secondary change lies in the elevation of the fertilization "membrane," which is impervious to spermatozoa. Removal of this "membrane" and reexposure to spermatozoa results in multiple refertilization (Tyler et al., 1956). The egg surface change, involving a loss of ability to fuse with the acrosome filament, could well have a widespread importance among marine invertebrates and fish. It appears to be more rapid than changes involving egg investments.

Elevation of the fertilization "membrane" in the sea urchin egg involves, first, the breakdown of cortical granules, the contents of which, together with portions of egg cortical cytoplasm, apparently unite with the tenuous chorion or vitelline "membrane." The composite structure is raised by fluid pressure in the perivitelline space, arising through change in the osmotic gradient. The chain of events is similar in the starfish egg, though the chorion starts off much thicker and elevation is not so dramatic (Monroy, 1965). In the polychaete worm *Nereis*, emission of a broad sperm-impermeable jelly coat occurs following the breakdown of large surface vesicles (Fallon and Austin, 1967).

Most mammalian eggs show two reactions: (1) a change in the surface properties of the egg cytoplasm precluding sperm attachment, and (2) a loss of sperm permeability in the zona pellucida (the "zona reaction") (Braden et al., 1954). From recent observations it appears possible that the zona reaction could in fact be attributable to the passage into the zona of an enzyme inhibitor released by the breakdown of cortical granules (Conrad et al., 1971). A change of this kind may also represent the protective device in the eggs of anurans. In mammals the chances of polyspermic fertilization are further reduced by the fact that only a small fraction of the deposited spermatozoa reach the site of fertilization (Fig. 3). In certain fish, such as the sturgeon, response to the penetration of the first spermatozoon seems to be mainly an emission of jellylike material from the vitellus that fills the perivitelline space and is apparently impermeable to spermatozoa (Ginsburg, 1957, 1961).

Anomalies of Fertilization

POLYSPERMY

The immediate consequences following the penetration of supernumerary spermatozoa differ in different species (Table 7). In the frog all spermatozoa give rise to male pronuclei, and cleavage centers develop in association with each. Only one male pronucleus enters into syngamy with a female pronucleus, but multiple cleavage occurs under the influence of the several spindles. Development is thus chaotic from the start, and no proper embryo is formed. In the sea urchin the tendency is for multipolar spindles to develop in association with the pronuclei as they gather in the center of the egg; the first cleavage is thus multiple, and, again, early development is chaotic and limited.

Polyspermy in mammals involves nearly always only one extra spermatozoon so that one female and two male pronuclei develop. Fertilization proceeds normally, and three groups of chromosomes become arranged on the first cleavage spindle, which is bipolar and apparently normal in structure and function. Cleav-

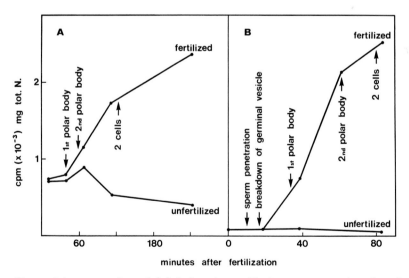

FIGURE 10. *Rate of incorporation of labeled amino acids into egg proteins after fertilization. (A)* ^{35}S-L-*methionine and eggs of* Asterias forbesii; *(B)* L-*valine-1-*^{14}C *and eggs of* Spisula solidissima. G. V., *germinal vesicle. (Modified from A. Monroy and H. Tolis, Biol. Bull. 126, 456, 1964.)*

TABLE 7

FORMS OF POLYSPERMY[a]

Type	Number of sperms entering	Fate of extra male pronuclei	First cleavage spindle	Outcome	Organism
Physiological	Several	Suppressed	Normal	Normal	Urodeles, bryozoans, angiosperms, some insects[b]
	Many	Repelled to periphery of blastodisc	Normal	Normal (possibly some mosaicism)	Elasmobranchs, reptiles, birds
	Several	Repelled to periphery of blastodisc	Normal	Normal	Arachnids
Pathological	Two; rarely three or more[c]	Polyandry	Multipolar	Random distribution of chromosomes during cleavage; development fails to pass blastula stage	Sea urchins, fish (sturgeons)
			Bipolar	Triploid embryo; full development	Angiosperms (occasionally)
		Fuse with synergids or antipodal cells		Triploid embryo; development fails to proceed beyond mid-pregnancy	Mammals
		Form separate division centers	Several	Haploid embryos of which about 10% develop into tadpoles	Frogs
				Multiple embryos	Angiosperms (occasionally)

[a] From C. R. Austin, "Fertilization," Prentice-Hall, Englewood Cliffs, New Jersey, 1965.
[b] It seems probable that polyspermy is pathological in most insects.
[c] Numbers are higher under some experimental conditions.

age is also normal, so that an embryo indistinguishable from a diploid embryo develops—although it is in fact triploid.

The state of triploidy is a highly lethal one for the mammalian embryo; in the different mammalian species studied, essentially all triploid embryos regressed and died during pregnancy—usually about halfway through. Four cases of apparently pure triploidy in man that have survived to birth have been described (Schindler and Mikamo, 1970); they displayed multiple defects and lived only a few hours. The reason for these results in mammals is not as yet known; triploids in nonmammalian vertebrates (fowl, newt, toad, frog) may survive to maturity and beyond. Mosaic 3n/2n children, on the other hand, can fare rather better, though with many abnormalities, perhaps because of diploidy in certain vital tissues, such as those concerned with blood cell formation (see Böök et al., 1961, 1970).

POLYGYNY (DIGYNY)

In the mammalian egg, under certain circumstances, the second polar spindle is prone to fail in its functions so that two female pronuclei develop. Consequently, upon fertilization a triploid embryo is initiated. The course and development of these embryos is similar to triploid embryos arising from polyspermy. Polyspermy and polygyny (digyny) vary in their relative frequency in different mammalian species: in the rabbit the two conditions are of about equal frequency; in the rat and mouse polyspermy is the more frequent occurrence; while in the hamster and the pig polygyny predominates. Both conditions are liable to show a large increase in frequency as a result of delayed mating or artificial insemination, whereby the egg undergoes aging before fertilization takes place (Table 8).

GYNOGENESIS

In some animal species gynogenesis represents the normal mechanism of development: sperm penetration occurs, but the male pronucleus fails to form, and development involves only the maternal genome (see Austin, 1969);

in a few instances, the species in question lacks males altogether, and the activating spermatozoon is supplied by males of a related species. In other animals, including mammals, gynogenesis is rare and apparently quite abnormal, giving rise only to rudimentary development.

DOUBLE FERTILIZATION

Occasionally, especially after experimental treatment such as centrifugation or the application of heat, cold, or drugs, the second maturation spindle moves inward from the surface of the egg and takes up a position more or less in the center; thus when the second meiotic division ensues, the egg is divided into two more or less equal parts. This phenomenon has been reported in several invertebrate and mammalian species (see Austin, 1969). One of the "halves" represents a polar body and the other is a matured egg—but they are more or less identical in size, and both are of course haploid; owing to the occurrence of crossing-over, the genomes of the two cells must differ.

Fertilization of both halves of an egg cleaved at maturation can occur, and development thus initiated could give rise to an apparently normal diploid embryo. If both spermatozoa carry the same sex chromosome, the resulting mature individual may be quite indistinguishable from a normal one. Degrees of hermaphroditism are likely to follow, however, if the spermatozoa carry different sex chromosomes. These types of mosaicism have been described in the mouse and in several human subjects (see De Gouchy et al., 1964; Brøgger and Gudersen, 1966; Corey et al., 1967; and Austin, 1969).

Parthenogenesis

A number of animal forms reproduce parthenogenetically, most often as a seasonal alternative to sexual reproduction or as part of a mechanism of sex determination (see Beatty, 1967). Thus in the aphid *Tetraneura ulmi* sexual reproduction occurs at the end of the summer, but throughout the rest of the year

TABLE 8

INCIDENCE OF POLYSPERMY AND POLYGYNY IN MAMMALIAN EGGS BASED ON PRESENCE OR ABSENCE OF SPERM TAILS

Animal	Polyspermy (%)	Polygyny (%)	Experimental conditions
Rat	0.3–1.8	—	control
	2.2–9.2[b]	—	d.f.[a]
	0, 0.9, 3.2[b]	—	control
	4.5, 6.6, 7.1[b]	—	d.f.
Mouse	0.9–1.2	0.1–4.5[b]	control
Hamster	1.4–3.8	—	control
	—	34	d.f.
Field vole	2	—	control
Rabbit	0	0	control
	5	10	d.f.
	0	50	d.f.[c]
Sheep	0	—	control
	6	—	d.f.
Cow	2.6	—	control
Pig	0–1.8	—	control
	12.0–29	21.0	d.f.

[a]d.f., delayed fertilization.
[b]Different stocks or strains of animals.
[c]Fertilized *in vitro*.

successive generations are produced parthenogenetically. In the honeybee the great majority of individuals (the workers) are females produced from fertilized eggs and are diploid, while the males (drones) arise parthenogenetically and are haploid. Parthenogenesis is abnormal and rare in most other groups—though occasionally it can give rise to mature animals. This is known in frogs and toads from the early experiments of Bataillon and curiously enough occurs with quite a high frequency as an apparently spontaneous process in turkeys (Olsen, 1960). The parthenogenetic turkeys were all diploid males, probably arising through suppression of the second polar division, and several survived to maturity and proved fertile. The incidence could be improved by selection and also by infection with fowl pox virus.

Full development beginning with parthenogenesis is not yet known in mammals; under experimental conditions rabbit, rat, and mouse embryos have been found to go at least halfway through pregnancy in an apparently normal manner (see Beatty, 1967; Graham, 1971). Certain breeding patterns in mice that tend to give litters with a preponderance of females have been investigated for the possibility that parthenogenesis was involved, but this was found not to be the explanation (Whitten, 1971).

Fertilization Under Experimental Conditions

Although the gametes of some of the animals that reproduce by means of external fertilization require for the fulfillment of their function conditions that are difficult to achieve in the laboratory, those of many other species in this group perform reliably in an artificially controlled environment and are thus extensively used for experimental purposes. By contrast, the success rate is quite low where fertilization is normally internal (Costello *et al.*, 1957; Austin, 1961). Especially challenging are the conditions required for mammalian gametes, because their normal environment of fertilization is the most highly specialized of all.

Attempts to promote the fertilization of mammalian eggs *in vitro* have been made since the latter part of the nineteenth century, but it was only in the early 1950s that claims for positive results began to be adequately supported. The reason for the change in fortune lay in the realization that capacitation had necessarily to precede fertilization and therefore that uterine rather than ejaculated or epididymal spermatozoa should be used—or that, in some species, spermatozoa should first undergo capacitation *in vitro*. Once this was

established, many investigations were carried out on the rabbit, then extended to the golden hamster, mouse, Chinese hamster, cat, man, sheep, pig, guinea pig, and rat, in that order.

A certain irony lay in the selection of the first experimental animal for *in vitro* fertilization; had the mouse or the hamster been used, success would probably have been achieved much sooner, in view of the fact that the spermatozoa of these species can undergo capacitation *in vitro*. These animals would also have provided a better guide to the behavior of human gametes *in vitro*.

The recent report by Yanagimachi (1972) of the fertilization of zona-free, but not of entire, guinea pig eggs with hamster spermatozoa reinforces the view that penetration of the zona pellucida is seldom possible by other than spermatozoa of the same species.

TABLE 9

PRINCIPAL RESULTS REPORTED FOR FERTILIZATION OF
MAMMALIAN EGGS IN VITRO

Animal	Source of spermatozoa	Stages seen
Rabbit	Uterus	Sperm penetration to birth of young after egg transfer
Hamster	Uterus and epididymis	Sperm penetration, pronuclei and cleavage
Mouse	Uterus and epididymis	Sperm penetration, pronuclei, and birth of young after egg transfer
Chinese hamster	Uterus of golden hamster	Sperm penetration and cleavage
Cat	Uterus	Cleavage
Man	Ejaculate	Sperm penetration, pronuclei, blastocyst
Sheep and pig	Uterus or oviduct	Sperm penetration, pronuclei, cleavage
Guinea pig	Epididymis	Sperm penetration, pronuclei and cleavage
Rat	Epididymis	Sperm penetration and cleavage

References

ALFERT, M. (1950). A cytochemical study of oogenesis and cleavage in the mouse. *J. Cell Comp. Physiol. 36*, 381.

ALLISON, A. C., AND HARTREE, E. F. (1970). Lysosomal enzymes in the acrosome and their possible role in fertilization. *J. Reprod. Fert. 21*, 501.

ALLISON, A. C., AND YOUNG, M. R. (1969). Vital staining and fluorescence microscopy of lysosomes. In "Role of Lysosomes in Biology and Pathology" (J. T. Dingle and H. B. Fell, eds.), Vol. 2, p. 600. North-Holland, Amsterdam.

ASHLEY, D. J. B. (1959). Are ovarian pregnancies parthenogenetic? *Amer. J. Hum. Genet. 11*, 305.

AUSTIN, C. R. (1951). The formation, growth and conjugation of the pronuclei in the rat egg. *J. Roy. Microsc. Soc. 71*, 295.

AUSTIN, C. R. (1961). "The Mammalian Egg." Blackwell Scientific Publications, Oxford, England.

AUSTIN, C. R. (1963). Fertilization in *Pectinari* (= *Cistenides*) Gouldii. *Biol. Bull. 124*, 115.

AUSTIN, C. R. (1968). "Ultrastructure of Fertilization." Holt, New York.

AUSTIN, C. R. (1969). Variation and anomalies in fertilization. In "Fertilization" (C. B. Metz and A. Monroy, eds.), Vol. 2, Chap. 10. Academic Press, New York.

AUSTIN, C. R. (1970a). Ageing and reproduction: post-ovulatory deterioration of the egg. *J. Reprod. Fert.*, Suppl. 12, 39.

AUSTIN, C. R. (1970b). Sperm capacitation—biological significance in various species. *Advan. Biosci. 4.*

AUSTIN, C. R., BAVISTER, B. D., AND EDWARDS, R. G. (1972). Components of capacitation. N.I.H. Conference on Regulation of Mammalian Reproduction, Sept. 1970. Washington.

AUSTIN, C. R., LUTWAK-MANN, C., AND MANN, T., (1964). Spermatophores and spermatozoa of the squid *Loligo pealii. Proc. Roy. Soc. B 161*, 143.

BARROS, C., AND AUSTIN, C. R. (1967). *In vitro* fertilization and the sperm acrosome reaction in the hamster. *J. Exp. Zool. 166*, 317.

BARROS, C., AND FRANKLIN, M. C. (1968). Behaviour of the gamete membranes during sperm entry into the mammalian egg. *J. Cell Biol. 37*, C13.

BAVISTER, B. D. (1969). Environmental factors important for *in vitro* fertilization in the hamster. *J. Reprod. Fert. 18*, 544.

BAVISTER, B. D. (1971). A study of *in vitro* fertilization and capacitation in the hamster. Ph.D. thesis, University of Cambridge.

BEATTY, R. A. (1967). Parthenogenesis in vertebrates. In "Fertilization" (C. B. Metz and A. Monroy, eds.), Vol. 1, Chap. 9. Academic Press, New York.

BEDFORD, J. M. (1965). Changes in fine structure of the rabbit sperm head during passage through the epididymis. *J. Anat. 99*, 891.

BEDFORD, J. M. (1966). Development of the fertilizing ability of spermatozoa in the epididymis of the rabbit. *J. Exp. Zool. 163*, 319.

BEDFORD, J. M. (1970). Sperm capacitation and fertilization in mammals. *Biol. Reprod.*, Suppl. 2, 128.

BEDFORD, J. M. (1972). An electron microscopic study of sperm penetration into the rabbit egg after normal mating. *Amer. J. Anat. 133*, 213.

BEDFORD, J. M., AND CHANG, M. C. (1962). Removal of decapacitation factor from seminal plasma by high-speed centrifugation. *Amer. J. Physiol. 202*, 178.

BISHOP, M. W. H. (1970). Ageing and reproduction in the male. *J. Reprod. Fert.*, Suppl. 12, 65.

BISHOP, M. W. H., AND AUSTIN, C. R. (1957). Mammalian spermatozoa. *Endeavour 16*, 137.

BISHOP, M. W. H., AND SMILES, J. (1957). Induced fluorescence in mammalian gametes with acridine orange. *Nature* (London) *179*, 307.

BLANDAU, R. J. (1969). Gamete transport—comparative aspects. In "The Mammalian Oviduct" (E. S. E. Hafez and R. J. Blandau, eds.), Chap. 5, pp. 129–162. University of Chicago Press, Chicago.

BÖÖK, J. A., SANTESSON, B., AND ZETTERQUIST, P. (1961). Association between congenital heart malformation and chromosomal variations. *Acta Paediat, 50*, 216.

BÖÖK, J. A., SANTESSON, B., AND ZETTERQUIST, P. (1970). XXY triploidy and sexual development. *Acta Medica Auxo 2*, 45.

BRADEN, A. W. H., AUSTIN, C. R., AND DAVID, H. A. (1954). The reaction of the zona pellucida to sperm penetration. *Australian J. Biol. Sci. 7*, 391.

BRØGGER, A., AND GUDERSEN, S. K. (1966). Double fertilization in Down's syndrome. *Lancet 1*, 1270.

CALVIN, H. I., AND BEDFORD, J. M. (1971). Formation of disulphide bonds in the nucleus and accessory structures of mammalian spermatozoa during maturation in the epididymis. *J. Reprod. Fert.*, Suppl. 13, 65.

CHANG, M. C. (1957). A detrimental effect of seminal plasma on the fertilizing capacity of sperm. *Nature (London) 179*, 258.

COHEN, J. (1971). The comparative physiology of

gamete populations. *Advan. Comp. Physiol. Biochem. 4*, 267.

COLWIN, L. H., AND COLWIN, A. L. (1964). Role of the gamete membranes in fertilization. In "Cellular Membranes in Development" (M. Locke, ed.), p. 233. Academic Press, New York.

CONRAD, K., BUCKLEY, J., AND STAMBAUGH, R. (1971). Studies on the nature of the block to polyspermy in rabbit ova. *J. Reprod. Fert. 27*, 133.

COREY, M. J., MILLER, J. R., MacLEAN, L. R., AND CHOWN, B. (1967). A case of XX/XY mosaicism. *Amer. J. Hum. Genet. 19*, 378.

COSTELLO, D. P., DAVIDSON, M. E., EGGERS, A., FOX, M. H., and HENLEY, C. (1957). "Methods for Obtaining and Handling Marine Eggs and Embryos." Marine Biological Laboratory, Woods Hole, Massachusetts.

DAN, J. C. (1967). Acrosome reaction and lysins. In "Fertilization" (C. B. Metz, and A. Monroy, eds.), Vol. 1, Chap. 6, Academic Press, New York.

DAVEY, K. G. (1965). "Reproduction in the Insects." W. H. Freeman, San Francisco.

DE GROUCHY, J., MOULLEC, J., SALMON, C., JOSSO, N., FREZAL, J., AND LAMY, M. (1964). Hermaphrodise avec caryotype XX/XY. Etude génétique d'un cas. *Ann. Génét. 7*, 25.

EDWARDS, R. G., BAVISTER, B. D., AND STEPTOE, P. C. (1969). Early stages of fertilization *in vitro* of human oocytes matured *in vitro*. *Nature (London) 221*, 632.

EPEL, D., PRESSMAN, B. C., ELSAESSER, S., AND WEAVER, A. M. (1969). The program of structural and metabolic changes following fertilization of sea urchin eggs. In "The Cell Cycle" (G. M. Padilla, I. L. Cameron, G. L. Whitson, eds.), Chap. 13. Academic Press, New York.

ERICSSON, R. J. (1967). Technology, physiology and morphology of sperm capacitation (rabbit, bull, man). *J. Reprod. Fert.*, Suppl. 2, 65.

FALLON, J. F., AND AUSTIN, C. R. (1967). Fine structure of the gametes of *Noreis limbata* before and after interactions. *J. Exp. Zool. 166*, 225.

FAWCETT, D. W. (1970). A comparative view of sperm ultrastructure. *Biol. Reprod.*, Suppl. 2, 90.

FAWCETT, D. W., AND HOLLENBERG, R. D. (1963). Changes in acrosome of guinea pig spermatozoa during passage through the epididymis. *Z. Zellforsch. 60*, 276.

FOOR, W. E. (1970). Spermatozoan morphology and zygote formation in nematodes. *Biol. Reprod.*, Suppl. 2, 177.

FRANKLIN, L. E. (1970). Fertilization and the role of the acrosomal region in non-mammals. *Biol. Reprod.*, Suppl. 2, 159.

GINSBURG, A. (1957). Monospermy in sturgeons in normal fertilisation and the consequences of penetration into the egg of supernumerary spermatozoa. *Dokl. Acad. Nauk S.S.S.R. 144*, 445.

GINSBURG, A. (1961). The block to polyspermy in sturgeon and trout with special reference to the role of cortical granules (alveoli). *J. Embryol. Exp. Morphol. 9*, 173.

GRAHAM, C. F. (1971). Experimental early parthenogenesis in mammals. *Advan. Biosci. 6*.

HINEGARDNER, R. T., RAO, B., AND FELDMAN, D. F. (1964). The DNA synthetic period during early development of the sea urchin egg. *Exp. Cell Res. 36*, 53.

IWAMATSU, T., AND CHANG, M. C. (1969). *In vitro* fertilization of mouse eggs in the presence of bovine follicular fluid. *Nature (London) 224*, 919.

KANATANI, H., SHIRA, H., NAKANISHI, K., AND KUROKAWA, T. (1969). Isolation and identification of meiosis-inducing substance in starfish *Asterias amurensis*. *Nature (London) 221*, 273.

KILLE, R. A. (1960). Fertilization of the lamprey egg. *Exp. Cell Res. 20*, 12.

LAGERLÖF, N. (1934). Morphologische Untersuchungen über Veränderungen in Spermabild und in den Hoden bei Bullen mit Verminderter oder aufgehobener Fertilität. *Acta Pathol. Microbiol. Scand.*, Suppl., 19.

LILLIE, F. R. (1919). "Problems of Fertilization." University of Chicago Press, Chicago.

LONGO, F. J., AND ANDERSON, E. (1968). The fine structure of pronuclear development and fusion in the sea urchin *Arbacia punctulata*. *J. Cell Biol. 39*, 339.

LONGO, F. J., AND ANDERSON, E. (1969a). Cytological events leading to the formation of the two-cell stage in the rabbit: association of the maternally and paternally derived genomes. *J. Ultrastruct. Res. 29*, 86.

LONGO, F. J., AND ANDERSON E. (1969b). Cytological aspects of fertilization in the lamellibranch *Mytilus edulis*. II. Development of the male pronucleus and the association of the maternally and paternally derived chromosomes. *J. Exp. Zool. 172*, 97.

LONGO, F. J., AND ANDERSON, E. (1970). An ultrastructural analysis of fertilization in the surf clam, *Spicula solidissima*. II. Development of the male pronucleus and the association of the maternally and paternally derived chromosomes. *J. Ultrastruct. Res. 33*, 515.

MANN, T. (1964). "The Biochemistry of Semen and of the Male Reproductive Tract." Methuen, London; Wiley, New York.

MANN, T., MARTIN, A. W., AND THIERSCH, J. B. (1970). Male reproductive tract, spermatophoric reaction in the giant octopus of the North Pacific, *Octopus dofleini martini. Proc. Roy. Soc. B 175*, 31.

MARSTON, J. H., AND KELLY, W. A. (1968). Time relationships of spermatozoon penetration into the eggs of the rhesus monkey. *Nature (London) 217*, 1073.

METZ, C. B. (1967). Gamete surface components and their role in fertilization. In "Fertilization" (C. B. Metz and A. Monroy, eds.), Vol. 1, Chap. 5. Academic Press, New York.

METZ, C. B., AND ANIKA, J. (1970). Failure of conception in rabbits inseminated with nonagglutinating, univalent antibody-treated semen. *Biol. Reprod. 2*, 284.

METZ, C. B., SEIGUER, A. C., AND CASTRO, A. E. (1972). Sperm antigens involved in fertilization. "Proceedings 2nd International Symposium on Immunology of Reproduction" September 13–16, 1971, Varna, Bulgaria (R. G. Edwards, ed.).

MILLER, R. L. (1966). Chemotaxis during fertilization in the hydroid *Campanularia. J. Exp. Zool. 162*, 23.

MONROY, A. (1965). "Chemistry and Physiology of Fertilization." Holt, New York.

MORTON, J. E. (1958). "Molluscs." Hutchinson University Library, London.

NELSON, L. (1967). Sperm motility. In "Fertilization" (C. B. Metz and A. Monroy, eds.), Vol. 1, Chap. 2. Academic Press, New York.

OLSEN, M. W. (1960). Nine-year summary of parthenogenesis in turkeys. *Proc. Soc. Exp. Biol. Med. 105*, 279.

ONUMA, H., AND NISHIKAWA, Y. (1963). Studies on the acrosomic system of spermatozoa of domestic animals. I. Cytochemical nature of the PAS positive material in the acrosomic system of boar spermatids. *Bull. Nat. Inst. Anim. Ind. (Japan)*, No. 1, 125.

O'RAND, M. G. (1971). *In vitro* fertilization and capacitation-like interaction in the hydroid *Campanularia flexuosa. Biol. Bull. 141*, 398.

ORGEBIN-CHRIST, M. C. (1967). Maturation of spermatozoa in the rabbit epididymis. Fertilizing ability and embryonic mortality in does inseminated with epididymal spermatozoa. *Ann. Biol. Animale Biochim. Biophys. 7*, 373.

ORGEBIN-CHRIST, M. C. (1969). Studies on the function of the epididymis. *Biol. Reprod. 1*, 155.

PARKES, A. S. (1960). The biology of spermatozoa and artificial insemination. "Marshall's Physiology of Reproduction" (A. S. Parkes, ed.), Chap. 9. Longmans, London.

PETHICA, B. A. (1961). The physical chemistry of cell adhesion. *Exp. Cell Res.*, Suppl. 8, 123.

PICKWORTH, S., AND CHANG, M. C. (1969). Fertilization of Chinese hamster eggs *in vitro. J. Reprod. Fert. 19*, 371.

PIKÓ, L. (1969). Gamete structure and sperm entry in mammals. In "Fertilization" (C. B. Metz and A. Monroy, eds.), Vol. 2, Chap. 8. Academic Press, New York.

PIKÓ, L., AND TYLER, A. (1964). Fine structural studies of sperm penetration in the rat. *Proc. 5th Int. Congr. Anim. Reprod., Trento 2*, 372.

POLGE, C. (1955). Artificial insemination in fowl. Ph.D. thesis, University of London.

SALISBURY, G. W., AND HART, R. G. (1970). Gamete aging and its consequences. *Biol. Reprod.*, Suppl. 2, 1.

SALISBURY, G. W., AND LODGE, J. R. (1962). Metabolism of spermatozoa. *Advan. Enzymol. 24*, 35.

SAVORY, T. (1964). "Arachnida." Academic Press, New York.

SCHINDLER, A. M., AND MIKAMO, K. (1970). Triploidy in man. Report of a case and a discussion on etiology. *Cytogenetics, 9*, 116.

SHIVERS, C. A., AND JAMES, J. M. (1971). Fertilization of antiserum-inhibited frog eggs with "capacitated" sperm. *Biol. Reprod. 5*, 229.

SNELGROVE, L. E. (1946). "Queen Rearing." Somerset.

SOUPART, P. (1971). "Sperm Capacitation: Methodology, Hormonal Control and the Search for a Mechanism." Invitational paper presented at the Harold C. Mack Symposium on the Biology of Fertilization and Implantation, October 28–29, 1970. Charles C Thomas, Springfield, Illinois.

SRIVASTAVA, P. N., ADAMS, C. E., AND HARTREE, E. F. (1965). Enzymatic action of acrosomal preparations on the rabbit ovum *in vitro. J. Reprod. Fert. 10*, 61.

STAMBAUGH, R., AND BUCKLEY, J. (1969). Identification and subcellular localization of the enzymes effecting penetration of the zona pellucida by rabbit spermatozoa. *J. Reprod. Fert. 19*, 423.

STAMBAUGH, R., AND BUCKLEY, J. (1970). Comparative studies of the acrosomal enzymes of rabbit, rhesus monkey, and human spermatozoa. *Biol. Reprod. 3*, 275.

STEFANINI, M., OURA, C., AND ZAMBONI, L. (1969). Ultrastructure of fertilization in the mouse. II.

Penetration of sperm into the ovum. *J. Submicrosc. Cytol. 1, 1.*

STRAUSS, F. (1950). Ripe follicles without antra and fertilization within the follicle: a normal situation in a mammal. *Anat. Res. 106, 251.*

SZOLLOSI, D. G. (1965). The fate of sperm middle-piece mitochondria in the rat egg. *J. Exp. Zool. 159, 367.*

TRINKAUS, J. P. (1969). "Cells into Organs." Prentice-Hall, Englewood Cliffs, New Jersey.

TYLER, A. (1965). The biology and chemistry of fertilization. *Amer. Natur. 99, 309.*

TYLER, A., MONROY, A., AND METZ, C. B. (1956). Fertilization of fertilized sea-urchin eggs. *Biol. Bull. 110, 184.*

VACQUIER, V. D., EPEL, D., AND DOUGLAS, L. A. (1972). Sea urchin eggs release protease activity at fertilization. *Nature (London) 237, 34.*

VAIDYA, R. A., BEDFORD, J. M., GLASS, R. H., AND MORRIS, J. M. (1969). Evaluation of the removal of tetracycline fluorescence from spermatozoa as a test for capacitation in the rabbit. *J. Reprod. Fert. 19, 483.*

WHITTEN, W. K. (1971). Parthenogenesis: does it oocur spontaneously in mice? *Science 171, 406.*

WITSCHI, E. (1969). Teratogenic effects from over-ripeness of the egg. "Congenital Malformations." Excerpta Medica International Congress Series No. 204. Proceedings Third International Conference on Congenital Malformations, The Hague. Excerpta Medica, Amsterdam.

WITSCHI, E., AND LAGUENS, R. (1963). Chromo-somal aberrations in embryos from overripe eggs. *Develop. Biol. 7, 605.*

WOLF, D. P., AND HEDRICK, J. L. (1971). A molecular approach to fertilization. II. Development of a bioassay for sperm. *Develop. Biol. 25, 360.*

YANAGIMACHI, R. (1957). Some properties of the sperm-activating factors in the micropyle area of the herring egg. *Anat. Zool. Japan 30, 114.*

YANAGIMACHI, R. (1969). *In vitro* acrosome reaction and capacitation of golden hamster spermatozoa bovine follicular fluid and its fractions. *J. Exp. Zool. 170, 269.*

YANAGIMACHI, R. (1972). Penetration of guinea-pig spermatozoa into hamster eggs *in vitro*. *J. Reprod. Fert. 28, 477.*

YANAGIMACHI, R., AND CHANG, M. C. (1964). *In vitro* fertilization of golden hamster ova. *J. Exp. Zool. 156, 361.*

YANAGIMACHI, R., AND NODA, Y. D. (1970a). Electron microscope studies of sperm incorporation into the golden hamster egg. *Amer. J. Anat. 128, 429.*

YANAGIMACHI, R., AND NODA, Y. D. (1970b). Ultrastructural changes in the hamster sperm head during fertilization. *J. Ultrastr. Res. 31, 465.*

YOUNG, W. C. (1931). A study of the formation of the epididymis. III. Functional changes undergone by spermatozoa during their passage through the epididymis and vas deferens in the guinea pig. *J. Exp. Biol. 8, 151.*

ZAMBONI, L. (1971). "Fine Morphology of Mammalian Fertilization." Harper & Row, New York.

CHAPTER THREE

Cleavage

R. RAPPAPORT

Introduction

The cleavage divisions that follow soon after fertilization rapidly convert the zygote into a population of cells. These divisions are mitotic and are fundamentally similar to those that will take place later. They differ, however, in the relation between cell growth and cell division, the factors determining division rates, and the size and proportion of the mitotic apparatus.

The pattern of cleavages is usually predictable. It is often correlated with yolk distribution, but the basis of this correlation is not well understood. Investigators of the turn of the century considered the cleavage patterns of mosaically developing forms a manifestation of egg "promorphology" and studied them exhaustively. At the same time, by reason of their size, shape, transparency, durability, and ease of synchronization, cleaving eggs were recognized as ideal subjects for study of the fundamental processes of cell division. Consequently, most of the information concerning cytokine-

sis in animal cells derives from studies on cleaving eggs; in this chapter we shall therefore emphasize current research on the physical process of cleavage and the events immediately preceding it. More extensive reviews of cytokinesis in animal cells will be found elsewhere (Wolpert, 1960; Rappaport, 1971).

Cleavage is easy to observe; but observations of the process do not often permit discrimination between cause and effect, and understanding has thus required experimentation and physical measurement. Investigators have attempted to provide answers to two general questions: First, what is the role of the visible parts of the cell in the cleavage process? And second: What is the nature of the physical process that cleaves the egg?

Since the contents of the cleaving egg may be removed, scrambled, or replaced without interfering with the progress of an established furrow, it follows that the division mechanism must lie at or near the surface. During cleavage, the cell surface increases and the pattern of increase is often predictable; but disruption of the pattern does not interfere with division. Three theories of division have been adduced: polar expansion, equatorial constriction, and polar relaxation. The polar expansion theory has been disproved by experiments demonstrating that cells divided when they were stretched by attached weights and that isolated bits of furrow also continued to divide. Measurements indicating tension at the cell surface during cleavage are also inconsistent with this hypothesis. Proponents of the equatorial constriction and polar relaxation theories are in agreement that tension at the equatorial surface exceeds that at the poles but differ as to the manner in which the position of the furrow is established. It is certain, however—from experimental analyses of the furrow establishment process—that only the asters of the mitotic apparatus are essential, and that they affect the equatorial cortex and surface. In addition, electron microscope studies of the cortex at the base of the furrow have demonstrated a 10 μm wide band of circumferentially oriented microfilaments that have been considered the visible manifestation of the constriction mechanism.

Before and during cleavage the mechanical properties of the egg change, and although ingenious methods for measuring these changing properties have been devised, the relation between the fluctuating properties and the cleavage process has not been clearly established. The chemical basis of the cleavage mechanism is also currently under investigation, and several hypothetical systems have been proposed. Actin has been suggested as the chemical basis of cleavage because it is present in cleaving cells and because the microfilaments at the furrow base resemble actin fibers and absorb heavy meromyosin particles. On the other hand, an −SS −SH exchange reaction between a cortical contractile protein and a protein associated with the mitotic apparatus has been proposed. It is also possible that local oxidation of the cortical−SH−rich protein may provide the force required for division.

Although the process of cleavage furrow establishment has been investigated, much remains to be learned. It seems very likely, however, that the mitotic apparatus first delivers a stimulus to the equatorial surface. The stimulus moves along straight lines that are probably associated with linear elements of the mitotic apparatus. Furrow establishment requires about 1 minute. The nature of the original stimulus is still unknown, but it will probably be revealed when the chemical basis of the constriction mechanism is understood.

Characteristics of Cleavage Division

Cleavage divisions are typical animal cell divisions in that mitosis is immediately followed by cytokinesis. Cleavage mitosis is similar to that occurring in adult cells, and present evidence indicates that the physical processes that bisect blastomere and tissue cell are fundamentally the same. In several other respects, however, the divisions that follow fertilization are different from subsequent divisions, and these differences may be used to characterize the cleavage period.

NONGROWTH

Although cleavages usually occur in rapid succession, the total cytoplasmic volume remains constant during this period. As a consequence, the cells of the embryo become smaller, and the ratio of nuclear to cytoplasmic volume changes. In sea urchin eggs the ratio

$$\frac{\text{volume of nucleus}}{\text{volume of cytoplasm}}$$

changes from 1/550 to $\frac{1}{6}$ between fertilization and the blastula stage (Brachet, 1950). The progressive diminution of cell size continues until it reaches the magnitude normal for the species, after which each division is followed by a growth period. The duration of the cleavage period depends upon the size of the egg, the division rate, and the typical cell size. The number of cleavage divisions appears to be correlated with the ploidy of the blastomeres. In parthenogenetically stimulated sea urchin eggs, the cells are haploid, and cleavage results in cells half the normal size. But in embryos with the tetraploid nuclei, the cells are twice the normal size. Embryos developing from eggs with experimentally reduced cytoplasmic volumes cleave until the cell volume is normal for the species; consequently, the embryo is composed of fewer cells than normally (Kühn, 1971).

RATES AS SPECIES-CHARACTERISTIC

In common with other biological processes the cleavage rate is temperature-dependent, being slower at lower temperatures. At any given temperature, however, the cleavage rate of a particular species is usually highly predictable and, when gametes are properly ripe, highly synchronous. There is considerable variation among the species: zebra fish eggs cleave every 15 minutes, while the interval between mouse cleavages may be 10 to 12 hours. Several experimental studies have revealed that the rate-determining factors are not located within the nucleus. It is possible to hybridize echinoderms with different characteristic cleavage rates and evaluate the role of nucleus and cytoplasm in determining cleavage frequency. A. R. Moore found that when the enucleated egg of a rapidly cleaving species is fertilized by a slowly cleaving species, the sperm's activities are accelerated, so that the cleavage times are characteristic of the egg (Moore, 1933). More recently it has been shown that the cleavage rate is independent of cytoplasmic volume or nucleocytoplasmic ratio. When nucleated and anucleate halves of sea urchin eggs are fertilized, they cleave at the same time as normal eggs. A cytoplasmic component essential for normal division rates appears to be associated with the "mitochondrial layer" of centrifuged *Arbacia* eggs (Rustad et al., 1970).

DIFFERENCES IN SIZE OF MITOTIC APPARATUS

The term *mitotic apparatus* includes the chromosomes, spindles, asters, and centrioles. The asters and spindle are composed of linear elements that line up visible cytoplasmic granules into radially oriented array. The linear elements are birefringent and at the ultrastructural level consist of microtubules and oriented vesicular elements. The size of the mitotic apparatus is proportional to the cytoplasmic volume it occupies, and when under abnormal circumstances, a single cell contains several, their size is inversely proportional to their number. These and similar observations suggest that during their formation, the components of the mitotic apparatus were drawn from a common pool. It is not surprising that the mitotic apparatus of a large blastomere is greater than that of a small one. The volume of the mitotic apparatus of cleaving cells consists largely of the asters that are centered upon the poles of the spindles. Although asters have been described in a great many dividing animal tissue cells, they achieve their maximum size during the cleavage period.

Cleavage Patterns

Despite their relatively similar origins and fates, animal eggs may superficially appear strikingly different and their cleavage patterns may vary widely.

CORRELATION WITH YOLK DISTRIBUTION

Egg cytoplasm may be roughly divided into two categories: active cytoplasm and yolk cytoplasm. Differences in egg volume are usually attributable to different amounts of yolk. The relative amounts of the two kinds of cytoplasm and the degree to which they are mixed or separated may be correlated with the rate of cleavage, and the rate of cleavage is usually inversely proportional to the concentration of yolk, even within the same egg. When the yolk is homogeneously distributed or scanty, as in sea urchin and mammalian eggs, the rate of cleavage is nearly uniform throughout the egg. In amphibian eggs the concentration of yolk is greater in the vegetal region, and the cleavage rate of vegetal cells is slower than that of animal cells. In the extreme cases of bird and teleost fish eggs, yolk and active cytoplasm are discrete and the yolk portion never cleaves but is later assimilated by developing embryonic tissues. The correlation between cleavage rate and yolk content (Balfour's law) is one of the oldest generalizations in developmental mechanics. The correlation is attributed to an impending effect that the yolk exerts upon mitosis and cytokinesis. As yet, however, no experimental analysis of the impeding effect has been made.

Although the patterns of yolk distribution and the cleavage patterns of embryos may be strikingly different, they leave no lasting imprint on larval or adult morphology. When, for example, we inspect a sturgeon and a shark, we can find no distinctive features revealing that the former developed from an egg that cleaved in a modified holoblastic pattern and the other in a meroblastic pattern.

CORRELATION WITH DEVELOPMENTAL PATTERN

In the early years of this century meticulous descriptive studies of the cleavage patterns of the eggs of many species revealed precise relations between the cells formed by cleavage and the parts of the adult. Further, such studies indicated that comparable cells in cleavage stages might play the same morphological roles in the development of groups with strikingly different adult forms, regardless of the relative sizes of the blastomeres. Very striking in this respect are comparisons of mollusk and annelid development. E. B. Wilson (1928) and others pointed out that such regularity and predictability in cleavage patterns were probably attributable to preexisting organization or "promorphology" of the egg. Since cleavage planes are always established in the plane previously occupied by the metaphase plate of the mitotic apparatus, the understanding of the basis of the precise cleavage pattern might be greatly increased if we knew why a mitotic apparatus is found in a particular place in a particular orientation at a particular point in developmental time. E. G. Conklin (1902) emphasized the importance of cytoplasmic currents in determining the positions of intracellular structures in molluskan cleavage. Berrill (1961) has pointed out that in ascidians the position of the mitotic

apparatus may be determined by the densities of the different egg plasms. Both of these intriguing hypothetical mechanisms are amenable to experimental analysis.

The Cleavage Process

For investigations of the cleavage mechanism, most researchers have usually selected as experimental material spherical eggs that survive mechanical and chemical manipulation and are sufficiently transparent to reveal distinctly the mitotic apparatus. For these and other reasons, we find that most studies have been accomplished on marine invertebrate eggs, notably those of the Echinodermata. It is highly probable that the lack of variety of experimental material has biased the outlook of many investigators.

VISIBLE EVENTS

Cytokinesis follows mitosis, and the analysis of the relationship between these two events has been one of the persistent themes of cytology. During the early stages of mitosis, cells round up so that their surfaces are minimal. In cleavage stages, cells that are flattened against each other at interphase reduce their contact areas and rise as individual protuberances from the embryonic mass. Immediately before cleavage in *Arbacia*, an equatorial surface band 15 μm wide contracts isotropically (Scott, 1960). At late anaphase, cells elongate parallel to the spindle axis so that the surface in the equatorial region (adjacent to the midpoint of the spindle) is flattened. This condition persists into telophase, when the diameter at the equator actively diminishes and the cleavage furrow develops. The relation between the position of the mitotic apparatus and that of the furrow appears to hold for all mitotically dividing cells. The time of furrow appearance is also correlated with the location of the mitotic apparatus. When the mitotic apparatus is centered in the equatorial region, the furrow appears simultaneously throughout the circumference; but when the mitotic apparatus is naturally or artificially displaced to an excentric position, the furrow appears first in the nearest surface and later in the more distant surface. If, however, the degree of excentricity is extreme, a unilateral furrow forms only on the near surface, and the more distant surface remains passive throughout the process.

When a single sphere is divided into two spheres, the surface area increases about 26 percent—even when the volume remains constant. The mode of origin of new surface has been very carefully studied by K. Dan and his coworkers (1937), and by Y. Hiramoto (1958); they measured the movements of particles embedded in the denuded surface of cleaving invertebrate eggs and calculated changes in regional areas—under the assumption that the surface is and remains smooth. According to their findings, about 80 percent of the 26 percent area increase occurring in echinoderm cleavage takes place in the polar and subpolar

surface. There also appears to be no net gain at the equator, as the local poleward stretching is compensated for by circumferential shrinkage.

ESSENTIAL ELEMENTS FOR DIVISION

Fundamental to an understanding of division is information as to the roles of the visible cell parts in the process. At this time it is not possible to describe exactly how any part of the cell plays its role, but it is possible to determine which parts are unnecessary. If cells cleave despite removal or disruption of a region or organelle, one may conclude that the structure affected is not essential to the mechanism. Hiramoto (1956) showed that removal of the entire mitotic apparatus of sea urchin eggs after metaphase does not block cleavage (Fig. 1). Since the egg was spherical at the time of

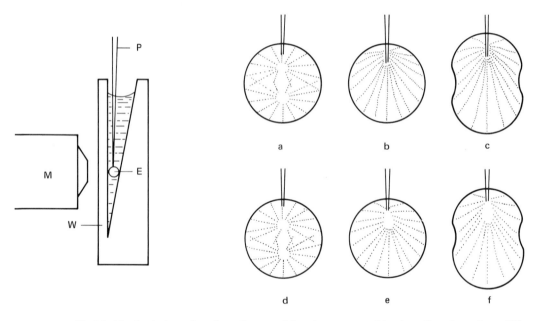

FIGURE 1. *(Left) Manipulation chamber. E, egg; M, microscope objective; P, micropipet; W, wall made of cover slip. (Right) Relation between position of mitotic figure and cleavage plane. (a–c) Successive stages of cleavage when the spindle is removed: (a) before removal of the spindle; (b) after removal; (c) furrow appears at a predetermined position. (d–f) Successive stages of cleavage when the mitotic figure is displaced by the removal of a part of egg protoplasm: (d) before the displacement of the mitotic figure; (e) after displacement (astral rays of the lower aster are much elongated); (f) furrow appears at a predetermined position. In both series, cleavage planes are independent of shifted position of asters. (From Y. Hiramoto, Exp. Cell Res. 11, 630–636, 1956.)*

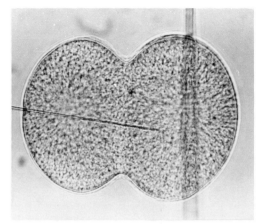

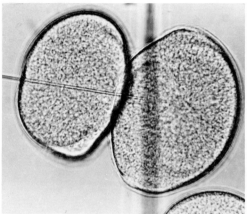

FIGURE 2. *Cleavage in a sand dollar egg with a moving needle inserted through the cleavage plane. The needle was swept back and forth during the period between the photographs. (From R. Rappaport, J. Exp. Zool. 161, 1–8, 1966.)*

the cell to look for the division mechanism is, then, in the cell surface.

EGG SURFACE AND PATTERNS OF NEW SURFACE FORMATION

The surface of eggs has special qualities not shared by the surface membranes that form when egg cytoplasm is exposed to the surrounding medium. Many years ago Chambers (1921b) found that nucleated portions of the egg could be fertilized and would cleave only if they contained some of the original egg sur-

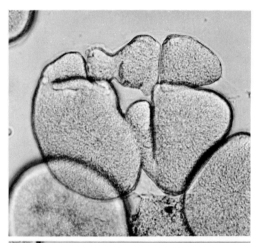

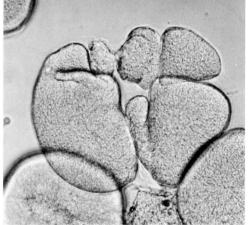

FIGURE 3. *Cleavage of an isolated furrow fragment of a sand dollar egg. (From R. Rappaport, Exp. Cell Res. 56, 87–91, 1969a.)*

the operation, it is apparent that the mitotic apparatus can play no essential physical role in division. As soon as the furrow is established, the region between the mitotic apparatus and the surface can be stirred and churned (Rappaport, 1966) or replaced with oil or sucrose solution (Hiramoto, 1965); but cleavage continues to completion (Fig. 2), and fragments of the furrow can continue division after isolation from the rest of the egg (Rappaport, 1969a) (Figs. 3 and 4). Such investigations indicate that after metaphase, division is independent of any subsurface structure or cytoplasmic arrangement. The most logical place in

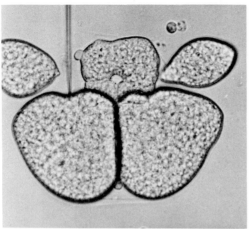

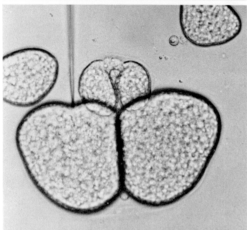

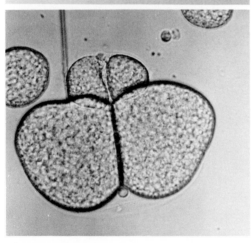

FIGURE 4. *Cleavage of an isolated furrow in* Aurelia aurita. *The furrow is normally unilateral. (From R. Rappaport,* Exp. Cell Res. *56, 87–91, 1969a.)*

face material—and that the degree of development of egg fragments was proportional to the amount of such material included. Chambers attributed the special qualities to a cortical region inseparable from and immediately beneath the surface. The cortex is demonstrable chiefly because it is denser and more viscous than the underlying cytoplasm. Its thickness has been estimated by microdissection methods to be about 3 μm in the polar region and 4 μm at the base of an active furrow in echinoderm eggs (Hiramoto, 1957). It has not, however, been possible to positively identify an ultra-structure that accounts for the demonstrated physical properties of the cortex.

The egg surface is usually covered with microvilli of varying dimensions and distribution. In some species fertilization membranes are absent, and tips of the villi may be embedded in an extracellular layer of chorionic material. The anchored villi constitute natural and convenient landmarks for estimating surface changes of cleaving eggs. Pasteels and de Harven (1962) observed the surface in the furrow and adjacent regions of the mollusk *Barnea candida* and found a pattern of new surface formation strikingly different from that of denuded echinoderm eggs. Electron microscope studies of attachments of the microvilli revealed that all the new surface formed at first cleavage originated from the stretching of an equatorial band 3.5 μm wide that must have increased its area 5- or 6-fold (Fig. 5). Similar new surface formation localized in the furrow region has been described in cleaving newt eggs (Selman and Perry, 1970).

Since normal cleavage of spherical eggs does not seem to be correlated with a particular pattern of new surface formation, it is logical to inquire whether an egg with a well-defined pattern of new surface formation can cleave when the pattern is disrupted. Denuded sea urchin and sand dollar egg surfaces stick to clean glass in calcium-free seawater, and this response has been used to constrain localized surface regions of cleaving sand dollar eggs. By sucking the polar and subpolar surfaces into opposed pipettes and similar manipulations, it was shown that the typical echinoderm pattern of new surface formation was by no means necessary for echinoderm cleavage (Rappaport and Ratner, 1967). All that appears necessary is

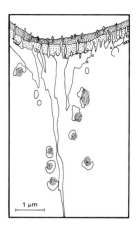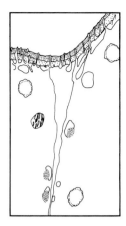

FIGURE 5. *Tracings of electron micrographs of cleaving* Barnea candida *eggs. The furrow on the left is partially complete and that on the right is complete. (From J. J. Pasteels and E. de Harven,* Arch. Biol. *73, 467–490, 1962.)*

sufficient surface next to the furrow to yield as the furrow deepens.

THEORETICAL CLEAVAGE MECHANISMS

With the above background it is possible to discuss the principal hypothetical mechanisms that have been considered in recent years. The experimental elimination of subsurface structures from consideration as active participants imposes the requirement that active furrowing be preceded by a kind of functional differentiation of the surface. The surface cannot bisect cells with the observed precision if it is a homogeneous structure with a uniform behavior pattern and physical properties. Rather, it is necessary to propose that, before cleavage, regional differences in surface behavior arise. The number of possible mechanisms is very limited, and the mechanism of cell division has been a subject of intense speculation for about a century. Thus it is not surprising that most of the hypothetical mechanisms have a long history.

According to the *polar expansion* theory (Swann and Mitchison, 1958), active growth and expansion of the polar regions causes a passive inward dipping of the equatorial surface that becomes the furrow. It is implicit

that change in polar surface properties elicits furrowing and that the equatorial surface retains the qualities that previously characterized the entire surface. An alternative theory of *equatorial constriction* readily suggests itself to any observer of the living process. According to this hypothesis, cells are divided by an equatorial ring of contractile material. This mechanism has been suggested many times (Ziegler, 1903; Lewis, 1939; Marsland, 1950; Marsland and Landau, 1954). It is implied that the contractility of the equatorial surface exceeds that which characterized the entire surface before functional differentiation. According to a third hypothesis, the *polar relaxation* mechanism, division occurs because the polar surfaces relax and allow the equatorial surface, which retains the degree of contractility that previously characterized the entire surface, to contract and bisect the cell like a tightening belt. Polar relaxation mechanisms have been frequently proposed (Conklin, 1902; Lillie, 1903; Chalkley, 1935), but Wolpert (1960) has presented the most thoughtful and detailed exposition. In both the equatorial constriction and polar relaxation theories, the contractility in the equatorial surface exceeds that at the poles. In the former the difference arises by increase at the equator, but in the latter it arises by decrease at the poles.

EXPERIMENTAL ANALYSIS OF THEORETICAL MECHANISMS

According to the polar expansion mechanism, stress in the cell surface during division is in the form of compression, and the mechanism cannot operate if there is tension at the surface. Proponents of the theory made physical measurements that they felt indicated no tension at the surface, but others have differed as to interpretation of the data. The theory was more directly tested by determining whether eggs can cleave in an imposed state of obvious tensile stress. Recently fertilized and denuded sand dollar eggs were made sticky at the surface by immersion in acidulated calcium-free seawater and pipetted to the surface of a submerged sheet of glass. Small glass beads were scattered over the eggs so that some eggs adhered to both the glass sheet and a bead. The egg was made to support the weight of the glass bead by inverting the glass sheet (Fig. 6). Despite the tensile stress imposed by this manipulation, the eggs cleaved (Rappaport, 1960).

The polar relaxation and the equatorial constriction mechanisms have much in common, and, in the final analysis, their operation is based upon an equatorial band of greater surface contractility. There is no doubt that the band exists, as measurements of the tension at the surface in the polar and equatorial zones have been made (Hiramoto, 1968). The crux of the difference between these two hypothetical mechanisms lies not so much with the mechanical nature of the cleavage event but rather with the preliminary events and the region where functional surface differentiation takes place. We must now consider the process whereby the cleavage furrow is established.

ROLE OF MITOTIC APPARATUS

Although the possibility that the mitotic apparatus may play a physical role in cleavage has been experimentally eliminated, there is abundant evidence that it is somehow involved in the process. The correlation between the position of the metaphase plate and the orientation of the subsequent cleavage plane was noted by many investigators. In 1917 E. G. Conklin centrifuged *Crepidula* eggs during the maturation, or meiotic, divisions and found that when the mitotic apparatus was moved to the center of the oocyte before metaphase, the division furrow appeared in the normal relation to it, resulting in an abnormally large polar body. Observations of this kind have been repeated on cells in mitotic as well as meiotic divisions. It is apparent that in addition to playing its well-known role in the sorting and transportation of chromosomes, the achromatic portion of the mitotic apparatus elicits the surface functional differentiation that necessarily precedes cleavage. Experimental studies of this aspect of division have involved several assumptions. After surface differentiation has taken place, it is assumed that some regions of the egg surface remain unchanged, while others are functionally altered. It is further assumed that the part of the surface acted upon by the mitotic apparatus is the part that changes and that the change

(or the direct consequence of it) is the event that precipitates cleavage. By learning more about the process of furrow establishment, we can better understand the nature of the physical division process; for it is apparent that the polar relaxation mechanism requires that the division-precipitating change in surface properties take place at the polar surface, but an equatorial constriction mechanism requires that it occur at the equator.

HYPOTHETICAL STIMULUS PATTERNS

Any hypothetical explanation of the relation between mitosis and cleavage must contain a provision for exposing only part of the cortex and surface to the influence of the mitotic apparatus. The influence that the mitotic apparatus exerts upon the surface may be considered a form of stimulus, and several possible stimulus patterns have been suggested. According to the polar stimulus pattern, the telophase chromosomes (Swann and Mitchison, 1958) or the asters (Wolpert, 1960) affect the adjacent polar surfaces but, by reason of the greater distance, do not affect the equatorial surface (Fig. 7). It is implicit in this mechanism that the furrow retains the qualities that previously characterized the entire egg surface and that it plays a passive role. Alternatively, in equatorial stimulation, the influence of the mitotic apparatus would be brought to bear upon the equatorial surface, and changes in the properties of the equatorial surface would precipitate cleavage. Events at the poles would consequently be passive. A logical first step in an analysis of the process of furrow establishment is to determine which parts of the mitotic apparatus are necessary for the appearance of normal-looking furrows at the normal time. By altering the egg geometry before the interaction of mitotic apparatus and surface, some information concerning the nature of their relationship may be obtained. When a sand dollar egg was changed from a sphere to a torus or doughnut by pressing a glass ball through it, the mitotic apparatus was crowded to one side and the first cleavage produced a horseshoe-shaped binucleate cell. Nuclear preparation for the second cleavage was normal, but subse-

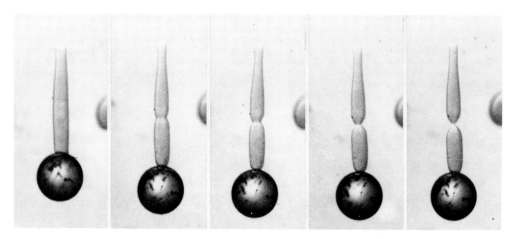

FIGURE 6. *Cleavage of a sand dollar egg under tensile stress. The dark sphere is a glass bead. (From R. Rappaport, J. Exp. Zool. 144, 225–231, 1960.)*

quently three furrows were formed, producing four uninucleate cells. Two of the furrows appeared adjacent to the spindles of the mitotic apparatuses, but the third furrow developed between asters in a region that had never been occupied by a spindle or chromosomes (Fig. 8). Asters alone, in this case, can provide whatever is essential for establishment of a temporally and morphologically normal furrow (Rappaport, 1961; Hiramoto, 1971). They might play the role ascribed to the mitotic ap-

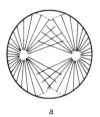 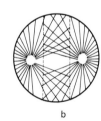

a   b

FIGURE 7. *Cleavage stimulus patterns. (a) Polar stimulation. The influence of the asters reaches the polar surfaces but fails to reach the equatorial surface peripheral to the dotted lines. The equatorial surface is not altered by stimulation. (b) Equatorial stimulation. The entire cell surface can be reached by influence from the asters or achromatic apparatus, but the equatorial surface between the dotted lines is subjected to greater stimulatory activity. The equatorial surface is altered by stimulation. (From R. Rappaport, Int. Rev. Cytol. 31, 169–213, 1971.)*

paratus in the equatorial stimulation hypothesis by jointly influencing the equatorial surface so that the level of stimulation is higher there than at the poles.

EXPERIMENTAL ANALYSIS OF STIMULUS PATTERN

The hypothetical stimulus patterns just described have been tested by altering the geometrical relation between the mitotic apparatus and the surface before the position of the furrow is established. According to the polar stimulation mechanism, functional surface differentiation is achieved because the stimulus fails to reach the equatorial surface; the failure is attributed to the inability of the stimulus to traverse the distance between the asters and the equatorial surface. The stimulus can, however, traverse the shorter distance from the asters to the polar surface. It follows that it should be possible to abolish cleavage or create extra furrows by manipulating the distance between the asters and the surface. If eggs are artificially constricted in the future cleavage plane before furrow establishment, it should be possible to prevent cleavage by equalizing the distance from the asters to the surface. In this circumstance, the entire surface would be stimulated (or unstimulated), and the functional differentiation necessary for cleavage would be prevented. However, eggs artificially con-

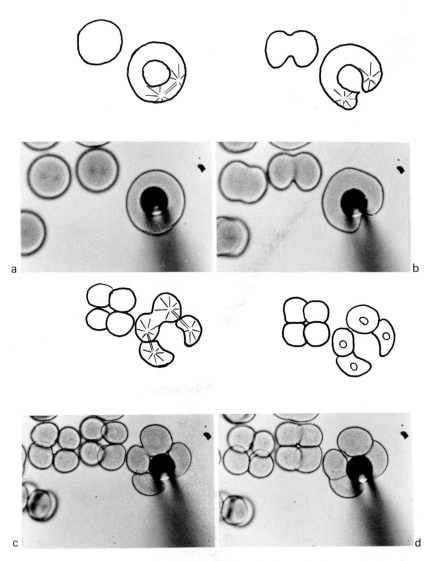

FIGURE 8. *Cleavage of torus-shaped cell. Condition of the mitotic apparatus shown in line drawings. The position of the spindle is indicated by a double line. Note synchrony with controls. Initial temperature 19.5°C. Timing begins at fertilization. (a) Immediately before furrowing; 69 minutes. (b) First cleavage completed, resulting in binucleate cell; 79 minutes. (c) Second cleavage. Two cells have divided from the free ends of the horseshoe and the binucleate cell is dividing between the polar regions of the asters of the second division; 142 minutes (d) Division completed; each cell contains one nucleus; 144 minutes. (From R. Rappaport, J. Exp. Zool. 148, 81–89, 1961.)*

stricted in the future plane of cleavage divide normally (Rappaport, 1964). Conversely, if the position of the furrow is determined by absence of stimulation and that absence is due to distance, it should be possible to create extra areas of furrowing activity by altering the cell geometry so that parts of the surface lie farther from the asters than does the equator. In cells stretched by glass weights, the polar regions lie farther from the asters than the equator but furrowing is confined to the surface adjacent to the spindle (Fig. 6). Also, in the first division of the torus-shaped cell there is an extensive region between the "backs" of the asters that lies at a considerable distance from them, but, again, furrowing is confined to the equatorial surface (Fig. 8). Measures that should, according to the polar stimulation hypothesis, abolish furrowing or produce extra furrowing activity fail to do so.

If, on the other hand, furrow establishment involved an equatorial stimulus pattern based upon joint action of asters upon the equatorial surface furrow, establishment would be affected by two dimensions. One is the distance be-tween the asters (interastral distance), and the other is the distance from a line joining the astral centers to the equatorial surface (spindle-to-surface distance) (Fig. 9). If the interastral distance is increased, the zone in which the asters can act jointly must be diminished and could be too small to affect the surface; thus no furrow would appear. It should, however, be possible to compensate for an abnormally large interastral distance by decreasing the spindle-to-surface distance. It has been possible to determine whether such compensation can be demonstrated in dispermic sea urchin eggs that contain four asters before the first cleavage. Sperm asters can elicit furrows as effectively as those of the mitotic apparatus. It was found that an increase in the interastral distance without changing the spindle-to-surface distance prevented furrow establishment, but furrows appeared in the normal relation to abnormally distant asters when the spindle-to-surface distance was reduced (Fig. 10). These experiments indicate that there exists between the asters and the equatorial surface a geometrical relationship that must be satisfied for a furrow to be established. No essential relationship between the mitotic apparatus and the polar surface has yet been demonstrated. The extensive series of experiments concerning furrow establishment were initially undertaken as a way of discriminating between the equatorial constriction hypothesis and the polar relaxation hypothesis. Since polar relaxation requires polar surface stimulation, equatorial constriction now appears the more likely possibility.

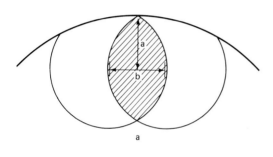

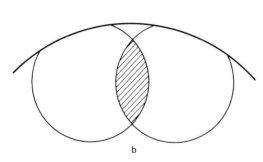

FIGURE 9. *Aster–equatorial surface relations that would obtain if furrow establishment were a consequence of joint action of the asters. In (a) the area affected by both asters reaches the surface. (b) shows the effect of moving the asters apart. Astral diameter and distance from the astral center to the surface are unchanged. (a) Spindle-to-surface distance measured parallel to the flattened surfaces; (b) interastral distance. (From R. Rappaport, J. Exp. Zool. 171, 59–68, 1969b.)*

Ultrastructure of Furrow Region

LOCATION OF CLEAVAGE MECHANISM

Cleavage is a relatively brief event, and the physical mechanism that accomplishes it appears to exist only during the period when it is being used. The mechanism must be established by reorganization of components already present in the egg, and one would expect that the effects of the reorganization could be studied

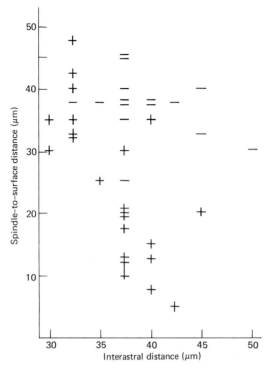

FIGURE 10. *Summary of data from perforated and unperforated cells. When the spindle-to-surface distance is 35 μm or more and the interastral distance is 35 μm or more, furrowing almost invariably fails. If the spindle-to-surface distance is reduced to 20 μm or less, furrowing occurs in conjunction with a 35 μm interastral distance. Plus indicates a furrow formed adjacent to the asters; minus indicates no furrow formation in that location. (From R. Rappaport, J. Exp. Zool. 171, 59–68, 1969b.)*

with the electron microscope. Results of experiments described in the previous section indicate fairly clearly where studies should be concentrated. The furrow's ability to continue functioning despite isolation and disruption or replacement of subsurface cytoplasm suggests that the mechanism lies at or immediately beneath and attached to the surface. And, again, if we assume that functional differentiation is a consequence of the stimulation process and that the part stimulated undergoes the changes that precipitate division, we would expect to find ultrastructural alterations in the equatorial region. Studies of the surface beginning in 1958 (Mercer and Wolpert, 1958) indicated the

presence of a well-developed electron-dense layer at the furrow base, but the present interpretation of its structure and function took some time to develop.

MICROFILAMENTOUS BAND

In 1968 Schroeder demonstrated that the electron-dense layer consists of many microfilaments 50 to 70 Å in diameter, oriented parallel to the equatorial surface and parallel to the cleavage plane. They form a layer about 0.1 μm thick immediately beneath the surface, and subsequent investigations indicate that they extend on either side of the tip of the furrow, forming a band about 10μm wide (Schroeder, 1969). The existence of a very similar structure was soon demonstrated at the furrow base in cleaving eggs of squid (Arnold, 1969), polychaetes (Szollosi, 1970), and newts (Selman and Perry, 1970). Microfilaments have also been found in maturation divisions (Longo and Anderson, 1969) and in dividing animal tissue cells (Fig. 11).

The microfilamentous band is regarded by many as the visible manifestation of the contractile ring that cuts the egg into blastomeres by actively decreasing its diameter. In addition to the correlation between active furrowing and the presence of the band, support for this interpretation also derives from studies involving cytochalasin B, a product of mold metabolism that in low concentrations causes the very rapid dissolution of microfilaments. Eggs exposed to this substance before cleavage undergo mitosis normally but fail to divide. Exposure after the beginning of cleavage may result in reversal or arrest of furrowing, depending upon how far advanced the process was at the time of exposure (Schroeder, 1969).

Little is known about the ultrastructure of the band assembly process. It has been found only at the base of already deepening furrows, even in eggs like those of the squid, where the furrow's progress across the egg surface can be conveniently studied (Arnold, 1969). Szollosi (1970) has proposed that filaments associated with surface microvilli constitute the pool from which the constituents of the band are recruited, but the actual process of band formation has hardly been studied. After the band

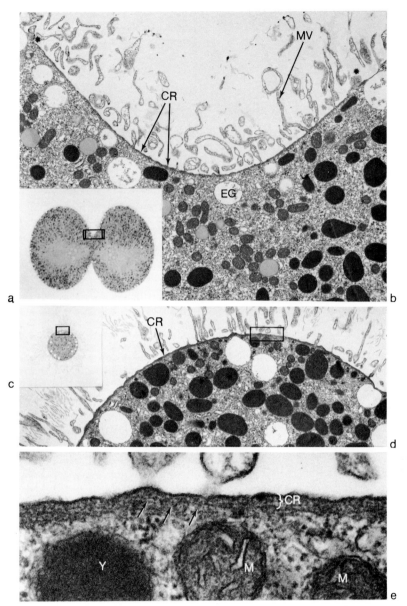

FIGURE 11. *Meridional and equatorial sections of cleaving sea urchin* (Arbacia) *eggs. (a) Meridional section about 4 minutes after initiation of furrowing.* × *365. (b) Meridional section of region comparable to that enclosed by the rectangle in (a). The contractile ring spans the concave surface between the asterisks.* × *7,100. (c) Equatorial section of the furrow of an egg comparable to that shown in (a).* × *485. (d) Equatorial section comparable to the region shown in the rectangle in (a). The contractile ring encircles the egg without interruption. (e) Equatorial section of the region enclosed by the rectangle in (d). Organization of the contractile ring and individual microfilaments (arrows) are visible.* × *75,000. CR, contractile ring; MV, microvillus-like projections; EG, echinochrome granule; M, mitochondria; arrows, individual microfilaments; asterisks, lateral limits of the contractile ring. (From T. E. Schroeder, J. Cell Biol. 53, 419–434, 1972.)*

appears, its volume decreases throughout the cleavage process (Schroeder, 1972).

Changes in Mechanical Properties Associated with Cleavage

The tendency of blastomeres to round up just before cleavage has long been attributed to changes in the mechanical properties and behavior of the surface that take place immediately before and during cleavage. Dividing tissue cells and cells in culture display similar changes that are rendered more visible when elongate cell processes are rapidly retracted as rounding up occurs. The widespread incidence of the phenomenon and its association with cytokinesis led to the assumption that these changes reflected some fundamental processes that are an integral part of the division mechanism. According to this interpretation, careful analysis of the nature of the physical changes should lead to a better understanding of the division mechanism. Danielli (1952) placed fragments of glass over cleaving eggs and found that the degree of flattening due to weight of the glass changed during the division cycle. The eggs were flattened to the greatest extent at interphase, but during mitosis they persisted in rounding up and actually lifted the pieces of glass. After cleavage, the blastomeres were again flattened, and the cycle was repeated at the next cleavage. These observations clearly showed the possibility that changes in resistance to deformation could be measured, and there followed a series of ingenious measurements of changes in the mechanical properties of cleaving eggs.

RESISTANCE TO DEFORMATION

When the tip of a pipette is held against the egg surface, gentle suction will cause a part of the surface to bulge into the orifice of the pipette. The magnitude of the bulge varies according to the amount of negative pressure applied and the phase of the cell cycle in which the measurement is made (Fig. 12a). Mitchison and Swann (1954), who devised this technique, interpreted the increase in resistance to deformation at the time of cleavage to changes in the thickness and modulus of elasticity of the surface.

Changes in resistance to deformation can also be measured by determining the amount of force necessary to produce a particular degree of flattening. This technique was used by K. S. Cole (1932), who flattened eggs under a gold beam and by Hiramoto (1963), who flattened eggs between glass plates while making the necessary measurements with a calibrated glass fiber. The eggs' resistance to this kind of deformational force varied in a fashion similar to that determined by the method of Mitchison and Swann (1955) (Fig. 12b), although Hiramoto attributed the changes to tension at the surface rather than to changes in elastic modulus or thickness.

Although fluctuating properties of the egg surface and other changes such as in birefringence have attracted attention primarily because of their presumed close relation to the cleavage mechanism, the nature of the relationship is presently unknown. As Hiramoto (1968, 1970) has pointed out, the connection is probably indirect at best. We have seen that the establishment of the cleavage mechanism requires the influence of the mitotic apparatus; yet anucleate pieces of egg (Bell, 1963) and eggs exposed to concentrations of colchicine that dissolve the mitotic apparatus (Monroy and Montalenti, 1947; Swann and Mitchison, 1953) display the same fluctuations synchronously with normal eggs.

CORTICAL GEL STRENGTH

Another property that changes during the division cycle is the firmness with which visible cell parts are held in their normal positions. The amount of centrifugation required to obtain a particular degree of stratification of egg contents is predictable and fairly easy to determine. Early studies were focused upon the deeper cytoplasm, and the changes probably reflected the state of the mitotic apparatus.

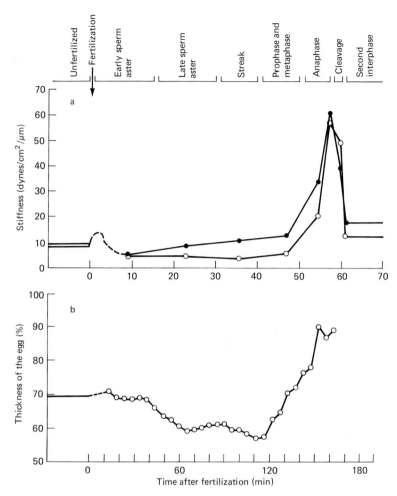

FIGURE 12. *Changes in the stiffness of the egg from fertilization to cleavage. (a) Stiffness change measured by the suction method. Solid circles, eggs with hyaline layer; open circles, eggs without hyaline layer. (After J. M. Mitchison and M. M. Swann, J. Exp. Biol. 32, 734–750, 1955.) (b) Change in thickness of egg by application of a definite force. (After Hiramoto, 1963.) The interval between fertilization and cleavage is adjusted to be the same in both figures. (From Y. Hiramoto, Biorheology 6, 201–234, 1970. Permission granted by Microforms International Marketing Corp.)*

Visible granules in the egg surface are, however, so firmly held that they can not be dislodged by simple centrifugation, and it was not until Marsland (1939) combined high hydrostatic pressures with centrifugal force that these properties of the cortex could be studied. The amount of centrifugal force and hydrostatic pressure required to produce a particular degree of stratification was considered a reflection of the cortical gel strength and, as may be seen in Fig. 13, the cortical gel strength fluctuates, reaching a maximum before cleavage. Cortical gel strength has been interpreted as a manifestation of the cleavage mechanism, and it has been shown that measures that decrease the cortical gel strength also reduce the egg's capacity for furrowing. It is unclear, however, how or why the tenacity with which granules are held in the cortex is related to the physical division process. It must also be pointed out that gel strength is at a peak well before the furrow is established and remains

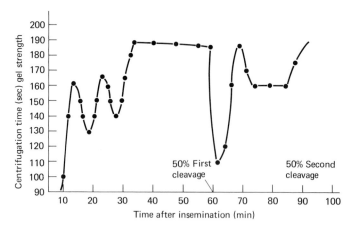

FIGURE 13. *Cortical sol-gel changes following insemination. Pressure-centrifuge measurements of the structural state of the plasmagel layer of* Arbacia punctulata *eggs made at 20°C and at high pressure (8000 lb/in.²) and high force (41,000 ×g). (From A. M. Zimmerman, J. V. Landau, and D. Marsland, J. Cell. Comp. Physiol. 49, 395–435, 1957.)*

fairly constant while establishment is taking place, suggesting that, contrary to the interpretation just described, formation of the cleavage mechanism is not reflected in gel strength determinations of the egg as a whole.

TENSION AT FURROW SURFACE

Many of the investigations we have just described were undertaken for the ultimate purpose of characterizing and identifying the physical mechanism that accomplishes cleavage. Other measurements have been made in order to determine the magnitude of the forces involved in the process. Direct measurements of the force generated in the furrows of cleaving sea urchin eggs were made by inserting a calibrated flexible needle and a stiff needle through the polar regions in opposite directions (Fig. 14). As the furrow constricted, the stiff needle did not move, but the flexible needle bent until its resistance to deformation equalled the force exerted by the furrow. A cleavage furrow halted in this fashion immediately resumes activity when the needles are removed. The forces measured averaged 2.5×10^{-3} dyne in two species of echinoderms (Rappaport, 1967). They represent the total available to the egg, rather than the amount actually required for

cleavage. Furrows exerting as little as 1.5×10^{-3} dyne promptly completed division after the needles were removed. Hiramoto (1968) calculated that the tension at the furrow is about 1.5 dynes/cm shortly before cleavage. His calculations were based on tension measured elsewhere on the surface and cell curvature. In order to divide the cell, the tension generated at the furrow surface need only

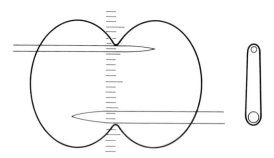

FIGURE 14. *(Left) Arrangement of cell and needles for determination. Upper, calibrated needle is deflected downward during cleavage. Lower, holding needle does not move. (Right) Schematic section through and parallel to the cleavage plane during isometric contraction. Diameter of the needles is exaggerated. (From R. Rappaport, Science 156, 1241–1243. Copyright 1967 by the American Association for the Advancement of Science.)*

exceed that existing at the surface in the other parts of the cell, where measurements by Hiramoto (1963) indicate a value of 0.2 dynes/cm.

Chemical Basis of Cleavage Mechanism

The chemical basis of the cleavage mechanism has yet to be determined, but investigators have described several likely possibilities, two of which we shall describe. Fluctuations in chemical composition correlated with the division cycle have been known for many years, but it has been difficult to determine their relation to the division process.

ACTOMYOSINOID SYSTEMS

Since a furrow decreases its circumference when it functions, and a muscle decreases its length, the possibility that furrow function may be related to the better-known process of muscle contraction has been recognized for some time. Actinoid proteins have been isolated from many kinds of cells, and their presence has been related to protoplasmic movement (Jahn and Bovee, 1969) and active form change. Hoffman-Berling (1964) found that cultured fibroblasts completed division in the presence of ATP, divalent ions, and "relaxing grana," a particulate fraction isolated from skeletal muscle. Kinoshita and Yazaki (1967) adapted the technique for use with cleaving sea urchin eggs. Miki-Noumura and Oosawa (1969) have isolated an actin-like protein that interacts with rabbit myosin from unfertilized sea urchin eggs. The sea urchin–rabbit actomyosin complex has many of the qualities of muscle actomyosin. More specifically, Perry et al (1971) reported that negatively stained filaments in newt egg furrows closely resemble those of actin from striated muscle and F-actin prepared from sea urchin eggs. Glycerination followed by treatment with heavy meromyosin coats the filaments with an array of fine hairs, some of which are arranged in the "arrow-head" configuration. The 50 to 70-Å-diameter microfilaments that have been described at the bases of the furrows in cleaving vertebrate and invertebrate eggs have characteristics similar to those of actin (Jahn and Bovee, 1969). A further resemblance concerns the contractility of furrows and actomyosin threads. By assuming that the microfilamentous band is the contractile structure, its capacity for isometric contraction can be calculated as 1.25 to 2.4 × 10^5 dynes/cm^2 (Rappaport, 1967); actomyosin threads exert a tension of 2.45 × 10^5 dynes/cm^2 in isometric contraction (Portzehl, 1951).

THIOL–DISULFIDE EXCHANGE SYSTEM

Sakai (reviewed, Sakai, 1968) has extracted a KCl-soluble protein from a water-insoluble residue of homogenized sea urchin eggs. The protein forms threads that contract and relax in the presence of metal ions and their chelating agents as well as oxidants of —SH groups; it is not affected by ATP. The contractility of these threads fluctuates during the division cycle, increasing as metaphase is approached and rapidly decreasing just before the furrow appears. The KCl-soluble protein may be obtained from the isolated egg cortex, and its contractility is correlated with fluctuation in —SH level, which increases toward metaphase and falls rapidly during cytokinesis. Since the thiol content of sea urchin egg protein remains constant during the division cycle, Sakai reasoned that there must be another fraction whose thiol content changes oppositely to that of the cortical protein. He found that the —SH content of a protein fraction which precipitates from the water-soluble fraction in the presence of calcium behaves according to the prediction (Fig. 15). Cortical contraction would in this mechanism be based upon *in vivo* oxidation of —SH groups facilitated by the exchange reaction between the two protein fractions. Sakai considers the calcium-insoluble protein to be associated with the mitotic apparatus and has demonstrated an exchange reaction *in vitro* between the contractile protein of the cortex and protein of the mitotic apparatus. The reaction requires several cofactors and an enzymatic protein fraction that precipitates at pH

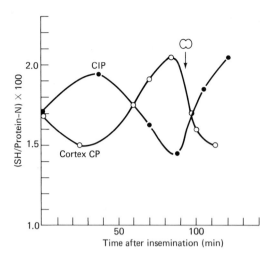

FIGURE 15. *Mirror image fluctuation of —SH groups between the contractile protein (CP) of the cortex and the calcium-insoluble protein fraction (CIP) during cleavage cycle (Hemicentrotus pulcherrimus). (From H. Sakai,* Int. Rev. Cytol. 23, *89–112, 1968.)*

5.0 from the calcium-insoluble protein-free water extract.

Establishment of Cleavage Mechanism

From the foregoing, it is apparent that the cleavage mechanism is established in the equatorial surface by the influence of the mitotic apparatus. It is logical to propose that the mitotic apparatus alters the cytoplasmic environment immediately beneath the surface, but the nature of the alteration and the manner in which it is accomplished are unknown. We are accustomed to attribute stimulatory activity of this kind to the addition of some substance to the altered environment, but existing information does not exclude the possibility that alteration could be accomplished by removal. It is, however, clear that, in eggs, the only parts of the mitotic apparatus necessary for this particular function are the asters. It must further be recognized that this stimulatory activity is not

restricted to the asters of the mitotic apparatus. Furrows form between sperm asters (or between a sperm aster and an aster of the mitotic apparatus) and between cytasters that may arise "spontaneously" in the cytoplasm following certain kinds of chemical treatment (Chambers, 1921a). In addition to their similarity of appearance, the chemical composition of these structures is indistinguishable.

MODE OF ACTION OF MITOTIC APPARATUS

The position of the furrow is established at anaphase, which is also the time when the asters attain their maximum size. In echinoderm eggs the mitotic apparatus is relatively large, and astral rays penetrate to all regions of the cell and frequently cross at the equator. Any mechanism that could move material along or within the astral structures could be involved in furrow establishment, and an association between their radiate appearance and cytoplasmic flow has been proposed many times. Some early cytologists considered the asters the visible consequences of complex flow patterns that periodically shifted from centrifugal to centripedal (Conklin, 1902); others suggested that the asters represented a cytoskeleton that regulated the flow pattern. This interpretation of the asters has not been substantiated. It has recently been shown, however, that particles move centripedally in asters, but the movement is not associated with cytoplasmic flow and no relation to establishment of the cleavage mechanism has been clearly demonstrated. The most specific attempts to relate cytoplasmic flow patterns to furrow establishment have concerned "fountain streaming." In this scheme a current flowing from the astral centers toward the polar surface was reflected toward the equator, where it meets the flow from the opposite pole and forms a vortex (Henley and Costello, 1965). By this mechanism, cleavage-inducing substances could accumulate under the equatorial surface. Cytoplasmic flow conforming to this pattern exists in some cleaving nematode eggs, but it is not visible in the eggs of other species. It has, however, been shown that geometrical alterations of the cell that would greatly disrupt the fountain streaming pattern do not interfere with furrow establishment (Rappaport, 1970).

It seems unlikely that the cleavage stimulus is associated with any simple fluid flow, as it appears to move only in straight lines. When cells are flattened to about the proportions of an English muffin, a perforation joining the opposed flattened surfaces is bounded by a cylindrical surface in which a furrow may be established. When perforations are made between the asters, the furrow forms only on the side closest to the asters and not on the opposite side, even though the diameter of the perforation may be very small. Further, no furrow forms at the cell margin distal to the perforation, even though it may lie within the normal spindle-to-surface distance. It is as though any object in the path of the astral rays throws a shadow within which furrows cannot form. Many different variations on this experiment have been performed (Dan, 1943; Rappaport, 1968), and their results provide a strong temptation to associate the cleavage stimulus with the linear elements of the mitotic apparatus. But whether these linear elements play their role by transporting substances or by propagating changes in molecular configuration is presently unknown.

NATURE OF STIMULUS

It is possible that the stimulus may consist of some missing element in the contractile system that is conveyed to the equatorial surface by the mitotic apparatus. The nature of the stimulus would therefore depend on the nature of the contractile system that is activated. Since the contractile system has not been positively identified, we cannot make definite statements concerning the nature of the stimulus. Several possibilities exist. If the thiol–disulfide exchange system is the basis of contractility (Sakai, 1968), then the enzyme-like protein that precipitates at pH 5.0 from the calcium-insoluble protein-free water extract could act as a stimulating substance by locally promoting the —SS —SH reaction. Should the contractile system prove to be actomyosinoid, local fluctuations of any of the necessary components of the system could constitute a stimulus mechanism. Calcium ions have a long history of

implication in cell division, although their role has not yet been identified. The activating effect of calcium on myosin ATPase has suggested the possibility that localization of calcium under the equatorial surface could facilitate cleavage (Roberts, 1961).

TIME RELATIONS OF CLEAVAGE STIMULUS

When the mitotic apparatus is moved relative to the surface, a new furrow is established within a few minutes in the surface adjacent to the zone between the asters in their new position. By manipuating the cell margin in echinoderm eggs with an excentric mitotic apparatus, it is possible to study precisely the time relations of the stimulation process (Rappaport and Ebstein, 1965). The surface must be exposed to the zone between the asters for about 1 minute in order to accomplish furrow establishment (Fig. 16). There follows a latent period of about 2½ minutes, during which the exposed part of the surface remains outwardly passive. At the end of the latent period, 3½ minutes after the surface is pushed inward, active furrowing begins in the newly exposed surface. During this relatively brief period, both the molecular reorganization required for establishment of the cleavage mechanism and the initiation of its function are accomplished.

To a considerable extent, the problems that intrigued classical cytologists have been answered in the terms of their time. The contractile nature of the cleavage mechanisms appears to be established. The visible cell parts that are essential to the process have been identified, and the nature of their function is known in a general way. This knowledge opens the way to more sophisticated analysis of the cleavage process.

References

ARNOLD, J. M. (1969). Cleavage furrow formation in a telolecithal egg (*Loligo pealii*). I. Filaments in early furrow formation. *J. Cell Biol. 41*, 894–904.

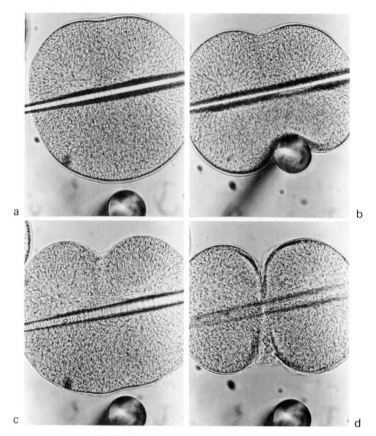

FIGURE 16. *Establishment of the furrow in a flattened sand dollar egg with an excentric mitotic apparatus. (a) Beginning of a unilateral furrow close to the excentric mitotic apparatus, which was held in position by a needle. (b) The more distant surface is held closer to the mitotic apparatus. (c) Immediately after the surface was released after about 1 minute, as shown in (b). (d) Active furrowing in the more distant surface. (Fom R. Rappaport and R. P. Ebstein,* J. Exp. Zool. 158, 373–382, 1965.)

BELL, L. G. E. (1963). Some observations concerning cell movement and cell cleavage. *In* "Cell Growth and Cell Division" (R. J. C. Harris, ed.), pp. 215–228. Academic Press, New York.

BERRILL, N. J. (1961). "Growth Development, and Pattern." W. H. Freeman, San Francisco.

BRACHET, J. (1950). "Chemical Embryology." Wiley-Interscience, New York.

CHALKLEY, H. W. (1935). The mechanism of cytoplasmic fission in *Amoeba proteus. Protoplasma 24,* 607–621.

CHAMBERS, R. (1921a). The formation of the aster in artificial parthenogenesis. *J. Cell. Comp. Physiol. 4,* 33–39.

CHAMBERS, R. (1921b). Studies on the organization of the starfish egg. *J. Cell. Comp. Physiol. 4,* 41–44.

COLE, K. S. (1932). Surface forces of the Arbacia egg. *J. Cell. Comp. Physiol. 1,* 1–9.

CONKLIN, E. G. (1902). Karyokinesis and cytokinesis in the maturation, fertilization and cleavage of *Crepidula* and other gasteropoda. *J. Acad. Nat. Sci. Philadelphia, Ser. 2 12,* Pt. 1, 1–21.

CONKLIN, E. G. (1917). Effects of centrifugal force on the structure and development of the eggs of *Crepidula. J. Exp. Zool. 22,* 311–419.

DAN, K. (1943). Behavior of the cell surface during cleavage. V. Perforation experiment. *J. Fac. Sci. Univ. Tokyo 6,* 297–321.

DAN, K., YANIGATA, T., AND SUGIYAMA, M. (1937).

Behavior of the cell surface during cleavage. I. *Protoplasma 28*, 66–81.

DANIELLI, J. F. (1952). Division of the flattened egg. *Nature (London) 170*, 496.

HENLEY, C., AND COSTELLO, D. P. (1965). The cytological effects of podophyllin and podophyllotoxin on the fertilized eggs of Chaetopterus. *Biol. Bull. 128*, 369–391.

HIRAMOTO, Y. (1956). Cell division without mitotic apparatus. *Exp. Cell Res. 11*, 630–636.

HIRAMOTO, Y. (1957). The thickness of the cortex and the refractive index of the protoplasm in sea urchin eggs. *Embryologia 3*, 361–374.

HIRAMOTO, Y. (1958). A quantitative description of protoplasmic movement during cleavage in the sea-urchin egg. *J. Exp. Biol. 35*, 407–424.

HIRAMOTO, Y. (1963). Mechanical properties of sea urchin eggs. I. Surface force and elastic modulus of the cell membrane. *Exp. Cell Res. 32*, 59–75.

HIRAMOTO, Y. (1965). Further studies on cell division without mitotic apparatus in sea urchin eggs. *J. Cell Biol. 25*, 161–167.

HIRAMOTO, Y. (1968). The mechanics and mechanisms of cleavage in the sea urchin egg. *Symp. Soc. Exp. Biol. 22*, 311–327.

HIRAMOTO, Y. (1970). Rheological properties of sea urchin eggs. *Biorheology 6*, 201–234.

HIRAMOTO, Y. (1971). Analysis of cleavage stimulus by means of micromanipulation of sea urchin eggs. *Exp. Cell Res. 68*, 291–298.

HOFFMAN-BERLING, H. (1964). Relaxation of fibroblast cells. *In* "Primitive Motile Systems in Cell Biology" (R. Allen and N. Kamiya, eds.), pp. 365–376. Academic Press, New York.

JAHN, T. L., AND BOVEE, E. C. (1969). Protoplasmic movements within cells. *Physiol. Rev. 49*, 793–862.

KINOSHITA, S., AND YAZAKI, I. (1967). The behavior and localization of intracellular relaxing system during cleavage in the sea urchin egg. *Exp. Cell Res. 47*, 449–458.

KÜHN, A. (1971). "Lectures on Developmental Physiology." Springer-Verlag, New York.

LEWIS, W. H. (1939). The role of a superficial plasmagel layer in changes of form, locomotion and division of cells in tissue cultures. *Arch. Exp. Zellforsch. 23*, 1–7.

LILLIE, R. S. (1903). Fusion of blastomeres and nuclear division without cell-division in solutions of non-electrolytes. *Biol. Bull. 4*, 164–178.

LONGO, F. J., AND ANDERSON, E. (1969). Cytological aspects of fertilization in the lamellibranch *Mytilus edulis*. I. Polar body formation and development of the female pronucleus. *J. Exp. Zool. 172*, 69–96.

MARSLAND, D. (1939). The mechanism of cell division. Hydrostatic pressure effects upon dividing egg cells. *J. Cell. Comp. Physiol. 13*, 15–22.

MARSLAND, D. (1950). The mechanisms of cell division; temperature–pressure experiments on the cleaving eggs of *Arbacia punctulata*. *J. Cell. Comp. Physiol. 36*, 205–227.

MARSLAND, D., AND LANDAU, J. V. (1954). The mechanics of cytokinesis: temperature–pressure studies on the cortical gel system in various marine eggs. *J. Exp. Zool. 125*, 507–539.

MERCER, E. H., AND WOLPERT, L. (1958). Electron microscopy of cleaving sea urchin eggs. *Exp. Cell Res. 14*, 629–632.

MIKI-NOUMURA, T., AND OOSAWA, F. (1969). An actin-like protein of the sea urchin eggs. *Exp. Cell Res. 56*, 224–232.

MITCHISON, J. M., AND SWANN, M. M. (1954). The mechanical properties of the cell surface. I. The cell elastimeter. *J. Exp. Biol. 31*, 443–472.

MITCHISON, J. M., AND SWANN, M. M. (1955). The mechanical properties of the cell surface. III. The sea urchin egg from fertilization to cleavage. *J. Exp. Biol. 32*, 734–750.

MONROY, A., AND MONTALENTI, G. (1947). Variations of the submicroscopic structure of the cortical layer of fertilized and parthenogenetic sea urchin eggs. *Biol. Bull. 92*, 151–161.

MOORE, A. R. (1933). Is cleavage a function of the cytoplasm or of the nucleus? *J. Exp. Biol. 10*, 230–236.

PASTEELS, J. J., AND DE HARVEN, E. (1962). Étude au microscope électronique du cortex de l'oeuf de *Barnea candida* (mollusque bivalve), et son évolution au moment de la fécondation, de la maturation et la segmentation. *Arch. Biol. 73*, 467–490.

PERRY, M. M., JOHN, H. A., AND THOMAS, N. S. T. (1971). Actin-like filaments in the cleavage furrow of the newt egg. *Exp. Cell Res. 65*, 249–253.

PORTZEHL, H. (1951). Muskelkontraction und Modelkontraction. *Z. Naturforsch. 66*, 355–361.

RAPPAPORT, R. (1960). Cleavage of sand dollar eggs under constant tensile stress. *J. Exp. Zool. 144*, 225–231.

RAPPAPORT, R. (1961). Experiments concerning the cleavage stimulus in sand dollar eggs. *J. Exp. Zool. 148*, 81–89.

RAPPAPORT, R. (1964). Geometrical relations of the cleavage stimulus in constricted sand dollar eggs. *J. Exp. Zool. 155*, 225–230.

RAPPAPORT, R. (1966). Experiments concerning the

cleavage furrow in invertebrate eggs. *J. Exp. Zool. 161*, 1–8.

RAPPAPORT, R. (1967). Cell division: direct measurement of maximum tension exerted by furrow of echinoderm eggs. *Science 156*, 1241–1243.

RAPPAPORT, R. (1968). Geometrical relations of the cleavage stimulus in flattened, perforated sea urchin eggs. *Embryologia 10*, 115–130.

RAPPAPORT, R. (1969a). Division of isolated furrows and furrow fragments in invertebrate eggs. *Exp. Cell Res. 56*, 87–91.

RAPPAPORT, R. (1969b). Aster-equatorial surface relations and furrow establishment. *J. Exp. Zool. 171*, 59–68.

RAPPAPORT, R. (1970). An experimental analysis of the role of fountain streaming in furrow establishment. *Devel. Growth and Diff. 12*, 31–40.

RAPPAPORT, R. (1971). Cytokinesis in animal cells. *Int. Rev. Cytol. 31*, 169–213.

RAPPAPORT, R., AND EBSTEIN, R. P. (1965). Duration of stimulus and latent periods preceding furrow formation in sand dollar eggs. *J. Exp. Zool. 158*, 373–382.

RAPPAPORT, R., AND RATNER, J. H. (1967). Cleavage of sand dollar eggs with altered patterns of new surface formation. *J. Exp. Zool. 165*, 89–100.

ROBERTS, H. S. (1961). Mechanisms of cytokinesis: a critical review. *Quart. Rev. Biol. 36*, 155–177.

RUSTAD, R. C., YUYAMA, S., AND RUSTAD, L. C. (1970). Nuclear-cytoplasmic relations in the mitosis of sea urchin eggs. II. The division times of whole eggs and haploid and diploid half-eggs. *Biol Bull. 138*, 184–193.

SAKAI, H. (1968). Contractile properties of protein threads from sea urchin eggs in relation to cell division. *Int. Rev. Cytol. 23*, 89–112.

SCHROEDER, T. E. (1968). Cytokinesis: filaments in the cleavage furrow. *Exp. Cell Res. 53*, 272–276.

SCHROEDER, T. E. (1969). The role of "contractile" ring" filaments in dividing *Arbacia* eggs. *Biol. Bull. 137*, 413–414.

SCHROEDER, T. E. (1972). The contractile ring II. Determining its brief existence, volumetric changes and vital role in cleaving *Arbacia* eggs. *J. Cell Biol. 53*, 419–434.

SCOTT, A. (1960). Surface changes during cell division. *Biol. Bull. 119*, 260–272.

SELMAN, G. G., AND PERRY, M. M. (1970). Ultrastructural changes in the surface layers of the newt's egg in relation to the mechanism of its cleavage. *J. Cell Sci. 6*, 207–227.

SWANN, M. M., AND MITCHISON, J. M. (1953). Cleavage of sea urchin eggs in colchicine. *J. Exp. Biol. 30*, 506–514.

SWANN, M. M., AND MITCHISON, J. M. (1958). The mechanism of cleavage in animal cells. *Biol. Rev. 33*, 103–135.

SZOLLOSI, D. (1970). Cortical cytoplasmic filaments of cleaving eggs: a structural element corresponding to the contractile ring. *J. Cell Biol. 44*, 192–209.

WILSON, E. B. (1928). "The Cell in Development and Heredity." Macmillan, New York.

WOLPERT, L. (1960). The mechanics and mechanism of cleavage. *Int. Rev. Cytol. 10*, 163–216.

ZIEGLER, H. E. (1903). Experimentelle Studien über Zelltheilung. IV. Die Zelltheilung der Furchungzellen bei *Beroë* und *Echinus*. *Wilhelm Roux Arch. Entwicklungmech. Organismem 16*, 155–175.

ZIMMERMAN, A. M., LANDAU, J. V., AND MARSLAND, D. (1957). Cell division: A pressure-temperature analysis of the effects of sulfhydryl reagents on the cortical plasmagel structure and furrowing strength of dividing eggs (Arbacia and chaetopterus). *J. Cell. Comp. Physiol. 49*, 395–435.

II

CELLS

CHAPTER FOUR

Molecular Mechanisms of Cellular Differentiation*

MARTIN NEMER

The mechanisms that call forth the production of specified populations of protein molecules determine at the same time the direction of embryological development. These mechanisms involve the selection and transfer of genetic information and consist of a sequence of events beginning with the specification of chromosomal sites of activity and the synthesis of RNA gene products. The newly synthesized RNA may require processing before a part of it is transferred to ribosomal machinery for translation of the genetic message and the synthesis of specific proteins. The transport of message from nucleus to cytoplasm and the translation of messenger RNA (mRNA) may be susceptible to regulation; in a differentiated cell some genes are ultimately expressed and others are not. In the process of development,

*Dedicated to Jack Schultz (1904–1971) whose imagination and compassion will be remembered by many who have been inspired by him to pursue such problems as are described here.

the achievement of selective genetic expression may depend upon a variable application of regulatory devices that suit the particular situations encountered.

Heterochromatin and Euchromatin as Components of Chromosomes

The morphology of chromatin divides into heterochromatin and euchromatin. The heterochromatic regions of chromosomes are highly compacted and do not disperse during the cell cycle, specifically during telophase, to form the highly diffuse interphase material that is characteristic of the rest of the chromatin. The chromosomal regions that maintain their compact status all the time are termed *constitu-*

tively *heterochromatic* and are usually considered genetically inert. Other chromosomal regions may become heterochromatic only in certain situations. This is a second type of heterochromatin, which has been termed *facultative* or *functional* and is often associated with the regulation of genes on the same chromosome. Constitutive heterochromatin, although in many situations appearing to be genetically inert, does show genetic activity in genetic position effects (Lewis, 1950; Baker, 1968) and in quantitative characteristics, such as body weight, size of organs, or rate of development (Mather, 1944). Schultz (1965) proposed that the position effect controlling adjacent genes operates by compaction of the chromosome in this adjacent region, which is then unable to respond to stimuli for activity.

Heterochromatin and Satellite DNA

A prominent example of constituitive heterochromatin is to be found in the centromeres, which are known to serve as sites of attachment for mitotic spindle proteins. Pardue and Gall (1970) and Jones (1970) have demonstrated that mouse satellite DNA is localized in the centromeric heterochromatin. These DNA sequences have defined properties and are highly repetitive. The specific character of this type of constitutive heterochromatin establishes that it is not a variable state of chromatin. The mouse satellite DNA sequences do not appear to code for a complementary fraction of RNA. Whereas Flamm et al. (1969) were unable to detect hybridization of pulse-labeled RNA from mouse liver, spleen, and kidney to satellite DNA, Harel et al. (1968). reported some hybridization with the same material.

The same specification of satellite DNA in heterochromatin (Yasmineh and Yunis, 1969, 1970; Yunis and Yasmineh, 1970) was made by a method for separating heterochromatin and euchromatin, developed by Frenster et al. (1963). The method was based on the differential solubility and sedimentation of chromatin extracts. The original observation made from the use of such extracts was that incorporation of RNA precursors into the RNA of nuclei of lymphocytes occurred predominantly in the diffuse, readily soluble euchromatin. The difficultly soluble heterochromatin was shown cytologically to resemble the compacted heterochromatin visible in chromosomes. Since the time of this demonstration there have been numerous uses of so-called chromatin extracts prepared from assorted nuclei.

Chromatin as Template for RNA Synthesis

When purified chromatin is used as template to synthesize RNA *in vitro* (Paul and Gilmour, 1966, 1968, 1969; Huang and Huang, 1969; Bekhor et al., 1969; Smith et al., 1969), the RNA populations synthesized on chromatin isolated from different tissues are distinguishable, and each chromatin preparation produces RNA that is apparently similar to that synthesized *in vivo*. These results indicate that the regulatory elements that influence selective transcription reside in the structure of the chromosomes themselves. The experiments on *in vitro* chromatin activity, particularly referring to template properties, have been designed to evaluate the nature of the synthesized RNA and to assess the role of chromatin proteins in the synthesis of specific RNA. The first object, RNA characterization, was approached by hybridization of the RNA with DNA to compare the results of competition with RNA extracted from the same tissue or different tissues. The RNA synthesized *in vitro* from a given tissue chromatin is characterized by its susceptibility to competition with homologous RNA synthesized by the same tissue *in vivo*, and by its limited susceptibility to competition by RNA from other tissues (Fig. 1). All the experiments have of necessity involved only the "redundant" sequences of the genome. Therefore, since related families of DNA sequences may not be distinguished, efficient competition between two RNA populations does not establish an identity between them. Thus we cannot

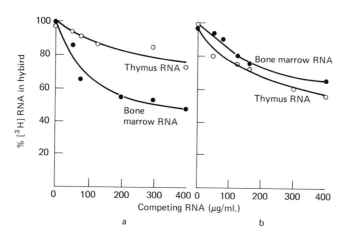

FIGURE 1. *Competition–hybridization experiments between (a)* [³H] *RNA made with a bone marrow chromatin template versus natural bone marrow RNA (–●–●–), and natural thymus RNA (–○–○–); (b)* [³H] *RNA made with a thymus chromatin template versus natural bone marrow RNA (–●–●–) and natural thymus RNA (–○–○–). All filters were loaded with 1.4 μg of rabbit embryo DNA. (From J. Paul and R. S. Gilmour, J. Mol. Biol. 34, 305–316, 1968.)*

conclude from successful competition that the *in vivo* and *in vitro* synthesized populations are the same but only that they are apparently similar. With this reservation, such competition has been the criterion for specific RNA synthesis in experiments that have attributed to histones a repressive role in *in vitro* transcription (Huang and Bonner, 1962) and that have assigned to some of the nonhistone chromosomal proteins a role in determining the specificity of transcription (Paul and Gilmour, 1966, 1968, 1969). Another chromosomal component implicated in the same type of experiment is the so-called "chromosomal RNA," which has been thought to be regulatory also. This RNA is of small and uniform size with a peculiarly high content of dihyrouridine. Artman and Roth (1971) have recently questioned whether or not this RNA is in fact an artifact of preparation.

In all the experiments designed to test the intervention of protein factors, chromatin was reconstituted by the reassembly of its parts in high salt concentration. Clark and Felsenfeld (1971) and Itzhaki (1970) have demonstrated considerable exchange of protein between DNA molecules in the presence of high salt or calcium concentrations. They have questioned the validity of chromatin treatment under such conditions—hence the results previously obtained on the effect of proteins on reconstituted chromatin. They have also concluded from deoxyribonuclease digestions of chromatin DNA that a considerable portion of the DNA is susceptible to digestion and thus not covered by protein. However, Mirsky (1971), in repeating and extending these studies, came to the opposite conclusion, that chromatin DNA *is* extensively covered by protein in complexes of various degrees of looseness. Experiments by Paul et al. (1972) indicate that chromatin transcribed in low salt yields a product that more closely resembles the natural RNA than that of chromatin transcribed in high salt concentrations, thus substantiating the crucial role of salt concentration. These investigators note also that if extensive stretches of free DNA do exist in chromatin, and are the sites of transcription as postulated by Clark and Felsenfeld (1971), then histones added to chromatin ought to combine with this free DNA and result in repression. Indeed, they find little effect of such addition on chromatin transcription. It is therefore possible that nonhistone proteins normally cover stretches of seemingly free DNA and serve to antagonize repressive effects of histones. Another aspect of the examination of the template activities of chromatin that needs to be further evaluated is the use of heterologous, bacterial RNA polymerase. Weaver et al.

(1971) have purified RNA polymerase (II) from calf thymus and rat liver and have found what they consider to be structural similarities between these eukaryotic RNA polymerases and those from prokaryotes. Although this finding suggests analogous modes of action and regulation, the eukaryotic RNA polymerase may utilize chromatin DNA template in a manner different from prokaryotic polymerases. Indeed, Butterworth et al. (1971) have found that RNA polymerase purified from rat liver and RNA polymerase from *Micrococcus lysodeikticus* bind to and transcribe from different sites on chromatin DNA.

The most definitive study of chromatin as template would involve the examination of the products of a single gene or a small number of genes, thereby eliminating the problems of RNA complexity. Such specifications have recently been fulfilled by Astrin (1973), who has utilized chromatin extracted from 3T3 mouse cells transformed by simian virus 40 (SV 40). Specific viral gene products were transcribed from this chromatin, which resembled viral RNA products *in vivo*, but differed from those products made on purified viral DNA templates and were not produced on chromatin from uninfected cells. The same sequences of the SV40 "—" strand (the DNA strand which serves as a template for early RNA) are transcribed *in vitro* from chromatin as are found in transformed cells.

Fractionation of Chromatin into Active and Inactive Components

The most useful goal in a biochemical approach to the analysis of chromatin would be the separation and purification of active genes from the bulk of the chromatin presumed to be inactive. The first step in this direction was made by Frenster et al. (1963). The readily soluble euchromatin in their experiments could be viewed as being enriched in the active portion of the genome. It was, at least, deficient in the relatively inert satellite-containing fraction of the genome (Yunis and Yasmineh, 1970). Solubility differences in physiological saline

have recently been employed by Marushige and Bonner (1971) to fractionate chromatin fragments into two components—the soluble one containing RNA polymerase and being enriched in nonhistone proteins, the insoluble component lacking RNA polymerase and diminishing in its content of nonhistone proteins. A most promising approach is the separation of sheared chromatin into active and inactive parts by the simple device of gel exclusion chromatography on agarose columns (Janowski et al., (1972). The active chromatin fragments are preferentially excluded on these columns. These authors refer to previous studies that noted the slower sedimentation of active compared to inactive chromatin and conclude that these differences in sedimentation and gel exclusion chromatography may be explained on the basis of the active fragments being more extended than the inactive fragments, as would be expected from cytological observations of extended euchromatin and compacted heterochromatin.

DNA–Redundant and in Excess

Eukaryotic genomes contain multiple copies of some DNA sequences. The widespread occurrence of redundant, or at least very similar, nucleotide sequences was observed by DNA renaturation kinetics (Britten and Kohne, 1968) and by the reannealing of DNA sequences to form circles (Thomas et al., 1970). The most highly specified category of multiple gene copies is the group of genes that code for the two large ribosomal RNAs. In *Xenopus laevis* several hundred of these genes are clustered at a single locus (Wallace and Birnstiel, 1966; Brown and Weber, 1968). A similar multiplicity of ribosomal genes was shown for *Drosophila* (Ritossa et al., 1966). The 5S ribosomal RNA (rRNA) genes are also present in multiple copies (Tartof and Perry, 1970). Further designation of other parts of the genome that might be included in the category of redundant DNA has not been so clear. For example, the genetic or physiological role of mouse satellite DNA in centromeric heterochromatin is unknown. However, the demonstration of high redundancy hybridization for histone mRNA from sea urchin embryos (Kedes and Birnstiel, 1971) suggests that some structural genes (coding for mRNA) may be repeated (Fig. 2). The well-characterized mRNA for hemoglobin arises from single copy genes (unique DNA sequences), according to Bishop et al. (1972); however, observations on the occurrence of redundancy in polysomal (messenger) RNA show that nuclear RNA is not alone in being highly enriched in sequences complementary to redundant DNA (Greenberg and Perry, 1971). A recent study by Dina et al. (1973) with mRNAs from developing *Xenopus* embryos goes much further, concluding that each mRNA molecule consists of a main part transcribed from unique DNA and a smaller part transcribed from homogeneously repeated DNA sequences. Therefore, in order to understand the significance of repetitious DNA, we shall have to test rigorously (1) the repetition of genes, (2) the existence of both unique and redundant stretches of nucleotides within

genes, and (3) the reiteration of nongene DNA sequences. Redundant DNA sequences, whether inside or outside genes, may have presently unknown functions, possibly in the regulation of gene expression. Gene repetition, it can be argued, imparts an advantage in evolution through the conservation of essential functions. The vital ribosomes and the highly conserved histone proteins may have been perpetuated with great certitude by coding in repetitious genes.

The very large genome sizes (DNA content per nucleus) of certain amphibians have long been a source of puzzlement (Mirsky and Ris, 1951). Indeed, there is a wide range of genome sizes among eukaryotes completely uncorrelated with presently conceivable differences in functional requirements of DNA among the different organisms. Britten and Davidson (1971) have noted that the sizes and complexities of genomes of higher organisms are much greater than those of microorganisms (Fig. 3), and yet the complexity of enzyme reactions in

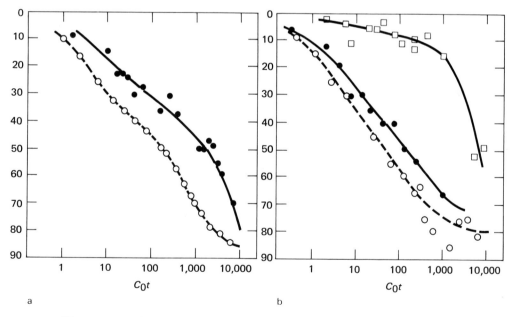

a   b

FIGURE 2.   *Time course of reassociation of sea urchin DNA, and sea urchin DNA in the presence of messenger RNA fractions. (a) DNA renaturation curves of sonicated sea urchin sperm DNA in 0.12 M phosphate buffer (–•–•–) or in 3xSSC and in 50 percent formamide (–o–o–). (b) Renaturation curves of 9S RNA and >25S RNA: of 9S RNA in 0.12 M PO$_4$ (–•–•–), 9S RNA in formamide (–o–o–), and >25S RNA (–□–□–). Ordinate is per cent reaction. (From L. H. Kedes and M. L. Birnsteil,* Nature *(London) 230, 165–176, 1971.)*

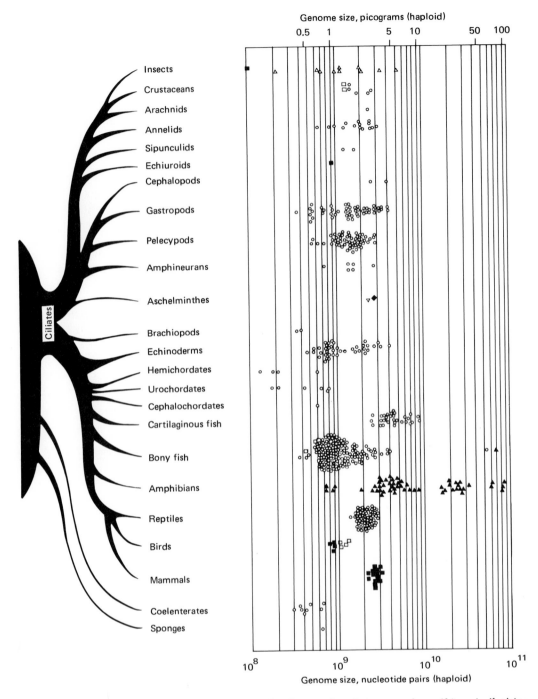

FIGURE 3. *Distribution of genome size in animals. Arranged to bring out the striking similarities in the range of genome sizes present in distantly related groups. The accuracy of all the measurements is not comparable. Phyla lacking data were not included. (From R. J. Britten and E. H. Davidson, Quart. Rev. Biol. 46, 111–133, 1971.)*

prokaryotes and eukaryotes are not much different. Thus there appears to be a vast excess of DNA in many eukaryotes beyond the essentials of specifying the known enzyme systems. Therefore, a considerable portion of the genomic DNA, either redundant or unique, may be regulatory of gene expression, functional in chromosomal structure, or simply inert.

Previous to the recent experiments on reannealing of DNA, the possibility of excessive genetic material and repetition of genes was raised by Callan (1967). He proposed that chromatids consist of *serial duplication units* along their lengths. This proposition was derived from observations by Callan and Lloyd (1960) on the structure of lampbrush chromosomes and by Keyl (1965) on the geometric doubling in DNA content of homologous bands of *Chironomous* chromosomes. However, serial repeats of genetic units would seem to be at variance with the fact that heterozygosity is normally restricted to not more than pairs of allelic alternatives. This objection is circumvented by Callan's hypothesis that each unit of genetic function consists of one *master copy* of coded information followed by a series of *slave repeats* that are specified by the master copy. Furthermore, a matching of slave to master sequences may be followed by correlation of base sequences in the slave wherever incongruities might appear. Thus mutation of the master copy would be propagated; alterations of the slaves would be corrected upon replication.

Models for Eukaryotic Gene Function

Britten and Davidson (1969) attempt to account for what seems in differentiated cells to be a coordinated activity of a number of genes. They illustrate that, for example, in liver, batteries of functionally linked enzyme systems must be maintained and presumably produced in an integrated way. They show that for some two dozen enzymes, batteries or groupings are made up in different tissues that consist of different but overlapping membership, thus mak-

ing it unlikely that each set of enzymes could have been generated from a set of contiguous genes. Furthermore, the alpha and nonalpha chains of hemoglobin in man are derived from unlinked genes, and the two enzymes glucose-6-phosphate dehydrogenase and 6-phosphogluconate dehydrogenase of the phosphogluconic acid oxidation pathway are from genes on separate linkage groups in *Drosophila melanogaster*. Thus genes whose activities must be intimately integrated may be located far apart in the genome. Finally, the response of uterine cells to estrogen is so physiologically varied that a pleiotropic effect on a number of genes is called for.

Britten and Davidson (1969) delineate *five* activities that logically describe an integrated gene activity: (1) response to an external signal; (2) production of a second signal; (3) transmission of the second signal to a number of receptors unresponsive to the original signal; (4) reception of the second signal, and (5) response to this event by activation of a gene and its transcription to provide the cell with the gene product (Fig. 4). These activities can be assigned to categories of agents as follows: (1) *sensor*, a site on the genome capable of receiving the external signal, directly or indirectly; (2) *integrator*, a sequence in the genome that synthesizes an activator agent, as a second signal; (3) *activator*, specifically "activator RNA," which is a specifically nuclear RNA, there being evidence that RNA sequences exist in the nucleus of presently unknown function that do not appear in the cytoplasm; (4) *receptor*, a site on the genome receiving the activator RNA in a sequence-specific manner, resulting in transcription of (5) the *structural* or *producer gene*, to yield a template RNA molecule or other RNA not involved in gene regulation. The authors utilize the observation of Britten and Kohne (1968) that groupings of extensive sequence similarities may exist in vast proportion in the genome of a higher organism. A proposed function of these sequence similarities is to serve as common production and receptor sites for integrative "activator RNA" or other signals. The recent report of Dina et al. (1973) adds still another possible scheme, whereby reiterated sequences may serve a regulatory function. In this case a portion of the mRNA molecule is

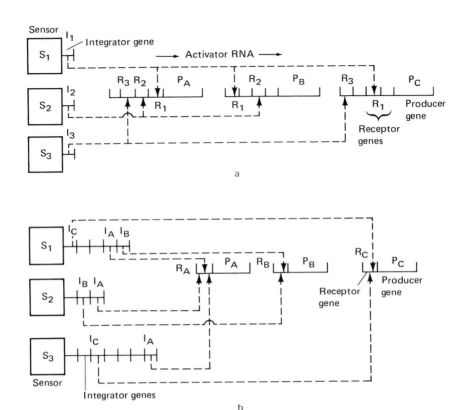

FIGURE 4. *Types of integrative system within the model of Britten and Davidson (1969). Integrative systems depending upon redundancy among (A) the regulator genes and (B) the integrator genes. These diagrams schematize the events that occur after the three sensor genes (in squares) have initiated transcription of their integrator genes. Activator RNAs diffuse (symbolized by dotted line) from their sites of synthesis—the integrator genes—to receptor genes. The formation of a complex between them leads to active transcription of the produced genes* P_A, P_B, *and* P_C. *(From R. J. Britten and E. H. Davidson,* Science 165, 349–358, 1969.)

complementary to redundant DNA sequences and linked to the larger portion of the mRNA molecule, which consists of unique sequences. Therefore, the producer or structural gene appears to consist of two parts, one of which may serve as a reiterated signal receptor. Regulation through such a device would be an alternative to the Britten–Davidson models, yet similarly would make use of reiterated DNA sequences.

Clustering of genes has been shown to afford coordinate regulation of gene expression in microorganisms, the most influential case in point being the lac operon of *Escherichia coli*. Whereas Britten and Davidson (1969) have contemplated the ways in which unlinked genes in eukaryotes could be coordinately expressed, Kabat (1972) has recently pointed out that there may be cases where the expression of one gene in a cluster precludes the expression of any other linked gene. Such close linkage of mutually exclusive genes occurs in the nonchain hemoglobin genes and in the immunoglobin genes of various mammals. Hemoglobin consists of two α chains and two non-α chains, that are called β, γ, δ, or ϵ. The ϵ chain is normally present in very early embryonic fetuses and is replaced by the γ (or fetal) chain and finally by the β chain in the adult. Only the β and δ chains appear coordinately (in a ratio of approximately 97:3). The appearance of each non-α chain is closely correlated with the disappearance of the more primitive chain. Various lines of evidence are cited by Kabat (1972) for the close linkage of the non-α genes. To account for the temporal sequence and the mutually exclusive expression of the non-α genes, Kabat (1972) proposes that these genes are aligned in series starting at a promoter site, which is the site of attachment of RNA polymerase. The first gene in the series would be ϵ. Its expression would be followed by termination of transcription. In order to make way for the expression of γ and then β, the ϵ gene would undergo intrachromosomal crossing-over, involving excision as an acentric loop. Thus the next gene, γ, would become adjacent to the promoter site.

Whereas this "looping out exclusion" hypothesis explains the sequential expression of clustered genes, there may be no necessary connection between the proximity of genes and their exclusive expressions. Certainly, common promoter gene or DNA sequence has yet to be demonstrated in eukaryotes. The alternative existence of separate promoter or regulatory sites would make this elaborate hypothesis superfluous. These mutually exclusively expressed genes may well be regulated by mutually exclusively expressed sequences. Each separate regulatory sequence might belong to a class of reiterated DNA, represented either within or external to the mRNA sequence.

Processing of Nuclear RNA

Something is known about the synthesis of large RNA precursors in the nucleus and their processing to the 18S and 28S forms in the ribosomes in the cytoplasm. Hardly anything is known about the steps that allow the transmission of template mRNA from its genic site of synthesis to the sites of action, the cytoplasmic polyribosomes. We do know that nuclear RNA molecules are very large, predominantly in a range from 5–10 $\times$ 10^6 daltons, that there are nucleotide sequences in the nuclear RNA that are not represented in the cytoplasmic RNA, and that turnover of the nuclear compartment is considerably more rapid than the cytoplasmic RNA. This exclusive character

of nuclear RNA was demonstrated by Shearer and McCarthy (1970) by means of hybridization of nuclear RNA with DNA and competition with either nuclear or cytoplasmic RNA (Fig. 5). Although the experiments involved only the redundant portion of the genome, the demonstration of a difference between the nuclear and cytoplasmic RNAs is unambiguous. Approximately 40 percent of the "redundant" nuclear RNA sequences escape competition by hybridization in the presence of cytoplasmic RNA (in the case of the mouse L cell). The strictly nuclear RNA has a very great turnover rate, accounts for the bulk of DNA transcription, and is, at present, of unknown function. By analogy with the synthesis of RNA from high molecular weight precursors and also the synthesis of transfer RNA (tRNA) from larger precursor molecules (Altman and Smith, 1971), we might speculate that a portion, at least, of the nuclear transcript serves as high-molecular-weight precursor that must be processed, perhaps by the excision of parts of the molecule, in order to furnish the completed mRNA for transfer to the cytoplasm. Still another portion of the nuclear transcript may serve a strictly nuclear function in the regulation of gene expression.

The processing of heterogeneous nuclear RNA, although it remains only a topic of speculation, may be elucidated by the recent demonstration that both the nuclear RNA and RNA from polyribosomes contain polyadenylic acid (poly A) segments covalently linked and in uniform amount 150 to 250 nucleotides, according to Edmonds et al. (1971). There is evidence (Darnell et al. 1971) that transcription may not be involved in the attachment of the poly A sequences to the RNA molecules—that is, that the attachment may be posttranscriptional. Kinetics of labeling indicates that the nuclear RNA has a higher content of poly A than cytoplasmic RNA after a short labeling period but that this reverses after longer periods, consistently with a precursor-product relationship. A posttranscriptional processing that involves the addition of poly A may be part of a process through which the appropriate mRNA sequences are selected from the total RNA transcript for subsequent transfer to the cytoplasm.

The most exact model so far used to solve

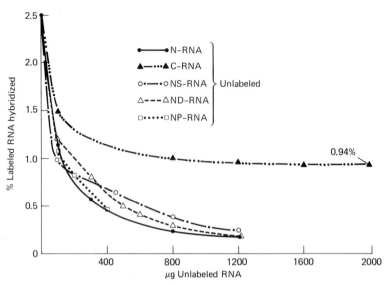

FIGURE 5. *Hybridization of L-cell nuclear RNA in the presence of competing RNAs from cytoplasm, chromatin, and nuclear sap. Five micrograms of 100-minute pulse-labeled RNA were incubated with 14 μg of filter-bound mouse DNA for 16 hours at 67°C in 0.5 ml of 2xSSC.* (From R. W. Shearer and B. J. McCarthy, J. Cell Physiol. 75, 97–106, 1970.)

the problem of the processing and transfer of RNA has been the cells transformed by DNA virus (Sambrook et al. 1968; Benjamin, 1966; Lindberg and Darnell, 1970). The viral DNA becomes covalently integrated into the cellular DNA, which is template for the synthesis of virus-specific nuclear and cytoplasmic RNA sequences. In the nucleus, the virus-specific mRNA's are part of large heterogeneous RNA molecules, but in the cytoplasm the molecules are much smaller and of uniform size. A precursor–product relationship here is strongly suggested.

Information Transfer to Cytoplasm

The transfer of mRNA from the nucleus to the cytoplasm is a problem that may have significance, particularly if such transport depended upon specific mechanisms that were capable of selecting RNA molecules for transport or if an accompanying agent (protein or membrane) were capable of specifically influencing template activity (Nemer, 1967). Several modes of transfer have been suggested:

1. Messenger RNA may diffuse into the cytoplasm as unattached nucleic acid or in nonspecific complex with protein. Its entry into the cytoplasm would then depend only upon its rate of transcription and processing and relative susceptibility to degradation (Aronson, 1971).

2. Transfer may depend upon associations with membranes, particularly those contiguous with the nuclear membrane (Faiferman et al., 1971).

3. Transfer may depend upon a mRNA complex forming either with ribosomal subunits or with specific proteins. There is evidence against ribosomal subunit complexing and for protein complexing from a number of sources: Spirin et al. (1964), Spirin and Nemer (1965), Infante and Nemer (1968), Spirin (1969), Perry and Kelly (1968), Spohr et al. (1970), and Henshaw (1968). Spirin proposed that a specific complex between RNA and protein, which he called the *informosome*, was indeed a special class of particles responsible for transfer. The possibility offered by one theory (Tomkins et al., 1969; see page 112) that repressor proteins may interact with mRNA suggests that specific mRNA-protein complexes may afford an in-

crease in either stability or lability of template activity.

4. Finally, a transfer of excised genes or DNA from nucleus to cytoplasm as so called *I-somes* has been proposed by Bell (1969). The data presented, however, were open to criticism based on the possibility of the cytoplasmic DNA being leaked from the nucleus (Fromson and Nemer, 1970). Bell (1971) has recently noted that the relative inhibitions of nuclear and cytoplasmic DNA incorporations varied considerably with the inhibitors 5-fluorodeoxyuridine, cytosine arabinoside, and hydroxyurea in chick muscle cells, thus affording a way to discriminate metabolically between nuclear and cytoplasmic forms. However, the mechanisms of inhibition are not known well enough to be able to infer validly what the differential effects of these inhibitors might mean. The association of DNA with cytoplasmic membranes of lymphocytes has been shown definitively to arise as a product of nuclear synthesis (Hall et al., 1971). A demonstration of valid cytoplasmic DNA of nuclear origin may have implications other than an involvement in information transfer for translational purposes.

The various theories for transfer of information from nucleus to cytoplasm are all controversial because they depend exclusively on biochemical fractionations and they lack specific markers. In the case of mRNA transfer, the specific properties of the informosomal proteins (proteins associated with mRNA) need to be delineated. If membranes are involved, then mere association between mRNA and membrane is not enough. The I-some theory proposed that DNA found in the cytoplasm is the agent for the transfer of genetic information from the nucleus. If this is so, it should be possible to demonstrate that cytoplasmic DNA base sequences are equivalent to those selectively represented in polysomal (messenger) RNA or, at least, a selection from and not fully representative of nuclear DNA. Williamson et al. (1972) have recently tested the "informational content" of cytoplasmic DNA from primary embryonic cell cultures, by observing its kinetics of hybridization with nuclear DNA, both for intermediate and slow

annealing fractions. The rate of reannealing of nuclear and cytoplasmic DNAs were indistinguishable. Therefore, the "cytoplasmic" DNA appears to contain all, rather than a selection of, the information in the genome, and, as such, cannot perform the informational role assigned to it by Bell (1969).

Again, as mentioned for the problem of RNA processing, the transport of a viral mRNA, transcribed from an integrated viral DNA in the host nucleus, presents the most promising model. Raskas (1971) has followed the release of Adenovirus mRNA from isolated nuclei, as promoted by incubation with ATP, and found the emerging mRNA to appear as a definite ribonucleoprotein or informosome. Ishikawa et al. (1969) previously showed that rat liver nuclei released ribonucleoprotein when incubated with ATP. These systems appear to hold great promise for analyses aimed at the specificity of the protein moiety of the informosome.

Untranslated Messengers

Cytoplasmic ribonucleoproteins containing messenger-like RNA appear to exist unassociated with ribosomes or ribosomal subunits (Spirin et al., 1964; Spirin and Nemer, 1965; Infante and Nemer, 1968; Spirin, 1969). These particles, or informosomes, are of unknown function but have been posited to be involved in either transfer or storage of messenger. Template activity for specific protein synthesis has been demonstrated for the RNA isolated from the informosomes of reticulocytes (Olsen et al., 1972; Jacobs-Lorena and Baglioni, 1972). In this case, 10S RNA, originating from a 20S ribonucleoprotein particle, was extracted from the postribosomal supernatant fluid and used to stimulate the synthesis of globin *in vitro*. It is interesting here that the distribution of mRNA for α and β globin chains in the informosomes does not reflect that in the polyribosomes. Jacobs-Lorena and Baglioni (1972) found about 20 percent of the α chain mRNA in informosomes, but none of the β chain mRNA—which was almost exclusively in the polyribosomes. Another case in which the informosomal compartment might reveal its function is that of the sea urchin egg. Mano and Nagano (1966) indicated that large particles in the egg contained messenger ribonucleoproteins and that the RNA of these particles had template activity. This work has yet to receive independent substantiation. Recently, Slater et al. (1973) have detected messenger-like RNA in the postribosomal (informosomal) compartment of sea urchin eggs. Its detection was based solely on hybridization of polyadenylic acid portions of the RNA with labeled polyuridylic acid. The criterion of such a hybridization for the presence of mRNA is not rigorous but should serve tentatively. According to Slater et al. (1973) the polyadenylic acid-containing RNA shifts in its distribution after fertilization with a substantial decrease in the informosomes and an accumulation in the polyribosomes. However, Wilt (1973), who also noted an increase in poly A content after fertilization, does not observe a shift in the cellular distribution. Since homogenization was effected in the presence of detergent by Slater et al. and in its absence by Wilt, poly A containing RNA may well be associated with large, detergent-susceptible particles in the egg.

The entry of polysomal mRNA into a messenger ribonucleoprotein compartment has been shown in temperature-shocked mouse L cells (Schochetman and Perry, 1972). Exposure of cells to 42°C results in the dissociation of polyribosomes and the release of mRNA in particles resembling informosomes. Restoration to normal temperature results in the formation of complexes between these particles and monoribosomes and the reformation of polyribosomes. The shift of mRNA between polyribosomes and messenger ribonucleoproteins may occur in various situations (see Schochetman and Perry, 1972, for references). In all these cases the mRNA in the informosome-like particles may be only temporarily suspended from translation. The dissociation of polyribosomes *in vitro* has allowed the analysis of the released messenger ribonucleoprotein, especially the protein part (Olsnes, 1970; Spohr et al. 1970), which may or may not be equivalent to the proteins of informosomes normally present *in vivo*. Indeed, specific proteins have been isolated. Kwan and Brawerman (1972) described a protein specifically associated with the polyadenylic acid portion of polysomal mRNA. Apparently the same polyadenylic acid-associated protein (of 78,000 molecular weight) was described by Blobel (1973) in several different cells, and, in addition, a second protein of 52,000 molecular weight was posited by him to be associated with some other portion of the mRNA. If the function of these proteins can be determined and if they are demonstrably present in informosomes, the role of informosomes or, generally, of structures containing untranslated mRNA may be explained.

Gene-Regulated Repression of mRNA Translation

Tomkins et al. (1969) proposed on the basis of a number of points of evidence that the induction of an enzyme in eukaryotes involves the production of mRNA for that enzyme as well as for the synthesis of a labile protein repressor of the mRNA activity. The system under study was tyrosine aminotransferase, which can be induced in liver cells or in rat hepatoma (HTC) cells by adrenal steroid or dexamethasone. The observations of Tomkins are as follows:

1. Steroid inducers stimulate the rate of enzyme synthesis.
2. Enzyme-specific mRNA accumulates in the presence of inducer, even when protein synthesis is inhibited.
3. The constant presence of an inducing steroid is required to maintain the induced rate of enzyme synthesis.
4. RNA synthesis is required for enzyme induction.
5. Continued RNA synthesis is not required to maintain enzyme synthesis at the basal or induced levels.
6. If RNA synthesis is blocked after induction, enzyme formation becomes constitutive.
7. Actinomycin D superinduces the synthesis of the enzyme and increases its intracellular concentration.
8. Induced enzyme synthesis, slowed by removing the inducer, may be reactivated by blocking RNA synthesis.

These observations are explained by the hypothesis that the steroid inducers are antagonistic to a posttranscriptional repressor, the so-called *R protein*, which both inhibits messenger translation and promotes messenger degradation. Thus we might suppose that a structural gene produces mRNA for enzyme synthesis and that a corresponding repressor gene simultaneously provides for the synthesis of a repressor R protein. The R protein might interact with the mRNA, thus affecting its template activity and its longevity, unless the R protein were itself acted upon by the inducer steroid. Such an interaction between R protein and inducer would allow increased mRNA activity and enhanced enzyme production.

Alternative explanations for this enzyme induction are possible without the introduction of a repressor protein at the level of translation. The super induction of tyrosine transaminase by actinomycin D need not occur through the inhibition of RNA synthesis. It now seems likely that this antibiotic has an effect on the initiation of protein synthesis (Greenberg, 1972; Singer and Penman, 1972; Schwartz, 1973). If, in fact, the mRNA for tyrosine transaminase were less susceptible to an actinomycin-linked blockage of initiation, then this mRNA would have a competitive advantage for translation. Ribosomes, elongation factors, and tRNA would become more abundantly available to it. The result would be a superinduction without regulation of translation. Indeed, ovalbumin synthesis in chick magnum is superinduced by actinomycin D treatment without an increase in translatable mRNA, but with an increase in the rate of mRNA translation (Palmiter and Schimke, 1973). These authors attribute this specific increase to an increase in the rate of initiation.

Selective Initiation of Protein Synthesis

A second scheme for controlling the activity of mRNA is derived from the analysis of cofactors involved in initiating translation. In prokaryotes three factors are required for initiating the translation of mRNA. Factors F_1 and F_2 are proteins required for the binding of the initiator formyl methionyl tRNA and the formation of the active 70S ribosome from the 30S and 50S subunits. In addition, a third protein, F_3, promotes the binding of mRNA to the 30S subunit (Iwasaki et al., 1968; Sabol et al., 1970) and also acts to dissociate the 70S ribosome into subunits (Subramanian and Davis, 1970). Sabol and Ochoa (1971) draw the following scheme from studies with labeled F_3 protein. The F_3 protein binds to a 30S subunit and consequently complexes with mRNA. This complex joins with a 50S subunit to form an active 70S ribosome, and the F_3 protein becomes displaced. The displaced F_3 protein may then interact with a 70S ribosome just released from a polyribosome and effect a dissociation into subunits and the formation again of a complex between the F_3 protein and the 30S subunit. Perhaps the most interesting property of the F_3 factor is that it may consist of a multiplicity of proteins, capable of discriminating between different recognition sites on mRNAs. Two classes of F_3 factors have been extracted from *Escherichia coli* (Hsu and Weiss, 1969; Berissi et al., 1971; Lee-Huang and Ochoa, 1971). One F_3 fraction allows the translation of *E. coli* messengers and MS2 phage RNA; another allows translation of T4-phage messengers. Each activity is mutually exclusive.

Heywood (1969) observed that the 25 to 27S mRNA for myosin synthesis will promote protein synthesis by ribosomes from reticulocytes as well as from muscle cells but that a salt-extract from muscle ribosomes was necessary to furnish cofactor (Table 1). This specific ribosome-associated factor is believed to be a specific recognition or initiation protein, as obtained from bacteria (Prichard et al., 1970). Thus it is possible that cell-specific recognition factors exist on ribosomes and that these factors may play a role in the regulation of translation bearing on the events of differentiation. Ilan and Ilan (1971) have reported that protein factors from ribosomal washes appear to be able to discriminate between mRNAs obtained from different stages of insect development. A complex (80S) between mRNA and ribosome is formed only when the mRNA is obtained from the same stage of development. Purified initiation factors from Krebs II ascites

TABLE 1

EFFICIENCY OF VARIOUS ASSAY SYSTEMS TO SUPPORT MYOSIN SYNTHESIS[a]

Assay system	25–27 S RNA	Myosin count/min[b]	% Myosin synthesis
Muscle ribosomes (NaF)	+	110	100
Muscle ribosomes (NaF)	–	20	18.1
Muscle ribosomes (S)	+	16	14.5
Muscle ribosomes (S) + CFm	+	68	61.8
Muscle ribosomes (S) + CFr	+	14	12.7
Retic. ribosomes (NaF)	+	0	0
Retic. ribosomes (S)	+	0	0
Retic. ribosomes (S) + CFm	–	0	0
Retic. ribosomes (S) + CFm	+	31	28.1
Retic. ribosomes (S) + CFr	+	0	0

[a]Composite data from average of at least four runs with each assay system. From S. M. Heywood, *Cold Spring Harbor Symp. Quant. Biol. 34*, 799–803, 1969.
[b]Radioactivity moving electrophoretically with the 200,000 mol wt subunit of myosin. Amino acid incorporating systems were as previously described. CFm, crude factor high-salt wash from muscle ribosomes; CFr, crude factor high-salt wash from reticulocyte ribosomes.

cells have been shown to distinguish between different messengers (Wigle and Smith, 1973). The translation of encephalomyocarditis viral RNA (EMC RNA) is absolutely dependent on the purified initiation factor "IF$_{EMC}$," whereas mouse 9S globin messenger is only slightly affected. Since this factor, IF$_{EMC}$, was purified from uninfected cells, it might be expected to have some role in the translation of some class of endogenous cellular message with recognition sites similar to those of EMC RNA. This remains to be seen. Kaempfer and Kaufman (1973) have recently indicated that double-stranded RNA interacts with the initiation factor IF-3 from rabbit reticulocytes. They propose that the initiation protein recognizes a double-stranded region in mRNA. This proposition brings us to a consideration of the role of mRNA structure per se in regulation.

Regulation Determined by the Intrinsic Properties of mRNA

Finally, we shall focus upon the intricate structure of mRNA to uncover signals that may elicit regulation of translation. There may be either secondary and tertiary configurations, such as helices and loops, or primary base sequences involved in furnishing signals. Partial nucleotide sequences of RNA of the bacteriophages R 17 (Steitz, 1969) and Qβ (Hindley and Staples, 1969) at the sites of polypeptide chain initiation reveal possible loops associated with these sections of the RNA. Furthermore, R 17 RNA has been shown to contain extensive regions of double helix (Adams et al., 1969). Such conformational aspects of the mRNA structure might impose restrictions particularly on binding to ribosomes. Berissi et al. (1971), who noted that a specific F$_3$ fraction was required for the recognition of the initiation sites on MS2 phage RNA by 30S subunits, found, in addition, that if the MS2 RNA is denatured, binding occurs in the absence of F$_3$ factor but not at the specific initiation sites of the mRNA.

Sussman (1970) presented a model whereby the primary structure of the mRNA molecule might serve as regulator of its activity. The observations from which the model was conceived involved the study of the accumulation of various enzymes at different stages of slime mold morphogenesis. A stochiometric relationship appeared to exist between the amount of transcription that was allowed and the amount of particular enzyme that subsequently accumulated. The RNA product of a transcription period usually remained stable for 4 to 5 hours before actual accumulation of enzyme. When a reversible inhibitor of protein synthesis, cycloheximide, was added, then removed, the RNA remained stable several hours longer and no reduction of enzyme activity occurred. Thus the cessation of protein synthesis stabilized mRNA activity, and a decline in mRNA activity seemed concomitant with protein synthesis. The model proposed was that at the 5-hydroxyl terminus, the initiator portion of the template, a series of redundant nucleotide sequences were attached and preceded the initiator codons (expected to be AUG). The ribosomes were supposed to attach to these parts of the mRNA before initiation of protein synthesis, and as they did, a portion of the redundant sequences became excised. After a given number of passes these 5-hydroxyl sequences are consumed; the mRNA may not become complexed with ribosomes thereafter but instead is susceptible to degradation. The poly A sequences known to be associated with mRNA cannot be implicated in such a model, since they have now been shown to be attached to the 3-hydroxyl terminus (Kates, 1970; Burr and Lingrel, 1971; Molloy et al., 1972).

References

ADAMS, J. M., JEPPESEN, P. G. N., SANGER, F., AND BARRELL, B. G. (1969). Nucleotide sequence from the coat protein cistron of R 17 bacteriophage RNA. *Nature (London) 223*, 1009–1014.

ALTMAN, S., AND SMITH, J. D. (1971). Tyrosine tRNA precursor molecule polynucleotide sequence. *Nature New Biol. 233*, 35–39.

ARONSON, A. I. (1972). Degradation products and a unique endonuclease in heterogeneous nuclear RNA in sea urchin embryos. *Nature New Biol. 235*, 40–44.

ARTMAN, M., AND ROTH, J. S. (1971). Chromosomal RNA: an artifact of preparations? *J. Mol. Biol. 60*, 291–301.

ASTRIN, S. M. (1973). *In vitro* transcription of the SV40 sequences in SV3T3 chromatin (transformed cells/Simian virus 40). *Proc. Nat. Acad. Sci. U.S.A. 70*, 2304–2308.

BAKER, W. K. (1968). Position-effect variegation. *Advan. Genet. 14*, 133–169.

BEKHOR, I., KUNG, G. G., AND BONNER, J. (1969). Sequence-specific interaction of DNA and chromosomal protein. *J. Mol. Biol. 39*, 351–364.

BELL, E. (1969). I-DNA: its packaging into I-somes and its relation to protein synthesis during differentiation. *Nature (London) 224* 326–328.

BELL, E. (1971). Informational DNA synthesis distinguished from that of nuclear DNA by inhibitors of DNA synthesis. *Science 174*, 603–606.

BENJAMIN, T. L. (1966). Virus-specific RNA in cells productively infected or transformed by polyoma virus. *J. Mol. Biol. 16*, 359–373.

BERISSI, H., GRONER, Y., AND REVEL, M. (1971). Effect of a purified initiation factor F3(B) on the selection of ribosomal binding sites on phage MS2 RNA. *Nature New Biol. 234*, 44–47.

BISHOP, J. O., PEMBERTON, R., AND BAGLIONI, C. (1972). Reiteration frequency of haemoglobin genes in the duck. *Nature New Biol. 235*, 231–234.

BLOBEL, G. (1973). A protein of molecular weight 78,000 bound to the polyadenylate region of eukaryotic messenger RNAs. *Proc. Nat. Acad. Sci. U.S.A. 70*, 924–928.

BRITTEN, R. J., AND DAVIDSON, E. H. (1969). Gene regulation for higher cells: a theory. *Science 165*, 349–358.

BRITTEN, R. J., AND DAVIDSON, E. H. (1971). Repetitive and non-repetitive DNA sequences and a speculation on the origins of evolutionary novelty. *Quart. Rev. Biol. 46*, 111–133.

BRITTEN, R. J., AND KOHNE, E. E. (1968). Repeated sequences in DNA. *Science 161*, 529–540.

BROWN, D. D., AND WEBER, C. S. (1968). Gene linkage by RNA-DNA hybridization. *J. Mol. Biol. 34*, 661–680.

BURR, H., AND LINGREL, J. B. (1971). Poly A sequences at the 3' termini of rabbit globin mRNAs. *Nature New Biol. 233*, 41.

BUTTERWORTH, P. H., COX, R. F., AND CHESTERTON, C. J. (1971). Transcription of mammalian chromatin by mammalian DNA-dependent RNA polymerase. *Eur. J. Biochem. 23*, 229–241.

CALLAN, H. G. (1967). The organization of genetic units in chromosomes. *J. Cell Sci. 2*, 1–7.

CALLAN, H. G., AND LLOYD, L. (1960). Lampbrush chromosomes of crested newts *Triturus cristatus* (Laurenti). *Phil. Trans. Roy. Soc. B 243*, 135–219.

CLARK, R. J., AND FELSENFELD, G. (1971). Structure of chromatin. *Nature New Biol. 229*, 101–106.

DARNELL, J. E., PHILIPSON, L., WALL, R., AND ADESNIK, M. (1971). Polyadenylic acid sequences:

role in conversion of nuclear RNA into messenger RNA. *Science 174*, 507–510.

DINA, D., CRIPPA, M., AND BECCARI, E. (1973). Hybridization properties and sequence arrrangement in a population of mRNAs. *Nature New Biol. 242*, 101–105.

EDMONDS, M., VAUGHN, M. H., AND NAKAZATO, H. (1971) Polyadenylic acid sequences in the hetergeneous nuclear RNA and rapidly-labeled polyribosomal RNA of HeLa cells. *Proc. Nat. Acad. Sci. U.S.A. 68*, 1136–1340.

FAIFERMAN, I., CORNUDELLA, L., AND POGO, A. O. (1971). Messenger RNA nuclear particles and their attachment to cytoplasmic membranes in Krebs tumour cells. *Nature New Biol. 233*, 234–237.

FLAMM, W. G., WALKER, P. N. B., AND McCALLUM, M. (1969). Some properties of the single strands isolated from the DNA of the nuclear satellite of the mouse (*Mus musculus*). *J. Mol. Biol. 40*, 423–443.

FRENSTER, J. H., ALLFREY, J. G., AND MIRSKY, A. F. (1963). Repressed and active chromatin isolated from interphase lymphocytes. *Proc. Nat. Acad. Sci. U.S.A. 56*, 1026–1032.

FROMSON, D., AND NEMER, M. (1970). Cytoplasmic extraction: polyribosomes and heterogeneous ribonucleoproteins without associated DNA. *Science 168*, 266–267.

GREENBERG, J. (1972). High stability of messenger RNA in growing cultured cells. *Nature (London) 240*, 102–104.

GREENBERG, J. R., AND PERRY, R. P. (1971). Hybridization properties of DNA sequences directing the synthesis of messenger RNA and heterogeneous nuclear RNA. *J. Cell Biol. 50*, 774–786.

HALL, M. R., MEINKE, W., GOLDSTEIN, D. A., AND LERNER, R. A. (1971). Synthesis of cytoplasmic membrane-associated DNA in lymphocyte nucleus. *Nature New Biol. 234*, 227–229.

HAREL, J., HANANIA, N., TAPIERO, H., AND HAREL, L. (1963). DNA replication by nuclear satellite DNA in different mouse cells. *Biochem. Biophys. Res. Comm. 33*, 696–701.

HENSHAW, E. C. (1968). Messenger RNA in rat liver polyribosomes: evidence that it exists as ribonucleoprotein particles. *J. Mol. Biol. 36*, 401–411.

HEYWOOD, S. M. (1969). Synthesis of myosin on heterologous ribosomes. *Cold Spring Harbor Symp. Quant. Biol. 34*, 799–803.

HINDLEY, J., AND STAPLES, D. H. (1969). Sequence of a ribosome binding site in bacteriophage Qβ-RNA. *Nature (London) 224*, 964–967.

HSU, W., AND WEISS, S. B. (1969). Selective translation of T⁴-infected *Escherichia coli*. *Proc. Nat. Acad. Sci. U.S.A. 64*, 345–351.

HUANG, R. C. C., AND BONNER, J. (1962). Histone, a suppressor of chromosomal RNA synthesis. *Proc. Nat. Acad. Sci. U.S.A. 48*, 1216–1222.

HUANG, R. C. C., AND HUANG, P. C. (1969). Effect of protein-bound RNA associated with chick embryo chromatin on template specificity of the chromatin. *J. Mol. Biol. 39*, 365–378.

ILAN, J., AND ILAN, J. (1971). Stage-specific initiation factors for protein synthesis during insect development. *Develop. Biol. 25*, 280–292.

INFANTE, A. A., AND NEMER, M. (1968). Heterogeneous ribonucleoprotein particles in the cytoplasm of sea urchin embryos. *J. Mol. Biol. 32*, 543–565.

ISHIKAWA, K., KURODA, C., AND OGATA, K. (1969). Release of ribonucleoprotein particles containing rapidly labeled ribonucleic acid from rat liver nuclei. *Biochim. Biophys. Acta 179*, 316–331.

ITZHAKI, R. (1970). Structure of deoxyribonucleoprotein as revealed by its binding to polylysine. *Biochem. Biophys. Res. Comm. 41*, 25–32.

IWASAKI, K., SABOL, S., WAHBA, A. J., AND OCHOA, S. (1968). Role of initiation factors in formation of the chain initiation complex with *Escherichia coli* ribosomes. *Arch. Biochem. Biophys. 125*, 542–547.

JACOBS-LORENA, M., AND BAGLIONI, C. (1972). Messenger RNA for globin in the postribosomal supernatant of rabbit reticulocytes. *Proc. Nat. Acad. Sci. U.S.A. 69*, 1425–1428.

JANOWSKI, M., NASSER, D. S., AND McCARTHY, B. J. (1972). Fractionation of mammalian chromatin. *In* "Gene Transcription in Reproductive Tissue" Karolinska Symposia on Research Methods in Reproductive Endocrinology, 5th Symposium (A. Diczfalusy, ed.), pp. 112–129.

JONES, K. W. (1970). Chromosomal and nuclear location of mouse satellite DNA in individual cells. *Nature (London) 225*, 912–915.

KABAT, D. (1972). Gene selection in hemoglobin and in antibody synthesizing cells. *Science 175*, 135–140.

KAEMPFER, R., AND KAUFMAN, J. (1973). Inhibition of cellular protein synthesis by double-stranded RNA: inactivation of an initiation factor. *Proc. Nat. Acad. Sci. U.S.A. 70*, 1222–1226.

KATES, J. (1970). Transcription of the vaccinia virus genome and the occurrence of polyadenylic acid sequences in messenger RNA. *Cold Spring Harbor Symp. Quant. Biol. 35*, 743–752.

KEDES, L. H., AND BIRNSTIEL, M. L. (1971). Reiteration and clustering of DNA sequences complementary to histone messenger RNA. *Nature (London) 230*, 165–176.

KEYL, H. G. (1965). A demonstrable local and geometric increase in the chromosomal DNA of *Chironomus. Experientia 21*, 191–193.

KWAN, S. W., AND BRAWERMAN, G. (1972). A particle associated with the polyadenylate segment in mammalian messenger RNA. *Proc. Nat. Acad. Sci. U.S.A. 69*, 3247–3250.

LEE-HUANG, S., AND OCHOA, S. (1971). Messenger discriminating species of initiation factor F^3. *Nature New Biol. 234*, 236–239.

LEWIS, E. B. (1950). The phenomenon of position effect. *Advan. Genet. 3*, 73–115.

LINDBERG, U., AND DARNELL, J. E. (1970). SV40-specific RNA in the nucleus and polyribosomes of transformed cells. *Proc. Nat. Acad. Sci. U.S.A. 65*, 1089–1096.

MANO, Y., AND NAGANO, H. (1966). Release of maternal RNA from some particles as a mechanism of activation of protein synthesis by fertilization in sea urchin eggs. *Biochem. Biophys. Res. Commun. 25*, 210–215.

MARUSHIGE, K., AND BONNER, J. (1971). Fractionation of liver chromatin. *Proc. Nat. Acad. Sci. U.S.A. 68*, 2941–2944.

MATHER, K. (1944). The genetic activity of heterochromatin. *Proc. Roy. Soc. (London) B 132*, 308–332.

MIRSKY, A. E. (1971). The structure of chromatin. *Proc. Nat. Acad. Sci. U.S.A. 68*, 2945–2948.

MIRSKY, A. E., AND RIS, H. (1951). The deoxyribonucleic acid content of animal cells and its significance in evolution. *J. Gen. Physiol. 34*, 451–462.

MOLLOY, G. R., SPORN, M. B., KELLEY, D. E., AND PERRY, R. P. (1972). "Localization of Polyadenylic Acid Sequences in Messenger RNA of Mammalian Cells." *Biochemistry 11*, 3256–3260.

NEMER, M. (1967). Transfer of genetic information during embryogenesis. *Progr. Nucleic Acid Res. Mol. Biol. 7*, 243–301.

OLSEN, G. D., GASKELL, P., AND KABAT, D. (1972). Presence of hemoglobin messenger ribonucleoprotein in a reticulocyte supernatant fraction. *Biochim. Biophys. Acta 272*, 297–304.

OLSNES, S. (1970). Characterization of protein bound to rapidly-labelled RNA in polyribosomes from rat liver. *Eur. J. Biochem. 15*, 464–471.

PALMITER, R. D., AND SCHIMKE, R. T. (1973). Regulation of protein synthesis in chick oviduct. III. Mechanism of ovalbumin "superinduction" by actinomycin D. *J. Biol. Chem. 248*, 1502–1512.

PARDUE, M. L., AND GALL, J. G. (1970). Chromosomal localization of mouse satellite DNA. *Science 168*, 1356–1358.

PAUL, J., CARROLL, D., GILMOUR, R. S., MORE, J.A.R., THRELFALL, G., WILKIE, M., AND WILSON, S. (1972). Functional studies on chromatin. "Gene Transcription in Reproductive Tissue." Karolinska Symposia on Research Methods in Reproductive Endocrinology, 5th Symposium (A. Diczfalusy, ed.), pp. 277–297.

PAUL, J., AND GILMOUR, R. S. (1966). Template activity of DNA is restricted in chromatin. *J. Mol. Biol. 16*, 242–244.

PAUL, J., AND GILMOUR, R. S. (1968). Organ specific restriction of transcription in mammalian chromatin. *J. Mol. Biol. 34*, 305–316.

PAUL, J., AND GILMOUR, R. S. (1969). RNA transcribed from reconstituted nucleoprotein is similar to natural RNA. *J. Mol. Biol. 40*, 137–139.

PERRY, R. P., and KELLEY, D. E. (1968). Messenger RNA-protein complexes and newly synthesized ribosomal subunits: analysis of free particles and compounls of polyribosomes. *J. Mol. Biol. 35*, 37–59.

PRICHARD, P., GILBERT, J., SHAFRITZ, D., AND ANDERSON, W. (1970). Factors for initiation of haemoglobin synthesis by rabbit reticulocyte ribosomes. *Nature (London) 226*, 511–514.

RASKAS, H. J. (1971). Release of adenovirus messenger RNA from isolated nuclei. *Nature New Biol. 233*, 134–136.

RITOSSA, F. M., ATWOOD, K. C., LINDSLEY, D. L., AND SPIEGELMAN, S. (1966). On the chromosomal distribution of DNA complementary to ribosomal RNA. *Nat. Cancer Inst. Monogr. 23*, 449–472.

SABOL, S., AND OCHOA, S. (1971). Ribosomal binding of labelled initiation factor F_3. *Nature New Biol. 234*, 233–239.

SABOL, S. SILLERO, M. A. G., IWASAKI, K., AND OCHOA, S. (1970). Purification and properties of initiation factor F_3. *Nature (London) 228*, 1269–1273.

SAMBROOK, J., WESTPHAL, H., SRINIVASAN, P. R., AND DULBECCO, R. (1968). The integrated state of viral DNA in SV40-transformed cells. *Proc. Nat. Acad. Sci. U.S.A. 60*, 1288–1295.

SCHOCHETMAN, G., AND PERRY, R. P. (1972). Characterization of the messenger RNA released from L cell polyribosomes as a result of temperature shock. *J. Mol. Biol. 63*, 577–590.

SCHULTZ, J. (1965). Genes, differentiation and animal development. *Brookhaven Symp. Biol. 18*, 116–147.

SCHWARTZ, R. J. (1973). Control of glutamine synthetase synthesis in the embryonic chick neural retina: a caution in the use of actinomycin D. *J. Biol. Chem. 248*, 6426–6435.

SHEARER, R. W., AND McCARTHY, B. J. (1970). Characterization of RNA molecules restricted to the nucleus in mouse L-cells. *J. Cell. Physiol. 75*, 97–106.

SINGER, R., AND PENMAN, S. (1972). Stability of HeLa cell RNA in actinomycin. *Nature (London) 240*, 100–102.

SLATER, I., GILLESPIE, D., AND SLATER, D. W. (1973). Cytoplasmic adenylylation and processing of maternal RNA. *Proc. Nat. Acad. Sci. U.S.A. 70*, 406–411.

SMITH, K. D., CHURCH, R. B., AND McCARTHY, B. J. (1969). Template specificity of isolated chromatin. *Biochemistry 8*, 4271–4277.

SPIRIN, A. S. (1969). Informosomes. *Eur. J. Biochem. 10*, 20–35.

SPIRIN, A. S. BELITSINA, N. V., AND AJTKHOZHIN, M. A. (1964). Messenger RNA in early embryogenesis. *Zh. Obstchey Biol. (Russian) 25*, 321–338.

SPIRIN, A. S., AND NEMER, M. (1965). Messenger RNA in early sea urchin embryos: cytoplasmic particles. *Science 150*, 214–217.

SPOHR, G., GRANBOULAN, N., MOREL, C., AND SCHERRER, K. (1970). Messenger RNA in HeLa cells: an investigation of free and polyribosomes-bound cytoplasmic messenger ribonucleoprotein particles by kinetic labelling and electron microscopy. *Eur. J. Biochem. 17*, 296–318.

STEITZ, J. A. (1969). Nucleotide sequences of the three ribosomal binding sites in bacteriophage R 17 RNA. *Nature (London) 224*, 957–964.

SUBRAMANIAN, A. R., AND DAVIS, B. D. (1970). Activity of initiation factor F₃ in dissociating *Escherichia coli* ribosomes. *Nature (London) 228*, 1273–1275.

SUSSMAN, M. (1970). Model for quantitative and qualitative control of mRNA translation in eukaryotes. *Nature (London) 225*, 1245–1246.

TARTOF, K. D., AND PERRY, R. P. (1970). The 5S RNA genes of *Drosophila melanogaster*. *J. Mol. Biol. 51*, 171–183.

THOMAS, C. A., JR., HAMKALO, B. A., MISRA, D. N., AND LEE, C. S. (1970). Cyclization of eucaryotic deoxyribonucleic acid fragments. *J. Mol. Biol. 51*, 621–632.

TOMKINS, G. M., GELEHRTER, T. D., GRANNER, D., MARTIN, D., JR., SAMUELS, H. H., AND THOMPSON, E. B. (1969). Control of specific gene expression in higher organisms. *Science 166*, 1474–1480.

WALLACE, H., AND BIRNSTIEL, M. L. (1966). Ribosomal cistrons and the nucleolar organizer. *Biochim. Biophys. Acta 114*, 296–310.

WEAVER, R. F., BLATT, S. P., AND RUTTER, W. J., (1971). Molecular structures of DNA dependent RNA polymerases (II) from calf thymus and rat liver. *Proc. Nat. Acad. Sci. U.S.A. 68*, 2994–2999.

WIGLE, D. T., AND SMITH, A. E. (1973). Specificity in initiation of protein synthesis in a fractionated mammalian cell-free system. *Nature New Biol. 242*, 136–140.

WILLIAMSON, R., McSHANE, T., GRUNSTEIN, M., AND FLAVELL, R. A. (1972). Cytoplasmic DNA from primary embryonic cell cultures is not informational. *FEBS Letters 20*, 108–110.

WILT, F. H. (1973). Polyadenylation of maternal RNA of sea urchin eggs following fertilization. *Proc. Nat. Acad. Sci. U.S.A. 70*, 2345–2349.

YASMINEH, W. G., AND YUNIS, J. J. (1969). Satellite DNA in mouse autosomal heterochromatin. *Biochem. Biophys. Res. Comm. 35*, 779–782.

YASMINEH, W. G., AND YUNIS, J. J. (1970). Socialization of satellite DNA in constitutive heterochromatin. *Exp. Cell Res. 59*, 69–75.

YUNIS, J. J., AND YASMINEH, W. G. (1970). Satellite DNA in constitutive heterochromatin of the guinea pig. *Science 168*, 263–265.

CHAPTER FIVE

Molecular Basis of Embryogenesis

MARTIN NEMER

Embryogenesis is the development of an egg into an embryo, a multitude of differentiated cells committed to highly elaborate programs of future change. The process results in the establishment of stem cells for a variety of future cell lines. These stem cells may arise in the blastula or gastrula and display little or no specific morphological characteristics. Indeed, they and their progenitors might be termed "undifferentiated." If such a designation is valid—an undifferentiated cell being one without immediate specialized physiological function—then during embryogenesis there is a transition from the undifferentiated to the differentiated. This transition may not occur at any other time in the life of an organism, since all subsequent developmental changes would involve transformations along cell lines and the emergence of further specialized cells from already differentiated cells. The central feature of embryogenesis is, then, the biological interphase between the undifferentiated and the differentiated state.

To understand the molecular basis of embryogenesis, we should inquire as to how the process is promoted by the utilization of genetic information. During embryogenesis the nature of the flow of genetic information changes markedly. Stored maternal messenger RNAs (mRNA) in the egg cytoplasm support a considerable portion of protein synthesis in the early period after fertilization, but transcription of embryo genes contributes increasing amounts and eventually all the genetic information for development. As this transition continues toward completion, a major step toward differentiation occurs, the formation of the primary germ layers of ectoderm, endoderm, and mesoderm, which may entail important changes in the regulation of the flow of genetic information. The cells that give rise to the primary germ layers may well qualify as being undifferentiated both morphologically and in their mode of genetic expression. We expect that in specialized cells the flow of genetic information supports, at least in part, the synthesis of special products and the maintenance of special functions. What can be said about the flow of genetic information in a cell with no apparent specialized functions, in a cell that may at most be preparing for future specializations?

Information in the Egg

The accumulation of preformed gene products appears to be an essential characteristic of gametogenesis, which is, further, a process that involves or leads to the programmed utilization of gene products. On the Y chromosome of the spermatocytes of *Drosophila hydei* lateral loops form, which resemble the lampbrush chromosome configuration demonstrable in *Triturus cristatus*. Hess (1969) and Hess and Meyer (1968) have pointed out that the loops appear to be structural devices for the accumulation of gene products—namely, ribonucleoproteins. It seems likely that these particles include mRNA for directing the later synthesis of various components of the maturing sperm cell, particularly the flagellum. Another aspect of sperm maturation is the actual replacement of preexisting histones by protamines, from combination with DNA at the terminal stage of spermatogenesis in salminoid fish (Marushige and Dixon, 1969). Protamines are synthesized in the cytoplasm (Ling et al., 1969), then transported into the nucleus, followed by displacement of histones and finally association of protamines with DNA.

The mature egg appears to contain an extensive array of mRNAs formed during the course of oogenesis, possibly much of this stored RNA dating back to the lampbrush stage of the oocyte. The complexity of this population of RNA molecules in the eggs of *Xenopus* has been well described by Davidson and Hough (1971). Gross and Cousineau (1963) first observed that sea urchin eggs did not require RNA synthesis in order that protein synthesis, which was very low in the mature egg, become activated after fertilization. Their experiments with actinomycin D led them to postulate that the egg contained a store of RNA templates, which were used in early development. Subsequent experiments have reinforced the idea that preformed mRNA exists in the sea urchin egg. Gross et al. (1965) were able to hybridize labeled egg

RNA with DNA, and Whiteley et al. (1966) and Glisin et al. (1966) obtained competition in RNA–DNA hybridization using egg RNA. Slater and Spiegelman (1966) used egg RNA as template in a bacterial *in vitro* protein-synthesizing system. Infante and Nemer (1967) demonstrated a definite class of polyribosomes that formed after fertilization in the absence of RNA synthesis. These polyribosomes of approximately 300S were appreciably different from the class whose formation depended upon new RNA synthesis. Identifications of the stored RNA templates still need to be made. Recent data suggest that possibly among other functions the synthesis of mitotic spindle protein may depend on this preformed mRNA (Raff et al., 1971). Also Barrett and Angelo (1969) furnish evidence that the synthesis of hatching enzyme in the blastula may depend on stored mRNA. We might ask, then, whether or not the stored mRNA templates in the egg are a highly selected population that serve a narrow physiological function—principally that of cell replication. The complexity of the RNA stored in the egg of *Xenopus* is such that it is capable of representing 40,000 different proteins the size of hemoglobin (Davidson and Hough, 1971), thus allowing for the support of a considerable physiological complexity. Crippa and Gross (1969) suggest that there may be an orderly selection among the population of mRNA molecules in the *Xenopus* egg for use in protein synthesis through the course of early development.

Activation of Protein Synthesis in the Egg

The mechanism of activation of protein synthesis in the sea urchin egg is a problem without a really satisfactory solution. It is not yet clear whether the repression of activity in the unfertilized egg is due to a limitation on (1) the capacity of the protein synthesizing machinery—that is, the ribosomes, transfer RNA, activating enzymes, and so on, or (2) the availability of mRNA template. Nemer and Bard (1963) offered data indicating that there was a major limitation on template RNA availability, as well as a minor limitation on the protein synthesizing machinery (Fig. 1). Mitochondria-free supernatant fluids of unfertilized eggs of *Arbacia punctulata* displayed a very low endogenous level of protein synthesis. However, after fertilization, this *in vitro* activity could be shown to increase 20-fold in 1 hour. On the other hand the unfertilized system responded very actively to the addition of polyuridylic acid (poly U) in synthesizing polyphenylalanine, but the system from 1-hour fertilized eggs showed an increased response of no more than 1.5-fold. Therefore, a minor activation of the protein synthesizing machinery (probably the ribosomes) is observed when template RNA is in excess, whereas a major activation is seen when template RNA is not in excess. Thus the major activation depends on the availability of

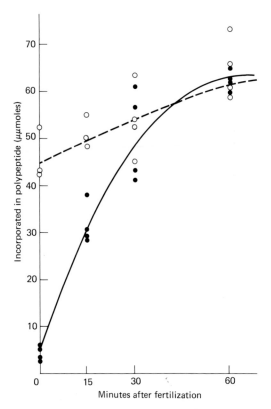

FIGURE 1. *Polyphenylalanine and protein synthesis by the postmitochondrial supernatant fluid of* Arbacia punctulata *before and after fertilization. (—●—●—) incorporation in protein; (—○—○—) PPA. (From M. Nemer and S. G. Bard, Science 140, 664–666, 1963.)*

mRNA. Stavy and Gross (1969) reached the same conclusion on the basis of more extensive experiments of the same general type with the species *Lytechinus pictus*. Humphreys (1969) measured the efficiency of translation in unfertilized and fertilized egg and concluded that they were similar as defined by the number of protein molecules produced per mRNA molecule per unit time. He concluded that the translation level control of protein synthesis cannot be a general change in activity of some component in the cellular synthetic machinery such as ribosomes, but must specifically control the activity of a defined population of mRNA molecules. Contrary to these observations, MacKintosh and Bell (1969) have presented data that indicate that unfertilized egg polyribosomes are not loaded to capacity but, unlike fertilized egg polyribosomes, may be increased in number by incubation of eggs with cycloheximide. From Humphreys' (1969) measurements we would conclude that the limitation on polyribosome formation is due not to a lack of initiation factors for the complexing of additional ribosomes but to the masking or repressing of mRNA (Tyler, 1967; Spirin, 1966).

In opposition to these assorted studies, the work in Monroy's laboratory (Monroy et al., 1965; Maggio et al., 1968) on the species *Paracentrotus lividus* indicates that the deficiency in the unfertilized egg system is to be found in the egg ribosomes. Monroy et al. (1965) were able to stimulate both endogenous activity and poly U-directed polypeptide synthesis in unfertilized egg ribosomes by treatment of microsomal preparations with trypsin. The destruction of a protein inhibitor apparently derepressed the system. In further studies Maggio et al. (1968) noted that egg ribosomes were more difficult to dissociate into subunits than were later stage ribosomes. They suggested that this "stickiness" might be related to reduced activity on the egg ribosomes, and more specifically to an inhibitory protein. Infante and Graves (1971), however, did not find that egg ribosomes of *Strongylocentrotus purpuratus* and *Arbacia punctulata* displayed this property. Also Kedes and Stavy (1969) were not able to activate protein synthesis by trypsin treatment of egg ribosomes from either *S. purpuratus* or *L. pictus*. In addition, they observed no differences in response to poly U by ribosomes reconstituted from various combinations of unfertilized egg and embryo ribosomal subunits. Recently, Metafora et al. (1971) reported that they were able to extract from egg ribosomes of *P. lividus* an extract of proteins that inhibited both binding of poly U to ribosomes and polyphenylalanine synthesis. Similar inhibition was obtained by addition of the extract to other protein synthesizing systems, such as those from *E. coli*, rat liver, and pea germ. Indirect assessments of nuclease activity in the inhibitory extract were made, but direct nuclease activity was not measured. No assessment of the effect of the extract on endogenous protein synthesis was given. The results reported in this case are consistent with the previous work on the species *P. lividus*. On the other hand, all the results reported using the other three (North American) species of sea urchin are also internally consistent. At present, the most likely resolution of these differences involves a species difference and the possibility that there is validity to both sets of claims. It is therefore possible that the activation of protein synthesis in the unfertilized egg involves both an unmasking of stored mRNA and a change in the activity of the ribosomes—but each of those to substantially different degrees in different species. The translational control of protein synthesis by stored egg mRNA would not be expected to manifest itself simply in a general inhibition of protein synthesis, as offered by the repression of ribosomal activity; rather a specific programming of protein synthesis seems to work, even in actinomycin D-blocked embryos, as deduced from the results of Terman (1970) and by the specific timing of the appearance of hatching enzyme synthesized on egg mRNA (Barrett and Angelo, 1969).

Proportion of Genome Transcribed During Embryogenesis

There are three types of changes in transcription that may occur as a function of embryonic development: (1) a change in pro-

portion of the genome that is transcribed, (2) a shift in the relative output of the different functional classes of RNA—informational, transfer, and ribosomal, and (3) a change in the population of newly synthesized informational RNA, characterized mainly on the basis of qualitative differences in base sequences and expected to consist of mRNA and RNA of some presently unknown regulatory functions.

A direct saturation of DNA by hybridization with nuclear RNA, labeled *in vitro*, was performed by Flickinger (1970) on several stages of *Rana pipiens*. The result was that in going from the neurula to tail bud to larval stage the percent of the genome hybridized decreased markedly. Although the base sequences represented only the redundant part of the genome, the experiment does indicate that widespread transcription occurs in the early embryos but that restriction sets in with development. The same series performed with total RNA, including, indeed, mostly the cytoplasmic, shows a reverse of this trend—namely, a greater representation of the genome, which can be interpreted as an increase of sequence representation principally in the cytoplasm, very likely of relatively stable mRNA.

A similar developmental trend may be the case for the sea urchin embryo. Brandhorst and Humphreys (1971) calculate that a blastula nucleus synthesizes RNA at the rate of 9.7×10^{-15} g/min, whereas the rate of the nucleus of a pluteus is 3.1×10^{-15} g/min. Roeder and Rutter (1970) have measured RNA synthesis in isolated nuclei of different embryonic stages. Observed under various cation conditions, the transcription per nucleus declined during development from the early blastula to the late gastrula stages.

DIFFERENTIAL OUTPUT OF RNA CLASSES

In the frog *Rana pipiens* (Brown and Caston, 1962a, b) and in the toad *Xenopus laevis* (Brown and Littna, 1964a, b), ribosomal RNA (rRNA) synthesis is not detectable in cleavage-stage embryos. The onset of RNA synthesis is correlated with the appearance of visible nucleoli, which in these animals is at the gastrula stage (Brown, 1966). During the early stages DNA-like RNA and soluble RNA are the predominant classes of RNA synthesized. The situation in the sea urchin embryo is roughly parallel. Nemer and Infante (1965, 1967) observed DNA-like RNA synthesis from the outset of sea urchin development, and Nemer (1963) noted that substantial rRNA synthesis could not be detected until the mesenchyme blastula stage. Transfer RNA synthesis is also not detectable in the early stages (Yang and Comb, 1968). The pattern for the sea urchin embryo seemed similar to what has been concluded for the regulation of rRNA synthesis in amphibians—namely, that little or no rRNA synthesis occurred before gastrulation and that there was activation of synthesis at that time. Recently, working at the limit of detectability for rRNA incorporation, Emerson and Humphreys (1970) reported that rRNA synthesis could be detected in late cleaving embryos. Furthermore, they estimated an output of rRNA per nucleus that differed only slightly from what it would be in the blastula and pluteus stages. Their conclusion was that the output of rRNA per nucleus is constant throughout development and that the relative output of DNA-like RNA shifts massively—being so high in the cleaving embryo and blastula that the less than 1 percent of the output attributable to rRNA would be difficultly detectable. Therefore, the shift in output of RNA classes would seem to be attributable solely to a less and less extensive transcription of DNA-like RNA.

Correlated with a shift in production of DNA-like RNA compared to rRNA, Roeder and Rutter (1970) have observed, first, that there are three discernible RNA polymerases: I, which is responsible for rRNA synthesis, and II and III, which are apparently responsible for DNA-like RNA synthesis; and second, that polymerase I appears to maintain a constant activity level throughout development, whereas polymerases II and III decline in activity. Assuming that the *in vitro* activities mirror the *in vivo* activities, the steady production of rRNA, advocated by Emerson and Humphreys (1970) would seem highly likely. An alternative explanation is that polymerase I activity is regulated *in vivo*. Another explanation is offered by Sconzo and Guidice (1971), who find that incorporation of methyl-labeled methionine is observed in rRNA in the gastrula but not in the blastula. Their results indicate that incorporation per nucleus in the gastrula is at least seven times that in the blastula, thus disputing the conclusions of Emerson and Humphreys (1970). The two sets of conclusions might be resolved, as Sconzo and Guidice suggest, if transcriptions of the ribosomal genes did indeed occur actively in the early embryo but that processing, including methylation, of the gene product were deficient. Therefore, the question of whether or not activation of ribosomal genes occurs at a discrete embryonic stage in the sea urchin has not yet been satisfactorily answered.

CHANGING OUTPUT OF INFORMATIONAL RNA

RNA–DNA hybridization, particularly involving competition between RNA populations, has been used to assess the possible qualitative changes during development. The protocol has been to compete unlabeled RNA of different stages with the hybridization of labeled RNA of a given stage. Under the conditions used, only the redundant portions of the genome appear to be involved; thus competition again cannot establish identities but only resemblances between families within the RNA population. An example of such a study is in the competition of various early stage RNAs with labeled RNA from stage-42 tadpoles (Denis, 1966). The RNA of stages 6 to 8 cleaving eggs does not compete with RNA labeled by pulsing stage-42 embryos with $^{14}CO_2$. However, the RNA from gastrula onward is increasingly more competitive with the late stage RNA. One interpretation is that an increasingly higher proportion of these nucleotide sequences are present in the mRNA of the progressively older embryos. Another interpretation, in addition to considerations about the pertinence here of only redundant DNA sequences, is that the qualitative nature of the informational RNA population does not change but that the quantity of informational RNA increases relative to the total, mostly ribosomal, RNA. Thus a given amount of extracted cellular RNA might simply contain progressively more informational RNA molecules of the same kind. This interpretation is, indeed, favored in a series of competitions of various RNAs with labeled stages 26 to 28 tail bud RNA (Denis, 1966). Stage-42 RNA competes better with stages 26 to 28 RNA than the latter does with itself.

A more promising protocol is offered by the approach of Davidson and Hough (1971). The complexity of information represented by the RNA population of a cell is most unambiguously approached by matching it with the non-repetitive or unique DNA sequences of the genome. Davidson and Hough (1971) have evaluated the RNA population of the *Xenopus* oocyte by saturating *in vitro*-labeled RNA with the purified unique portion of *Xenopus* DNA. The result was that a saturation level was reached that was equivalent to 20×10^6 nucleotide pairs (2×10^4 genes), or 4.5 times the size of the *E. coli* genome (4×10^3 genes). Furthermore, they estimate that the RNA population consisted of an average of 10^4 copies. This population consists of an accumulation of molecules synthesized during oogenesis. What the complexity is of the RNA in any given somatic cell type has still to be determined. Nor do we know entirely what the significance of this complexity may be.

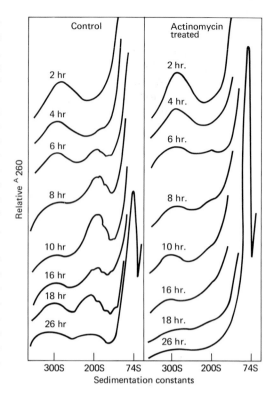

FIGURE 2. *Sedimentation profiles of polyribosomes from embryos of* Strongylocentrotus purpuratus *in the presence and absence of actinomycin D. Postmitochondrial supernatant fluids used for sedimentation were from embryos harvested at given times after development at 18°C. Cleavage to morula (2–6 hours), early blastula (8–10 hours), hatched blastula (16–18 hours), mesenchyme blastula to early gastrula (26 hours). (From A. A. Infante and M. Nemer, Proc. Nat. Acad. Sci. U.S.A. 58, 681–688, 1967.)*

Messenger RNAs During Early Development

New RNA synthesis after fertilization of the sea urchin egg results in the accumulation of a class of polyribosomes containing 3 to 7 ribosomes per mRNA (Infante and Nemer, 1967). These polyribosomes increase in concentration from barely perceptible amounts in the early cleaving embryo to over 70 percent of the polysomal ribosomes in the 10-hour, 200-cell early blastula of *S. purpuratus* (Fig. 2). The new polysomal RNA is predominantly of a 9 to 10S class (Nemer and Infante, 1965; Kedes and Gross, 1969). Similarly to their suspected function in mammalian tissue culture cells (Borun et al., 1967), this narrow class of mRNAs has been implicated in histone synthesis in the early sea urchin embryo (Nemer and

Lindsay, 1969; Kedes and Gross, 1969; Moav and Nemer, 1971). Moav and Nemer (1971) demonstrated that the nascent proteins of these small-sized polyribosomes were histone-like according to a number of criteria. Using the relative incorporation of tryptophan, which is lacking in histones, and arginine and lysine, they estimated the relative output of histone-like protein by the free polyribosomes during the course of early development (Fig. 3). Of the total output by these polyribosomes, histone-like protein amounted to approximately

50 percent at the 10-hour, 200-cell blastula stage. Histone synthesis apparently reached a peak at this stage and was found throughout this early period of development to be parallel to the rate of DNA synthesis. It therefore seems that the cleaving embryo and early blastula expend a considerable effort in support of chromosomal replication and cell division, and little toward net growth and immediate differentiation.

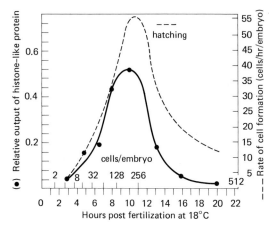

FIGURE 3. *Relative output of histone-like proteins by free polyribosomes during embryonic development of* S. purpuratus. *(From B. Moav and M. Nemer,* Biochemistry 10, *881–888, 1971.)*

The activation of protein synthesis after fertilization is accompanied by polyadenylation of the stored mRNAs of the egg (Slater et al., 1972). The poly A polymerase activity involved is believed to be cytoplasmic (Slater et al., 1973; Wilt, 1973) and the effect is to increase the stretch of adenylate residues from approximately 100 to 200. Thus, previously polyadenylated mRNAs are readjusted to some significantly fixed length of poly A size. We do not know the significance of this adenylate addition. Slater et al. (1972) discount the possibility that the polyadenylation is the mechanism through which stored mRNAs are primed for activity in protein synthesis. Their argument is that the timing and magnitude of the change in protein synthesis are much different from those of the addition of poly A. Furthermore, such prim-

ing would not be expected for the mRNAs coding for histones, recently shown to be among the mRNAs stored in the egg (Gross et al., 1973) and known to lack poly A (Adesnik and Darnell, 1972). Nevertheless, the poly A moiety of the mRNA may serve some regulatory function, if not in translation *per se*, perhaps through its specific associations with proteins (Kwan and Brawerman, 1972; Blobel, 1973).

The association of poly A has been made up until now only with the mRNAs for highly specialized proteins, such as hemoglobin and immunoglobulins, or with specific viral RNAs. In addition, evidence is accumulating that in tissue culture cells, highly differentiated cells and tumor cells, the mRNAs are essentially all polyadenylated, except the histone's mRNAs (Adesnik et al., 1972). However, in studies of sea urchin embryos at cleavage and blastula stages (Nemer et al. unpublished) newly synthesized nonhistone mRNAs lacking poly A have been shown to exist together with polyadenylated mRNAs. These two classes occur in approximately equal amounts in these early stage embryos. One might speculate that the appearance of the mRNAs lacking poly A may have something to do with the relatively undifferentiated status of the cleavage and blastula cells.

References

ADESNIK, M., AND DARNELL, J. E. (1972). Biogenesis and characterization of histone messenger RNA from HeLa cells. *J. Mol. Biol. 67,* 397–406.

ADESNIK, M., SALDITT, M., THOMAS, W., AND DARNELL, J. E. (1972). Evidence that all messenger RNA molecules (except histone messenger RNA) contain poly (A) sequences and that the poly (A) has a nuclear function. *J. Mol. Biol. 71,* 21–30.

BARRETT, D., AND ANGELO, G. M. (1969). Material characteristics of hatching enzymes in hybrid sea urchin embryos. *Exp. Cell Res. 57,* 159–166.

BLOBEL, G. (1973). A protein of molecular weight 78,000 bound to the polyadenylate region of eukaryotic messenger RNAs. *Proc. Nat. Acad. Sci. U.S.A. 70,* 924–928.

BORUN, T. W., SCHARFF, M. D., AND ROBBINS, E. (1967). Rapidly labeled polyribosome-associated

RNA having the properties of histone messenger. *Proc. Nat. Acad. Sci. U.S.A. 58,* 1977-1983.

BRANDHORST, B. P., AND HUMPHREYS, T. (1971). Synthesis and decay rates of major classes of deoxyribonucleic acid like ribonucleic acid in sea urchin embryos. *Biochemistry 10,* 877-881.

BROWN, D. D. (1966). The nucleolus and synthesis of ribosomal RNA during oogenesis and embryogenesis of *Xenopus laevis. Nat. Cancer Inst. Monogr. 23,* 297-309.

BROWN, D. D., AND CASTON, J. D. (1962a). Biochemistry of amphibian development. I. Ribosome and protein synthesis in early development of *Rana pipiens. Develop. Biol. 5,* 412-434.

BROWN, D. D., AND CASTON, J. D. (1962b). Biochemistry of amphibian development. II. High molecular weight RNA. *Develop. Biol. 5,* 435-444.

BROWN, D. D., AND LITTNA, E. (1964a). RNA synthesis during development of *Xenopus laevis,* the South African clawed toad. *J. Mol. Biol. 8,* 669-687.

BROWN, D. D., AND LITTNA, E. (1964b). Variations in the synthesis of stable RNA's during oogenesis and development of *Xenopus laevis. J. Mol. Biol. 8,* 688-695.

CRIPPA, M., AND GROSS, P. R. (1969). Maternal and embryonic contributions to the functional messenger RNA of early development. *Proc. Nat. Acad. Sci. U.S.A. 62,* 120-127.

DAVIDSON, E. H., AND HOUGH, B. R. (1971). Genetic information in oocyte RNA. *J. Mol. Biol. 56,* 491-506.

DENIS, H. (1966). Gene expression in amphibian development. II. Release of the genetic information in growing embryos. *J. Mol. Biol. 22,* 285-304.

EMERSON, C. P., AND HUMPHREYS, T. (1970). Regulation of DNA-like RNA and the apparent activation of ribosomal RNA synthesis in sea urchin embryos: quantitative measurements of newly synthesized RNA. *Develop. Biol. 23,* 86-112.

FLICKINGER, R. (1970). The role of gene redundancy and number of cell divisions in embryonic determination. *Develop. Biol., Suppl. 4,* 1-41.

GLISIN, V. R., GLISIN, M. V., AND DOTY, P. (1966). The nature of messenger RNA in the early stages of sea urchin development. *Proc. Nat. Acad. Sci. U.S.A. 56,* 285-289.

GROSS, K. W., JACOBS-LORENA, M., BAGLIONI, C., AND GROSS, P. R. (1973). Cell-free translation of material messenger RNA from sea urchin eggs. *Proc. Nat. Acad. Sci. U.S.A. 70,* 2614-2618.

GROSS, P. R., AND COUSINEAU, G. H. (1963). Effects of actinomycin D on macromolecule synthesis and early development in sea urchin eggs. *Biochem. Biophys. Res. Comm. 10,* 321-326.

GROSS, P. R., MALKIN, L. I., AND HUBBARD, M. (1965). Synthesis of RNA during oogenesis in the sea urchin. *J. Mol. Biol. 13,* 463-481.

HESS, O. (1969). Chromosome structure and activity. "Congenital Malformations." Excerpta Medica International Congress Series No. 204, pp. 29-41. Excerpta Medica, Amsterdam.

HESS, O., AND MEYER, G. F. (1968). Genetic activities of the Y chromosome in *Drosophila* during spermatogenesis. *Advan. Genet. 14,* 171-223.

HUMPHREYS, T. (1969). Efficiency of translation of messenger RNA before and after fertilization in sea urchins. *Develop. Biol. 20,* 435-458.

INFANTE, A. A., AND GRAVES, P. N. (1971). Stability of free ribosomes, derived ribosomes and polysomes of the sea urchins. *Biochim. Biophys. Acta 246,* 100-110.

INFANTE, A. A., AND NEMER, M. (1967). Accumulation of newly synthesized RNA templates in a unique class of polyribosomes during embryogenesis. *Proc. Nat. Acad. Sci. U.S.A. 58,* 681-688.

KEDES, L. H., AND GROSS, P. R. (1969). Identification in cleaving embryos of three RNA species serving as templates for the synthesis of nuclear protein. *Nature (London) 223,* 1335-1339.

KEDES, L. H., AND STAVY, L. (1969). Structural and functional identity of ribosomes from eggs and embryos of sea urchins. *J. Mol. Biol. 43,* 337-340.

KWAN, S. W., AND BRAWERMAN, G. (1972). A particle associated with the polyadenylate segment in mammalian messenger RNA. *Proc. Nat. Acad. Sci. U.S.A. 69,* 3247-3250.

LING, V., TREVITHICK, J. R., AND DIXON, G. H. (1969). The biosynthesis of protamine in trout testis. I. Intracellular site of synthesis. *Canad. J. Biochem. 47,* 51-60.

MACKINTOSH, F. R., AND BELL, E. (1969). Regulation of protein synthesis in sea urchin eggs. *J. Mol. Biol. 41,* 365-380.

MAGGIO, R., VITORELLI, M. L., CAFFARELLI-MORMINO, I., AND MONROY, A. (1968). Dissociation of ribosomes of unfertilized eggs and embryos of sea urchin. *J. Mol. Biol. 31,* 621-626.

MARUSHIGE, K., AND DIXON, G. H. (1969). Developmental changes in chromosomal composition and template activity during spermatogenesis in trout testis. *Develop. Biol. 19,* 397-414.

METAFORA, S., FELICETTI, L., AND GAMBINO, R. (1971). The mechanism of protein synthesis activation after fertilization of sea urchin eggs. *Proc. Nat. Acad. Sci. U.S.A. 68,* 600–604.

MOAV, B., AND NEMER, M. (1971). Histone synthesis: assignment to a special class of polyribosomes in sea urchin embryos. *Biochemistry 10,* 881–888.

MONROY, A., MAGGIO, R., AND RINALDI, A. M. (1965). Experimentally induced activation on the ribosomes of the unfertilized sea urchin egg. *Proc. Nat. Acad. Sci. U.S.A. 54,* 107–111.

NEMER, M. (1963). Old and new RNA in the embryogenesis of the purple sea urchin. *Proc. Nat. Acad. Sci. U.S.A. 50,* 230–235.

NEMER, M., AND BARD, S. G. (1963). Polypeptide synthesis in sea urchin embryogenesis: an examination with synthetic polyribonucleotides. *Science 140,* 664–666.

NEMER, M., AND INFANTE, A. A. (1965). Messenger RNA in early sea urchin embryos: size classes. *Science 150,* 217–221.

NEMER, M., AND INFANTE, A. A. (1967). Early control of gene expression. "The Control of Nuclear Activity" (L. Goldstein, ed.), pp. 101–127. Prentice-Hall, Englewood Cliffs, New Jersey.

NEMER, M., AND LINDSAY, D. T. (1969). Evidence that the s-polysomes of early sea urchin embryos may be responsible for the synthesis of chromosomal histones. *Biochem. Biophys. Res. Commun. 35,* 156–160.

RAFF, R. A., GREENHOUSE, S., GROSS, K. W., AND GROSS, P. R. (1971). Synthesis and storage of proteins by sea urchin embryos. *J. Cell Biol.* 516–527.

ROEDER, R. G., AND RUTTER, W. J. (1970). Multiple ribonucleic acid polymerases and ribonucleic acid synthesis during sea urchin development. *Biochemistry 9,* 2543–2553.

SCONZO, G., AND GUIDICE, G. (1971). Synthesis of ribosomal RNA in sea urchin embryos. V. Further evidence for an activator following the hatching blastula stage. *Biochim. Biophys. Acta 254,* 447–451.

SLATER, I., GILLESPIE, D., AND SLATER, D. W. (1973). Cytoplasmic adenylation and processing of maternal RNA. *Proc. Nat. Acad. Sci. U.S.A. 70,* 406–411.

SLATER, D. W., SLATER, I., AND GILLESPIE, D. (1972). Postfertilization synthesis of polyadenylic acid in sea urchin embryos. *Nature 240,* 333–337.

SLATER, D. W., AND SPIEGELMAN, S. (1966). An estimation of genetic messages in the unfertilized echinoid egg. *Proc. Nat. Acad. Sci. U.S.A. 56,* 164–170.

SPIRIN, A. S. (1966). On "masked" forms of messenger RNA in early embryogenesis and in other differentiation systems. *Curr. Top. Develop. Biol., 11.*

STAVY, L., AND GROSS, P. R. (1969). Availability of mRNA for translation during normal and transcription-blocked development. *Biochim. Biophys. Acta 182,* 203–213.

TERMAN, S. A. (1970). Relative effect of transcription-level and translation-level control of protein synthesis during early development of the sea urchin. *Proc. Nat. Acad. Sci. U.S.A. 65,* 985–992.

TYLER, A. (1967). Masked messenger RNA and cytoplasmic DNA in relation to protein synthesis and processes of fertilization and determination in embryonic development. *Develop. Biol.,* Suppl. 1, 170–226.

WHITELEY, A. H., MCCARTHY, B. J., AND WHITELEY, H. R. (1966). Changing populations of messenger RNA during sea urchin development. *Proc. Nat. Acad. Sci. U.S.A. 55,* 519–525.

WILT, F. H. (1973). Polyadenylation of maternal RNA of sea urchin eggs after fertilization. *Proc. Nat. Acad. Sci. U.S.A. 70,* 2345–2349.

YANG, S. S., AND COMB, D. G. (1968). Distribution of multiple forms of lysyl transfer RNA during early embryogenesis of sea urchin *Lytechinus variegatus. J. Mol. Biol. 31,* 139–142.

CHAPTER SIX

Gastrulation and Cell Interactions

KURT E. JOHNSON

During the early development of any metazoan organism, there is a period of rapid cell division when the large mass of the fertilized zygote is cleaved into many cells. During this cleavage period, cells arrange themselves to form a structure of many cells, called the *blastula,* which surrounds or lies upon a cavity called the *blastocoel* (see Fig. 1). Gastrulation, a developmental process that often involves the movement of the primordia of internal structures, such as muscle and gut, from the surface of the embryo to their proper position within the embryo, begins after the blastula stage is completed. Cells move about in the embryo in such a fashion that groups of cells are brought into new relations with new neighbors. During this series of morphogenetic cell movements and the subsequent events of neurulation, the basic body plan of the embryo (and frequently of the adult) is established. The cells that induce the central nervous system and other axial structures are brought into position adjacent to the cells that will respond to inductive stimuli and became the central nervous system. The complex and delicate inductive interactions that occur as an immediate consequence of gastrulation require that morphogenetic cell movements be accomplished with a high degree of precision.

In our consideration of gastrulation, we shall first describe its morphology in a number of representative chordate forms. Next, we shall examine well-characterized examples of the roles in gastrulation of changes in cell shape, cell contact behavior, and cell motility. After that, we shall outline and discuss several analytical approaches that have been employed in attempts to unravel the secrets of morphogenesis. Several models of cell behavior that seem to be relevant to the study of morphogenesis will be presented. In the final portion of the chapter we shall suggest some directions for future investigation of morphogenetic cellular interactions in embryos.

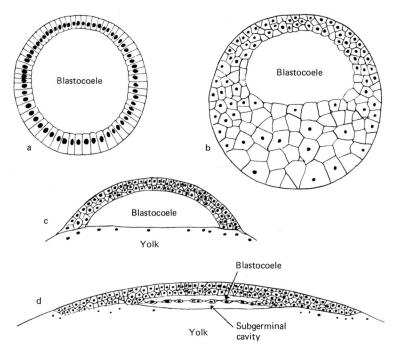

FIGURE 1. *Comparison of blastulae of amphioxus (a), frog (b), teleost fish (c), and bird (d). (From B. I. Balinsky, "An Introduction to Embryology," 3rd ed., Saunders, Philadelphia, 1970, p. 154.)*

Cleavage and Blastula Formation

After fertilization and union of the male and female pronuclei (see Chapter 1), there is invariably a period of rapid cell division with little dramatic change in the overall size or external shape of the embryo. The cytoplasmic mass of the zygote is repeatedly cleaved into progressively smaller cells (blastomeres), presumably serving as more convenient and manageable packages for the cell movements that will soon begin. As more and more cells are produced during cleavage, a large internal cavity is formed between them. This cavity, the blastocoel, is located centrally or is displaced away from the yolkier portion of the embryo (Fig. 1). The embryo at this stage is called the blastula.

Recent electron microscope observations in amphibian embryos have shown that a small but detectable "blastocoel" is evident even in the first cleavage furrow (Kalt, 1971a,b). This cavity enlarges rapidly with each successive division. It also appears that there is some selective modification or differentiation of the cell surface membrane in cleavage furrows, since some portions form close adhesions while other portions do not. Those parts of the cell surface that do not form adhesions become part of the wall of the blastocoel. Later in this chapter, more will be said about the role of surface modifications in gastrulation.

Amphioxus Gastrulation

Amphioxus is a common name used for members of the genus *Branchiostoma*. These small, worm-like, marine, primitive chordates are members of the subphylum Cephalochordata, whereas other forms that we will consider (frog and chick) are members of the subphylum Vertebrata (see Ballard, 1964, Chap. 2 for details). The morphology of gastrulation in amphioxus is easy to visualize and therefore will be considered first. The oligolecithal egg (i.e., an egg in which a relatively small number

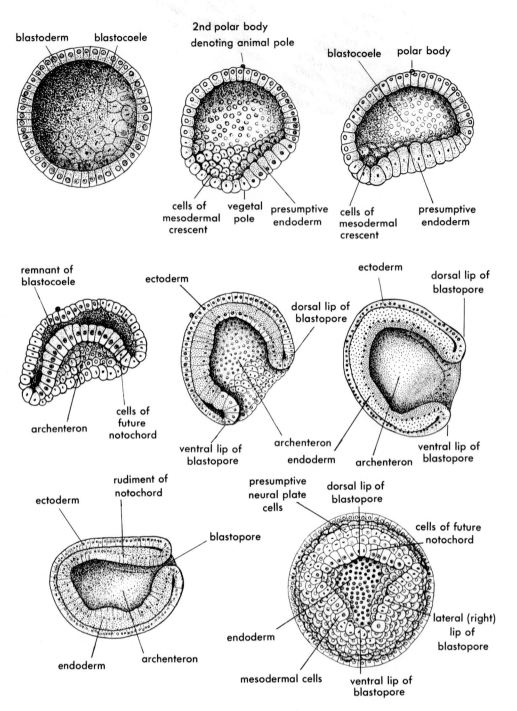

FIGURE 2. *Invagination in* Branchiostoma. *(From B. I. Balinsky, "An Introduction to Embryology," 3rd ed., Saunders, Philadelphia, 1970, p. 196.)*

of yolk granules are distributed homogene-ously) begins rapid cleavage soon after fertilization. The second polar body is shed after formation of the fertilization membrane and does not move during subsequent cleavages. Consequently, the animal pole of the embryo is marked for a relatively long time during development, serving as a guide for orientation of the embryonic axes. Because certain early cleavages are unequal, the ventral blastomeres are slightly larger than the dorsal blastomeres. This is the beginning of a trend that continues into the later phases of cleavage, when a hollow spherical blastula is formed whose dorsal (animal) blastomeres are somewhat smaller than the ventral (vegetal) blastomeres. The *Branchiostoma* blastula is composed of approximately 1000 blastomeres, an order of magnitude less than in amphibian or chick embryos. Cell division proceeds at a rapid pace all through blastula and gastrula stages and may play an important morphogenetic role. The beginning of gastrulation is marked by a flattening of the endoderm (Fig. 2B). These cells become increasingly columnar as flattening continues. Meanwhile, the presumptive ectodermal cells become *less* columnar, presumably because of stretching. It is not known, however, if the blastoderm is under tension during gastrulation. As the embryo progresses in its development, the flattened endodermal plate becomes invaginated toward the animal pole, and finally the blastocoelic surfaces of the animal and vegetal poles come together (Figs. 2C–E). At this stage, the spherical embryo has become cup-shaped, and the inner blastocoelic cavity has been all but obliterated. A small remnant can be seen at the ventral lip of the blastopore. During the indentation of the endodermal plate, the number of cells in the embryo is increasing. Also, there is a pronounced elongation parallel to the anterior–posterior axis.

A second asymmetry is introduced when the dorsal surface of the embryo becomes more flattened than the ventral surface (Fig. 2G). At this stage, the amphioxus gastrula consists of two nested cup-like structures that are joined together into a single sheet of cells over the lips of the blastopore. The topology of the entire gastrulation process so far can be visualized nicely by imagining the result of indenting and stretching a beach ball. Confluent sheets of cells now become discontinuous and migrate over one another during the final phases of gastrulation and the establishment of the primary organ rudiments. The outer ectodermal layer of the tubular gastrula divides into (1) a neural plate, which subsequently rolls up on itself to form a neural tube, and (2) neural folds, which migrate over the neural plate and fuse at the embryonic midline to form a continuous sheet. In the meantime, the inner tube is developing (1) twin constrictions on each side of the dorsal midline, which later will become mesodermal somites, and (2) a single median dorsal constriction, which is destined to become notochord (Fig. 3). When the notochord and the mesodermal somites have separated from the inner tubular structure, a tube still remains, which is the endoderm, or primitive gut. These processes are illustrated in detail in Figs. 2 and 3 and should be studied carefully.

In a detailed and careful study that has since become a classic, Conklin (1932) described the embryology of *Branchiostoma lanceolatum*, a Mediterranean species once plentiful around Naples. He observed that the flattening and infolding of the endodermal plate occurred with dramatic changes in constituent cell shape. Interestingly enough, he also felt that the endodermal plate was overgrown and engulfed by rapid cell divisions of presumptive mesodermal and ectodermal cells near the lips of the blastopore. This assertion bears investigation, since in most forms cell division plays a minor role in animal morphogenesis (see, for example, Trinkaus, 1969, pp. 169–170).

Amphibian Gastrulation

At the end of the cleavage–blastula period, an amphibian embryo is 1 to 2 mm in diameter and is composed of approximately 15,000 cells. Animal pole cells are smaller than vegetal pole cells and contain less yolk. The blastocoel is located eccentrically toward the animal pole (Fig. 1b). The center of gravity is located toward the vegetal pole away from the geometric center of the spherical or ellipsoidal embryo. Thus amphibian embryos rest with the vegetal portion located downward with respect

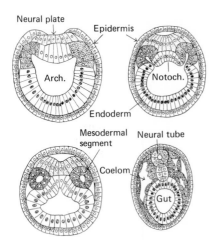

Neural plate

Epidermis

Arch.

Notoch.

Endoderm

Mesodermal segment

Neural tube

Coelom

Gut

FIGURE 3. *Establishment of the primary organ rudiments in* Branchiostoma. *Drawings are diagrammatic transverse sections. Development proceeds left to right and top to bottom. (From B. I. Balinsky, "An Introduction to Embryology," 3rd ed., Saunders, Philadelphia, 1970, p. 198.)*

to gravity. The upper animal portion is frequently heavily pigmented, whereas the lower vegetal portion is usually less pigmented. The intersection between the dark upper and light lower portions of the embryo is not abrupt; rather, there is an intermediate zone, called the *marginal zone*, where cell size, yolkiness, and pigmentation are all intermediate between the extremes of the animal and vegetal poles. This marginal zone occupies a portion of the surface of the spherical embryo that lies below the equator toward the vegetal pole. It is here that the first clearly recognizable changes of gastrulation are detectable.

It should be realized that the processes of cleavage, blastula formation, and gastrulation overlap during early development. For convenience, we shall take the first appearance of the blastopore as the beginning of gastrulation. It is clear that the changes that cause the initiation of gastrulation must have occurred prior to gastrulation itself. For example, certain factors in oocyte nuclei appear to make some essential contribution for gastrulation in salamanders (Briggs and Cassens, 1966; Briggs and Justus, 1967). Unfortunately, at present there is not much solid information about the chemi-

cal control of morphological changes in early morphogenesis.

When first visible, the blastopore appears as a faint streak of pigmentation near the marginal zone. Certain cells at the blastopore begin to elongate and burrow into the surrounding presumptive endodermal mass. At the same time, these migratory cells maintain a firm connection with the surface by way of an attenuated cytoplasmic neck (Fig. 4). Because of their shape, these cells are commonly referred to as *bottle cells*. The blastopore grows by "recruiting" other cells and soon becomes an indentation in the surface of the embryo. Next, the blastopore widens and becomes crescent-shaped, then semicircular, and finally circular. It is not known what stimuli cause bottle-cell formation or how these cells are coordinated with one another.

Sheets of cells invaginate into the embryo by migrating over the rim of the blastopore. As invagination proceeds, the archenteron increases in size at the expense of the blastocoel (Fig. 5c). There is no gross change in embryonic volume during gastrulation. The archenteron grows by the invagination and spreading of presumptive mesodermal cells and presumptive endodermal cells. Presumptive mesodermal cells were once on the outer surface of the embryo. As a result of gastrulation, presumptive mesodermal cells come to lie under the presumptive medullary plate. There they interact with overlying cells and cause primary embryonic induction, which leads to neurulation and other events of embryonic axiation. As presumptive mesoderm and presumptive endoderm leave the embryonic surface and move into the embryo, presumptive ectoderm spreads to occupy a greater surface area. This spreading of presumptive ectoderm is called *epiboly*. Invagination, which leads to formation of the archenteron, and epiboly are coupled to one another. The morphogenetic movements occur in a *sheet* of cells, just as is the case in *Branchiostoma*. Comparison of Figs. 2 and 5 will allow one to see both the similarities and the differences between gastrulation in *Branchiostoma* and in amphibians. The major difference between the two is that amphibian embryos are larger and have a larger amount of intracellular yolk to contend with. The similarities are easier to appreciate than the differences. Al-

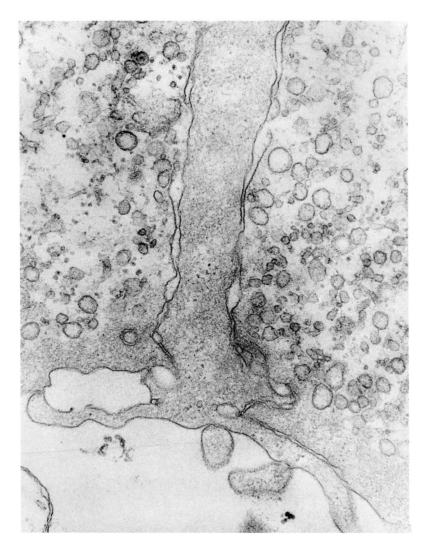

FIGURE 4. *Electron micrograph of the neck of a bottle cell of a tree frog early gastrula. The open space at the bottom of the photograph is the blastoral groove. Approximate magnification × 48,500. (From P. C. Baker, J. Cell. Biol. 24, 101, 1965.)*

though the morphological events of amphibian gastrulation have been carefully studied and analyzed, we do not know how this (or any other) embryonic morphogenetic process is initiated or controlled.

Avian Gastrulation

Avian gastrulation is quite different from that seen in *Branchiostoma* or in amphibians (although there are, of course, a great number of fundamental similarities among the morphogenetic cell movements of the different forms). The avian embryo forms on the top of a large, telolecithal ovum as a result of meroblastic cleavages (only the surface cytoplasm at the animal pole is cleaved by each furrow). Consequently, early cleavage blastomeres are not separated from the yolky mass below by a plasma membrane. Later, cleavage furrows form that produce cells with a complete plasma membrane as well as cells without a complete

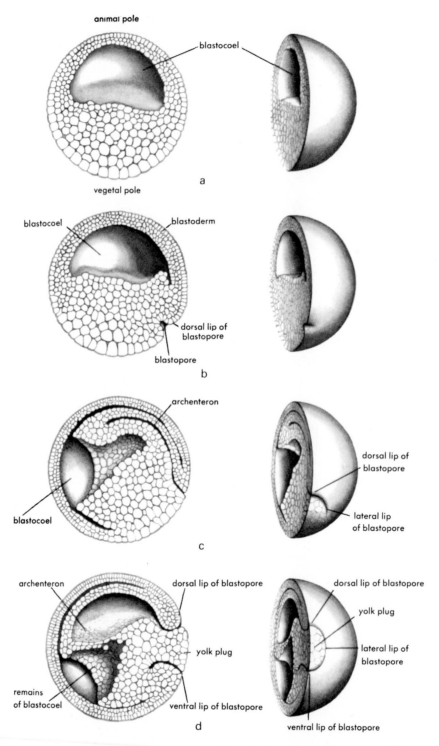

FIGURE 5. *Four stages in frog embryos. a, late blastula; b, early gastrula; c, middle gastrula; d, late gastrula. Drawings on the left represent sections in the median plane; drawings on the right represent the same embryos viewed at an angle from the dorsal side (a, b, and c) or from the posterior end (d). (From B. I. Balinsky, "An Introduction to Embryology," 3rd ed., Saunders, Philadelphia, 1970, p. 206.)*

plasma membrane. After further cleavage, a segmentation cavity forms beneath the blastoderm (Fig. 1d) that is homologous with the blastocoel in *Branchiostoma* and in amphibians. Gastrulation is more difficult to visualize in birds. Indeed, it has not been until very recently that different authors have agreed about the morphological details of avian gastrulation (Nicolet, 1971; Bellairs, 1971). We shall divide gastrulation into three different processes for descriptive convenience (although, in reality, the three overlap one another in time): (1) separation of epiblast and hypoblast, (2) formation of primitive streak and mesoblast, and (3) movement of Hensen's node and notochord formation. Each will be described separately.

SEPARATION OF EPIBLAST AND HYPOBLAST

After segmentation is completed, the blastoderm is a multilayered sheet of cells separated from the yolk by a segmentation cavity (blastocoel). In the future caudal region of the embryo, there is an accumulation of cells called the *embryonic shield*. Cells migrate away from this region of apparent cell proliferation anteriorly along the midline of the embryo as well as away from the midline in a fan-shaped pattern. In this manner, the single-layered blastoderm becomes a double-layered structure. The upper layer is referred to as the *epiblast*, and the lower layer is referred to as the *hypoblast*. As in many other morphogenetic cell movements, the mechanism of hypoblast establishment is poorly understood, although most workers now agree that hypoblast forms by some combination of invagination, proliferation, and emigration in and around the region of the embryonic shield. The establishment of the hypoblast can be considered to be a part of gastrulation that precedes the formation of the primitive streak.

FORMATION OF PRIMITIVE STREAK AND MESOBLAST

While the hypoblast is being formed by an anterior and lateral set of cell movements on the underside of the blastoderm, there is a simultaneous posterior and medial flow of cells in the epiblast followed by a sharp turning toward the anterior part of the embryo at the midline. These movements result in the formation of a primitive streak. There is also substantial invagination occurring at the primitive streak. Epiblast cells move toward the primitive streak, invaginate, and migrate laterally away from the primitive streak, largely between the epiblast and the hypoblast. While these cells are migrating, they maintain their contacts with one another via delicate cell processes (Trelstad et al., 1967). As more and more cells invaginate and move laterally from the primitive streak, a new layer, called the *mesoblast*, is formed (see Bellairs, 1971, for more details). This three-layered blastoderm rests upon the immense yolk of the hen's egg.

MOVEMENT OF HENSEN'S NODE AND NOTOCHORD FORMATION

When the primitive streak has attained its maximum length, an indented nodule becomes prominent at its anterior end. This depression, called *Hensen's node*, undergoes a dramatic posterior movement, forming notochord and somites in its wake. It may be helpful to think of Hensen's node as a traveling organizational state in the cells of the primitive streak that plays some obscure role in notochord and somite formation. With the completion of nodal movement, chick gastrulation is completed. It is important to realize that the entire sequence of gastrulation as described here occurs first in the anterior portion and later in the posterior portion of the embryo. Thus, at certain stages, it is possible to find, in one embryo, regions where (1) mesoblast formation is in progress (for the most part, posteriorly), (2) Hensen's node is migrating (in the middle of the embryo), and (3) neural induction is in progress (anteriorly) (Fig. 6). It should be clear by now to the reader that gastrulation is a complex series of mass cell movements that are integrated and controlled to an impressive degree. We do not know what the control mechanisms are. Nevertheless, there are changes in

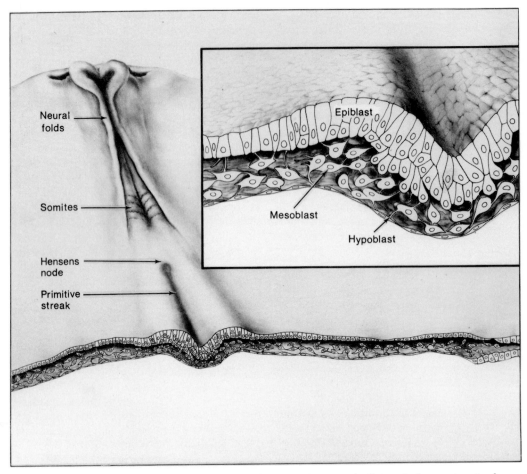

FIGURE 6. *Chick embryo sectioned through the primitive streak. Part of the section is shown at higher magnification in the inset. The anterior portion of the embryo is at the top. Neurulation is in progress. Hensen's node has already laid down four somites. Lateral migration of mesoblast can also be seen in the cross section. [From E. D. Hay, in "Epithelial–Mesenchymal Interactions" (R. Fleishmajer and R. E. Billingham, eds.), p. 32, Williams & Wilkins, Baltimore, 1968. Copyright © 1968, The Williams & Wilkins Co., Baltimore.]*

cellular properties that are common to all morphogenetic systems and are undoubtedly important for the control of gastrulation.

Role of Changes in Cell Shape in Gastrulation

Extensive rearrangements of cells occur by translocation and by changes in the shapes of individual cells. The best studied change in cell shape with obvious relevance to gastrulation is *bottle cell formation*. In frog and bird embryos, a collection of bottle cells forms at the site of invagination. Most attention has been focused on bottle cells in amphibian embryos. These cells can be as long as 200 μm. They have firm connections with neighboring cells at the blastopore, long necks, and an enlarged, bulbous portion located away from the blastopore. The necks are packed with microtubules oriented parallel to the long axis of the cell (Perry and Waddington, 1966). At present, it is still uncertain whether or not microtubules are responsible for active cell elongation (and neck contraction) or whether they are simply arranged passively in response to vectorial cytoplasmic streaming. In short, there is still considerable controversy concerning the role of microtubules in cytoplasmic movement. Often the blastoporal ends of bottle cells appear dense in the electron microscope. Frequently, a filamentous substructure is visible in these regions. There is some evidence that these microfilaments may have a contractile function. In some ways, however, the opposite end of the cell is more interesting. This part is probably burrowing into and through neighboring presumptive mesodermal and endodermal cells. One wonders what kind of contacts

are formed between cells that are burrowing and cells that are being burrowed through. We also do not know how bottle cells move in the embryo or even in culture. Finally, it would be fascinating to know why cells that are contacting so many other cells are at the same time so highly motile (see the later discussions of contact inhibition of motility). Based on an extensive morphological study, but little behavioral evidence, a general mechanism for bottle cell formation and movement has been proposed (Baker, 1965). This model is plausible enough and has some strong evidence in its favor (see Fig. 7). However, the actual behavior of bottle cells in embryos and cultures is not well understood.

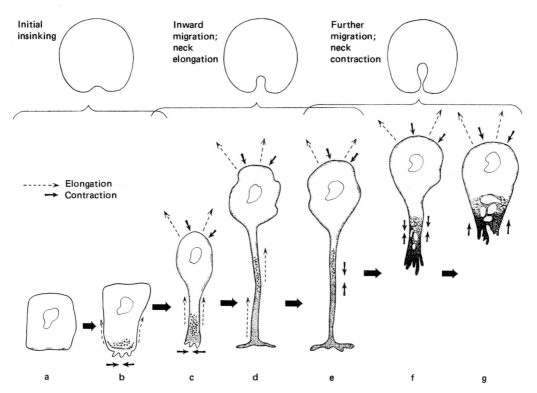

FIGURE 7. *Theoretical scheme for the transformations of a bottle cell during gastrulation, based on the hypothesis of contraction and expansion of the dense layer. The end of the cell neck remains securely adhered to adjacent cells. (a) The cell as it appears on the surface prior to gastrulation. (b) The distal surface contracts, causing initial insinking of the blastoporal pit. (c) and (d) The proximal cell surface begins to migrate inward, deepening the groove; the distal surface continues to contract and the neck elongates. (e)–(g) The neck contracts after it reaches maximum elongation. This shortening, together with the pull exerted by the migrating inner ends, furthers invagination and pulls adjacent cells into the groove. (From P. C. Baker, J. Cell. Biol. 24, p. 115, 1965.)*

In addition to such well characterized examples of changes in cell shape, there are a host of other examples that are less dramatic but probably of great significance for morphogenesis. For example, blastula stage cells in many embryos are often more or less spherical or at least fairly regular in shape. In contrast, cells in gastrulating embryos become stretched and distorted into a bewildering array of shapes. Many cells may become highly polygonal, form numerous amorphous surface blisters, or throw out long exploratory processes from their surfaces. Still other cells can be seen to pulsate and change shape rapidly with no apparent organization. Although it is not clear how these varied modes of cell shape change are related to gastrulation, it is widely believed that they are important, largely on the evidence that they often commence in close synchrony with gastrulation.

Role of Changes in Cell Contact Behavior in Gastrulation

There is a sound theoretical reason for believing that intercellular adhesions become more resistant to disruption prior to the onset of gastrulation. As masses of cells begin to push and pull on one another as they attempt to readjust cell contact relations, the contacts that hold the cells together could reasonably be expected to become more resistant to breakage. It would be ineffective to create a tension in a sheet by invagination in order to move cells into the embryo if the adhesions that connected the cells of the sheet were inadequate to resist disruption by these tensions. It has also been pointed out by a number of workers that changes in areas of cell contact

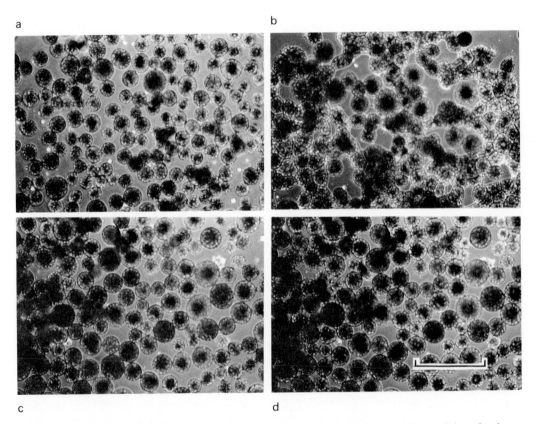

a

b

c

d

FIGURE 8. *Reaggregation of presumptive mesodermal cells. Normal cells after 0 (a) and 3 hours (b) of reaggregation. Hybrid cells after 0 (c) and 3 hours (d) of reaggregation. × 93. Scale equals 200 μm. (From K. E. Johnson,* J. Exp. Zool. 179, *230, 1972.)*

or kinds of cell contacts may be an important morphogenetic mechanism in itself (see Gustafson and Wolpert, 1967, for a comprehensive discussion of this subject). One need only imagine a sheet of cuboidal cells that could increase their area of lateral cell contact and thus make themselves taller, increasing the height of the sheet. Although this mechanism, and variants of it, has been presented as a possible morphogenetic agent, there is as yet no convincing evidence that it is a real morphogenetic mechanism.

Several lines of experimental evidence support the notion that changes in cell contact behavior make an important contribution to gastrulation. Blastula cells are often spherical, contacting neighbors over a small percentage of their cell surface area. In contrast, gastrula cells are often polygonal with adjacent cells sharing broad areas of contact. Blastula cells also have a low ability to attach to glass and other cells in culture, whereas gastrula cells have a high ability to attach to glass and other cells in culture (Trinkaus, 1963). Finally, many changes in cell contact behavior seen in normal gastrulating amphibian embryos are not seen in interspecific hybrid embryos, which undergo developmental arrest at the onset of gastrulation (Johnson, 1970, 1972). If one isolates presumptive mesodermal cells from normal *Rana pipiens* early gastrulae, they will reaggregate *in vitro* in a short time. Similar cells taken from arrested hybrid embryos fail to reaggregate (see Fig. 8).

Although we have treated changes in cell shape and changes in cell contact behavior under separate headings, it is well known that the two phenomena are closely interrelated. For example, highly stellate cells with cell processes radiating in several directions will quickly round up when deprived of a substratum for adhesion. Furthermore, if a cell changes its contacts with its neighbors in some asymmetric fashion, its cell shape will quickly change to accommodate new tensions.

Loss of cell contacts can also serve as a powerful morphogenetic mechanism. We have seen several instances of mass cell movements where cells appear to be released from a coherent mass and then migrate. Recall, for example, how chick hypoblast cells appear to originate in and migrate away from the embryonic shield or how mesoblast cells migrate away from the primitive groove. These cells may in fact become more migratory as a result of changes in cell contacts. This possible morphogenetic mechanism will be considered more fully in the next section.

Role of Changes in Cell Motility in Gastrulation

Gastrulation involves mass movements that are accomplished by the coordinated motile interaction of collections of individual cells. We can think of the problem of gastrulation metaphorically by imagining how we could move all the residents of one city to another city. We might instruct each person individually, or we might link them together physically—instructing a few as to how to go and instructing the rest to follow. Bottle cells elongate and migrate away from the blastopore or primitive streak, producing an invagination as they go and carrying other cells with them. Are bottle cells leaders in some sense?

One of the most difficult morphogenetic problems is how individual cell behavior can be coordinated to give mass cell movements. Most workers have tried to approach this problem by studying individual cell behavior in live embryos or by studying behavior of dissociated cells in short-term cultures. The former technique is preferable but is not often practical. Some embryos are opaque, are protected with various investments, or contain collections of intracellular inclusions that make direct observations difficult or impossible. Consequently, it sometimes becomes necessary to disrupt embryos with chelating agents or proteolytic enzymes. Both treatments break cell contacts and can be used to produce a cell suspension that may be cultured on an artificial substratum. This is a potentially misleading approach because one may end up studying a phenomenon that has nothing to do with morphogenesis but that results from disruption of the normal developmental process or from culture conditions. Finally, cells are known to show behavioral alterations, depending on the number of cells they are touching and on their environment. One might wonder whether it is possible *in principle* to study morphogenesis with cells taken from embryos. This rather gloomy viewpoint may be correct, but strict adherence to it would make it difficult to introduce any probe into a developing system. It should also be pointed out that experiments on cells taken from embryos and studied in isolation have been used to discover behavioral traits that are widely thought to have great significance for an understanding of morphogenetic cell movements. There are a number of embryos that are transparent and that develop under conditions suitable for microscope examination. It would be a good idea to study the behavior of well-defined groups of cells in such embryos both *in vivo* and *in vitro* in an attempt to discover if dissociation and culturing do indeed generate modes of behavior that have no consequence for morphogenesis or if *in vitro* behavior resembles that *in vivo*.

There is a growing body of knowledge on the mechanism of locomotion of a variety of tissue culture cells. Unfortunately, at the present time, there is only the most rudimentary kind of information available about how bottle cells or spreading ectodermal cells move. More is known about movement of morphogenetically active cells in sea urchins (Gustafson and Wolpert, 1967) and fish (Trinkaus, 1973). In addition, there is almost no information about how cell contact in embryos alters motile behavior. Concerted studies in those embryos which are favorable material for following cell movements are needed before we will be able to understand the role of cell motility in gastrulation.

In the preceding sections of this chapter, we have considered changes in cell shape, changes in cell contact behavior, and changes in cell motility as if they were separate phenomena. In fact, they are inextricably interrelated. For example, a cell must adhere to a substratum (or gain some traction) for it to be able to move. When a cell moves, it frequently changes shape dramatically. A round cell might throw out a long process (thereby changing shape), which then adheres to a substratum (changing contact), contracts, and pulls the cell to a new place (and thus moves). As this imaginary cell moves, the process is taken back

into the cell and the shape is changed again. In addition, for a cell to leave one location, it must release any contacts that it has in that location. A cell could have a fully developed motile capacity but be unable to get a grip on the substratum. Or, a cell might have the ability to grip the substratum so firmly that it is frozen in one location. One might also imagine a situation in which a cell had an invariant ability to grip the substratum but an altered motile capacity in such a way that it could break existing contacts. It is extremely important for the reader to realize that morphogenetic cell movements involve a dynamic interplay between cell shape, cell contact, and cell motility. Gastrulation will not be explained until a complete picture of this interplay can be constructed.

Causal Analysis of Gastrulation

Much of the early work on gastrulation consisted of attempts to clarify the morphological details of the process. At first glance, this might appear to be a relatively simple matter. In fact, this effort was difficult even for skilled and competent researchers. Most of the morphological details of gastrulation are now well understood; we know where the cells go. Unfortunately, however, we are largely ignorant about what causes gastrulation. We are unable to answer fundamental questions about gastrulation. Why do only certain cells initiate invagination? Why do cells move in their particular directions? What signals initiate and terminate changes in shape, contact, and motility? What forces drive and guide morphogenesis?

The work of Johannes Holtfreter (1939, 1943, 1944) pioneered and developed much of our current view of gastrulation. He observed live embryos, fragments of embryos, individual cells in culture, and reaggregated suspensions of embryonic cells. As a result of these studies, he introduced a new concept, *cellular affinities*, in the first comprehensive attempt to account for morphogenesis. His experimental investigations led him to believe that the inherent associative properties of cells and cell clusters could go a long way toward explaining gastrulation. For example, when he isolated a fragment of presumptive endoderm from a blastula stage amphibian embryo and cultured it in a simple salt solution, he was able to show that it first rounded up into a compact mass and later spread on the bottom of the culture dish. Spreading occurred at the same time as the presumptive endoderm of control embryos was also spreading. Holtfreter was clever enough to realize that part of gastrulation could be accounted for by the inherent tendencies of the different parts of the embryo. In another elegant experiment, he showed that an isolated dorsal lip of the blastopore could still burrow into a fragment of endoderm to form a small invagination (Fig. 9).

Although Holtfreter did not uncover the "causes" of these inherent behavior patterns, he had made an important discovery. By combining fragments of embryos *in vitro*, he was able to demonstrate how selective affinities might work as a morphogenetic agent. When he combined equal-sized fragments of presumptive ectoderm and presumptive endoderm,

they would at first adhere to one another to form a ball. Later, the two tissues would undergo spontaneous self-isolation. In contrast, if fragments of presumptive ectoderm and presumptive endoderm were combined with a little bit of presumptive mesoderm, the ectoderm and endoderm still separated from one another but now were held together by an intermediate layer of mesoderm. These experiments demonstrated once again how one important part of gastrulation might be the expression of cellular affinities. It is also apparent from a close reading of Holtfreter that he was aware of the importance of changes in surface free energy in governing cellular affinities. Later work by Townes and Holtfreter (1955) showed that type-specific segregation of coaggregated embryonic amphibian cells resulted in final configurations similar to those seen earlier with fragment combinations. Holtfreter can be credited with a clear demonstration that preferential associative behavior among cells individually or as a group in a fragment was an important morphogenetic mechanism.

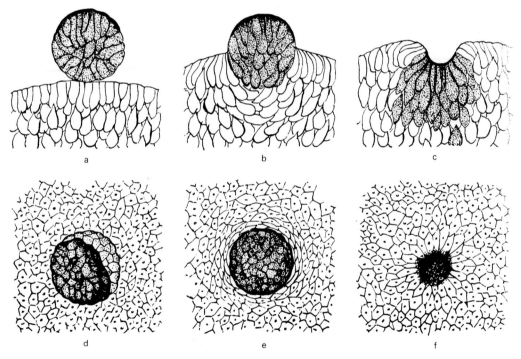

FIGURE 9. *A fragment of dorsal lip of the blastopore will sink into a fragment of endoderm and form a definite invagination. a, b, and c are sections; d, e, and f are surface views of a, b, and c. (From J. Holtfreter, J. Exp. Zool. 95, 194, 1944.)*

Recently, Steinberg (1964, 1970) contributed to our present understanding of morphogenesis by extensive studies of the behavior of embryonic chick cells in mixed aggregates. A brief description of this experimental system will be helpful. Different organs or regions of a chick embryo are dissociated into cell suspensions with proteolytic enzymes. Under appropriate conditions, the cells in these suspensions will reassociate to form aggregates. If two or more organs—for example, heart and liver—are dissociated and the cell suspensions are mixed and allowed to reaggregate, heart and liver cells will be mixed randomly in these aggregates at first. In a short while, however, cell sorting occurs. Heart and liver cells segregate into separate phases, most often as a central mass of heart cells surrounded by an outer layer of liver cells. In addition, whole fragments of liver usually engulf whole fragments of heart when the two are confronted in culture. Sorting out and spreading is seen with many tissue combinations. Steinberg also made the important discovery that there is a *hierarchy* of "preference" for the external position (or internal position) in combinations of tissues. This hierarchy follows a transitive rule: that is, if A engulfs B and B engulfs C, then A will engulf C. The existence of a hierarchy that follows a transitive rule and the attainment of the same equilibrium configuration of cells from two completely different initial states (mixed aggregates versus confronted fragments) are both predicted by Steinberg's *differential adhesion hypothesis*. This hypothesis is somewhat complicated and is illustrated in Fig. 10.

Let us consider only case 2, where complete segregation (or engulfment) occurs. Steinberg would say that this equilibrium configuration will result when the work (strength) of adhesion between two *a* cells (W_a) is greater than the average of the work (strength) of adhesion between two *b* cells (W_b), which in turn is greater than the work (strength) of adhesion between an *a* and a *b* cell (W_{ab}), which in turn is greater than or equal to the work (strength) of adhesion between two *b* cells. That is,

$$W_a > \frac{W_a + W_b}{2} > W_{ab} \geqslant W_b$$

Recently, Phillips and Steinberg (1969) made direct measurements of the interfacial free energies of adhesion for a number of embryonic chick tissues. These measurements support the validity of the differential adhesion hypothesis. It is also widely held that differences in the strengths of intercellular adhesions may be a driving force in morphogenesis. Sorting out is a kind of morphogenetic process that has been used as a model for morphogenetic processes that occur in real embryos. It is quite easy to imagine ways in which changes could occur in the surface of an embryonic cell that would result in changes in the strengths of the adhesions it makes with neighboring cells. In addition, masses of cells could change their properties to invaginate into or spread upon other masses of cells. A recent demonstration of sorting out while cell motility is inhibited (Armstrong and Parenti, 1972) is an additional indication that adhesive differentials may be important as a driving force for sorting out and morphogenesis.

There is another impressive body of evidence that suggests that changes in cell adhesiveness and cell motility may play a central part in animal morphogenesis. Using clear sea urchin embryos and time-lapse cinemicrography, Gustafson, Wolpert, and their associates have made many observations that suggest that changes in cell behavior might account for morphogenetic cell movements (see reviews by Gustafson and Wolpert, 1963, 1967). For example, certain cells in the vegetal plate of late blastula stage embryos appear to lose their intimate contacts with their neighbors and migrate freely into the blastocoel. These primary mesenchymal cells later participate in the formation of the larval skeleton. During this dramatic behavioral transition, there are concomitant changes in cell contact relations and cell motility. Gustafson and Wolpert have proposed that changes in cellular adhesiveness cause changes in cell behavior. Other cells, called *secondary* mesenchymal cells, remain intimately associated with their neighbors in the invaginating vegetal plate and extend long filopodia. These long, slender processes seem to "explore" and selectively adhere to portions of the inner blastocoel wall. Gustafson and Wolpert have also presented convincing descriptions that suggest that the filopodia are contractile and are instrumental

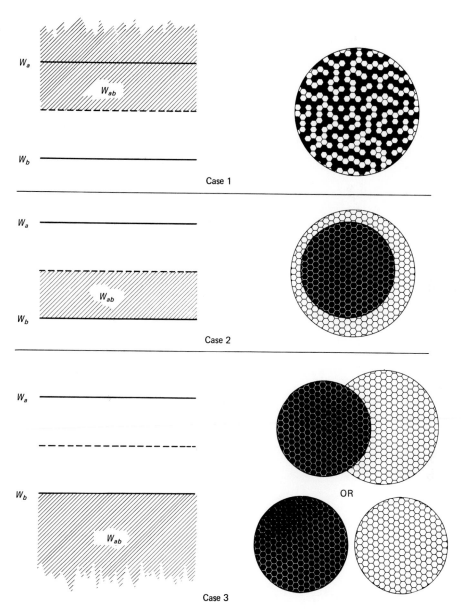

FIGURE 10. *Theoretical correspondence between intercellular adhesive strengths and equilibrium configuration after sorting out. The work (strength) of adhesion between two a cells (black), between two b cells (white), and between an a and a b cell are* W_a, W_b, *and* W_{ab}, *respectively.* [*From M. S. Steinberg, in "Cellular Membranes in Development" (M. Locke, ed.), p. 328, Academic Press, New York, 1964.*]

in pulling the vegetal plate toward the animal pole. Although much of their evidence is descriptive, it is important to realize that we know more about the mechanism of sea urchin gastrulation than about most other forms. Any student with a serious interest in animal morphogenesis should study the work of Gustafson and Wolpert in detail.

We now know that cell shape, cell motility, and cell adhesiveness are intimately interrelated. New research on the relationship between cell contact and cell motility has been done in recent years. Much of this work has come from the laboratory of Abercrombie and his associates. Several years ago, Abercrombie (Abercrombie and Heaysman, 1953, 1954; Abercrombie et al., 1957) was responsible for coining the phrase "contact inhibition of cell movement." He gave this term a rigorous operational definition that has been largely ignored ever since. Originally, Abercrombie and his associates observed that there was a relationship between the rate of cells emigrating from an explanted fragment of embryonic chick heart and the number of cells any individual cell touched. In short, the more cells that another cell contacted, the slower it moved. They also observed that when two explants are placed close together in culture, a halo of cells would grow out from each explant, and, although these halos would eventually meet, they would not overlap. Prior to meeting, cells in the peripheral portion of the halo would move rapidly, but, after contacting cells in the other halo, they would increase their number of contacts with other cells and consequently decrease their rate of movement. Abercrombie and his associates also showed that cell overlapping was minimal in the interzone between explants. From these results, they reasoned that cell movement was restricted by contact and that cells would not move over one another. This behavior pattern was called "contact inhibition of cell movement." More recently, fibroblastic cell motility has been subjected to intense scrutiny by many workers, and we now have a more comprehensive notion of the mechanism of tissue-culture cell motility and its contact inhibition (Abercrombie et al., 1970a, b, c, 1971, 1972; Trinkaus et al., 1971; Harris and Dunn, 1972; Harris, 1973). Fibroblastic cells in culture may not be the same as embryonic cells in culture. There is a pressing need for more information about the motile behavior of embryonic cells.

It is clear that we need to learn a good deal more about cell movements in embryos. Such studies, however, will entail certain difficulties. Many embryos are opaque or rest on a bothersome yolk mass. Some are transparent but rather small, making it difficult to isolate and study the behavior of the same cells *in vivo*

and *in vitro*. Trinkaus has made excellent use of the eggs of *Fundulus heteroclitus*, the killifish, for studies of cell behavior *in vitro* (Trinkaus, 1963) and *in vivo* (Trinkaus, 1973). He has been able to show that contact inhibition of cell movement is an important morphogenetic mechanism in a spreading cell sheet in this embryo (the enveloping layer that engulfs the yolk mass). Detailed analysis of other cells deep within this transparent embryo, using time-lapse cinemicrography, have demonstrated that there are dramatic changes in cell motile–adhesive behavior that may in part account for gastrulation.

Contact inhibition of cell movement has also been evoked as a morphogenetic mechanism to explain lateral migration of chick mesoblast cells (Trelstad et al., 1967; Hay, 1968). This is plausible, and it would be most interesting to study chick mesoblast cells *in vitro* when they confront other mesoblast cells as well as epiblast and hypoblast cells to see if this hypothesis is really valid. This would be a challenging and difficult study, since it would be a demanding task to isolate these cells. In a similar vein, it would be desirable to have a detailed account of the behavior of bottle cells of various sorts to try to get some understanding of how they move and how they interact with neighboring cells. It is only necessary to point out that these cells move rapidly while in contact with a large number of neighboring cells, apparently lacking contact inhibition of cell movement. In short, there is a pressing need for a more detailed characterization of the behavior of "morphogenetically interesting" cells.

Future Directions in Study of Gastrulation and Cell Interactions

In all gastrulating embryos there are clear examples where cells move with precision from one location to another in the embryo. What is the chemical nature of the control mechanisms that guide cell movements? Do cells sense their positions in an embryo? Can they detect and respond to their proper pathway of movement through the embryo? Do embryonic cells have the capacity to read and respond to some chemical road map to guide their movements? These questions cannot be answered at present. It is not even known if they are meaningful questions. There is very little information available about the chemical composition of the surface of early embryonic cells. Recent advances in membrane structure and surface chemistry in nonembryonic systems, however, have pointed to several directions of potentially fruitful investigation for future developmental biologists.

There is an increasingly impressive body of evidence that indicates that carbohydrate-containing molecules, located in or near the outer surface of cells, play an important role in a wide variety of phenomena in which cell behavior is modified by the recognition of and response to something in the extracellular environment. Often, monosaccharides are covalently linked to protein (glycoproteins) or lipid (glycolipids) molecules known to be plasma membrane constituents. In the case of glycoproteins, side chains of variable numbers of covalently bonded monosaccharides are linked to polypeptide chains. The oligosaccharide side chains can be of variable length, degree of branching, or residue sequence. Many workers now feel that the oligosaccharide side chains of cell surface glycoproteins are projecting into the environment surrounding the cell and are free to interact with extracellular constituents of the organism and, most particularly, with *other cells* (Burger, 1971; Marchesi et al., 1972; Warren et al., 1973). Complex information could be encoded in the structure of the oligosaccharide side chains of glycoproteins. For example, if a cell had a large number of oligosaccharide chains projecting from its surface with the sequence protein–sugar A–sugar B–sugar B–sugar C and another cell had a way of *recognizing* this sequence, we would have a way of understanding how cell A differs from cell B (and other cell types) and how cells could recognize these differences. The carbohydrate sequences of glycoproteins are potential candidates for carrying the kind of specific surface chemical information that may be involved in guiding morphogenetic cell movements.

We have alluded to the need for recognition of differences in surface chemical composition. There is a class of enzymatic proteins that are specialized for recognizing carbohydrate sequences. These enzymes, the glycosyltransferases, catalyse the addition of "activated" sugars to glycoprotein "acceptors." These enzymes often have a dual specificity: (1) they add only certain activated sugars, and (2) they add these sugars only to certain acceptors. Let us consider, for example, the addition of galactose to a glycoprotein. Galactose (Gal) is first phosphorylated by adenosine triphosphate (ATP) to form galactose-1-phosphate (Gal-1-P). This compound is then coupled with uridine triphosphate (UTP) to form an activated sugar-nucleotide, uridine diphosphogalactose (UDP-Gal). Finally, a galactosyltransferase catalyses the addition of the Gal from the UDP-Gal to an acceptor, which is modified by the addition of Gal and is therefore no longer an acceptor (Fig. 11).

Some galactosyltransferases show specificity for both sugar nucleotides and acceptors. Once galactose had been added to the acceptor protein-A-B-C, a new sugar would have to be added to protein-A-B-C-Gal by a different glycosyltransferase and a different nucleotide sugar. Roseman (1970) has used these facts to propose a model for the role of glycosyltransferases in intercellular adhesion. His diagrammatic presentation of this model can be seen in Fig. 12.

Recent experiments have shown that there are indeed glycosyltransferases on embryonic cell surfaces and that these enzymes may really play the role proposed in the Roseman model (Roth et al., 1971). This model is important whether or not it is true in detail, since it is the first time that a detailed chemical mechanism of specific cell interaction has been proposed. It contains four attractive features. First, it allows specific recognition and interaction between cells. Second, it could provide a relatively simple device for rapidly and reversibly changing cell adhesions. Third, this model could be used to account for control of changes in cell adhesion. Fourth, it allows for control of adhesion by controlling insertion of transferases into membranes or by controlling the level of sugar nucleotides and divalent cations at the active site of the transferase. This sort of approach seems to be one promising future direction that may lead to a better understanding of gastrulation.

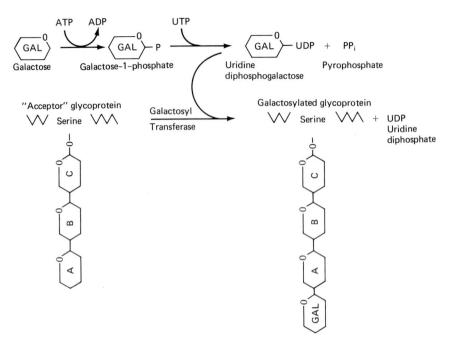

FIGURE 11. *Biochemistry of the addition of a sugar residue to a glycoprotein oligosaccharide side chain.*

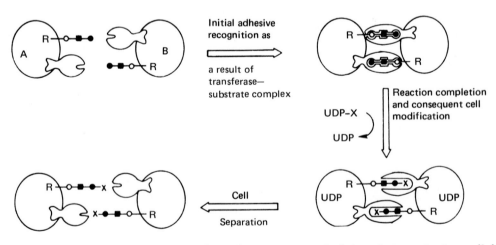

FIGURE 12. *Possible role of surface glycosyltransferases and their substrates in intercellular adhesive recognition. Cells A and B have both substrates (R-containing groups) and enzymes with active sites exposed. Roseman's theory allows for surface modification when the new sugar (X) is added to the acceptor molecule. (From S. Roth, E. J. McGuire and S. Roseman, J. Cell Biol. 51, 546, 1971.)*

References

ABERCROMBIE, M., AND HEAYSMAN, J. E. M. (1953). Observations on the social behavior of cells in human tissue culture. I. Speed of movement of chick heart fibroblasts in relation to their mutual contacts. *Exp. Cell Res. 5*, 111–131.

ABERCROMBIE, M., AND HEAYSMAN, J. E. M. (1954). Observations on the social behavior of cells in tissue culture. II. "Monolayering" of fibroblasts. *Exp. Cell Res. 6*, 293–306.

ABERCROMBIE, M., HEAYSMAN, J. E. M., AND KARTHAUSER, H. M. (1957). Social behavior of cells in tissue culture. III. Mutual influence of sarcoma cells and fibroblasts. *Exp. Cell Res. 13*, 276–291.

ABERCROMBIE, M., HEAYSMAN, J. E. M., AND PEGRUM, S. M. (1970a). The locomotion of fibroblasts in culture. I. Movements of the leading edge. *Exp. Cell Res. 59*, 393–398.

ABERCROMBIE, M., HEAYSMAN, J. E. M., AND PEGRUM, S. M. (1970b). The locomotion of fibroblasts in culture. II. "Ruffling." *Exp. Cell Res. 60*, 437–444.

ABERCROMBIE, M., HEAYSMAN, J. E. M., AND PEGRUM, S. M. (1970c). The locomotion of fibroblasts in culture. III. Movements of particles on the dorsal surface of the leading lamella. *Exp. Cell Res. 62*, 389–398.

ABERCROMBIE, M., HEAYSMAN, J. E. M., AND PEGRUM (1971). The locomotion of fibroblasts in culture. IV. Electron microscopy of the leading lamella. *Exp. Cell Res. 67*, 359–367.

ABERCROMBIE, M., HEAYSMAN, J. E. M., AND PEGRUM, S. M., (1972). Locomotion of fibroblasts in culture. V. Surface marking with concanavalin A. *Exp. Cell Res. 73*, 536–539.

ARMSTRONG, P. B., AND PARENTI, D. (1972). Cell sorting in the presence of cytochalasin B. *J. Cell Biol., 55*, 542–553.

BAKER, P. C. (1965). Fine structure and morphogenetic movements in the gastrula of the tree frog, *Hyla regilla*. *J. Cell Biol. 24*, 95–116.

BALLARD, W. W. (1964). "Comparative Anatomy and Embryology." Ronald Press, New York.

BELLAIRS, R. (1971). "Developmental Processes in Higher Vertebrates." University of Miami Press, Coral Gables, Florida.

BRIGGS, R., AND CASSENS, G. (1966). Accumulation in the oocyte nucleus of a gene product essential for embryonic development beyond gastrulation. *Proc. Nat. Acad. Sci. U.S.A. 55*, 1103–1109.

BRIGGS, R., AND JUSTUS, J. T. (1967). Partial characterization of the component from normal eggs which corrects the maternal effect of gene O in the Mexican axolotl (*Ambystoma mexicanum*). *J. Exp. Zool. 167*, 105–116.

BURGER, M. M. (1971). Cell surfaces in neoplastic transformation. *In* "Current Topics in Cellular Regulation" (B. L. Horecker and E. R. Stadtman, eds.), pp. 135–193. Academic Press, New York.

CONKLIN, E. G. (1932). The embryology of amphioxus. *J. Morphol. 54*, 69–151.

GUSTAFSON, T., AND WOLPERT, L. (1963). The cellular basis of morphogenesis and sea urchin development. *Int. Rev. Cytol. 15*, 139–214.

GUSTAFSON, T., AND WOLPERT, L. (1967). Cellular movement and contact in sea urchin morphogenesis. *Biol. Rev. 42*, 442–498.

HARRIS, A. (1973). Behavior of cultured cells on substrata of variable adhesiveness. *Exp. Cell Res. 77*, 285–297.

HARRIS, A., AND DUNN, G. (1972). Centripetal transport of attached particles on both surfaces of moving fibroblasts. *Exp. Cell Res. 73*, 519–523.

HAY, E. D. (1968). Organization and fine structure of epithelium and mesenchyme in the developing chick embryo. *In* "Epithelial-Mesenchymal Interactions" (R. Fleischmajer and R. E. Billingham, eds.), pp. 31–55. Williams & Wilkins, Baltimore.

HOLTFRETER, J. (1939). Gewebeaffinität, ein mittel der embryonalen formbildung. *Arch. Exp. Zellforsch. 23* 169–209. Available in translation *in* "Foundations of Experimental Embryology" (B. H. Willier and J. M. Oppenheimer, eds.), pp. 186–225. Prentice-Hall, Englewood Cliffs, New Jersey.

HOLTFRETER, J. (1943). A study of the mechanics of gastrulation, I. *J. Exp. Zool. 94*, 261–318.

HOLTFRETER, J. (1944). A Study of the mechanics of gastrulation, II. *J. Exp. Zool. 95*, 171–212.

JOHNSON, K. E. (1970). The role of changes in cell contact behavior in amphibian gastrulation. *J. Exp. Zool. 175*, 391–428.

JOHNSON, K. E. (1972). The extent of cell contact and the relative frequency of small and large gaps between presumptive mesodermal cells in normal gastrulae of *Rana pipiens* and the arrested gastrulae of the *Rana pipiens* ♀ X *Rana catesbeiana* ♂ hybrid. *J. Exp. Zool. 179*, 227–238.

KALT, M. R. (1971a). The relationship between cleavage and blastocoel formation in *Xenopus laevis*. I. Light microscopic observations. *J. Embryol. Exp. Morphol. 26*, 37–49.

KALT, M. R. (1971b). The relationship between cleavage and blastocoel formation in *Xenopus laevis*. II. Electron microscopic observations. *J. Embryol. Exp. Morphol. 26*, 51–66.

MARCHESI, V. T., TILLACK, T. W., JACKSON, R. L., SEGREST, J. P., AND SCOTT, R. E. (1972). Chemical characterization and surface orientation of the major glycoprotein of the human erythrocyte membrane. *Proc. Nat. Acad. Sci. U.S.A. 69*, 1445–1449.

NICOLET, G. (1971). Avian gastrulation. *Adv. Morphogen. 9*, 231–262.

PERRY, M. M., AND WADDINGTON, C. H. (1966). Ultrastructure of the blastopore cells in the newt. *J. Embryol. Exp. Morphol. 15*, 317–330.

PHILLIPS, H. M., AND STEINBERG, M. S. (1969). Equilibrium measurements of embryonic chick cell adhesiveness. I. Shape equilibrium in centrifugal fields. *Proc. Nat. Acad. Sci. U.S.A. 64*, 121–127.

REVERBERI, G. (1971). Amphioxus. *In* "Experimental Embryology of Marine and Freshwater Invertebrates" (G. Reverberi, ed.), pp. 551–572. North-Holland, Amsterdam.

ROSEMAN, S. (1970). The synthesis of complex carbyhydrates by multiglycosyltransferase systems and their potential function in intercellular adhesion. *Chem. Phys. Lipids 5*, 270–297.

ROTH, S., MCGUIRE, E. J., AND ROSEMAN, S. (1971). Evidence for cell-surface glycosyltransferases. Their potential role in cellular recognition. *J. Cel Biol. 51*, 536–547.

STEINBERG, M. S. (1964). The problem of adhesive selectivity in cellular interactions. *In* "Cellular Membranes in Development" (M. Locke, ed.), pp. 321–366. Academic Press, New York.

STEINBERG, M. S. (1970). Does differential adhesion govern self-assembly processes in histogenesis? Equilibrium configurations and the emergence of a hierarchy among populations of embryonic cells. *J. Exp. Zool. 173*, 395–434.

TOWNES, P. L., AND HOLTFRETER, J. (1955). Directed movements and selective adhesion of embryonic amphibian cells. *J. Exp. Zool. 128*, 53–120.

TRELSTAD, R. L., HAY, E. D., AND REVEL, J. P. (1967). Cell contact during early morphogenesis in the chick embryo. *Develop. Biol. 16*, 78–106.

TRINKAUS, J. P. (1963). The cellular basis of *Fundulus* epiboly. Adhesivity of blastula and gastrula cells in culture. *Develop. Biol. 7*, 513–532.

TRINKAUS, J. P. (1969). "Cells into Organs." Prentice-Hall, Englewood Cliffs, New Jersey.

TRINKAUS, J. P. (1973). Surface activity and locomotion of *Fundulus* deep cells during blastula and gastrula stages. *Develop. Biol. 30*, 68–103.

TRINKAUS, J. P., BETCHAKU, T., AND KRULIKOWSKI, L. S. (1971). Local inhibition of ruffling during contact inhibition of cell movement. *Exp. Cell Res. 64*, 291–300.

WARREN, L., FUHRER, J. P., AND BUCK, C. A. (1973). Surface glycoproteins of cells before and after transformation by oncogenic viruses. *Fed. Proc. 32*, 80–85. This volume of *Fed. Proc.* contains a large number of papers that stemmed from the Conference on Membranes in Growth, Differentiation, and Neoplasia held in April, 1972.

CHAPTER SEVEN

Erythroid Cell Differentiation

RICHARD A. RIFKIND

Among the many differentiated cells found in higher organisms those of the hematopoietic lineages, and principally the erythroid cells, have proved particularly valuable in the study of regulatory mechanisms that operate during normal and pathological cell differentiation (Marks and Rifkind, 1972; Gordon, 1970a,b). At the present time, although considerable progress has been made in the study of leukocyte development (Gordon, 1970b), it is erythropoiesis that provides the most comprehensive and integrated picture of a developmental hematopoietic system. In particular, study of erythroid cell differentiation, both in the fetal and adult organism, has contributed to the elucidation of a number of basic problems in developmental biology, including:

1. The cellular and molecular basis for changes in protein synthesis during development.
2. The relationship between cell division, DNA synthesis, and the synthesis of specialized proteins by differentiating cell populations.
3. The nature and stability of RNA synthesized during cell differentiation.
4. The mechanism of action of hormonal regulators in cell differentiation.
5. The special features which characterize terminally differentiated cell types.

The advantages offered by erythropoiesis as a paradigm model for the study of cell differentiation derive from certain unique features of erythropoiesis in both the adult and the embryonic organism. Hemoglobin, a biochemically and genetically well-characterized protein, is the principal differentiated cell product and comprises over 90 percent of the protein synthesized during differentiation. Relatively large numbers of differentiating cells in discrete and accessible anatomic sites are readily available for experimental manipulation. An established body of cytological, biochemical, and immunological information permits identification and separation of erythroid cells at their several stages of differentiation. A well-documented humoral (i.e., hormonal) regulatory system is responsible for control of the rate of erythroid cell production in response to the physiological requirements of the organism. Finally, there exist a large number of genetic and acquired disorders of erythroid cell development both in man and in experimental animals, which help define and clarify normal biological mechanisms. Although a detailed treatment of the physiology and pathophysiology of the erythroid cell system is beyond the scope of this chapter (see Harris and Kellermeyer, 1970, for an exhaustive review of this subject), some basic aspects of red blood cell function are essential to our understanding of the developmental biology of this cell system.

Red blood cells (erythrocytes) constitute the population of specialized cells designed for the transport of respiratory gases (oxygen and carbon dioxide) between metabolizing body cells and the external environment. This function is accomplished by virtue of the oxygen-binding properties of hemoglobin, the porphyrin-containing protein synthesized by erythroid cell precursors, which constitutes over 95 percent of the protein content of the mature erythrocyte. The mature red cell, then, consists of a solution of hemoglobin, packaged within a plasma membrane, in association with the minimal energy sources and enzymatic machinery required to maintain the osmotic properties of the cell (ionic transport mechanisms) and to protect both the hemoglobin and structural elements of the cell from denaturation in the blood stream. The production of circulating erythrocytes (erythropoiesis) comprises a cytological and biochemical sequence of events whereby a hemopoietic precursor cell is induced to initiate the synthesis of hemoglobin and other cell-specific components of erythroid cells, to undertake a definitive number of sequential cell divisions, and to undergo characteristic morphological transformations. In virtually all species of the higher vertebrates the final product of erythroid cell differentiation, the circulating mature erythrocyte, is extremely restricted with respect to further macromolecular synthesis and is incapable of cell division. The genetic apparatus of such terminally differentiated cells is either eliminated or severely repressed. Indeed, in adult mammals the erythroid cell nucleus is actually extruded from the developing precursor cell at the end of the process of differentiation. Both the cytological and the biochemical details of erythropoietic differentiation will be examined in detail in the following sections.

Under normal circumstances, throughout postpartem life and possibly during embryogenesis as well, the rate of erythropoiesis (that is, the number of erythrocytes and the amount of hemoglobin produced by the differentiating precursor cell population) is regulated by the titer of a hormone, erythropoietin, which is produced, released, or activated in response to the level of tissue oxygenation (Reissmann, 1950; Jacobson and Goldwasser, 1957; Gordon, 1959; Gordon and Zanjani, 1970; Erslev, 1953). The flexibility of this control system is considerable. In the adult human approximately 6 billion new erythrocytes are produced hourly to meet the normal requirements for replacement of effete red cells (normal human red cells circulate for about 110 to 120 days before they are destroyed by the reticuloendothelial cell system of macrophages; Rifkind, 1966) and the normal tissue requirement for oxygen delivery. Low but demonstrable levels of erythropoietin are responsible for sustaining this production rate. Should the circulating erythrocyte mass (that is, oxygen-delivery capacity) exceed the physiological requirements, erythropoiesis may shut down virtually to nil as a result of a drop in erythropoietin activity. In the presence of an inadequate red cell mass (that is, anemia—due, for example, to accelerated red cell destruction, bleeding or other causes) or tissue hypoxia for any reason, erythropoietin activity rises, and red cell production may increase manyfold (as much as 6- to 8-fold increase in bone marrow erythropoiesis can be documented in the presence of certain severe and chronic hemolytic anemias in man). The present status of the chemistry and physiology of erythropoietin, a glycoprotein of less than 70,000 molecular weight and the hypoxia-erythropoiesis regulatory mechanism, lies outside the scope of this chapter but has been the subject of extensive and recent review (Camiscoli and Gordon, 1970; Lowy, 1970).

The presently available data suggest that in response to tissue oxygen tension the kidney elaborates either erythropoietin itself (Kuratowska et al, 1960) or a renal activator of a plasma erythropoietin precursor (Gordon et al., 1967) at a rate appropriate for maintaining or stimulating erythropoiesis according to metabolic demands. Clearly this feedback system is sluggish but useful in maintaining the long-range respiratory homeostasis of the organism. More rapid physiological adaptation to acute changes in the demand for oxygen delivery are mediated by other mechanisms, operating directly on the oxygen-carrying properties of red cell hemoglobin (Oski and Gottlieb, 1971) and on the cardiovascular system.

In summary, the system of red blood cells (the *erythron*) may be considered to have two major portions: (1) a proliferating and differentiating compartment responsible for the synthesis of hemoglobin and for producing the required numbers of mature erythrocytes, at a rate responsive to the level of the erythropoiesis-stimulating hormone erythropoietin; and (2) a circulating erythrocyte compartment containing cells of finite life-span, incapable of either cell division or macromolecular synthesis but responsible for the physiological functions of the erythron. The detailed cellular biology of this system, particularly with reference to its embryological development in mammalian species, will constitute the substance of the remainder of this chapter.

Fetal Erythropoiesis

Although studies on adult organisms provide the most complete body of information concerning the physiology of erythropoiesis, the study of erythroid cell differentiation during embryogenesis has provided much of the present understanding of the molecular and cellular regulatory mechanisms that are involved in this process. In particular, the fetal mouse has proved an attractive and useful model for experimental study. For these reasons, the review that follows will concentrate on a set of observations made in several laboratories with the C57BL/6J strain of mouse. Nevertheless, an increasing body of information suggests strongly analogous developmental processes in other species and, wherever appropriate, reference to comparative studies will be provided.

During the course of embryogenesis and early postnatal life there occur, in virtually all species so far studied in sufficient detail, a sequence of events that involve changes in the sites of erythroid differentiation, in the cytology of the erythrocyte population, in the types of hemoglobins synthesized, in the proportion of heme- and nonheme proteins being synthesized, in the types of RNA produced in erythroid cell precursors, and in the stability of the protein-synthetic apparatus. Each of these events or changes casts light upon the fundamental regulatory mechanisms involved in erythropoietic differentiation, as well as upon more general problems in developmental biology. Each of these will be dealt with in subsequent sections.

Primitive (Yolk Sac) Erythropoiesis

ANATOMY AND CYTOLOGY

In those species studied to date, including mouse, man, tadpole (frog), chick, and other species (Marks and Rifkind, 1972; Ingram, 1972; Maniatis and Ingram, 1971b), embryonic erythropoiesis is characterized by the sequential appearance of two distinct populations of erythroid cells (erythrocytes and their differentiating precursors). In the mouse, with a gestation period of roughly 21 days, the first population of erythroid cells appears, at the end of the seventh gestational day, in the mesenchymal layer of cells lying between the endodermal and ectodermal epithelial layers of the embryonic yolk sac (Attfield, 1951; Craig and Russell, 1964; Bank et al., 1970; Moore and Metcalf, 1970). In this site groups of mesenchymal cells round up, lose their typical mesenchymal intercellular junctional complexes, proliferate and form discrete foci of hemoglobinizing erythroid cells, termed *blood islands*, in a fashion entirely analogous to primitive yolk sac erythropoiesis in other vertebrate species (Wilt, 1965; Lillie, 1908). At their inception blood islands are essentially extravascular. Cells at the periphery of the blood islands rapidly transform into recognizable capillary endothelial cells so that by the tenth gestational day the differentiating and maturing erythroid cells are enclosed within sinusoidal vascular spaces communicating with the fetus's systemic circulation. Yolk sac-derived differentiating erythroid cells (erythroblasts) actually enter the fetal circulation before completing their cytological and biochemical development. Nucleated erythroblasts of yolk sac origin enter the circulation between the ninth and tenth gestational days and continue their development while circulating in the fetal bloodstream. Cytological evidence of maturation consists of progressive condensation of nuclear chromatin, shrinkage of the nucleolus, and nuclear pycnosis (Kovach et al., 1967). Mitotic yolk sac erythroblasts are seen through the thirteenth fetal day but not subsequently. Ribosomes and polyribosomes are abundant in the cytoplasm during early stages of yolk sac erythroblast development, but their concentration decreases progressively as the cells accumulate hemoglobin. These erythroid cells reach maturity by about the fifteenth gestational day. The mature yolk sac erythrocyte is a nucleated cell, resembling, in this respect, the definitive (adult) erythrocytes of many lower vertebrate species (i.e., birds and amphibia). The nucleus is, however, highly pycnotic and inactive with regard to the normal nuclear biosynthetic functions. Unlike subsequent phases of erythropoiesis both in the embryo and in the adult animal, yolk sac erythropoiesis is not self-sustaining; yolk sac-derived erythroid cells mature as a relatively homogeneous cohort. At each time point, all the cells are at approximately the same stage of differentiation (de la Chapelle et al., 1969). The circulating life-span of these primitive erythrocytes is short; by the sixteenth day cells of this lineage are already found undergoing phagocytosis by cells of the fetal reticuloendothelial system and they are virtually all destroyed prior to birth, as the circulation fills with erythrocytes of a subsequent (adult or definitive) lineage.

PROTEIN AND NUCLEIC ACID SYNTHESIS DURING YOLK SAC ERYTHROPOIESIS

Mouse yolk sac erythroblasts synthesize three hemoglobins, designated E_I, E_{II}, and E_{III}, none of which is found in the adult animal. Distinct "embryonic" hemoglobin of this sort is characteristic of virtually all vertebrate species studied (Marks and Rifkind, 1972; Ingram, 1972). The embryonic mouse hemoglobins are readily distinguished by electrophoresis and chromatography, and the globin chains comprising these hemoglobins have likewise been analyzed (Fantoni et al., 1967). Whereas adult C57BL/6J mouse hemoglobin is composed of two distinct globin chains (designated α and β) in the form of a tetramer ($\alpha_2\beta_2$), yolk sac erythroblasts synthesize four globin chains, only one of which (the α chain) corresponds to an adult globin. The globin chains of mouse yolk sac erythroblasts are designated α, x, y, and z. Hemoglobin E_I consists of x and y chains, E_{II} of α and y chains, E_{III} of α and z chains. The best available data suggests that all three hemoglobins (hence all four globin chains) are synthesized in each of the yolk sac erythroblasts. Nevertheless, it seems likely that the rate of synthesis of each of these embryonic hemoglobins is subject to independent control. From fetal days 10 to 14 in the C57BL/6J mouse, the relative rates of synthesis of hemoglobins E_I, E_{II}, and E_{III} change and the most abundant hemoglobin on day 10 (E_I) becomes a minor component of the total on day 14, when hemoglobin E_{II} is the predominant embryonic hemoglobin. Indeed, studies on the synthetic rates for each of these hemoglobins indicate a changing pattern during this period of embryogenesis (Fantoni et al., 1968).

In addition to a change in the relative rates of synthesis of the several embryonic hemoglobins there also occur significant changes in the rates of synthesis of nonhemoglobin proteins during yolk sac erythropoiesis (Fantoni et al., 1968b; Terada et al., 1971). On the eleventh gestational day about one third of the protein synthesized is hemoglobin, while two thirds is nonheme protein. By the thirteenth fetal day nonhemoglobin protein synthesis has virtually ceased, while hemoglobin synthesis persists at an essentially unchanged rate. Between days 13 and 15 hemoglobin synthesis likewise declines as the cells mature to the erythrocyte stage.

Yolk sac erythropoiesis in the mouse, as well as in other species, proceeds in a relatively synchronous fashion during embryogenesis (de la Chapelle et al., 1969; Weintraub et al., 1971).

That is, at each time point during gestation virtually all the yolk sac erythroid cells are at the same stage of maturation. (This is not the case during definitive erythropoiesis later in embryonic life and during adult erythropoiesis, as we shall show.) This situation affords an attractive opportunity for examining some important questions with regard to the factors that regulate the observed changes in protein synthesis during differentiation. In particular, it has been possible to consider the relationship between protein synthesis and (1) the synthesis and stability of RNA, and (2) the synthesis of DNA and cell division.

As already noted, differentiation of yolk sac erythroblasts, particularly during the phase of intravascular development, is accompanied by a progressive decrease in the concentration of cytoplasmic ribosomes, as visualized by electron microscopy. Simultaneously there is progressive nuclear pycnosis. Biochemical studies demonstrate a sharp fall in the capacity for RNA synthesis between the eleventh and thirteenth gestation days accompanied by a progressive decrease in the content of RNA between days 11 and 15. Although the rate of hemoglobin synthesis remains unchanged while RNA synthesis is falling, nonheme protein production falls rapidly to nil (Fantoni et al., 1968b). Thus, there is a marked difference in the rate of decay in the capacity to synthesize hemoglobin and nonheme proteins. This suggests that these two classes of protein may be formed on messenger RNA (mRNA) of distinctly different stabilities. This possibility is further substantiated by the differential effect of actinomycin D on protein synthesis in differentiating yolk sac erythroid cells. Thus, incubation of day 11 cells with the antibiotic has an insignificant effect on the rate of hemoglobin synthesis while profoundly suppressing nonhemoglobin protein production (Fantoni et al., 1968b). Over 90 percent of the nonhemoglobin proteins synthesized by these cells are nuclear proteins; 50 percent of these are acid-insoluble proteins (Terada, 1971). Taken together, the available data suggest that in yolk sac erythroid cells from at least the eleventh gestational day onward, nonhemoglobin (predominantly nuclear) proteins are synthesized on relatively unstable, short-lived mRNA, while synthesis of the specialized protein hemoglobin is directed by a relatively stable template. This feature is not unique to erythropoiesis; similar observations have been made in the case of such diverse differentiating systems as lens (Stewart and Papaconstantinou, 1967; Reeder and Bell, 1967), pancreas (Wessells, 1964), and silkworm galea (Kafatos and Reich, 1968).

The possibility that changes in the pattern of protein synthesis in differentiating yolk sac erythroblasts might be associated with alterations in the types of RNA produced has been explored by Terada and colleagues. In these studies newly synthesized mRNA, associated with cytoplasmic polyribosomes, was examined by agarose acrylamide gel electrophoresis and a 9S RNA fraction identified whose rate of synthesis declines precipitously between days 11 and 13, concomitant with the decrease in nonhemoglobin protein synthesis. Coelectrophoresis with reticulocyte polysomal mRNA demonstrates that this mRNA fraction is not an identifiable hemoglobin mRNA. Rather, it is suggested, it represents mRNA for one or more nonhemoglobin proteins. The mRNA for yolk sac hemoglobin is apparently produced, in a stable form, early in the course of yolk sac erythropoiesis, certainly prior to day 11. This interpretation is consistent with observations on yolk sac erythropoiesis in the chick as well (Miura and Wilt, 1971).

It has been suggested that under some circumstances the production of differentiated (e.g., specialized) proteins is restricted to nonproliferating cells (Ebert and Kaighn, 1966; Holtzer, 1970). This does not appear to be the case for yolk sac erythroid cells of the several species studied (de la Chapelle et al., 1969; Campbell et al., 1971; Maniatis and Ingram, 1971a) nor for certain other developmental systems (Kafatos and Feder, 1968; Davies et al., 1968). In yolk sac erythroblasts of the 11-day fetus, for example, at least 80 percent of the cells incorporate tritiated thymidine into DNA; all these cells contain histochemically demonstrable hemoglobin and are actively synthesizing this protein (de la Chapelle et al., 1969).

Yolk sac erythroblasts undergo about three cell divisions between the tenth and fourteenth gestational days. With each successive mitosis the rate of hemoglobin synthesis declines, although it continues, as a low level, past the final cell division until the fifteenth fetal day. Similar observations have been reported for the primitive (yolk sac) lineage of chick erythroblasts in which the declining rate of hemoglobin synthesis is accompanied by a progressive prolongation of the generation time for these differentiating erythroblasts (Hagopian and Ingram, 1971; Hagopian et al., 1972; Weintraub et al., 1971; Campbell et al., 1971). Taken together, these observations suggest to Holtzer (1970) that the final cell divisions in terminally differentiated cells such as erythroblasts differ from the self-renewing mitoses of precursor cells, in that the differentiating cell is irreversibly programmed both for the synthesis of its specialized proteins and for a finite and predetermined number of proliferative cell divisions. That cell division that marks the transition from self-renewal to terminal proliferation and differentiation is called a "*quantal*" mitosis (Holtzer, 1970). The precise moment for such a hypothetical "*quantal*" mitosis in yolk sac erythropoiesis has not been identified. Studies with metabolic inhibitors (Miura and Wilt, 1971; Hagopian et al., 1972) suggest that in the chick, at least, it would have to be a relatively early event.

Definitive Erythropoiesis and Synthesis of Adult Hemoglobin

Definitive erythroid cells in the mouse (as in all mammals) differ from erythrocytes of the yolk sac-derived primitive lineage with respect to cellular morphology, site of production, the type of hemoglobin synthesized, and the fact that this is a perpetuating or self-renewing population capable of sustained erythropoiesis throughout the adult life of the organism (Kovach et al., 1967; Craig and Russell, 1964; Rifkind et al., 1969; see also Bloom, 1938; Grasso et al., 1962; Sorenson, 1963; Zamboni, 1965, for studies of other species). Indeed, the transition from embryonic hemoglobin to hemoglobin of the definitive adult type is a consequence of the substitution of the adult (or definitive) erythroid cell lineage for the yolk

sac (primitive) population. This sequence appears to be the rule in all vertebrate species studied so far (Marks and Rifkind, 1972; Ingram, 1972).

In common with definitive erythrocytes of other mammalian species the definitive mouse erythrocyte is a nonnucleated cell, the progeny of nucleated erythroblasts that, in the mouse, make their first appearance in the fetal liver during the tenth gestational day (Rifkind et al., 1969; Moore and Metcalf, 1970). Definitive erythroid cell precursors are first recognized among those mesenchymal cells of the transverse septum adjacent to the cords of hepatic epithelial cells that are growing radially from the endodermal gut diverticulum at this stage of embryogenesis. It is of some interest that both the primitive (yolk sac) and definitive (adult, hepatic) erythroid cell precursors originate in a mesenchymal tissue in the immediate vicinity of an endodermal derivative. The precise nature of the influence of endodermal tissue on the differentiation of the erythropoietic cells has yet to be defined (Wilt, 1965). Whether the definitive erythroid cell line arises *de novo* or represents a colonization by precursors from sites of primitive erythropoiesis (i.e., yolk sac) is at present an active question (see below).

The definitive hepatic erythroid cell precursors observed on the tenth gestational day undergo both self-renewal and differentiation. Unlike the situation in the yolk sac, hepatic erythropoiesis is self-sustaining at least until the time of birth, when the liver is supplanted as the principal erythropoietic organ by spleen and bone marrow (Bloom, 1938; Djaldetti et al., 1972).

Differentiation in the liver (and in subsequent sites of definitive erythropoiesis) is characterized by a series of morphologic transformations that correlate closely with the biochemical features of erythropoietic development. The earliest recognizable precursor found in the liver on the late tenth and early eleventh fetal days may be arbitrarily termed a hemocytoblast (Rifkind et al., 1969). By the eleventh day, cells displaying clear features of erythroid cell differentiation (*proerythroblasts* and subsequent stages) are found within the epithelial cords of the fetal liver (Fig. 1). Cytochemical staining fails to reveal hemoglobin in the cytoplasm of these precursor cells, and there is little or no synthesis of globin by these cells (Terada et al., 1972). The proerythroblast nucleus contains principally noncondensed euchromatin, and the nucleolus is large and active in the synthesis of ribosomal RNA. The ribosomes and perhaps other nuclear RNA products of this cell type are transported to the cytoplasm and account for the intense basophilia characteristic of the next sequential stage of erythroid cell differentiation, the *basophilic erythroblast*. At this stage the nucleus first begins to show the heterochromatinization characteristic of maturing erythroblasts; the nucleolus begins to regress as RNA synthesis declines. Cytoplasmic hemoglobin is still undetectable, but, presumably, hemoglobin synthesis commences during this stage, and the first tinctorial evidence of accumulation of this protein (positive benzidine reaction) ushers in the *polychromatophilic* stage. Further nuclear condensation and reduction in the rate of RNA

synthesis occur during this phase, which is also the last stage capable of cell division. Hemoglobin synthesis is most active in cells of this type, although the concentration of cytoplasmic RNA (ribosomes and other) gradually decreases as cell divisions and accumulating hemoglobin produce dilution in the face of declining nuclear synthesis. The final nucleated stage, the *orthochromic* erythroblast, displays a highly pycnotic nucleus and intense cytoplasmic staining for hemoglobin. The nucleus is expelled along with a thin rim of cytoplasm (Bank et al., 1970) to form the anucleate reticulocyte. Reticulocytes enter the circulation by the twelfth fetal day and are produced continuously throughout the animal's life by differentiating erythroblasts in the liver, spleen, or bone marrow. Reticulocytes, lacking a nucleus, are incapable of cell division or RNA synthesis, whereas they continue to produce hemoglobin for a brief period (24 to 48 hours) by means of their residual content of polyribo-

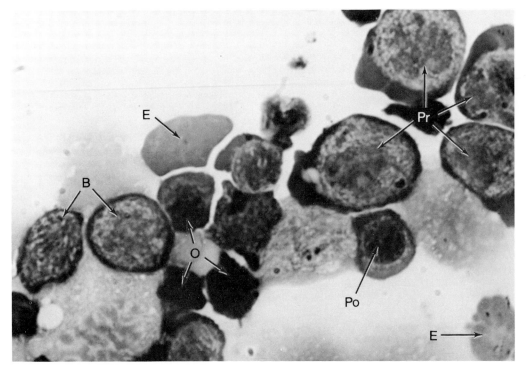

FIGURE 1. *Erythroid cell precursors obtained by disaggregation of 12-day fetal mouse livers. The cells are stained with benzidine (orange color indicates hemoglobin) and Wright-Giemsa. Proerythroblasts (Pr), basophilic erythroblasts (B), polychromatophilic erythroblasts (Po), orthochromic erythroblasts (O), and nonnucleated reticulocytes and erythrocytes (E) are observed.*

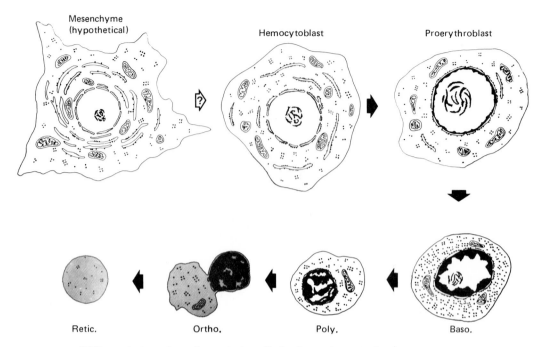

Mesenchyme (hypothetical) Hemocytoblast Proerythroblast

Retic. Ortho. Poly. Baso.

FIGURE 2. *Differentiation of erythropoietic cells in the embryonic liver and in all adult erythro-poietic tissues. The relationship between mesenchymal cells and the hemopoietic stem cell (hemocytoblast) is hypothetical. The morphologic features of each of the stages of erythroid cell development (proerythroblasts, basophilic, polychromatophilic, orthochromic erythroblasts, reticulocyte, and erythrocyte) are described in the text.*

somes (Marks et al., 1963; Rifkind et al., 1964). Cessation of hemoglobin synthesis and the disappearance of ribosomes define the transition from reticulocyte to mature erythrocyte. This morphogenetic sequence, which characterizes definitive erythropoiesis, is illustrated schematically in Fig. 2.

Protein and Nucleic Acid Synthesis in Definitive Erythroid Cells

It will be recalled that in the case of yolk sac erythropoiesis hemoglobin synthesis is directed by a relatively stable template, as demonstrated by resistance to inhibition by actinomycin D. Hepatic erythropoiesis has provided an opportunity to determine whether template stability is a feature of erythropoiesis at all stages of embryogenesis. These studies indicate that during hepatic erythropoiesis there is, indeed, a transition from an unstable to a stable hemoglobin-synthesizing apparatus (Fantoni et al., 1968; Djaldetti et al., 1970).

Erythroblast RNA synthesis is profoundly inhibited by actinomycin D on all gestational days. Nevertheless, whereas on fetal day 12 the antibiotic markedly inhibits hemoglobin synthesis, by day 13 the antibiotic has little or no effect on either hemoglobin synthesis or on the uptake of isotopic labeled leucine or iron by polychromatophilic or orthochromic erythroblasts. These observations suggest that fetal development is accompanied by a stabilization of the hemoglobin synthetic capacity. Although a change in the stability of globin mRNA could explain these observations, the molecular mechanisms that govern this transition and, indeed, the identity of the actinomycin-sensitive step in hemoglobin synthesis remain to be defined.

Hematopoietic Stem Cell and Fetal Erythropoiesis

It is apparent from the observations on fetal hepatic erythropoiesis, as well as from the physiology of erythropoietic homeostasis in the adult organism, that definitive erythropoiesis is a self-renewing as well as differentiating cell system. These dual capacities provide a strong theoretical argument for a stem cell compartment in erythropoiesis. Arguments developed elsewhere (McCulloch, 1970) may be adduced in support of a stem cell compartment that is a precursor for the three major hematopoietic cell lines. The existence of such a cell type, having the capacity for differentiation into erythropoietic, granulopoietic, and megakaryocytic cell lines, has now been definitely established by an extensive body of experimental evidence (McCulloch, 1970; Moore and Metcalf, 1970), although morphologic identity has yet to be definitively established (Niewisch et al., 1967; van Bekkum et al., 1971).

The process of differentiation of pluripotent stem cells into specific hemopoietic cells with a restricted capacity for differentiation (e.g., an erythroid precursor cell) implies acquisition of responsiveness to those selective physiological regulatory mechanisms characteristic of the differentiated cell lineage (e.g., erythropoietin, in the case of erythropoiesis, and less well-defined humoral factors in the case of granulocytopoiesis and platelet production). Substantial evidence now distinguishes the population of hematopoietic pluripotent stem cells (operationally termed *colony-forming cells*) from their progeny, which express the capacity for monopotent differentiation and responsiveness to physiologic stimuli (Bruce and McCulloch, 1964; Till et al., 1967; Fowler et al., 1967; Stephenson and Axelrad, 1971; Worton et al., 1969).

Critical to an understanding of factors implicated in the initiation and regulation of fetal erythropoiesis is the nature and source of fetal hematopoietic stem cells. Although yolk sac erythropoiesis (there is as yet no evidence for either granulopoiesis or megakaryocytopoiesis in the yolk sac) is not self-renewing, there is conflicting evidence as to the presence, at least transiently, of pluripotent stem cells in that organ. Employing both the *in vivo* spleen colony assay for pluripotent stem cells and the *in vitro* agar colony technique for the assay of granulopoietic precursors, Moore and Metcalf (1970) report detection of small numbers of stem cells in the fetal mouse yolk sac. Absence of granulopoietic or megakaryocytic differentiation in this site is attributed to an as yet undefined nonsupportive microenvironment in the yolk sac. Matioli and coworkers (1968), on the other hand, were unable to detect stem cells with the spleen colony assay. The implications of a clear resolution of this problem are considerable. Moore and Metcalf (1970) speculate, on the basis of colony assays and organ cultures, that yolk sac stem cells colonize the developing fetal liver and thereby provide the precursors not only for primitive erythropoiesis and embryonic hemoglobin synthesis but for definitive (hepatic and subsequent) erythropoiesis and adult hemoglobin production as well.

Erythropoietin and Regulation of Erythropoiesis

Normal homeostasis of erythrocyte production in response to physiologic demands for delivery of oxygen to the tissues is maintained through the mediation of the hormone erythropoietin—synthesized, released, or activated in response to tissue hypoxia at oxygen-sensitive sites as yet only tentatively identified (Gordon and Zanjani, 1970). Pivotal to an understanding of this regulatory process is elucidation of the nature of the erythropoietin-sensitive erythroid cell (target-cell) and of the cellular and molecular events that characterize the response to the hormone. Although cells of the adult spleen and marrow are capable of responding to erythropoietin with accelerated erythropoiesis, the fetal mouse has provided a cell system that has proved especially valuable in the pursuit of some of these mechanisms. Although it is as yet uncertain whether yolk sac erythropoiesis is responsive to erythropoietin (Cole and Paul, 1966; Bateman and Cole, 1971), fetal hepatic erythroid cells demonstrate unequivocal sensitivity to the hormone (Cole and Paul, 1966; Chui et al., 1971). The primary effect of erythropoietin is exerted on an immature erythroid precursor (the erythropoietin-sensitive cell), which is induced to multiply, to synthesize RNA species, and to produce hemoglobin in response to the hormone (Filmanowicz and Gurney, 1961; Hodgson, 1970; McCool et al., 1970; Cantor et al., 1972; Djaldetti et al., 1972b). The possibility that the hormone may accelerate the rate of erythropoiesis among already differentiating and hemoglobin-producing erythroblasts has been proposed (Stohlman, 1968; Borsook et al., 1968), but the best available data suggest that such an effect makes a relatively minor contribution to the overall erythropoietin-mediated stimulation of erythropoiesis (McCool et al., 1970; Chui et al., 1971), if it makes one at all.

The fetal hepatic erythroid cell population on the twelfth and thirteenth gestational day contains a high proportion of immature precursors and for this reason appears uniquely attractive for studies designed to identify and characterize the erythropoietin-sensitive cell. By means of immunological techniques, Cantor and colleagues (1972) have concentrated a population of precursor cells from this source and demonstrated their responsiveness to erythropoietin. In particular, the hormone appears to mediate replication of the immature precursors and the differentiation of a portion of these into hemoglobin-producing cells.

The molecular responses of erythropoietin-responsive cells to the hormone have yet to be fully explored. Both in marrow (Krantz and Goldwasser, 1965) and in erythroid cells of the fetal liver (Djaldetti et al., 1972; Paul and Hunter, 1969), the earliest response to erythropoietin detected at present is an increase in the rate of RNA synthesis. The full significance of this erythropoietin-stimulated RNA synthesis is not yet understood. Evidence for stimulation of a variety of size classes of RNA has been reported (Gross and Goldwasser, 1969). Employing purified preparations of hormone-responsive cells, Terada and coworkers (1972) have demonstrated that unstimulated precursors contain little or no globin mRNA activity and are virtually inactive in globin synthesis. Within 10 hours of incubation with erythropoietin, globin mRNA activity can be detected and

hemoglobin synthesis is initiated. These observations suggest that biologically active globin mRNA is not available for translation prior to stimulation by erythropoietin and that globin mRNA may prove to be at least a component of the RNA produced in response to the hormone. The nature of other RNA species produced and their relationship to other features of erythroid cell differentiation, such as the appearance of erythrocyte cell surface components (Dukes et al., 1964; Minio et al., 1972), the synthesis of nucleoproteins (Malpoix, 1971), and the accelerated proliferation of erythroid elements, have yet to be elucidated. The availability of a number of highly sensitive methods for biological assay (Mathews et al., 1971; Housman et al., 1971; Forget and Benz, 1971; Nienhuis and Anderson, 1971; Gurdon et al., 1971; Metafora et al., 1972) and molecular assay by hybridization (Kacian et al., 1972; Verma et al., 1972; Ross et al., 1972a,b) of mRNA species suggests that resolution of these problems may be anticipated in the not-too-distant future.

On the basis of accumulated evidence, a tentative and hypothetical model suitable for the design of future experimental studies can be proposed. Erythropoietin acts upon an erythroid precursor, itself the differentiated progeny of a pluripotent hemopoietic stem cell. This level of differentiation must include the presence of hormone-receptor sites (as yet undefined) and the program of molecular events to be triggered by the hormone itself. Erythropoietin induces the transition from this hormone-responsive state to the stage of the fully committed erythroblast capable of hemoglobin synthesis and of a restricted number of proliferative but terminal cell divisions terminating in enucleation and the release of reticulocytes and erythrocytes into the circulation.

Summary and Conclusions

Several important aspects of both cell differentiation and of embryogenesis are studied to advantage employing the erythroid cell system and, in particular, erythropoiesis in the fetal mouse. The change in pattern of hemoglobin produced during embryogenesis, found in all higher vertebrates so far studied, is shown to be the result of substitution of a primitive erythroid cell lineage by a definitive cell line. The characteristic template stability of hemoglobin synthesis is a developmental feature, but the molecular basis of the switch from unstable to stable globin synthesis remains obscure. Hemoglobin (a highly specialized protein) is synthesized in cells capable of cell division yet programmed for a severely restricted number of divisions (terminal differentiation). The rate of erythropoiesis is regulated by a hormone, erythropoietin, responsive to body oxygen requirements. The hormone target cell is an immature precursor that responds by initiating a program of events that includes the synthesis of hemoglobin and the proliferation of erythroblasts. A pluripotent hemopoietic stem cell feeds this hormone-responsive compartment, which is itself capable of some degree of self-renewal.

References

ATTFIELD, M. (1951). Inherited macrocytic anemias in the house mouse. III. Red blood cell diameter. J. Genet. 50, 250–263.

BANK, A., RIFKIND, R. A., AND MARKS, P. A. (1970). Regulation of globin synthesis. In "Regulation of Hematopoiesis" (A. S. Gordon, ed.), Vol. 1, pp. 701–729. Appleton-Century-Crofts, New York.

BATEMAN, A. E., AND COLE, R. J. (1971). Stimulation of haem synthesis by erythropoietin in mouse yolk-sac-stage embryonic cells. J. Embryol. Exp. Morphol. 26, 475–480.

BLOOM, W. (1938). Embryogenesis of mammalian blood. In "Handbook of Hematology" (H. Downey, ed.), Vol. 2, pp. 865–922. Harper & Row, New York.

BORSOOK, H., RATNER, K., TATTRIE, B., AND TIEGLER, D. (1968). Effect of erythropoietin in vitro which stimulates that of a massive dose in vivo. Nature (London) 217, 1024–1026.

BRUCE, W. R., AND MCCULLOCH, E. A. (1964). The effect of erythropoietin stimulation on the hemopoietic colony-forming cells of mice. Blood 23, 216–232.

CAMISCOLI, J. F., AND GORDON, A. S. (1970). Bioassay and standardization of erythropoietin. In "Regulation of Hematopoiesis" (A. S. Gordon, ed.), Vol. 1, pp. 369–393. Appleton-Century-Crofts, New York.

CAMPBELL, G. LE M., WEINTRAUB, H., MAYALL, B. H., AND HOLTZER, H. (1971). Primitive erythropoiesis in early chick embryogenesis. II. Correlation between hemoglobin synthesis and the mitotic history. J. Cell Biol. 50, 669–681.

CANTOR, L. N., MORRIS, A. J., MARKS, P. A., AND RIFKIND, R. A. (1972). Purification of erythropoietin-responsive cells by immune hemolysis. Proc. Nat. Acad. Sci. U.S.A. 69, 1337–1341.

CHUI, D. H. K., DJALDETTI, M., MARKS, P. A., AND RIFKIND, R. A. (1971). Erythropoietin effects on fetal mouse erythroid cells. I. Cell population and hemoglobin synthesis. J. Cell Biol. 51, 585–595.

COLE, R. J., AND PAUL, J. (1966). The effects of erythropoietin on haem synthesis in mouse yolk sac and cultured foetal liver cells. J. Embryol. Exp. Morphol. 15, 245–260.

CRAIG, M. L., AND RUSSELL, E. S. (1964). A developmental change in hemoglobins correlated with an embryonic cell population in the mouse. Develop. Biol. 10, 191–201.

DAVIES, L. M., PRIEST, J. H., AND PRIEST, R. E. (1968). Collagen synthesis by cells synchronously replicating DNA. Science 159, 91–93.

DE LA CHAPELLE, A., FANTONI, A., AND MARKS, P. A. (1969). Differentiation of mammalian somatic cells: DNA and hemoglobin synthesis in fetal mice. Proc. Nat. Acad. Sci. U.S.A. 63, 812.

DJALDETTI, M., CHUI, D., MARKS, P. A., AND RIFKIND, R. A. (1970). Erythroid cell development in fetal mice: stabilization of the hemoglobin synthetis capacity. J. Mol. Biol. 50, 345–358.

DJALDETTI, M., BESSLER, H., AND RIFKIND, R. A. (1972a). Hematopoiesis in the embryonic mouse spleen: an electron microscopic study. Blood 39, 826–841.

DJALDETTI, M., PREISLER, H., MARKS, P. A., AND RIFKIND, R. A. (1972b). Erythropoietin effects on fetal mouse erythroid cells. II. Nucleic acid synthesis and the erythropoietin-sensitive cell. J. Biol. Chem. 247, 731–735.

DUKES, P., TAKAKU, F., AND GOLDWASSER, E. (1964). In vitro studies of erythropoietin on ^{14}C-glucosamine incorporation into rat bone marrow cells. Endocrinology 74, 960–967.

EBERT, J. D., AND KAIGHN, M. E. (1966). In "Major Problems in Developmental Biology" (M. Locke, ed.), pp. 29–84. Academic Press, New York.

ERSLEV, A. J. (1953). Humoral regulation of red cell production. Blood 8, 349–357.

FANTONI, A., BANK, A., AND MARKS, P. A. (1967). Globin composition and synthesis of hemoglobins in developing fetal mice erythroid cells. Science 157, 1327–1329.

Fantoni, A., de la Chapelle, A., and Marks, P. A. (1968a). Synthesis of embryonic hemoglobins during erythroid cell development in fetal mice. *J. Biol. Chem. 244*, 675–681.

Fantoni, A., de la Chapelle, A., Rifkind, R. A., and Marks, P. A. (1968b). Erythroid cell development in fetal mice: synthetic capacity for different proteins. *J. Mol. Biol. 33*, 79–91.

Filmanowicz, E., and Gurney, C. W. (1961). Studies on erythropoiesis. XVI. Response to a single dose of erythropoietin in polycythemic mouse. *J. Lab. Clin. Med. 57*, 65–72.

Forget, B. G., and Benz, E. J., Jr. (1971). Messenger RNA for human globin synthesis in β thalassemia. *Blood 38*, 796.

Fowler, J. H., Till, J. E., McCulloch, E. A., and Siminovitch, L. (1967). The cellular basis for the defect in haematopoiesis in flexed-tailed mice. II. The specificity of the defect for erythropoiesis. *Brit. J. Haematol. 13*, 256–264.

Gordon, A. S. (1959). Hemopoietine. *Physiol. Rev. 39*, 1–40.

Gordon, A. S. (1970a). "Regulation of Hematopoiesis." Vol. 1. Appleton-Century-Crofts, New York.

Gordon, A. S. (1970b). "Regulation of Hematopoiesis." Vol. 2. Appleton-Century-Crofts, New York.

Gordon, A. S., Cooper, G. W., and Zanjani, E. D. (1967). The kidney and erythropoiesis. *Sem. Hematol, 4*, 337–358.

Gordon, A. S., and Zanjani, E. D. (1970). Some aspects of erythropoietin physiology. *In* "Regulation of Hematopoiesis" (A. S. Gordon, ed.), Vol. 1, pp. 413–457. Appleton-Century-Crofts, New York.

Grasso, J. A., Swift, H., and Ackerman, G. A. (1962). Observations on the development of erythrocytes in mammalian fetal liver. *J. Cell Biol. 14*, 235–254.

Gross, M., and Goldwasser, E. (1969). On the mechanism of erythropoietin-induced differentiation. V. Characterization of the ribonucleic acid formed as a result of erythropoietin action. *Biochemistry 8*, 1795–1805.

Gurdon, J. B., Lane, C. D., Woodland, H. R., and Marbaix, G. (1971). The use of frog eggs and oocytes for the study of messenger RNA and its translation in living cells. *Nature (London) 233*, 177–182.

Hagopian, H. K., and Ingram, V. M. (1971). Developmental changes of erythropoiesis in cultured chick blastoderms. *J. Cell Biol. 51*, 440–451.

Hagopian, H. K., Lippke, J. A., and Ingram, V. M. (1972). Erythropoietic cell-cultures from chick embryos. *J. Cell Biol. 54*, 98–106.

Harris, J. W., and Kellermeyer, R. W. (1970). "The Red Cell. Production, Metabolism, Destruction: Normal and Abnormal." Harvard University Press, Cambridge, Massachusetts.

Hodgson, G. S. (1970). Mechanism of action of erythropoietin. *In* "Regulation of Hematopoiesis" (A. S. Gordon, ed.), Vol. 1, pp. 459–469. Appleton-Century-Crofts, New York.

Holtzer, H. (1970). Proliferative and quantal cell cycles in the differentiation of muscle, cartilage, and red blood cells. *Symp. Int. Soc. Cell Biol. 9*, 69–88.

Housman, D., Pemberton, R., and Taber, R. (1971). Synthesis of α and β chains of rabbit hemoglobin in a cell-free extract from Krebs II ascites cells. *Proc. Nat. Acad. Sci. U.S.A. 68*, 2716–2719.

Ingram, V. M. (1972). Embryonic red blood cell formation. *Nature (London) 235*, 338–339.

Jacobson, L. O., and Goldwasser, E. (1957). The dynamic equilibrium of erythropoiesis. *Brookhaven Symp. Biol. 10*, 110–131.

Kacian, D. L., Spiegelman, S., Bank, A., Terada, M., Metafora, S., Dow, L. W., and Marks, P. A. (1972). *In vitro* synthesis of DNA components of human genes for globins. *Nature New Biol. 235*, 167–169.

Kafatos, F. C., and Feder, N. (1968). Cytodifferentiation during insect metamorphosis: the galea of silkmoths. *Science 161*, 470–472.

Kafatos, F. C., and Reich, J. (1968). Stability of differentiation-specific and nonspecific messenger RNA in insect cells. *Proc. Nat. Acad. Sci. U.S.A. 60*, 1458–1465.

Kovach, J. S., Marks, P. A., Russell, E. S., and Epler, H. (1967). Erythroid cell development in fetal mice: ultrastructural characteristics and hemoglobin synthesis. *J. Mol. Biol. 25*, 131–142.

Krantz, S. B., and Goldwasser, E. (1965). On the mechanism of erythropoietin-induced differentiation. II. The effect on RNA synthesis. *Biochim. Biophys. Acta 103*, 325–332.

Kuratowska, Z., Lewartowski, B., and Michalak, E. (1960). Studies on the production of erythropoietin by the isolated hypoxic kidney. *Bull. Acad. Polon. Sci. Ser. Biol. 8*, 77.

Lillie, F. R. (1908). "Development in the Chick," p. 117. Holt, New York.

Lowy, P. H. (1970). Preparation and chemistry of erythropoietin. *In* "Regulation of Hematopoiesis" (A. S. Gordon, ed.), pp. 395–412. Appleton-Century-Crofts, New York.

MALPOIX, P. J. (1971). Stimulation by erythropoietin of histone and nonhistone chromatin protein synthesis in disaggregated foetal mouse liver. *Exp. Cell Res. 65*, 393–400.

MANIATIS, G. M., AND INGRAM, V. M. (1971a). Erythropoiesis during amphibian metamorphosis. I. Site of maturation of erythrocytes in *Rana Catesbeiana. J. Cell Biol. 49*, 372–379.

MANIATIS, G. M., AND INGRAM, V. M. (1971b). Erythropoiesis during amphibian metamorphosis. III. Immunochemical detection of tadpole and frog hemoglobins (*Rana Catesbeiana*) in single erythrocytes. *J. Cell Biol. 49*, 390–404.

MARKS, P. A., AND RIFKIND, R. A. (1972). Protein synthesis: its control in erythropoiesis. *Science 175*, 955–961.

MARKS, P. A., RIFKIND, R. A., AND DANON, D. (1963). Polyribosomes and protein synthesis during reticulocyte maturation *in vitro. Proc. Nat. Acad. Sci. U.S.A. 50*, 336–342.

MATHEWS, M. B., OSBORN, M., AND LINGREL, J. (1971). Translation of globin messenger RNA in a heterologous cell-free system. *Nature (London) 233*, 206–208.

MATIOLI, G. T., NIEWISCH, H., AND VOGEL, H. (1968). Distribution of hematologically competent stem cells in tissues of mice (fetus and newborn). *Fed. Proc. 27*, 672.

McCOOL, D., MILLER, R. J., PAINTER, R. H., AND BRUCE, W. R. (1970). Erythropoietin sensitivity of rat bone marrow cells separated by velocity sedimentation. *Cell Tissue Kinet. 3*, 55–65.

McCULLOCH, E. A. (1970). Control of hematopoiesis at the cellular level. *In* "Regulation of Hematopoiesis" (A. S. Gordon, ed.), pp. 133–159. Appleton-Century-Crofts, New York.

METAFORA, S., TERADA, M., DOW, L. W., MARKS, P. A., AND BANK, A. (1972). Increased efficiency of exogenous messenger RNA translation in a Krebs ascites cell lysate. *Proc. Nat. Acad. Sci. U.S.A. 69*, 1299–1303.

MINIO, F., HOWE, C., HSU, K. C., AND RIFKIND, R. A. (1972). Antigen density on differentiating erythroid cells. *Nature New Biol. 237*, 187–188.

MIURA, Y., AND WILT, F. H. (1971). The effects of 5-bromodeoxyuridine on yolk sac erythropoiesis in the chick embryo. *J. Cell Biol. 48*, 523–532.

MOORE, M. A. S., AND METCALF, D. (1970). Ontogeny of the haemopoietic system: yolk sac origin of *in vivo* and *in vitro* colony-forming cells in the developing mouse embryo. *Brit. J. Haematol. 18*, 279–296.

NIENHUIS, A. W., AND ANDERSON, W. F. (1971). Isolation and translation of hemoglobin messenger RNA from thalassemia, sickle cell anemia and normal human reticulocytes. *J. Clin. Invest. 50*, 2458–2460.

NIEWISCH, H., VOGEL, H., AND MATIOLI, G. (1967). Concentration, quantitation, and identification of hemopoietic stem cells. *Proc. Nat. Acad. Sci. U.S.A. 58*, 2261–2267.

OSKI, F. A., AND GOTTLIER, A. J., (1971). The interrelationships between red blood cell metabolites, hemoglobin, and the oxygen-equilibrium curve. *In* "Progress in Hematology" (E. B. Brown, and C. V. Moore, eds.), Vol. 7, pp. 33–67. Grune & Stratton, New York.

PAUL, J., AND HUNTER, J. A. (1969). Synthesis of macromolecules during induction of haemoglobin synthesis by erythropoietin. *J. Mol. Biol. 42*, 31–41.

REEDER, R., AND BELL, E. (1967). Protein synthesis in embryonic chick lens cells. *J. Mol. Biol. 23*, 577–585.

REISSMANN, K. R. (1950). Studies on the mechanism of erythropoietic stimulation in parabiotic rats during hypoxia. *Blood 5*, 372–380.

RIFKIND, R. A. (1966). Destruction of injured red cells *in vivo. Amer. J. Med. 41*, 711–723.

RIFKIND, R. A., CHUI, D., AND EPLER, H. (1969). An ultrastructural study of early morphogenetic events during the establishment of fetal hepatic erythropoiesis. *J. Cell Biol. 40*, 343–365.

RIFKIND, R. A., DANON, D., AND MARKS, P. A. (1964). Alterations in polyribosomes during erythroid cell maturation. *J. Cell Biol. 22*, 599–611.

ROSS, J., AVIV, H., SCOLNICK, E., AND LEDER, P. (1972a). *In vitro* synthesis of DNA complementary to purified rabbit globin mRNA. *Proc. Nat. Acad. Sci. U.S.A. 69*, 264–268.

ROSS, J., IKAWA, Y., AND LEDER, P. (1972b). Globin messenger-RNA induction during erythroid differentiation of cultured leukemia cells. *Proc. Nat. Acad. Sci. U.S.A. 69*, 3620–3623.

SORENSON, G. D. (1963). Hepatic hemocytopoiesis in the fetal rabbit: a light and electron microscopic study. *Ann. N. Y. Acad. Sci. 111*, 45–69.

STEPHENSON, J. R., AND AXELRAD, A. A. (1971). Separation of erythropoietin-sensitive cells from hemopoietic spleen colony-forming stem cells of mouse fetal liver by unit gravity sedimentation. *Blood 37*, 417–427.

STEWART, J. A., AND PAPACONSTANTINOU, J. (1967). Stabilization of mRNA templates in bovine lens epithelial cells. *J. Mol. Biol. 29*, 357–370.

STOHLMAN, F., JR. (1968). Current concepts: the kidney and erythropoiesis. *New Eng. J. Med. 279*, 1437–1439.

TERADA, M., BANKS, J., AND MARKS, P. A. (1971). RNA synthesized during differentiation of yolk sac erythroid cells. *J. Mol. Biol. 62*, 347–360.

TERADA, M., CANTOR, L., METAFORA, S., RIFKIND, R. A., MARKS, P. A. AND BANK, A. (1972). Globin mRNA activity in erythroid precursor cells and the effect of erythropoietin. *Proc. Nat. Acad. Sci. U.S.A. 69*, 3575–3579.

TILL, J. E., SIMINOVITCH, L., AND McCULLOCH, E. A. (1967). The effect of plethora on growth and differentiation of normal hemopoietic colony-forming cells transplanted in mice of genotype W/W^v. *Blood 29*, 102–113.

VAN BEKKUM, D. W., VAN NOORD, M. J., AND DICKE, K. A. (1971). Attempts at identification of hemopoietic stem cells in mouse. *Blood 38*, 547–558.

VERMA, I. M., TEMPLE, G. F., FAN, H., AND BALTIMORE, D. (1972). *In vitro* synthesis of DNA complementary to rabbit reticulocyte 10S RNA. *Nature New Biol. 235*, 163–169.

WEINTRAUB, H., CAMPBELL, G. LE M., AND HOLTZER, H. (1971). Primitive erythropoiesis in early chick embryogenesis. I. Cell cycle kinetics and the control of cell division. *J. Cell Biol. 50*, 652–668.

WESSELLS, N. K. (1964). DNA synthesis, mitosis, and differentiation in pancreatic acinar cells *in vitro*. *J. Cell Biol. 20*, 415–433.

WILT, F. H. (1965). Erythropoiesis in the chick embryo: the role of endoderm. *Science 147*, 1588–1590.

WORTON, R. G., McCULLOCH, E. A., AND TILL, J. E. (1969). Physical separation of hemopoietic stem cells from cells forming colonies in culture. *J. Cell. Physiol. 74*, 171–182.

ZAMBONI, L. (1965). Electron microscopic study of blood embryogenesis in humans. II. The hemopoietic activity in the fetal liver. *J. Ultrastruct. Res. 12*, 525–541.

CHAPTER EIGHT

Aspects of Differentiation and Determination in Pigment Cells

J. R. WHITTAKER

Pigment cells and tissues have been a favored material in cell differentiation studies for many years because they possess a clearly visible marker of their differentiated condition—namely, the black or brown melanin pigment. There is also a great deal of information available about the biochemistry of melanin formation, and this has been useful in studying the acquisition of the melanotic phenotype in differentiating cells. In addition, the melanin marker allows one to examine the behavior of these cells in various natural and artificial interactions with other cells and tissues, and thereby to explore a whole range of other phenotypic properties of the cells.

Cellular differentiation is a sequential process consisting of many parallel intra- and intercellular events that ultimately lead to the production of unique structural and functional features of the cell. These unique features distinguish a cell from other kinds of cells that perform different or slightly different functions. Since no cell is ever completely unspecialized, differentiation must be regarded as the

general process leading to *new* specializations. There is no truly appropriate definition of differentiation other than a full explanation of the events that occur during it. Operationally, "differentiation" is regarded as the acquisition of cell or tissue products characteristic of the functional end state of the cell—usually proteins or other complex macromolecules. Unfortunately, this pragmatic viewpoint does much to impede the search for important earlier events in the history of the cell that undoubtedly specify the nature of the later changes.

At the present time, most differentiation research is focused on discovering the regulatory mechanisms controlling the expression of specific end products in the differentiating cell. Many of these end products can be fully characterized and the stages in their formation described. The prevailing opinion among investigators of differentiation is that working backward from selected phenotypic properties of cells will eventually lead to an uncovering of earlier events in the sequence and to the elaboration of mechanisms controlling these earlier events. Such optimism might prove un-justified, but there is currently no more suit-able way of approaching the problem of dif-ferentiation.

The pigment cells of animals are derived mainly from the embryonic neural ectoderm. This includes the black pigment cells found in the eyes of both vertebrates and invertebrates, since the eye is usually a neural derivative. The widely distributed peripheral body cells of var-ious animals, which make up the different black or brown body color patterns, are also ectodermal in origin. In vertebrates, these pe-ripheral cells, called *melanocytes*, migrate from the neural crest region of the closing neural tube. Most of the discussion in this chapter concerns the expression of melanotic pheno-type in developing melanocytes and in the retinal pigment epithelium (*tapetum nigrum*) of vertebrates. Some studies of the pigment cells in the brain of ascidian (protochordate)

larvae are also described, since the develop-ment of these giant cells illustrates many gen-eral problems and mechanisms of pigment cell differentiation.

The Melanotic System

Brown or black melanin pigments (called (*eumelanins*) occur within the vertebrate mel-anocyte cytoplasm exclusively in small organ-elles, the melanosomes. These melanosomes be-come sufficiently numerous to give a distinct coloration even to isolated single cells. The eumelanins are formed in the melanosome from the oxidation of the amino acid L-tyrosine by the enzyme tyrosinase. A number of interme-diate oxidation products occur (Fig. 1), but several of these, the quinones, are highly reac-

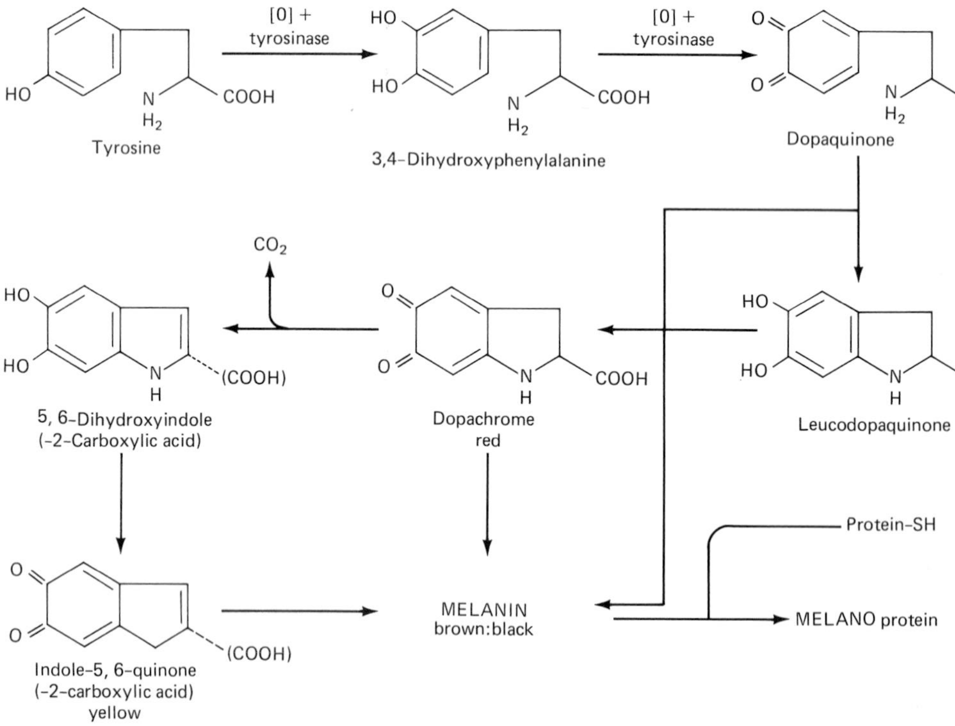

FIGURE 1. *Steps in the Raper–Mason pathway for the conversion of tyrosine to melanin. (Mod-ified according to K. Hempel, in "Structure and Control of the Melanocyte" (G. Della Porta and O. Muhlbock, eds.). pp. 162–175, Springer-Verlag, New York, 1966, and R. A. Nicolaus, "Melanins," Hermann, Paris, 1968.)*

tive and appear to condense directly to form a melanin polymer. The melanin polymer is apparently a random arrangement of the quinones into long chains with complex cross-linkages, possibly with some loss of parts of the aromatic ring from a few of the quinone subunits in the chain (Hempel, 1966; Nicolaus, 1968).

Some kinds of pigment cells synthesize a red or yellow form of melanin, which has been called *pheomelanin*. It is also produced in the melanosomes through the action of tyrosinase, and probably involves a secondary reaction of the quinone subunits with cysteine or some other sulfhydryl compound (Misuraca et al., 1969; Whittaker, 1971).

Tyrosinase (*o*-diphenol: oxygen oxidoreductase) seems to be the only enzyme necessary for melanin synthesis, since tyrosinase *in vitro* with L-tyrosine or the first intermediate L-3,4-dihydroxyphenylalanine (L-dopa), will readily produce a melanin. Other enzymes might be involved in the melanosomes *in vivo*, but the rapid rates of the reactions make it most difficult to identify these enzymes, and none are yet known. The instability and reactivity of the Raper–Mason intermediates (Fig. 1) render it unlikely that significant pools of these occur in embryonic cells; there is no experimental evidence for such pools (Whittaker, 1971). Melanogenesis is probably not regulated at the level of any of these intermediates. The key elements are apparently the presence of tyrosinase and the accessibility of tyrosine. It is generally believed that the conversion of L-tyrosine to L-dopa is the rate-limiting reaction in melanogenesis (Lerner and Fitzpatrick, 1950), although this point has never actually been established *in vivo*.

Recent investigations of vertebrate tyrosinases (e.g., Holstein et al., 1971) illustrate that there are three electrophoretically distinct forms of the enzyme in most animals. All three forms have been purified from mouse melanoma by Burnett (1971) and have been shown to act on both L-tyrosine and L-dopa. Agents that will dissociate proteins into their separate polypeptide chains failed to dissociate any of these tyrosinases (Holstein et al., 1971), thereby suggesting that tyrosinase does not consist of monomeric subunits stabilized by noncovalent bonds, as reported for lactic dehydrogenase and various other isozymes.

The C locus in the mouse probably contains the structural gene for tyrosinase (Wolfe and Coleman, 1966; Holstein et al., 1971), and possibly the multiple forms of tyrosinase result from secondary modifications of the product of the C locus. The expression of tyrosinase activity is subject to complex genetic influences; alleles at various other loci dramatically affect the occurrence and activity of the multiple forms of tyrosinase. Multiple forms of tyrosinase in mammals may result from the union of tyrosinase produced by the C locus, with various structural polypeptides specified by other gene loci (Holstein et al., 1971). However, in chick retinal pigment epithelium, tyrosinase occurs as a single monomer; electrophoretic variants seem to be simple aggregates of this monomer (Doezema, 1973a).

At the time of melanogenesis, tyrosinase functions exclusively within the melanosomes. The sequence of events involved in the formation of the oblong melanosomes of the mouse retinal pigment epithelium has been described by Moyer (1966). Thin "unit fibers" (35 Å in diameter) aggregate to form "compound fibers" (approximately 90 Å in diameter), which in turn aggregate in parallel array into a cross-linked matrix; the whole structure is membrane-bounded. The final matrix on which melanin is deposited appears to be a lattice of parallel compound fibers that are spaced approximately 100 Å apart and cross-linked by other fibers approximately 35 Å in diameter. Cross sections of melanin granules show that the flat lattice-like sheet is rolled concentrically in a loose cylinder to give the mature granule a three-dimensional form. Melanin is deposited on the matrix until all space in the granule is obscured by electron-opaque melanin. The early granule stages, prior to melanization, are called *premelanosomes*. Premelanosome formation in other species follows a similar sequence.

Studies of pigmentation mutants by Moyer (1966), Holstein et al. (1971), and others suggests indirectly that a number of different proteins are probably involved in melanogenesis. Doezema (1973b) has shown directly that isolated chick and mouse melanosomes contain 6 different major polypeptide components and about 15 minor ones. Many of these are undoubtedly part of the melanosome matrix, but others (in addition to tyrosinase) may be involved in melanin synthesis. Melanin has long been thought to occur as a melanoprotein complex within the pigment granules; this complex may result from a secondary coupling of melanin to granule proteins via their sulfhydryl groups (Nicolaus, 1968). Specific *melano*proteins might be real entities but strong evidence suggests that melanin synthesis per se is not immediately dependent on a coordinate synthesis of proteins (Whittaker, 1966; Kitano and Hu, 1971), and some kinds of melanin granules seem to have no significant amount of protein associated with the melanin (Whittaker, 1966).

Initiation of Melanotic Expression

The various proteins involved in melanogenesis, including tyrosinase, are made from mRNA templates transcribed from the pigment cell genome, but essentially nothing is yet known about these genetic control mechanisms. Some initial progress in understanding this control has been made from cell hybridization studies. Cells treated with inactivated Sendai virus will fuse in cell culture to form cells with "hybrid" nuclei. When cells from melanoma cell lines are hybridized with cells from a nonmelanotic line of fibroblasts, the resulting cells with hybrid nuclei and both "parental" sets of chromosomes, lose their melanotic expression and their tyrosinase activity (Davidson et al., 1966; Silagi, 1967). There is some kind of negative control exerted by the nonmelanotically differentiated cell, perhaps in the form of a repressor molecule that inactivates the genes for tyrosinase and melanogenesis. Whatever the nature of this control, it seems to be influenced by gene dosage, since about half the hybrid cells remain pigmented when tetraploid (4N) melanoma cells are hybridized with normal fibroblasts (Davidson, 1972), and there is an extra pair of pigmentation genes in the resulting hybrid cells. This suggests an interaction with the genome of some quantitatively limited chemical substance produced by the fibroblast cell.

Since tyrosinase can be readily identified at both the cellular and subcellular levels by histochemical means, one way of investigating the acquisition of the melanotic program is to see when tyrosinase activity first appears in the developmental time sequence. There are a number of examples that illustrate the occurrence of tyrosinase some time before the pigment cell actually begins to produce melanin. One of these examples concerns the two giant pigment cells of the ascidian larva. Ascidian larval development (*Ciona intestinalis*) is quite rapid (18 hours from egg to swimming larva at 18°C). The two pigment cells, the otolith and ocellus pigment cell, become melanized at 12 and 15 hours of development, respectively, but the first histochemically detectable tyrosinase in these cells occurs at 9 hours of development, when the embryo is in a late neurula stage (Whittaker, 1973a). Another example of this time difference occurs in the chick retinal pigment epithelium. Ide (1972) has shown that tyrosinase first occurs quite early in the development of the eye, at embryonic stage 16 (incubation age: 51 to 56 hours), but the first traces of melanin synthesis do not occur until stages 19 to 20 (age about 72 to 84 hours), at least 24 hours later.

Recent studies of melanosome formation using a histochemical reaction for tyrosinase (dopa oxidase) at the ultrastructural level have indicated a dual origin for these organelles. Premelanosomes and their matrices are formed from dilations of the rough endoplasmic reticulum, but tyrosinase originates in the Golgi apparatus and is then transferred to the premelanosomes from Golgi vesicles that fuse with the premelanosomes (Maul and Brumbaugh, 1971), or through tubules directly connecting the Golgi apparatus and the premelanosomes (Eppig and Dumont, 1972). The time lag between first occurrence of tyrosinase activity and subsequent melanin synthesis is not caused by a difference in the times at which tyrosinase and the premelanosomes are formed. Ide (1972) has found that tyrosinase-containing vesicles and premelanosomes both occur in stage-16 chick embryos.

The different components of the melanogenic system seem to be so closely integrated in their formation that the early occurrence of tyrosinase activity is probably indicative of the presence of the entire system. This is suggested not only by Ide's (1972) observation but also by experiments with the thymidine analogue, 5-

bromodeoxyuridine (BrdU). BrdU preferentially suppresses the differentiated functions of many different cell types, including melanoma cell lines and cultured chick retinal pigment cells, without adversely affecting the short-term growth and metabolism of the cells. It seems to do this by becoming incorporated into the cellular DNA in place of thymidine, and the effect is reversible after one or two cell divisions in the absence of BrdU. In one example by Wrathall et al. (1973), loss of tyrosinase activity from cultured melanoma cells was accompanied by a coordinate loss of premelanosome structures.

Experiments with ascidian embryos show that all the necessary protein components of melanogenesis are present long before melanin is synthesized. However, it is not known whether synthesis of all the proteins begins simultaneously with tyrosinase synthesis. When tyrosinase-containing cells (at $10\frac{1}{2}$ hours of development) were treated with puromycin, an inhibitor of protein synthesis, small melanized granules still developed in the otolith cell (Whittaker, 1973a).

In the ascidian system one can also get additional information on how the initiation of melanin synthesis is regulated by using phenylthiourea, an inhibitor of tyrosinase activity (Whittaker, 1966). In the absence of functional tyrosinase (i.e., in embryos exposed to phenylthiourea), the pigment cells continue to differentiate and the pigment granule of the otolith

cell is observed to accumulate both tyrosinase and substrate (L-tyrosine) in a large granule-sized vacuole (Fig. 2). This tyrosine accumulation is indicative of a mechanism that makes substrate available to the enzyme at a given time and that begins to function at the appropriate time whether or not melanin is actually being made. There is reasonable evidence that the tyrosinase is fully functional prior to melanogenesis but that it does not initially have access to substrate (Whittaker, 1973a).

However, there are biochemical events related to melanogenesis that occur even earlier in embryonic development than the first synthesis of tyrosinase. Actinomycin D experiments have shown that there is a period during ascidian (*Ciona*) embryogenesis between 6 and 7 hours of development when RNA related to tyrosinase synthesis is made (Whittaker, 1973b). The antibiotic actinomycin D is an effective *in vivo* inhibitor of RNA synthesis. Presumably the messenger RNA (mRNA) for tyrosinase and other melanogenic components is synthesized during this late gastrula stage of development, since *Ciona* embryos treated continuously with actinomycin D after 7 hours of development will synthesize tyrosinase and eventually some melanin as well. Embryos exposed to actinomycin D from 6 hours or earlier do not synthesize tyrosinase.

Recent studies by Johnson and Pastan (1972) have shown that adenosine 3',5'-cyclic monophosphate induces an increased rate of melanin

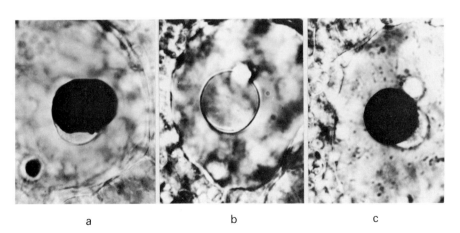

a   b   c

FIGURE 2. *Otolith pigment cells of ascidian*, Styela partita, *larvae. (a) Control. (b) Otolith cell of a larva reared in 0.35 mM phenylthiourea. (c) Otolith cell of a larva reared to hatching in phenylthiourea and then allowed to become pigmented in inhibitor-free seawater.* × 850.

synthesis in mouse melanoma cells, and a number of earlier investigations have indicated that certain peptide hormones can likewise stimulate melanogenesis in melanocytes and melanoma cells. There is undoubtedly a level of control over melanotic expression that is responsive to essentially organismal demands.

In the differentiation of pigment cells, and indeed of other cell types as well, one is faced with a perplexing series of regulatory events at different levels that control the expression of phenotypic characters. A decision is made at some point in early development to transcribe particular parts of the genome. Once a mRNA is transcribed from the genome this RNA may not be translated into protein for hours or perhaps even days afterward; nevertheless, translation occurs at a very specific later time in development. Even after certain histospecific proteins are present in the cytoplasm, additional time-dependent regulatory mechanisms govern when and if final expression of the phenotype occurs. The full range of possibilities is only just beginning to be discovered. One of the most appropriate aims of research in cellular differentiation is still a *description* of what the regulatory problems are and what, in fact, is being regulated.

Terminal Adjustment of Melanogenesis

Another interesting regulatory problem is the ability of embryonic pigment cells to shut off their melanin synthesis when some predetermined amount of melanin has been made. This limitation of phenotypic expression is particularly evident in the development of the ascidian otolith pigment cell (Fig. 2), where the melanin body stops increasing in volume after it reaches a certain definite size (Whittaker, 1966). The size regulation in this cell is almost certainly controlled primarily by the tyrosine accumulation mechanism, which begins to operate at the time of initial melanin synthesis, although this in turn is probably regulated by granule membrane permeability changes. Embryos developing in phenylthiourea produce a nonpigmented vacuole in the otolith cell (Fig. 2) that is almost exactly the same size as the normal pigmented otolith granule. Puromycin studies in this system (Whittaker, 1966) suggest that tyrosinase is probably not synthesized during the later part of granulogenesis.

Active melanogenesis in the chick retinal pigment epithelium also terminates during embryogenesis. After the twelfth day of incubation the cells have less demonstrable tyrosinase activity each successive day up to hatching (Miyamoto and Fitzpatrick, 1957; Doezema, 1973a), and they gradually cease producing melanin. Reduced tyrosinase synthesis is probably the primary event here, and the remaining enzyme is either gradually encapsulated in melanin (Seiji and Fitzpatrick, 1961) or eventually stops functioning because of a reaction-inactivation process. Even if the cells were producing no additional premelanosomes after day 12, one must assume that tyrosinase synthesis stops. Eppig and Dumont (1972) have shown in their studies of frog retinal pigment epithelium development that premelanosomes accept tyrosinase from the Golgi apparatus and that oocyte melanosomes still present in the eyes can also act as premelanosomes and accept additional tyrosinase. Accordingly, one must assume that mature melanosomes will go on accepting tyrosinase and synthesizing melanin as long as tyrosinase continues to be produced. In order to limit the extent of melanogenesis, pigment cells must cease making tyrosinase, stop the flow of tyrosine into the melanosomes, or possibly do both, depending on how rapidly the change must take place.

Pigment Cell Interactions: Pattern Formation

There are many studies of pigment cell behavior *in vitro* and *in situ* that could be cited to illustrate some of the complex cellular activities of pigment cells and that would appropriately be classified as differentiation of phenotype. One of the more interesting aspects of melanocyte activity is the generation of gross body pigment patterns, and the patterns on smaller epidermal structures such as scales, feathers, and hairs. In many cases, pigmentation patterns depend largely on variations in the movements of melanocytes, although sometimes these variations are imposed by the tissue environments in which the cells migrate. In some cases, the tissue environment appears to suppress the differentiation of melanocytes rather than their migration (Mayer, 1967).

Melanocyte movements in pattern organization are not directly dependent on melanotic expression, since melanocytes move to their final destinations during development even when melanotic expression is suppressed by an inhibitor (phenylthiourea) of tyrosinase activity (Lehman, 1957). For the sake of brevity, this discussion will be limited to a consideration of pigment cell autonomy in pattern formation and to several examples involving pattern formation that seem to be regulated by melanotic expression.

Melanocytes from various breeds of fowl reproduce the pigment pattern of the donor breed when grafted into nonpigmented host breeds. The details of the pattern are modified according to the particular feather and tract that the melanoblasts invade; yet the basic patterning seems to be an almost autonomous property of the pigment cells themselves rather than of the epidermal substratum in which they interact.

A most striking demonstration of this autonomy is Rawles' (1939) experiment in which she grafted pieces of skin from the posterolateral head region of robin (*Turdus migratorius*) embryos, containing neural crest melanoblasts, into the base of the wing bud of 72-hour White Leghorn chick embryos. The robin pigment cells proliferated and invaded adjacent chick feather germs and, at hatching, the chick hosts showed areas of colored down (pale cinnamon brown to blackish-brown) covering part or all of the wing and contiguous body regions. In one bird that attained sexual maturity, the colored down was replaced by juvenile contour feathers of mosaic coloration—the distal parts tawny, the proximal parts gray—identical in color to the breast feathers of the robin.

Nickerson (1944) investigated banding in the Barred Plymouth Rock and Silver Campine chicken breeds, which have alternate black pigmented and unpigmented (white) transverse bands throughout the feather vane. By grafting ectodermal melanoblasts from each breed to the wing bud of White Leghorn host embryos, he learned that the particular periodicity of the banding obtained in host feathers was characteristic of the donor breed. His most interesting observation, however, was that tissue isolates taken from white interband regions of the developing feather germ of either banded breed could develop pigmented melanocytes when grafted to the coelom or limb bud of White Leghorn embryos. Obviously, the banding pattern was not caused by a differential migration of the pigment cells, since the nonpigmented region of the developing feather germ also contained melanoblasts.

The pigment bands in a feather germ become differentiated in the nonkeratinized lower part of a small epidermal cylinder; the inner core of the cylinder is a dermal papilla from which all substances entering or leaving the developing feather parts must pass by diffusion. This cylindrical epidermal casing, devoid of circulation, seems likely to be an ideal structure for the establishment of some kind of diffusion gradient (Crick, 1970) to regulate melanotic expression in band formation. A diffusion gradient of some regulatory molecule might occur in this situation or possibly even a gradient of L-tyrosine.

Another interesting example of melanotic activity in producing banding patterns is the black-yellow-black pattern in agouti mouse hairs. This pattern is thought to be the result of a rapid shift from deposition of black pigment (eumelanin) to deposition of yellow pigment (pheomelanin) and back to black again in the hair shaft. Many investigators feel that there is only one population of melanocytes in the hair follicle and that the progeny of these cells alternate in the production of eumelanin and pheomelanin. Unlike the relative autonomy of the pigment cells in feather germs, the details of the agouti pattern are controlled by the genotype of the dermal (mesodermal) component of the hair follicle as shown by recombination skin grafts (Mayer and Fishbane, 1972).

In the light of the previous example, it is interesting that the pattern in agouti hairs can be modified experimentally by pigmentation substrates. Galbraith (1964) shifted the pattern of regenerating agouti mouse hairs in the direction of excessive eumelanin production by intradermal infusion with L-tyrosine or L-dopa at the time when pheomelanin production in the hair follicles was at its maximum (7 days after plucking out of the original hair). Also, glutathione, which might be one of the sulfhydryl compounds participating in normal pheomelanin production, can cause a pheomelanin phase in the black hair follicles of embryonic mouse skin explants that are exposed to glutathione in organ culture (Cleffman, 1963). These substrate effects suggest that a system involving some kind of substrate diffusion gradient might normally be operating in the agouti hair shaft. If so, the dermal component is probably directly involved in establishing the gradient.

Stability and Modulation

There are circumstances both natural and experimental under which pigment cells lose their melanotic phenotype. One naturally wonders how stable are certain differentiation pathways in cells, and whether anything important about the mechanisms controlling differentiation can be learned by investigating these stability phenomena. *In vitro* studies, especially, have illustrated how exceptionally stable are the programs for melanotic differentiation in pigment cells, even after the overt expression has disappeared.

Over the years, a number of investigators have studied cells of the retinal pigment epithelium and of pigmented iris from chick embryos in cell and tissue culture (see Whittaker, 1968). These cells are obtained free of other contaminating cell types with relative ease by simple surgical procedures. A common observation is that such cells, while rapidly growing in culture, fail to maintain their initially high level of melanotic activity (Fig. 3), and frequently (depending on the culture conditions) seem to lose their melanotic expression completely. In one example (Whittaker, 1967) of this so-called *dedifferentiation* in growing chick retinal pigment cells: (1) cell pigmentation and pigment-synthesizing components were diluted by growth, (2) the essential enzyme, tyro-

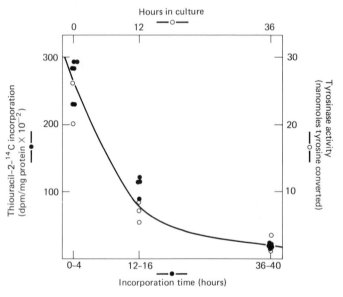

FIGURE 3. *Melanotic activity in cell cultures of 7-day chick embryo retinal pigment cells showing changes in activity up to 36 hours in culture. Rates of melanin biosynthesis were measured as thiouracil-2-¹⁴C incorporation by the live cells (Whittaker, 1971), and tyrosinase activity was measured in homogenates of the cultured cells on a per milligram of protein basis (Whittaker, 1967).*

sinase, underwent decay of activity, and (3) the synthesis of tyrosinase virtually stopped. All three activities occurred within a relatively short time, and can account for the rapid loss of melanotic expression depicted in Fig. 3.

An important observation in connection with these pigment cell culture experiments is that the loss of melanotic activity is essentially a modulation. Altered culture conditions, usually in the direction of diminished growth, will cause many of the cells to become strongly pigmented again. There is no evidence that the cells had truly lost their ability to express the melanotic phenotype; they have apparently only reduced the level of that expression. Modulation of this kind has also been observed by numerous investigators in cultures of established lines of melanoma cells (e.g., Oikawa et al., 1972), which are neoplastic melanocytes. A modulation can also occur *in situ* in some transplantable melanomas (Gray and Pierce, 1964). There is at least one example of normal melanocytes *in situ* undergoing modulation cycles of dedifferentiation and redifferentiation in the mammalian hair follicle germ (Silver et al., 1969). Pigment cell modulations are not solely a tissue-culture phenomenon.

It is possible to rule out least one potential explanation of the stability of phenotypic potential. There is no irreversible alteration of the genome nor any selective loss of genetic material. Nuclei from differentiated cells appear to be totipotent, and in the frog *Xenopus laevis* can support total development when transplanted into the activated ovum. This has recently been verified in part using nuclei of *Xenopus* melanocytes (Kobel et al., 1973). The nuclei of melanophores from tissue cultures set up from hatching tadpoles proved capable of supporting development as far as the heartbeat larval stage. This stage has a great diversity of differentiated tissue types.

Metaplasia and Determination

Regeneration experiments with vertebrate eye tissues also clearly illustrate that pigment cells of the retinal pigment epithelium and iris are capable of dedifferentiating. However, these melanotically dedifferentiated pigment cells can also develop into other histospecific cell types. There are many examples in fishes, amphibians, birds, and mammals in which retinal pigment cells of the embryo or adult (in urodele amphibians) can become depigmented and differentiate into neural (sensory) retina tissues of the eye to replace tissue that was surgically removed (see Coulombre and Coulombre, 1965). The classic example of such transformation is the case of Wolffian regeneration in larval and adult urodeles, where pigment cells from the dorsal part of the iris will dedifferentiate and, after proliferation, develop into a replacement lens for one that has been surgically removed (Yamada, 1967).

Various investigators, including Dumont and Yamada (1972), have examined the ultrastructural sequence of dedifferentiation in the iris pigment cells; a substantial part of the cell surface and cytoplasmic matrix is lost during an active extrusion of the melanosomes. The physical mechanism here is quite different from that which occurs during the modulation of retinal pigment cells in culture where melanosomes are not actively extruded from the cells (Whittaker, 1967). There is another interesting difference. Modulating cells shut down their synthesis of tyrosinase. Lens-regenerating dorsal iris tissues have higher tyrosinase activity than normal iris, and, in addition, the early lens regenerate, itself, has an even higher level of tyrosinase activity (Achazi and Yamada, 1972). These same authors have observed that normal lenses developing in frog embryos from head ectoderm have relatively high tyrosinase activity in the early stages of lens formation.

Pathologists originally coined the word *metaplasia* to refer to unusual or unexpected differentiative changes, particularly in tumors where these changes are not uncommon. One wonders if the pigment cell metaplasias noted above involve very far-reaching reprogramming of the pigment cells' differentiation potential. The older literature in experimental embryology used the term *determination* to describe the early restrictions or channeling of developmental potential that embryonic cells obviously undergo. Cells lose their ability to transform into a great variety of tissues and structures (totipotency) very early in development, and become limited to certain phenotypic programs. Determination is used in the subsequent discussion to denote these early restrictions of potential.

Most embryonic tissues have only a limited range of possible expression but probably a far wider range than they are called upon to exhibit during normal development. Possibly the regenerating eye tissues have remained within the natural boundaries of their determination. Since neural retina and retinal pigment epithelium have a common immediate origin from the wall of the developing optic vesicle, it is not surprising that retinal pigment cells are also determined for neural retina expression. Conversely, embryonic neural retina has a melanotic determination and can transform into pigment cells under certain simple cell and organ culture conditions (Dorris, 1938; Peck, 1964). There is also a rare human eye disease, retinitis pigmentosa, in which patches of pigment cells appear in the adult retina. Metaplasia of neural retina cells to pigment cells probably occurs.

It is more surprising that iris pigment cells have the ability to become lens cells, since lens ordinarily develops from embryonic head ectoderm, which is only distantly related to the neural ectoderm from which other eye tissues develop. Interestingly, chick embryo iris contains low levels of antigens that are immunologically identical to the crystallin proteins of adult lens (Brahma et al., 1971). The chick embryo iris may have a limited ability to restore a missing lens (van Deth, 1940), although this remains a controversial point.

On the other hand, presumptive amphibian epidermis (ectoderm) undoubtedly has been determined for melanotic expression, since simple stimuli *in vitro* (exposure to phenylalanine or lithium chloride) will readily induce melanotic differentiation in these cells (Wilde, 1955; Barth and Barth, 1959). The reasons for such determination are more mysterious, since epidermal melanocytes originate enclusively from the neural crest, at least in vertebrates above the evolutionary level of fishes. However, these cases do suggest that ectodermal tissues share a very wide common determination.

An interesting group of tissues that presumably share a broad determination are the neural crest derivatives. It is thought likely, but re-

mains completely unproven, that one population of migrating neural crest cells gives rise to melanocytes, other kinds of pigment cells (xanthophores and guanophores), various sensory and autonomic ganglia and some of their supportive cells, adrenal medulla, and certain elements of the skeletal and connective tissues. If so, then subsequent cell and tissue interactions are critically important in selecting the appropriate phenotypic expression.

Peterson and Murray (1955) and Cowell and Weston (1970) have provided evidence that spinal ganglion cells might remain melanotically determined, at least during a part of their early development. Some cells of the sensory ganglion explanted *in vitro* from 4-day chick embryos differentiated into melanin-producing cells, but ganglia explanted from embryos older than 6 days did not develop metaplastic melanocytes. Unfortunately, these experiments do not dispose of the possibility that explants of younger ganglia may contain prospective melanocytes en route to other parts of the embryo where pigment is normally produced. Peterson and Murray (1955) favor this second explanation.

Although it has been interesting to speculate on the possibility of multiple determinations in certain tissues, the metaplasias of ectodermal tissues (e.g., eye, lens, epidermis, and neural crest) might really involve changes of determination. Weiss and James (1955) concluded that the famous example of vitamin A-induced metaplasia of embryonic chick skin to ciliated epithelium is a real change of determination, because transformation was not dependent on the continued presence of the stimulus (vitamin A). Likewise, metaplasia of presumptive epidermis of frog gastrula to pigment cells requires only a brief treatment with lithium chloride (Barth and Barth, 1959). Hadorn would probably regard metaplasias as *transdeterminations*, according to his terminology (1965).

Hadorn (1965) transplanted imaginal disk tissues from larval *Drosophila melanogaster* serially through adult flies, where they proliferate without undergoing further differentiation. At each serial transplantation, part of the accumulated tissue can be returned to larval hosts, where it will develop into adult tissues at pupation. In this way one can see that imaginal tissues of genital disk, for example, will eventually, as time and transfers continue, develop into leg parts or head parts in addition to genital tissues. Some of the cells in the transplants have unquestionably changed their determination. However, unlike the transdetermination in insect tissues, vertebrate cells in regenerating eye tissues begin to undergo their metaplasia within one or two cell generations, and in some cases all cells in a particular location alter their phenotype.

In certain regenerating eye tissues (lens and iris), another interesting behavior of the cells was noted: low levels of histospecific proteins (tyrosinase and lens crystallins) were expressed in tissues that do not ordinarily differentiate in that direction. Rutter et al. (1968) have demonstrated that pancreas has very low levels of the histospecific proteins a long time before *active* accumulation of these proteins begins, and they have called this a *protodifferentiated* state. In the future, more sensitive analytic techniques may show that various tissues commonly reach a protodifferentiated state. At some point in their developmental sequence many kinds of cells may protodifferentiate elements of the alternative programs of their determination as well as of the major pathways that they will ordinarily follow. If this proves generally true, then the final phenotypic commitment of a cell is regulated at several levels of cellular activity beyond the selective activation of certain genes.

When Determination Occurs

A knowledge of the earliest events of differentiation in which phenotypic programs are selected and stabilized is crucial to any real understanding of cellular differentiation. Yet the details of the gene regulatory mechanisms in eukaryotic cells are still sufficiently vague and speculative (e.g., Davidson and Britten, 1971) that perhaps none of the right questions can even be posed at this time. Meanwhile, it is possible to gain some insight into the so-called determination problem by attempting to learn how soon in development the selection of the melanotic program occurs.

During the development of the mouse embryo, determination of primordial melanoblast clones in the neural crest seems to occur very early. Mintz (1967) fused cleavage-stage embryos *in vitro* from parents of different coat color genotypes and reimplanted the mosaic blastocysts that had developed from the fused embryos back into the uterus of foster mothers (Fig. 4). The tetraparental mice produced by fusing embryos with genotypes for two distinct coat colors—for example, black (*C/C*) and albino (*c/c*)—gave rise at birth to mice with a *standard pattern* of broad transverse bands (up to 17) of alternating color along the body and tail. This small number of bands, and the middorsal separation of occasionally mismatching bands permits one to speculate on the timing of mammalian melanocyte determination.

The coat color patterns of the banded tetraparental mice must originate from a small number of primordial melanoblasts, each of which develops into all the melanocytes invading a body segment on one side. The simplest view (favored by Mintz, 1967, 1971) is that two longitudinal rows of 17 cells each become clonally determined along the length of the neural crest at some early point in development (probably between 5 and 7 days). The 34 primordial melanoblasts had to be determined when the embryo had more than 34 cells. Day 5 of embryonic life is the earliest time that determination could have occurred, since the embryo has too few cells during the first 4 days to account for the pigment cell clones and those of other cell types as well. Since the clonal patterns show a bilateral symmetry, day 7 is probably the latest time at which the melanoblast loci might have become active. Fusion of the neural folds at day 7 might permit cells from the neural crest to cross over to the opposite side and prevent pattern autonomy.

The relatively small band number and the alternating arrangement of bands that make up the standard pattern imply a mechanism for alternating cells of different genotypes within the rows (Mintz, 1967). This seems very unlikely. However, according to probability calculations made by Wolpert and Gingell (1970), a *random* assortment of 64 cells in two longitudinal rows of 32 each on either side of the dorsal midline could be the basis of the observed standard pattern, if each cell gave rise to a single stripe by clonal proliferation.

Whichever explanation of the primordial clone number is correct, the melanocyte determination has probably occurred no later than the neurulation stage of development. This is the time at which the neural crest is becoming morphologically distinct. Experiments by Deol and Whitten (1972) on the timing of X-chromosome inactivation in mosaic mice suggest indirectly that the determination of melanotic expression in retinal pigment epithelium cells probably occurs no later than the neural plate stage. Neurulation is still sufficiently late in development that it might actually be the time when these presumptive mouse pigment cells begin to synthesize their phenotype-specific mRNA molecules. Perhaps this is a marking event of determination, and as suggested earlier, there may also be low-level synthesis of molecules concerned with other pathways of a cell's limited range of phenotypic potential.

As interesting as these conclusions about

time of determination might be, one must emphasize, as a note of caution, that the clonal theory of mammalian development advanced by Mintz (1967, 1971) is only one possible interpretation of her data. Her interpretation has been vigorously challenged by other investigators (McLaren, 1972; Lewis et al., 1972), and the same conclusions about time of determination would not necessarily follow from different interpretations of the patterns. In addition, others have found little evidence so far of her standard pattern of banding in their own tetraparental mice (McLaren and Bowman, 1972).

By using very different techniques, it is possible in ascidian embryos (*Ciona intestinalis*) not only to visualize microscopically the primordial melanoblasts that give rise to the two pigmented brain cells of the larva but also to visually follow their complete linear history from first cleavage onward (Whittaker, 1973b). *Ciona* eggs which have been arrested in cell division with cytochalasin B and other cleavage inhibitors were able to differentiate tyrosinase

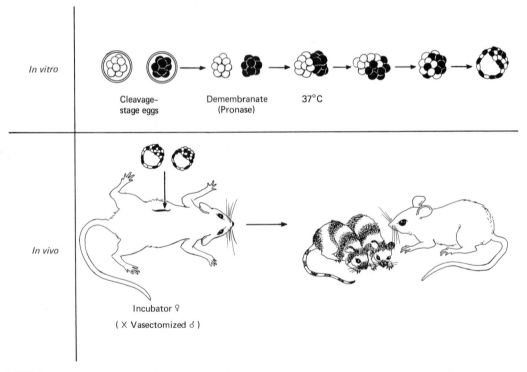

FIGURE 4. *Experimental procedures used by Mintz to produce tetraparental mice. (From B. Mintz, Proc. Nat. Acad. Sci. U.S.A. 58, 344–351, 1967.)*

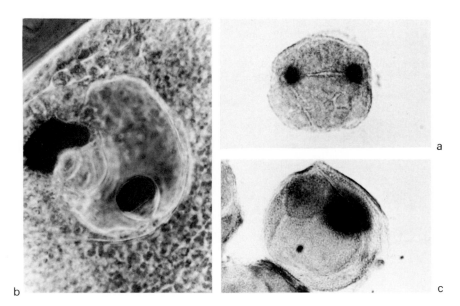

FIGURE 5. *(a)* Ciona intestinalis *embryo cleavage-arrested in cytochalasin B at the 64-cell stage, and reacted histochemically for tyrosinase (dopa oxidase) at 14 hours of development.* × *245. (b) Brain cavity of normal* Ciona *larva showing the two pigment cells (otolith cell lower right and ocellus pigment cell upper left).* × *850. (c)* Ciona *embryo cleavage- arrested in colchicine at the 8-cell stage, and reacted histochemically for tyrosinase activity at 14 hours.* × *320.*

(as detected histochemically) in the appropriate lineage blastomeres at a time corresponding to that of normal tyrosinase synthesis (9 to 10 hours). The majority of the embryos that were cleavage-arrested at the 32-cell stage or later developed tyrosinase in two blastomeres (Fig. 5a). By stopping cleavage at progressively later stages up to the time of normal differentiation, the lineage of these two cells could be followed. Each of the giant pigment cells of the larva (Fig. 5b) is a linear descendant of one of the bilaterally located cells seen in Fig. 5a. Tyrosinase determination is transferred to one daughter cell at each cleavage.

If cleavage arrest is begun earlier than the 32-cell stage, progressively fewer embryos eventually develop tyrosinase, but a few embryos can be found at each stage (Fig. 5c), including even the 2-cell stage, in which enzyme differentiation occurs in one or two of the blastomeres. The implication of these observations is that the fate of the cells is fixed as early as first cleavage, in the sense that something is segregated into one of the daughter cells at each subsequent cleavage. The parental cells can be regarded as *determined*

at each stage for tyrosinase production, since they themselves are capable of differentiating tyrosinase, at least under these experimental conditions. Cleavage-arrested embryos at the 64-cell stage and later are also capable of producing melanin in the two lineage blastomeres.

Ascidian embryos are perhaps highly unusual in the extremity of their determination, since their determination pattern is probably built into the structure of the egg during oogenesis. Experiments on cleavage-arrested and normal *Ciona* embryos with inhibitors of protein synthesis (puromycin) and RNA synthesis (actinomycin D) indicate that what is being segregated during cleavage is neither tyrosinase itself nor a template RNA for tyrosinase. Both enzyme and tyrosinase-related RNA are synthesized at essentially normal times in cleavage-arrested embryos (Whittaker, 1973b). It seems more likely that some regulatory material (Davidson and Britten, 1971) concerned with the (later) activation of tyrosinase genes is located in specific regions of the egg and is apportioned by cleavage into the appropriate tissue areas where the genes will become activated. Determination as a general mechanism could be the elaboration of such gene regulatory substances in certain cells and the transmission of these materials (and maintenance of their synthesis) through the cell progeny. Ascidian embryos are very "mosaic" in the sense that early blastomeres that are isolated surgically develop only into partial embryos. In embryos that are more plastic (regulative) and less "mosaic" than ascidians, these factors are probably produced later in development and not during oogenesis.

Conclusions

For such a relatively simple phenotypic expression as melanogenesis, the components of the system are remarkably complex. There is only one enzyme, tyrosinase, known to be essential, and much is known about it and the reaction product, melanin. Melanogenesis occurs exclusively in a special organelle, the melanosome. However, the composition and synthesis of the other melanosome constituents remain virtually unexplored, and there is still a paucity of information on what is being regulated in the phenotypic expression of a pigment cell. From the little that is presently known about regulation of melanogenesis, regulation of various kinds occurs at different levels. There is no simple answer to what controls differentiation of the melanotic phenotype in a pigment cell.

The program for melanotic expression appears to be selected by the pigment cell very early in development. Some embryonic pigment cells also seem to be programmed for other expression as well, since retinal pigment epithelium and iris tissue of the eye can readily transform during regeneration into other tissues (sensory retina and lens). Investigation of the occurrence and timing of these determination processes remains a very important and promising area for future research.

References

ACHAZI, R., AND YAMADA, T. (1972). Tyrosinase activity in the Wolffian lens regeneration system. *Develop. Biol.* 27, 295–306.

BARTH, L. G., AND BARTH, L. J. (1959). Differentiation of cells of the *Rana pipiens* gastrula in unconditioned medium. *J. Embryol. Exp. Morphol.* 7, 210–222.

BRAHMA, S. K., BOURS, J., AND VAN DOORENMAALEN, W. J. (1971). Immunochemical studies of chick iris. *Exp. Eye Res.* 12, 194–197.

BURNETT, J. B. (1971). The tyrosinases of mouse melanoma. *J. Biol. Chem.* 246, 3079–3091.

CLEFFMAN, G. (1963). Agouti pigment cells *in situ* and *in vitro*. *Ann. N. Y. Acad. Sci.* 100, 749–760.

COULOMBRE, J. L., AND COULOMBRE, A. J. (1965). Regeneration of neural retina from the pigmented epithelium in the chick embryo. *Develop. Biol.* 12, 79–92.

COWELL, L. A., AND WESTON, J. A. (1970). An analysis of melanogenesis in cultured chick embryo spinal ganglia. *Develop. Biol.* 22, 670–697.

CRICK, F. H. C. (1970). Diffusion in embryogenesis. *Nature (London)* 225, 420–422.

DAVIDSON, E. H., AND BRITTEN, R. J. (1971). Note on the control of gene expression during development. *J. Theoret. Biol.* 32, 123–130.

DAVIDSON, R. L. (1972). Regulation of melanin synthesis in mammalian cells: effect of gene dosage on the expression of differentiation. *Proc. Nat. Acad. Sci. U.S.A.* 69, 951–955.

DAVIDSON, R. L., EPHRUSSI, B., AND YAMAMOTO, K. (1966). Regulation of pigment synthesis in mammalian cells, as studied by somatic hybridization. *Proc. Nat. Acad. Sci. U.S.A.* 56, 1437–1440.

DEOL, M. S., AND WHITTEN, W. K. (1972). Time of X chromosome inactivation in retinal melanocytes of the mouse. *Nature New Biol.* 238, 159–160.

DOEZEMA, P. (1973a). Tyrosinase from embryonic chick retinal pigment epithelium. *Comp. Biochem. Physiol. B* 46, 509–517.

DOEZEMA, P. (1973b). Proteins from melanosomes of mouse and chick pigment cells. *J. Cell. Physiol.* 82, 65–74.

DORRIS, F. (1938). Differentiation of the chick eye *in vitro*. *J. Exp. Zool.* 78, 385–415.

DUMONT, J. N., AND YAMADA, T. (1972). Dedifferentiation of iris epithelial cells. *Develop. Biol.* 29, 385–401.

EPPIG, J. J., JR., AND DUMONT, J. N. (1972). Cytochemical localization of tyrosinase activity in pigmented epithelial cells of *Rana pipiens* and *Xenopus laevis* larvae. *J. Ultrastruct. Res.* 39, 397–410.

GALBRAITH, D. B. (1964). The agouti pigment pattern of the mouse: a quantitative and experimental study. *J. Exp. Zool.* 155, 71–90.

GRAY, J. M., AND PIERCE, G. B. (1964). Relationship between growth rate and differentiation of melanoma *in vivo*. *J. Natl. Cancer Inst.*, 32, 1201–1210.

HADORN, E. (1965). Problems of determination and transdetermination. *Brookhaven Symp. Biol. 18*, 148–161.

HEMPEL, K. (1966). Investigation of the structure of melanin in malignant melanoma with ^{3}H- and ^{14}C-dopa labeled at different positions. *In* "Structure and Control of the Melanocyte" (G. Della Porta and O. Muhlbock, eds.), pp. 162–175. Springer-Verlag, New York.

HOLSTEIN, T. J., QUEVEDO, W. C., JR., AND BURNETT, J. B. (1971). Multiple forms of tyrosinase in rodents and lagomorphs with special reference to their genetic control in mice. *J. Exp. Zool. 177*, 173–184.

IDE, C. (1972). The development of melanosomes in the pigment epithelium of the chick embryo. *Z. Zellforsch. 131*, 171–186.

JOHNSON, G. S., AND PASTAN, I. (1972). $N^6,O^{2'}$-Dibutyryl adenosine 3',5'-monophosphate induces pigment production in melanoma cells. *Nature New Biol. 237*, 267–268.

KITANO, Y., AND HU, F. (1971). Melanin versus protein synthesis in melanocytes *in vitro*. *Exp. Cell Res. 64*, 83–88.

KOBEL, H. R., BRUN, R. B., AND FISCHBERG, M. (1973). Nuclear transplantation with melanophores, ciliated epidermal cells, and the established cell-line A-8 in *Xenopus laevis*. *J. Embryol Exp. Morphol. 29*, 539–547.

LEHMAN, H. (1957). The developmental mechanics of pigment pattern formation in the black axolotl, *Amblystoma mexicanum*. I. The formation of yellow and black bars in young larvae. *J. Exp. Zool. 135*, 355–386.

LERNER, A. B., AND FITZPATRICK, T. B. (1950). Biochemistry of melanin formation. *Physiol. Rev. 30*, 91–126.

LEWIS, J. H., SUMMERBELL, D., AND WOLPERT, L. (1972). Chimaeras and cell lineage in development. *Nature (London) 239*, 276–279.

MAUL, G. G., AND BRUMBAUGH, J. A. (1971). On the possible function of coated vesicles in melanogenesis of the regenerating fowl feather. *J. Cell Biol. 48*, 41–48.

MAYER, T. C. (1967). Pigment cell migration in piebald mice. *Develop. Biol. 15*, 521–535.

MAYER, T. C., AND FISHBANE, J. L. (1972). Mesoderm-ectoderm interaction in the production of agouti pigmentation patterns in mice. *Genetics 71*, 297–303.

McLAREN, A. (1972). Numerology of development. *Nature (London) 239*, 274–276.

McLAREN, A., AND BOWMAN, P. (1972). Mouse chimaeras derived from fusion of embryos differing by nine genetic factors. *Nature (London) 224*, 238–240.

MINTZ, B. (1967). Gene control of mammalian pigmentary differentiation. I. Clonal origin of melanocytes. *Proc. Nat. Acad. Sci. U.S.A. 58*, 344–351

MINTZ, B. (1971). Clonal basis of mammalian differentiation. *Symp. Soc. Exp. Biol. 25*, 345–370.

MISURACA, G., NICOLAUS, R. A., PROTA, G., AND GHIARA, G. (1969). A cytochemical study of phaeomelanin formation in feather papillae of New Hampshire chick embryos. *Experientia 25*, 920–922.

MIYAMOTO, M., AND FITZPATRICK, T. B. (1957). On the nature of the pigment in retinal pigmented epithelium. *Science 126*, 449–450.

MOYER, F. H. (1966). Genetic variations in the fine structure and ontogeny of mouse melanin granules. *Amer. Zool. 6*, 43–66.

NICKERSON, M. (1944). An analysis of barred pattern formation in feathers. *J. Exp. Zool. 95*, 361–397.

NICOLAUS, R. A. (1968). "Melanins." Hermann, Paris. 310 pp.

OIKAWA, A., NAKAYASU, M., NOHARA, M., AND TCHEN, T. T. (1972). Fate of L-[3,5 −^{3}H]tyrosine in cell-free extracts and tissue cultures of melanoma cells: a new assay method for tyrosinase in living cells. *Arch. Biochem. Biophys. 148*, 548–557.

PECK, D. (1964). The role of tissue organization in the differentiation of embryonic chick neural retina. *J. Embryol. Exp. Morphol. 12*, 381–390.

PETERSON, E. R., AND MURRAY, M. R. (1955). Myelin sheath formation in cultures of avian spinal ganglia. *Amer. J. Anat. 96*, 319–355.

RAWLES, M. E. (1939). The production of robin pigments in White Leghorn feathers by grafts of embryonic robin tissue. *J. Genet. 37*, 517–532.

RUTTER, W. J., KEMP, J. D., BRADSHAW, W. S., CLARK, W. R., RONZIO, R. A., AND SANDERS, T. G. (1968). Regulation of specific protein synthesis in cytodifferentiation. *J. Cell. Physiol. 72*, Suppl. 1, 1–18.

SEIJI, M., AND FITZPATRICK, T. B. (1961). The reciprocal relationship between melanization and tyrosinase activity in melanosomes (melanin granules). *J. Biochem. (Japan) 49*, 700–706.

SILAGI, S. (1967). Hybridization of a malignant melanoma cell line with L cells *in vitro*. *Cancer Res. 27*, 1953–1960.

SILVER, A. F., CHASE, H. B., AND POTTEN, C. S. (1969). Melanocyte precursor cells in the hair follicle germ during the dormant stage (telogen). *Experientia 25,* 299–301.

VAN DETH, J. H. M. G. (1940). Induction and régénération du crystallin chez l'embryon de la poule. *Acta Neerland Morphol. 3,* 151–169.

WEISS, P., AND JAMES, R. (1955). Skin metaplasia *in vitro* induced by brief exposure to vitamin A. *Exp. Cell Res.,* Suppl. 3, 381–394.

WHITTAKER, J. R. (1966). An analysis of melanogenesis in differentiating pigment cells of ascidian embryos. *Develop. Biol. 14,* 1–39.

WHITTAKER, J. R. (1967). Loss of melanotic phenotype *in vitro* by differentiated retinal pigment cells: demonstration of mechanisms involved. *Develop. Biol. 15,* 553–574.

WHITTAKER, J. R. (1968). The nature and probable cause of modulations in pigment cell cultures. *In* "Results and Problems in Cell Differentiation" (H. Ursprung, ed.), Vol. 1, pp. 25–36. Springer-Verlag, New York.

WHITTAKER, J. R. (1971). Biosynthesis of a thiouracil pheomelanin in embryonic pigment cells exposed to thiouracil. *J. Biol. Chem. 246,* 6217–6226.

WHITTAKER, J. R. (1973a). Tyrosinase in the presumptive pigment cells of ascidian embryos: tyrosine accessibility may initiate melanin synthesis. *Develop. Biol. 31,* 441–454.

WHITTAKER, J. R. (1973b). Segregation during ascidian embryogenesis of egg cytoplasmic information for tissue-specific enzyme development. *Proc. Nat. Acad. Sci. U.S.A. 70,* 2096–2100.

WILDE, C. E., JR. (1955). The urodele neuroepithelium. II. The relationship between phenylalanine metabolism and the differentiation of neural crest cells. *J. Morphol. 97,* 313–344.

WOLFE, H. G., AND COLEMAN, D. L. (1966). Pigmentation. *In* "Biology of the Laboratory Mouse." 2nd ed. (E. L. Green, ed.), pp. 405–425. McGraw-Hill, New York.

WOLPERT, L., AND GINGELL, D. (1970). Striping and the pattern of melanocyte cells in chimaeric mice. *J. Theoret. Biol. 29,* 147–150.

WRATHALL, J. R., OLIVER, C., SILAGI, S., AND ESSNER, E. (1973). Suppression of pigmentation in mouse melanoma cells by 5-bromodeoxyuridine. Effects on tyrosinase activity and melanosome formation. *J. Cell Biol. 57,* 406–423.

YAMADA, T. (1967). Cellular and subcellular events in Wolffian lens regeneration. *Cur. Top. Develop. Biol. 2,* 247–283.

CHAPTER NINE

Regulation of the Cell Cycle and Myogenesis by Cell-Medium Interaction

IRWIN R. KONIGSBERG AND
PATRICIA A. BUCKLEY

The differentiation of skeletal muscle cells from embryonic myoblasts, isolated and grown in cell culture, closely parallels the developmental sequence observed in the tissue of origin (Konigsberg et al., 1960; Konigsberg, 1961b, 1964). As an experimental system, however, cell culture offers distinct advantages that cannot be attained by studying embryonic muscle *in vivo*. For example, using cell suspensions of embryonic muscle, selective techniques such as cloning (Konigsberg, 1961c, 1963) or differential cell attachment (Kaighn et al., 1966) can be employed to either entirely eliminate or at least greatly reduce contamination by non-myogenic cellular components of the tissue of origin. Not only do culture techniques afford the opportunity of working with relatively homogeneous populations of myoblasts, but culture conditions can be arranged to give a greater degree of synchrony of development than is ever observed *in vivo* (see Fig. 1).

The initial period after the culture is established is marked, simply by cell proliferation at a constant exponential rate (Konigsberg,

1961b; Coleman and Coleman, 1968). At a predictable time, which can be controlled to a large extent, long multinucleated cells form in culture by the fusion of mononucleated myogenic cells (Konigsberg, 1965). The number and size of these syncytia increase rapidly during the ensuing 24 to 48 hours and then appear to level off.

One can demonstrate in multinucleated cells, shortly after they form, the presence of typical cross-striated myofibrils by a variety of cytological techniques (Konigsberg, 1961b; Okasaki and Holtzer, 1965). In time, spontaneous, spasmatic contractions can also be observed in the syncytial fibers.

By the methods applied thus far the presence of contractile proteins, myosin in particular, cannot be detected in the proliferating myoblasts. Measurements of the rate of myosin synthesis indicate, moreover, not only that the initiation of rapid synthesis is correlated with the occurrence of fusion in culture (Patterson and Strohman, 1972) but that a 200-fold increase in rate occurs as a step function during

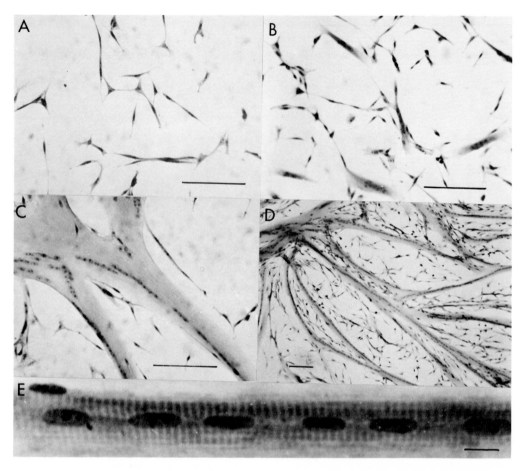

FIGURE 1. *Photomicrographs of quail muscle cells in culture. All cultures were inoculated with 31,250 cells. On representative days, cultures were fixed, stained, and photographed. (a) Day 2 culture consisting of mononucleated myoblasts recognizable by their characteristic bipolar shape. Bar = 0.1 mm. (b) Day 3 culture in which fusion is first observed as the formation of short, "stubby" multinucleated myotubes. Bar = 0.1 mm. (c) Day 5 culture demonstrating the progressive increase in the number of nuclei within multinucleated myotubes. Bar = 0.1 mm. (d) Lower magnification photograph of a day 5 culture demonstrating the extensiveness of fusion. Bar = 0.1 mm. (e) Higher magnification photograph of a multinucleated myotube demonstrating the characteristic cross striations. Bar = 0.01 mm. (From P. A. Buckley 1973.)*

the first 12 hours of fusion (Emerson, 1973). Other proteins, characteristic of mature skeletal muscle, although detectable in small amounts prior to the fusion of the mononucleated myoblasts also accumulate differentially after fusion (Shainberg et al., 1971). Whether specific biosynthesis is regulated at the translational level, as has been suggested (Yaffe and Dym, 1973), or at the transcriptional level is not presently known. It is clear, however, that such regula-

tion is linked to some aspect or aspects of myogenic fusion.

Our notions of what may signal the initiation of myogenic fusion are conditioned to a large extent by observations made after the fact—that is, by an examination of the properties of the multinuclear cells rather than the process of fusion per se.

The nuclei of these syncytia do not divide and, except under certain abnormal conditions

(Yaffe and Gershon, 1967; Fogel and Defendi, 1967; Stockdale, 1971; Ebert and Kaighn, 1966), do not incorporate tritiated thymidine (Firket, 1958; Stockdale and Holtzer, 1961; Bassleer, 1962). In cultures containing extensive networks of multinucleated myotubes, DNA polymerase activity is also reported to be greatly reduced (O'Neil and Strohman, 1970; Stockdale, 1970). The fact that DNA replication occurs only in the mononucleated cells is confirmed by comparing microspectrophotometric measurements of the DNA content of nuclei in syncytial cells to that of the mononucleated cells in culture (Strehler et al., 1961).

Although the distribution of DNA per nucleus is bimodal in populations of mononucleated cells, the values of the two peaks corresponding to 2N and 4N amounts of DNA, the DNA content of syncytial nuclei cluster around a single peak (see Fig. 2). In addition, since the single peak of DNA values observed in syncytial nuclei corresponds to the amount of DNA (viz., 2N) found in mononucleated cells during the G_1 phase of the cell cycle, it suggests that myoblasts when they fuse do so while traversing G_1.

All these data prove quite clearly that the nuclei of multinucleated muscle cells are no longer in the proliferative pool, but they do no more than that. They do not indicate whether myoblasts withdraw from the cycle *before* fusing or are withdrawn as a consequence of the fusion process.

It has been assumed, however, based on an observation of the shortest interval between pulse-labeling and fusion that myoblasts do, in fact, actually withdraw from the cell cycle before fusion occurs (Bischoff and Holtzer, 1969). The rationale underlying this assumption seems to be that, since the first labeled nucleus within a multinucleated myotube appears after a postmitotic gap period approxi-

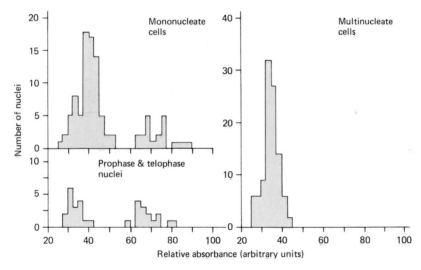

FIGURE 2. *Histograms showing the distribution of DNA per nucleus in cultures of embryonic chick skeletal muscle cells, measured by Feulgen microspectrophotometry.* Prophase and telophase nuclei: *Each chromosome set of the telophase configurations was measured separately to establish the diploid (2N) DNA content (m = 33.4 units). Measurements of prophase nuclei give the 4N value (m = 68.2).* Mononucleate cells: *The values of the 90 nuclei measured fall into a bimodal distribution. The first peak of the distribution (d = 39.4) which includes 79 of the measurements, corresponds to cells with the 2N value of DNA. The second peak of 21 values (m = 73.4) coincides with the 4N level of DNA.* Multinucleate cells: *In contrast to the mononucleate cells, the distribution of the 102 measurements of nuclei in multinucleate cells is unimodal (m = 34.5) and corresponds to the 2N value of DNA. None of the values fall in the range expected of nuclei (4N) which have doubled their DNA content. (From B. L. Strehler, I. R. Konigsberg, and J. E. T. Kelley,* Exp. Cell Res. 32, 232, 1963.)

mately twice as long as the *median* G_1 of the population (by cell cycle analysis), this cell must have permanently withdrawn from the cycle *before* it fused. The interpretation suggests further that, preceding fusion, a qualitatively different mitotic (or premitotic) event yields two daughter cells one or both being now not only permanently withdrawn from the cell cycle but programmed to fuse and to synthesize contractile proteins. The occurrence of this critical or "quantal" division is assumed to be an obligatory event (Bischoff and Holtzer, 1969).

This hypothesis can not be tested directly since it is not possible to determine whether any particular cell would or would not have divided had it not fused. Aside from that, however, the interpretation turns on the comparison of a single measurement (the shortest interval between labeling and fusion) with the median value of a large number of measurements (the median G_1 time). Without knowing the distribution of G_1 times or, more specifically, the fraction of cells that spend an equal or greater amount of time (than the single event measured) in the postmitotic gap and then reenter the S phase, we cannot assume that a G_1 period of even twice the median suggests that a cell is actually withdrawn from the cycle.

Autoradiographic data presented in two more recent publicaitons have, in fact, been interpreted by one group to indicate that fusion can occur early in G_1 (O'Neil and Stockdale, 1972) and alternatively, by the other, that although G_1 is "particularly favorable for fusion," fusion can occur in G_2 as well (Martinucci et al., 1970). If this last conclusion is correct, however, fusion in G_2 must be rather rare since three studies, in which cytophotometric measurements of feulgen DNA were made all indicate that the nuclei in syncytia constitute a single class having values corresponding to the 2N, not the 4N, amount of DNA (Firket, 1958; Strehler et al., 1961; Bassleer, 1962).

Earlier cytophotometric measurements of regenerating mouse muscle (Lash et al., 1957) are in essential agreement with these *in vitro* studies. Although a small percentage (1%) of the DNA values fall above the diploid distribution, these could be due, as the authors point out, to measuring errors inherent in sectioned material as well as to biological variation.

Although the three estimates of the shortest interval between the DNA synthetic period and fusion disagree, not enough measurements are presented in any of these reports to attempt to resolve the conflict.

Recent studies, in our laboratory, suggest that rather than some qualitatively different mitotic event, the initiation of fusion is a response to changes in the composition of the medium, generated by the metabolic activities of the proliferating culture population (Konigsberg, 1971). Furthermore, our data are consistent with the concept that the initial response to these changes is a protraction of the G_1 phase of the cycling myoblasts (Buckley, 1973).

Cell Density and Initiation of Fusion

During the course of an examination of the role of collagen substrata in promoting the development of muscle clones (Hauschka and Konigsberg, 1966), observations were made that suggested that, in the absence of collagen, daughter myoblasts form a more compact colony that fuses prematurely (Konigsberg, 1970). The obvious interpretation is that physical proximity promotes early fusion. Such an interpretation is entirely compatible with two observations made much earlier (Konigsberg et al., 1960; Konigsberg, 1961b, 1961a)—first, that there is a progressive increase in cell density preceding fusion, and, second, that when the inoculum size used to set up a series of cultures is increased, those cultures started with larger cell numbers fuse earlier. Two alternative explanations were offered to explain these earlier observations: (1) that higher cell densities might increase the probability that competent cells would contact one another and fuse, or (2) that increased cell density might accelerate metabolically generated alterations of the medium that could, in turn, provide a microenvironment, in some way, more favorable for the initiation of fusion.

To discriminate between these two possibilities we started with the premise that if increased cell density promoted fusion by alter-

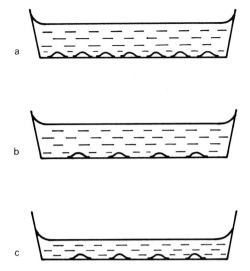

FIGURE 3. *Test to distinguish between alternative explanations of the cell density effect. (See the text.) a◄—►b: Inoculating a larger number of cells (a) into the same volume of medium used in b results in earlier fusion. This experiment in itself does not discriminate between (1) increased cell–cell encounters and (2) more rapid modification of the medium by the metabolic activities of the larger cell population. b◄—►c: If using the same inoculum size in c as in b but reducing the volume of medium results in earlier fusion in c it would strongly suggest alternative (2).*

ing the medium, the same end result—namely, early fusion—should be achieved whether we increased the number of cells inoculated into a petri plate or, keeping the cell number constant, decreased the volume of medium in contact with the cells (see Fig. 3). If, on the other hand, intercellular distance was the critical parameter, decreasing the volume of medium should be without effect.

In order to facilitate scoring these experiments, we scaled down our usual culture procedures, using lower numbers of cells confined to a small area, of constant size, on the petri plate surface (see Fig. 4).

Figure 5 illustrates the results that were obtained when our original premise was tested. Microcultures labeled A, B, and C were established with 12,500, 25,000, and 50,000 cells, respectively. Both the upper and lower series of microcultures are identical with respect to the progression of cell number. In the upper

series, however, the petri plates were flooded with 1 ml of medium, while the lower tier received 3 ml. Comparing A_3 with C_3 one can observe that, using the same volume of medium, fusion is well advanced at the highest inoculum size used (C_3), while at the lowest inoculum (A_3) fusion has barely started. Similar differences in the time of initiation of fusion can be obtained, even when the same inoculum size is employed, if the volume of medium in equilibrium with the cells is different (compare A_1 and A_3). Here, although the number of cells per unit area was initially identical in both cultures, fusion was initiated earlier when a smaller volume of medium was used, as in A_1. Since in this situation there would be no initial difference in the average intercellular distances in the two cultures, the result suggests that cell density controls the time of initiation of fusion, not by promoting cell contact, but by some diffusion-dependent metabolic processing of the medium.

Irrespective of whether such processing of the medium involves the accumulation of cell products, or the depletion of constituents of

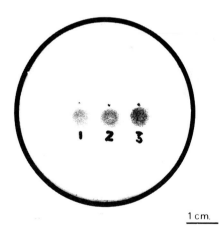

FIGURE 4. *Petri plate containing three microcultures inoculated with different cell numbers (1 = 12.5 × 10³; 2 = 25.0 × 10³; and 3 = 50.0 × 10³ cells) cultured in 1 ml of medium and fixed after 24 hours of incubation. Each culture is 6 mm in diameter. (Apparent differences in size, at this magnification, reflect the fact that the periphery of the lower density cultures consists chiefly of mononucleated cells. See Fig. 5, top tier.) (From I. R. Konigsberg, Develop. Biol. 26, 133, 1971.)*

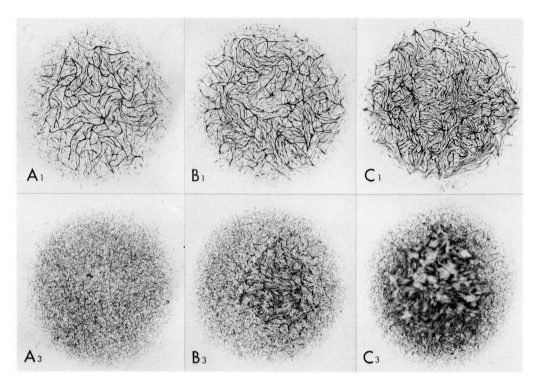

FIGURE 5. *Individual microcultures such as those in Fig. 4 in different volumes of media. Inoculum sizes used were: A = 12.5 × 10³; B = 25.0 × 10³; and C= 50.0 × 10³. Cultures in the upper horizontal row (subscript "1") received 1 ml of medium; those in the lower row (subscript "3") were fed 3 ml. All cultures were fixed after 24 hours of incubation. In addition to the differences in the time of initiation of fusion discussed in the text, the morphological pattern of the multinucleated cells is altered when fusion is delayed (compare C_1 and C_3). The massive fusion observed in C_3 is most probably due to the larger number of cells present when fusion does occur. (From I. R. Konigsberg, Develop. Biol. 26, 133, 1971.)*

the medium, or both, one would predict that if the medium were continuously circulated, fusion would be delayed by preventing the establishment of local inhomogeneities in the medium (due to processing) in the immediate vicinity of the cells.

Continuous Circulation of Medium

To test this prediction, single microcultures established with small numbers of cells (6,250 cells per cylinder) were placed at the edge of each petri plate. All of the petri plates were filled with 2 ml of medium and were then incubated for 42 hours. The control group, however, was kept stationary, while the medium in the experimental group was continuously circulated by fastening the petri plates to the edge of a slowly rotating turntable tilted at a slight angle with the horizontal. The results of such an experiment are shown in Fig. 6. At the end of the experimental period, although a well-developed network of multinucleated cells could be observed in the stationary cultures (B), the cultures maintained on the turntable consisted of a dense population of mononucleated cells. As we had predicted, circulating the medium *delays* fusion. Fusion occurs, despite continuous circulation of the medium, if the experimental period is extended for an additional 12 hours or more, and, as one might expect, with larger numbers of cells, fusion is observed even within the shorter experimental period. Photomicrographs C and D of Fig. 6 illustrate, in fact, that large populations of cells can influence the fusion of less dense "test" microcultures, even when separated by considerable distances. In D the microculture labeled "3" was seeded with the same number of cells as those in A and B; in this experiment, however, three microcultures (marked "5"), each established with four times the number of cells used in the test microculture, were placed in the same petri plate. At the end of the experimental period, not only did massive fusion occur in these three high-density cultures, but their presence actually promoted fusion in the low-density test microculture (compare C to the control microculture in A). In a similar fashion, the delay in the initiation of fusion can be overcome by circulating medium withdrawn from cultures in which fusion has already occurred, indicating that the alterations to the medium are, to some extent, stable in nature.

We conclude from these results that the time of initiation of fusion of myoblasts in cell culture is controlled by some metabolic processing of the medium that is diffusion-mediated and dependent upon population size. Since circulating the medium, which should accelerate gas exchange, and increasing the volume in stationary culture, which would impede gas exchange, both delay fusion, this parameter can play no significant role.

Although we do not as yet know the nature of the changes in the medium, we do know

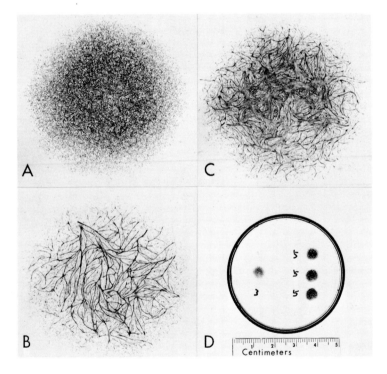

FIGURE 6. *Experimental and control cultures testing the effect of continuous circulation of the medium on the time of initiation of fusion. All test cultures (A, B, and C) were inoculated with the same number of cells (6250), cultured in 2 ml of medium, and fixed after 42 hours. (A) Test culture over which medium was circulated throughout the test period. (B) Control microculture; petri plate remained stationary throughout the test period. (C) Microculture treated as in A; that is, the medium was continuously circulated. However, three additional microcultures, each seeded with a large number of cells (25.0 × 10³), were also included in the same petri plate. (D) Photograph of the petri plate described in C (above). Microculture labeled "3" is the test culture shown in C at higher magnification. The numeral "5" identifies the "feeder" microcultures. (From I. R. Konigsberg, Develop. Biol. 26, 133, 1971.)*

Regulation of the Cell Cycle and Myogenesis by Cell-Medium Interaction 185

that the effective alterations reside in that fraction of the medium larger in molecular weight than 300,000 daltons. This finding would rule out trivial changes such as pH or the depletion of low-molecular-weight nutrients. Also, whatever the changes are, they cannot be reproduced by any of the three other cell types that we have assayed: fibroblasts, cardiac muscle cells, or liver cells.

Whenever fusion is delayed by either increasing the volume or circulating the medium, a comparison of experimental and control cultures suggests that a larger population of cells accumulates under the experimental conditions (compare Figs. 5A$_1$ and 5A$_3$ and Figs. 6A and 6B).

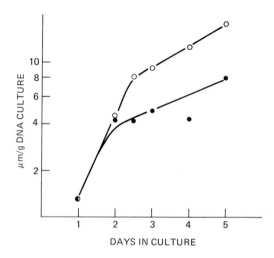

FIGURE 7. *DNA per culture as a function of time in cultures growing in different volumes of medium. All cultures were set up with 31,250 cells per 5-cm-diameter Falcon TC petri dish (collagen-treated) in 3 ml of medium. After 24 hours (day 1 on abscissa) medium was replaced with either 1 ml (closed circles) or 3 ml (open circles). (From I. R. Konigsberg, Develop. Biol. 26, 133, 1971.)*

Culture Growth and Cell Fusion

To examine this impression on a quantitative basis, growth in terms of DNA per culture was measured as a function of time in an experiment in which differences in the volume of medium used was employed to control the time of initiation of fusion. The results are presented in Fig. 7. All the cultures were established with 31,250 cells distributed over the entire surface of a 5 cm diameter petri dish (no retaining cylinder was employed) in 3 ml of medium. After 24 hours (day 1 in Fig. 7) one group was fed 1 ml of medium (closed circles) and the other 3 ml of medium (open circles). Both groups grew at the same rate during the first 24 hours following the change of medium. After this time the rate declined rather sharply, with the decline occurring 12 hours earlier in those cultures that had received the smaller volume of medium. Clear-cut differences in the time of initiation of fusion were also observed between these two sets of cultures corresponding to the difference in time when the break occurred. Following the break both groups continued to grow at essentially the same, slower rate. These data indicate that, concomitant with delayed fusion, cell division continues unabated for a longer period of time. Since the average generation time for these cultures is 10.0 hours (Buckley, 1973), the 12-hour difference in change of growth

rate would permit an additional round of cell division in those cultures in which fusion is delayed, which is in agreement with the approximately 2-fold difference observed in DNA per culture between the two groups.

It seems quite clear that the devices that we have employed to delay fusion operate by promoting cell proliferation. Here equal aliquots of cells from the same cell suspension were used to set up a series of replicate cultures treated in an identical manner until the first medium change. Since simply culturing in a larger volume of medium permits the same cells that would have fused in one third that volume to continue through another cell division, we conclude that fusion must be a response to an environmental cue rather than an intrinsically programmed event.

The simplest explanation of the abrupt change in growth concomitant with the appearance of multinucleated cells is that it merely reflects the progressive withdrawal from the proliferating fraction of those cells whose nuclei are recruited into the syncytial network. These syncytial nuclei, of course, are no longer capable of DNA synthesis (Firket, 1958; Stockdale and Holtzer, 1961; Bassleer, 1962). Such an explanation would not necessarily require assuming that any change has occurred in either the average generation time or the distribution of generation times of those cells that remain in the proliferating fraction. However, since cultures of all types of cells, not only muscle cells, exhibit a similar break in growth rate at high cell density, it seemed more reasonable to expect that some more general mechanism exclusive of fusion per se might be involved as well.

Cell Density and Mitotic Cycle

In order to determine whether any change in cell generation time does, in fact, occur concomitant with the change in growth rate, the cell cycle was measured (Buckley, 1973) using the pulse-chase method of Quastler and Sherman (1959).

This procedure exploits the fact that a cell in either of two different phases of the cell cycle can be recognized cytologically. Cells in any portion of the DNA synthetic phase can be recognized (after autoradiographic processing), since only these cells will have incorporated labeled precursors into DNA. The presence of the mitotic spindle would, of course, identify any cell in mitosis (M).

In any asynchronously growing population pulsed with tritiated thymidine, every cell which was in S during the brief pulse will have incorporated label. If, after replacing the medium containing isotope with "cold" medium, sample cultures are fixed and processed at regular intervals, every mitotic cell observed that is also labeled represents a cell that was in S during the pulse, traversed the premitotic gap (G_2), and entered M. Plotting the percentage of these labeled mitotic cells gives us a time-frequency distribution (see Fig. 8) from which the average (more correctly, the median) time spent in each cell cycle phase for the whole population can be calculated. Not only can mean values be determined, but the shape of the distribution itself gives a graphic

representation of the variability of certain parameters of the total cycle.

Two points in time were examined using cultures set up under conditions similar to those represented in the upper growth curve of Fig. 7. It is important to bear in mind that Quastler and Sherman developed their approach in order to study the cell cycle characteristics of crypt cells of the mouse duodenum *in vivo*. In the intact animal, a brief pulse can be achieved by the rapid clearing and excretion of isotope injected into the circulation. In culture, however, although the pulse is initiated by simply adding the labeled precursor to the medium, terminating the pulse can only be accomplished by withdrawing the labeling medium and replacing it with "chase" medium (containing an excess of "cold" precursor to rapidly dilute the intracellular pool). Preliminary studies were performed to see what effect, if any, replacing the medium would have on cell division. In retrospect, it should have been no surprise to find that although there was no effect up to day 2 in culture, thereafter (day 5, for example) simply replacing the medium produced a spurt of proliferation measured as a significant increase in the rate of DNA accumulation. This stimulatory effect could be avoided if at the later stages the "chase" was prepared from medium withdrawn from sister cultures of the same age.

The data for day 2 cultures (24 hours prior to the appearance of the first multinucleated cells) are shown on the graph in Fig. 8A, on which the percentage of labeled mitotic figures is plotted against time, following the termination of a 30-minute pulse with tritiated thymidine. The relatively small variance of the first peak of this curve (and the initial peak on the three other curves, plotted on this graph) reflects the fact that the labeled cohort, by virtue of the fact that they were all in S phase during the pulse are, to that extent, synchronized. In the curve for day 2 cultures, it appears that the initial synchronization is maintained through at least another complete round of division (note the similarity in shape of the two peaks). A plot of similar data, however, obtained from the as yet unfused single cells in day 5 cultures (in which fusion is well advanced), shows that this initial synchrony is maintained only through the G_2

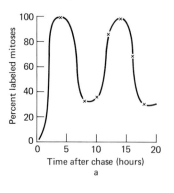

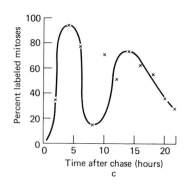

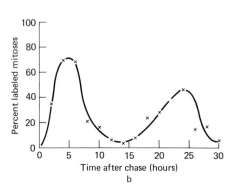

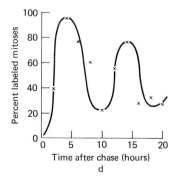

FIGURE 8. *Progression through mitosis of cohorts of cells pulse-labeled with tritiated thymidine. At zero minus 30 minutes a small volume of tritiated thymidine (s.a. = 6.7 C/mM) was added to the culture and mixed with the medium to give a final concentration of 0.5 μC, ml. At zero time the medium was withdrawn and replaced with chase medium supplemented with "cold" thymidine to a final concentration of 10⁻⁷ M. Sample cultures were fixed immediately after the pulse and at subsequent 2-hour intervals. Following autoradiography and staining, the percentage of labeled mitoses was scored. Each point represents 100–200 scored mitoses. (A). Day 2 cultures pulsed and chased using fresh medium (day 2, FM). The curve exhibits two homologous and symmetrical peaks which reach a maximum of 100 percent. (B) Day 5 cultures pulsed and chased in conditioned medium pooled from equivalent day 5 cultures (day 5, CM₅). Under these conditions, the two peaks are no longer homologous. The second peak appears later than that in the day 2, FM curve and is also much broader than the first peak and consequently considerably dampened. (C) Day 2 cultures pulsed and then chased using conditional medium collected from day 5 cultures (day 2, CM₅). The homology between the two peaks (see 8A) is lost. The second peak is broader and dampened. The curve appears more like the day 5, CM₅ curve (D) Day 5 cultures pulsed and then chased with fresh medium (day 5, FM). The two peaks exhibit the symmetry observed in day 2 cultures chased with fresh medium. [From P. A. Buckley, 1973.]*

and M phases immediately following the pulse (see Fig. 8B). Asynchrony appears during the second mitotic wave, which is markedly broader and flatter than the first.

The first wave of labeled mitosis, it should be recalled, represents cells that were labeled during the brief pulse, completed S and tra-

versed G_2 being caught in mitosis when the cultures were fixed at each time period. On the other hand, the mitotic figures that make up the second peak represent daughter cells of the first cell division after the pulse. These daughter cells traversed one complete cell cycle (*including* G_1) between birth and the time of fixation. Any variability due to the distributions of G_1 times in the population would show up in the second peak but would be absent from the first. The asynchrony that appears in the second mitotic wave of the day 5 curve must indicate, therefore, that between day 2 and day 5 the G_1 phase of the cell cycle has become more variable, many cells spending more time in G_1 before reentering S.

If one compares the median time spent in each of the cell cycle phases of day 2 cultures to day 5 cultures, it is obvious that, not only is the median total generation time increased at day 5, but virtually the entire increase can be accounted for by the protraction of G_1— from 3 hours on day 2 to 12 hours on day 5 (see Table 1).

It is clear from our data that the break in the rate of DNA accumulation concomitant with fusion is *not* simply a reflection of a progressive decrease in the numbers of proliferating cells due to their recruitment into multinucleated fibers. On the contrary, in cultures engaged in intensive fusion the population of mononucleated cells that have not as yet fused

proliferate at a significantly reduced rate, lagging in the G_1 phase specifically.

This observation raises two immediate questions. First, how early can this change in the length of the cell cycle be detected? In other words, does it occur before or after fusion is initiated in culture? Of equal importance, is the question of whether the change in the cell cycle is a response to metabolic alteration of the medium (like the cell-density dependent initiation of fusion) or simply due to time in culture (for example, the mitotic history of the myoblasts).

If the stage specific differences in cell cycle characteristics that we observed were due simply to differences in the medium, then we should be able to elicit the day 2 pattern from day 5 cultures by substituting the same medium routinely used to chase day 2 cultures. Figure 8D shows that by employing this stratagem we are able to mimic the day 2 pattern using older, day 5 cultures. The converse—that is, the day 5 response from day 2 cultures—can also be reproduced by chasing the younger cultures in medium withdrawn from older, day 5 cultures (see Fig. 8C). Also, as one might

TABLE 1

MATHEMATICAL ANALYSIS OF CELL CYCLE CURVES[a]

Mean Cell Cycle Times (hr) [Half Maximum Method of Quastler and Sherman (1959)]

	day 2, FM	day 5, CM
$G_2 + M$[b]	2.0	2.0
S	4.8	5.2
G_1	3.2	12.0
T_G	10.0	19.2

Standard Deviations (hr) [Calculated as per Takahashi (1966)]

	day 2, FM	day 5, CM
$G_2 + M$	±0.20	±0.60
S	±0.35	±0.30
T_G	±2.68	±8.48
$G_1 + M$	±1.32	±4.20

[a]The curves illustrated in Fig. 8 were analyzed according to the methods indicated.
[b]The average duration of M, measured by time-lapse cinematography, was calculated to be 20 minutes. Owing to its relatively short duration, it may be considered together with the G_2 phase.

predict, fusion is initiated earlier when day 2 cultures are fed with medium collected from the older cultures, and cell proliferation is stimulated when day 5 cultures are fed with fresh medium (Buckley, 1973).

From these experiments it is clear that the protraction of G_1, observed in the as yet unfused proliferating cells in fusing cultures is not simply due to some intrinsic, time-related cellular mechanism. It appears, rather, that the protraction of G_1 as well as the initiation of fusion, are both controlled by some progressive change or changes in the composition of the culture medium.

Since we only had compared the cell cycle in exponentially growing, prefusion cultures (day 2) with advanced fused cultures (day 5), it seemed of some importance to choose intervals between days 2 and 5 in order to determine how early we could detect the lengthening of the cell cycle, so pronounced on day 5. Measurements of cell cycle times were therefore made at regular intervals between days 2 and 5, using a simpler, continuous labeling technique.

At the beginning of each day, starting on day 2 and concluding on day 5, a group of cultures was fed medium withdrawn from sister cultures of the same age to which tritiated thymidine had been added (0.5 μC/ml). Cultures were fixed at regular intervals during the ensuing 24-hour period and the percentage of unfused single cells whose nuclei had incorporated label was determined for each interval. In Fig. 9 one sees two curves that reach the same maximum but at different rates. On day 2, maximum labeling is reached after approximately 6 to 8 hours. The values on days 3, 4, and 5 are all coincident and show that from day 3 on, progression of the population into the S phase is slower, maximum labeling not occurring until some 14 to 16 hours have elapsed. In the continuous presence of tritiated thymidine the period of time required for the percentage of labeled single cells to reach a constant maximum is equivalent to the sum of the longest G_2, M, and G_1 phases in the population. Since our cell cycle measurements indicate that G_1 is the only phase that lengthens significantly between days 2 and 5 the continuous labeling data indicate that protraction of G_1 occurs as early as day 3. On day 3,

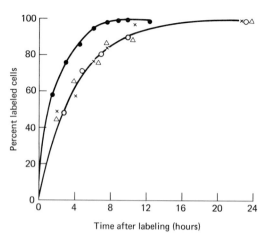

FIGURE 9. *Graph of the increase of percent labeled cells under conditions of continuous labeling. On successive days (days 2 to 5), the medium was withdrawn from a set of cultures and replaced with an equal volume of medium supplemented with tritiated thymidine (0.5 μC/cc). At intervals throughout the day, sample cultures were fixed. Following autoradiography and staining, the percentage of mononucleated cells with label was scored. Each point represents at least 200 mononucleated cells scored. The points on the graph represent the following values: solid circles, day 2; crosses, day 3; triangles, day 4; and open circles, day 5. The rate of increase of the percentage of labeled cells on days 3, 4, and 5 is much slower than that on day 2. [From P. A. Buckley, 1973.]*

under the conditions of inoculum size, medium volume, and replacement that we normally employ, the initiation of fusion is first observed. Thus not only are the protraction of G_1 and the initiation of fusion controlled by cell density-dependent changes in medium composition, but the two events occur concomitantly. While this correlation does not, in itself, establish any causal relationship, it indicates that such a relationship does deserve serious consideration.

One other piece of information obtained from the continuous labeling studies, which could not be obtained from the pulse-chase measurements of cell cycle times, is of considerable interest. Not only is the same maximum percentage of labeled, unfused cells reached both in prefused cultures (day 2) and in fusing cultures (days 3, 4, and 5), but this maximum is the same at all time intervals—virtually 100 percent. The overwhelming majority of these cells are indeed myoblasts. At clonal density approximately 95 percent of the colonies formed in cultures prepared from sieved suspensions of day 3, 4, or 5 cultures differentiate into muscle clones.

Although the "quantal" mitosis notion suggests that myoblasts actually withdraw from the cell cycle before fusing, our data indicate that, in healthy muscle cultures, no cells are withdrawn for any significant period, at any time.

Mitotic Cycle and Fusion: A Hypothesis

In summary, our experiments demonstrate that during the initial, purely proliferative, phase of culture development, the cells alter the medium in their immediate vicinity in such a manner that, although they continue to divide, the rate at which they traverse the cell cycle becomes highly variable, many cells spending more time in G_1. The simplest hypothesis that adequately accounts for these results is that, since myogenic cells, when they fuse, do so in G_1, it is not unreasonable to assume that some activity or activities required for fusion occurs during a portion, at least, of every G_1 period. If G_1 were protracted, the probability would be increased that these G_1-restricted activities attain some critical threshold. This hypothetical threshold might be reached either by cumulative processes occurring throughout every G_1 phase or by events which, since they are initiated only after some minimum time in G_1, are not expressed *except* during a protracted G_1 phase.

Based on this hypothesis one would predict that fusion is initiated when the G_1 time of a large-enough fraction of the cell population exceeds some minimum value. The fact that fusion does not occur simultaneously in the population may be a reflection of the consider-

able variability of G_1 times seen in older cultures.

If our formulation is correct, the interval between mitosis and fusion (the G_1 preceding fusion) should be long, certainly longer in duration than the mean G_1 time of exponentially growing myoblast populations. Although some data suggesting the validity of this assumption have been presented (Bischoff and Holtzer, 1969), conflicting observations have appeared more recently (O'Neil and Stockdale, 1972; Marinucci et al., 1970). In each of these studies, the objective was to determine the shortest G_1 interval preceding fusion. Perhaps more meaningful would be the measurements of the distribution of G_1 times preceding fusion. Viewing myogenic fusion as a probability function, as we do, we would be inclined to predict that such measurements might be continuous but skewed toward the longer values. Alternatively, if preparation for fusion is initiated only in late G_1, we might anticipate a region of excluded values preceding such a skewed distribution. The existence of a "window" late in G_1 has, in fact, been demonstrated in one study in which the synthesis of what is unequivocally the characteristic product of a differentiating cell was limited to a specific phase of the cell cycle. In proliferating lymphoid cells (Buell et al., 1971; Buell and Fahey, 1969) found that the production of immunoglobulin was restricted to late G_1 and early S.

In any event, none of the data, either of our own or of others, necessitate postulating a unique, critical or "quantal" cell cycle, particularly if one accepts the fact that the longer any cell remains in G_1 the greater is the probability that it will not reenter S (Whitmore and Till, 1964, but see also Smith and Martin, 1973). As we pointed out earlier (see page 182), the concept of an obligatory critical mitotic cycle preceding fusion can not be tested directly. However, our data, in addition to suggesting a simple, testable alternative, include two observations that are not compatible with this concept. First, we have shown that replicate cultures fed 3 ml of medium undergo another round of cell division prior to fusing, after sister cultures in a smaller volume of medium have initiated fusion. The implication is that fusion is illicited by changes in the external milieu rather than intrinsically programmed.

The second observation, contributed by our continuous labeling studies, is that even in cultures in which fusion is well in progress all of the mononucleated myoblasts are cycling (see page 191). We have no doubt that the "quantal" mitosis concept could be modified and expanded, as it already has been, to accommodate these observations, and others. In the evolution of any hypothesis, however, as it becomes more intricate and convoluted it loses its heuristic value, which is of course to serve as a framework upon which to base experiments.

References

BASSLEER, R. (1962). Étude de l'augmentation du numbre de noyaux don des bourgeons musculaires cultivés in vivo. Observations sur le vivant dosages cytophometriques et histoautoradiographies. Z. *Anat. Entwickl. Gesch. 123*, 184–205.

BISCHOFF, R., AND HOLTZER, H. (1969). Mitosis and the processes of differentiation of myogenic cells in vitro. *J. Cell Biol. 41*, 188–200.

BUCKLEY, P. A. (1973). Cell proliferation and myogenic differentiation. Ph.D. Dissertation, University of Virginia, Charlottesville, Virginia.

BUELL, D. N., AND FAHEY, J. L. (1969). Limited periods of gene expression in immunoglobulin synthesizing cells. *Science 164*, 1524–1525.

BUELL, D. N., SOX, H. C. AND FAHEY, J. L. (1971). Immunoglobulin production in proliferating lymphoid cells. In "Developmental Aspects of the Cell Cycle" (I. L. Cameron, G. M. Padilla, and A. M. Zimmerman, eds.), pp. 279–296. Academic Press, New York.

COLEMAN, J. R. AND COLEMAN, A. W. (1968). Muscle differentiation and macromolecular synthesis. *J. Cell. Physiol. 72*, (Suppl. 1), 19–24.

EBERT, J. D. AND KAIGHN, M. E. (1966). The keys to change: factors regulating differentiation. In "Major Problems in Developmental Biology" (M. Locke, ed.), pp. 29–84. Academic Press, New York.

EMERSON, C. P., JR. (1973). Control of myosin production during myogenesis *(in preparation)*.

FIRKET, H. (1958). Recherches sur la synthèse des acides désoxyribonucléiques et la préparation a la mitose dans des cellules cultivés in vitro. *Arch. Biol. Liege 69*, 3–166.

FOGEL, M. AND DEFENDI, V. (1967). The infection of muscle cultures from various species with oncogenic DNA viruses (SV40 and polyoma), *Proc. Nat. Acad. Sci. U. S. A. 58*, 967–973.

HAUSCHKA, S. D. AND KONIGSBERG, I. R. (1966). The influence of collagen on the development of muscle clones. *Proc. Nat. Acad. Sci. U.S.A. 55*, 119–126.

KAIGHN, M. E., EBERT, J. D. AND SCOTT, P. M. (1966), The susceptibility of differentiating muscle clones to Rous sarcoma virus. *Proc. Nat. Acad. Sci. U.S.A. 56*, 133–140.

KONIGSBERG, I. R. (1961a). "Carnegie Institution of Washington Yearbook," p. 399. Carnegie Institute of Washington, Washington, D.C.

KONIGSBERG, I. R. (1961b). Some aspects of myogenesis in vitro. *Circulation 24*, 447–457.

KONIGSBERG, I. R. (1961c). Cellular differentiation in colonies derived from single cell platings of freshly isolated chick embryo muscle cells. *Proc. Nat. Acad. Sci., U.S.A. 47*, 1868–1872.

KONIGSBERG, I. R. (1963). Conal analysis of myogenesis. *Science 140*, 1273–1284.

KONIGSBERG, I. R. (1964). The embryological origin of muscle. *Sci. Amer. 211*, 61–66.

KONIGSBERG, I. R. (1965). Aspects of cytodifferentiation of skeletal muscle. In "Organogenesis" (R. L. DeHaan and H. Ursprung, eds.), pp. 337–358. Holt, New York.

KONIGSBERG, I. R. (1970). The relationship of collagen to the clonal development of embryonic skeletal muscle. In "Chemistry and Molecular Biology of the Intercellular Matrix" (E. A. Balazs, ed.). Vol. 3, pp. 1779–1810. Academic Press, New York.

KONIGSBERG, I. R. (1971). Diffusion-mediated control of myoblast fusion. *Develop. Biol. 26*, 133–152.

KONIGSBERG, I. R. AND HAUSCHKA, S. D. (1966). Cell interactions in the reproduction of cell type. In "Reproduction: Molecular, Subcellular and Cellular" (M. Locke, ed.), pp. 243–289. Academic Press, New York.

KONIGSBERG, I. R., McELVAIN, N., TOOTLE, M., AND HERRMANN, H. (1960). The dissociability of deoxyribonucleic acid synthesis from the development of multinuclearity of muscle cells in culture. *J. Biophys. Biochem. Cytol. 8*, 333–343.

LASH, J. W., HOLTZER, H., AND SWIFT, H. (1957). Regeneration of mature-skeletal muscle. *Anat. Rec. 128*, 679–698.

MARTINUCCI, G., LEWIS, A. G., PICINNI, E., AND SAGGIORATO, R. (1970). Myoblast fusion and the mitotic cycle during muscle differentiation *in vitro. Acta Embryol. Exp. 1*, 59–77.

OKASAKI, K., AND HOLTZER, H. (1965). An analysis

of myogenesis *in vitro* using flourescein labelled antimyosin. *J. Histochem. Cytochem. 13*, 726–739.

O'NEIL, M., AND STOCKDALE, F. E. (1972). A kinetic analysis of myogenesis—*in vitro*. *J. Cell Biol. 52*, 52–65.

O'NEIL, M., AND STROHMAN, R. C. (1970). Studies of the decline in deoxyribonucleic acid polymerase activity during embryonic muscle cell fusion in vitro. *Biochemistry 9*, 2832–2839.

PATTERSON, B., AND STROHMAN, R. C. (1972). Myosin synthesis in cultures of differentiating chicken embryo skeletal muscle. *Develop. Biol. 29*, 113–138.

QUASTLER, H., AND SHERMAN, F. G. (1959). Cell population kinetics of the intestinal epithelium of the mouse. *Exp. Cell Res. 17*, 420–438.

SHAINBERG, A., YAGIL, G., AND YAFFE, D. (1971). Alterations of enzymatic activities during muscle differentiation *in vitro*. *Develop. Biol. 25*, 1–29.

SMITH, J. A., AND MARTIN, L. (1973). Do cells cycle? *Proc. Nat. Acad. Sci. U.S.A. 70*, 1263–1267.

STOCKDALE, F. E. (1970). Changing levels of DNA polymerase activity during development of skeletal muscle tissue *in vivo*. *Develop. Biol. 21*, 462–474.

STOCKDALE, F. E. (1971). DNA synthesis in differentiating skeletal muscle cells: initiation by ultraviolet light. *Science 171*, 1145–1147.

STOCKDALE, F. E., AND HOLTZER, H. (1961). DNA synthesis and myogenesis. *Exp. Cell Res. 24*, 508–520.

STREHLER, B. L., KONIGSBERG, I. R., AND KELLEY, F. E. I. (1963). Ploidy of myotube nuclei developing *in vitro* as determined with a recording double-beam microspectrophotometer. *Exp. Cell Res. 32*, 232–241.

WHITMORE, G. F., AND TILL, T. E. (1964). Quantitation of cellular radiobiological responses. *Ann. Rev. Nucl. Sci. 14*, 347–374.

YAFFE, D., AND DYM, H. (1973). Gene expression during differentiation of contractile muscle fibers. *Cold Spring Harbor Symp. Quant. Biol. 37*, 543–547.

YAFFE, D., AND GERSHON, D. (1967). Multinucleated muscle fibres: induction of DNA synthesis and mitosis by polyoma virus infection. *Nature 215*, 421–424.

III

TISSUES

CHAPTER TEN

Tissue Interactions and Related Subjects

JAY LASH

Introduction

Cell and tissue interactions originated in ancient evolutionary history, with the appearance of the first multicellular organisms [an example of such an organism in the alga *Volvox* (chlorophyceae), a simple vesicle of cells]. Individual cells came into contact with neighboring cells thus creating a totally new environment. It was now possible for cells to interact—and by so doing to influence each other's behavior.

With increasing biological complexity there is a corresponding increase in the array of cellular and tissue interactions. Interactions among cells, tissues, and organs occur not only during development but throughout the life of the organism. These interactions may involve trophic influences (e.g., nerves), metabolic influences over great or small distances (e.g., hormones and prostaglandins), or intimate relations through junctional surfaces (e.g., gap junctions).

Studies of development have concentrated primarily on an analysis of the *results* of such interactions: *the criterion used to determine developmentally significant interactions has been subsequent cyto- or histodifferentiation.* These studies, however, have usually been couched in the form of theories, laws, or dicta —from which experiments have been designed to confirm these concepts. Furthermore, the bewildering complexity of development is such that it has even been difficult to agree on general concepts. In the first section, therefore, we shall review several of the concepts of development that are of particular concern to embryologists today, along with the experimental evidence for their validity or invalidity.

Current embryological research in cell and tissue interactions deals with the acquisition, maintenance, and modulation of the differentiated state. These three areas are being approached at several levels, from molecular to morphological. Therefore, after our discussion of the prevalent concepts regarding cell and tissue differentiation, we shall review the more fruitful methods of analysis currently in use in the field.

General Concepts of Development

The first concept that we shall discuss may be stated as follows: there is a strong possibility that an antagonism may exist between growth and differentiation. *Growth* can be defined, in this instance, as primarily an increase in mass, such as the building of sand dunes through the action of winds and currents. In biological organisms this increase in mass occurs primarily through cell proliferation. *Differentiation* can be considered as the acquisition of distinctive morphology and function. To continue the analogy of dunes, the succession of dune communities would be the differentiation of subunits massed together to form the original dune. (The interaction of the constituent parts in dune succession is no less complicated than the interaction of constituent parts in organismic development.) The acquisition of distinctive morphology, or expression, may be at the intracellular or supracellular level. In reality, all expression is ultimately genotypic in origin, but, for the sake of convenience, intracellular and supracellular expression may be categorized as being either genotypic (intracellular) or phenotypic (supracellular). Figure 1 portrays this definition of differentiation.

In the past, embryologists were primarily concerned with histodifferentiation and its phenotypic, or easily visible, expression. With the improvement of techniques of biochemical analysis and more indirect methods of assay (e.g., electronmicroscopy, radioisotope incorporation, and various methods of chromatography and electrophoresis), it has become possible to study genotypic expression by assaying directly for the appearance of distinctive molecules or distinctive metabolic processes. Indeed, with the visualization of many processes that were previously studied only by inference, the boundary between genotypic and phenotypic expression is becoming blurred. The scheme of differentiation shown in Fig. 1 is not meant to be all-inclusive but is only an

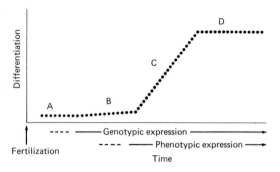

FIGURE 1. *Differentiation is the acquisition of distinctive metabolic (genotypic) and morphological (phenotypic) expressions. A represents the period shortly after fertilization when detection of distinctive products is uncommon. B represents the period during early development when specific products may be detected, but morphological manifestation is not evident. C is the period when organ and tissue structures become evident. D is the period for the differentiated state, which is usually stable.*

attempt to emphasize the different levels at which differentiation is being studied. Differentiation, after all, is primarily an operational term in the hands of the experimenter.

Another major area of concern in contemporary research has been the role of diffusion of substances between cell aggregates in growth and differentiation. Since the discovery of primary induction in amphibians by Spemann and Mangold in 1924, many investigators have sought the substance or substances, possibly transmitted from tissue to tissue, that are responsible for induction. Because the amphibian chordamesoderm influenced the overlying neural ectoderm to undergo neurulation, it was inferred that informational molecules were being transmitted. A great effort was made to isolate and characterize these molecules. In retrospect, it seems slightly naïve to have approached the problem from this viewpoint—rather than to have devoted energy to searching for the responding tissue—which is the current approach. Nevertheless, for the techniques available at the time, significant work was done in the 1930s by what might be called the Cambridge school (see Needham, 1950).

Thus, in spite of the excellent work done toward the development of the concept of "the inducer," or "the organizer," attempts to isolate the chemical inducer led to a dead end as far as understanding the mechanism of induction was concerned. Many and various bewildering compounds were isolated and characterized, and indeed some of them mimicked tissue induction. Confusion reached a height when it was found that tissue such as that from boiled beef heart had inductive properties when placed in combination with amphibian tissue. The classical work in this field is excellently reviewed in the books of Needham (1950) and Saxén and Toivonen (1962). We shall be dealing later in this chapter mainly with current approaches to the problems of tissue induction.

A third area of concern is whether or not differentiation is irreversible. The ultimate stage of differentiation is, naturally, senility and death. During the life of any organism, however, differentiation is invariably stable if we exclude from normality such matters as metaplastic conditions. If only for homeostasis, it is a good thing for the organism that liver cells remain liver cells, that muscle cells remain muscle cells, and that nerve cells remain nerve cells. The imbroglio resulting from anything but stability would not be tolerated by normal evolutionary processes. Even slow changes associated with aging should be considered as normal differentiative events and not as evidence of reversibility.

Before the time of tissue culture, the concept of reversibility was used in the study of such phenomena as amphibian lens regeneration (see Weiss, 1939). With the popularization of tissue culture, the possible reversibility of the differentiated state became a matter of serious concern. It is now generally agreed that most, if not all, cells will maintain a differentiated state in culture provided a suitable environment is maintained. Much of the work to be discussed in the second part of this paper is predicated upon the fact that the differentiated state, once attained, is perpetuated.

The concept of modulation is also primarily a result of *in vitro* manipulations. Modulation, in this sense, is an apparent change in form or function due to environmental conditions but not necessarily a change of genetic potential (see Weiss, 1939). When a cell loses or changes its character after being considered "differentiated," does this mean that it has

dedifferentiated, or that it had not differentiated in the first place? The concept of modulation, or dedifferentiation, has caused much controversy and, until recently, has tended to obscure more basic problems. Recent work on amphibian regeneration has done a great deal to clear up this controversy (see the chapter by Hay in this book).

Another current belief is that growth and differentiation may depend upon the flow of materials from one cell that is interior to another through junctional cell surfaces or connections. This concept has received impetus in recent years from the discovery that there is, between embryonic cells, electric coupling through low-resistance junctions (gap junctions). This very exciting problem is still at the descriptive stage and has not yet reached the point where developmental events can be explained or predicted. Nevertheless, this research has great potential in our efforts to understand certain types of cell-to-cell interactions.

The last concept that we shall mention is one that concerns the molecular level. Although the molecular aspects of development are extensively treated elsewhere in this book (see the chapters by Nemer, Rifkind, and Wilde), the area that concerns us here is the manner in which we obtain evidence of gene activity in differentiating systems. Two alternative views may be stated in one sentence: Is differentiation due to the progressive activation and/or restriction of gene activity? Our discussion of this concept will be based on the examination of distinctive gene products during differentiation.

Methods of Analysis

Needless to say, the most common method of analysis is one incorporated into all of modern experimentation—that of observation and measurement. Historically, this method predates record keeping and is the beginning of embryology as a descriptive science. Scientists, endowed with an inquisitive nature, continued investigating and exploring problems of development as a primarily descriptive science until the late nineteenth and early twentieth centuries, when experimental embryology had its beginning.

In what may be considered the birth of modern experimentation, Chabry in 1887, working with eggs of the ascidian *Ascidiella*, performed experiments leading to the concept of mosaic development. By surgically destroying selected blastomeres, he observed that the resulting embryo was defective. These observations led naturally to the concept of cytoplasmic localization of developmental information. In contrast to the ascidians, which show a marked cytoplasmic localization and "mosaic" development, the other extreme is seen in the "regulative" development of the echinoderms, where localization is a comparatively late event with respect to the rate of cleavage. As with most embryological concepts, upon further analysis extreme differences among groups gives way to intergradations, so that the dichotomy between mosaic and regulative development, although useful, is no longer the valuable concept it once was.

This brief discourse on the beginning of experimental embryology (or developmental biology), and the subsequent alteration of the conceptual content of its original findings concerning cytoplasmic localization, is given to show a certain continuity in the approach to developmental problems. Almost invariably, original discrete observations, upon which general principles can be constructed, become useful but by no means all-encompassing rules that tissues and cells must follow.

CELL REAGGREGATION

An important early concept in experimental embryology (that of the mechanisms of tissue construction) was derived from observations on the aggregation of component parts to yield a functional tissue or structure. This phenomenon has been studied by watching cells come together originally during development, or, in particular, by studying reaggregation after artificial dissociation. Reaggregation experiments go back to the turn of the century (1907), when H. V. Wilson first studied the dissociation and reaggregation of cells of marine sponges. Wilson found that he could take a sponge and disaggregate the constituent cells

by pressing the sponge through a bolting cloth of the proper mesh. The disaggregated, separate cells would then reaggregate to form a functional sponge. Obviously such experiments have limitations. It is unlikely, for example, that they can be performed using mouse embryos [although the experiments of Tarkowski (1961) and Mintz (1965) on creating mouse chimeras are a form of reaggregation experiment in which blastomeres are used instead of cells from later stages of development].

Thus there are vastly different levels of complexity when studying the phenomenon of self-assembly or reaggregation. Simple experiments with relatively few parameters can frequently give useful information about more complex systems, and this has been the case in studies of the interactions during cell reaggregation. Using Wilson's results as their experimental model, Moscona (1963) and Humphreys (1963), at the Marine Biological Laboratory, studied the dissociation and reaggregation of two sponges, *Haliclona occulata* and *Microciona prolifera*. Conveniently, these two sponges have different colorations: *Haliclona* is bluish purple and *Microciona* is reddish orange. If these two sponges are dissociated separately through bolting cloth, and the isolated cells placed in a dish together at room temperature, they reaggregate separately and sort out as to species type. The aggregates are made up of either *Haliclona* cells or *Microciona* cells—and never a mixture of the two. If the two sponges are dissociated "chemically," by means of agitating them in calcium- and magnesium-free seawater and then mixing them together in regular seawater, they reaggregate and sort out, but at a much slower rate. At lower temperatures (5°C), the chemically dissociated cells do not reaggregate, whereas "mechanically" dissociated cells do—though at a very slow rate. These experiments imply that the "chemical" dissociation removed something that was necessary for the cells to reaggregate.

The mode of reaggregation as reported by Sindelar and Burnett (Burnett, 1971) is by way of filopodal extensions. The sponge cells send out long filopods that contact neighboring cells. If the contacted cell is of the same species, then the two cells are drawn together. Cell–cell recognition thus occurs through filopodal contact alone, in that the filopods do not recognize cells of different species. The reaggregation experiments at different temperatures strongly suggest that underlying this phenomenon is a metabolic process rather than a purely physical one. Indeed, Moscona and Moscona (1963) have shown (using vertebrate tissues) that the process of reaggregation is inhibited by puromycin, an inhibitor of protein synthesis. This implies that the synthesis of specific "cell-ligands" (Moscona, 1968) was inhibited. Humphreys (1965), however—using sponge cells—showed that puromycin was not inhibitory to reaggregation. These conflicting interpretations have not yet been completely resolved.

There is general, though not unanimous (see Curtis, 1970; Maclennan, 1970), agreement that a glycoprotein from the extracellular matrix mediates cell-specific recognition and aggregation in some sponges, though not by any means all sponges. If this factor is removed by "chemical" dissociation, such dissociated cells will aggregate very slowly unless the factor is re-added to the seawater. It thus fosters, or stimulates, species-specific reaggregation. The extracellular substance of *Haliclona* (H factor) influenced *Haliclona* reaggregation but not that of *Microciona*, and the *Microciona* factor (M factor) only influenced *Microciona* behavior (Humphreys, 1970).

Thus the story is far from complete. There is no doubt that for these two species of sponges, an extracellular substance is formed that is strongly correlated with the ability of these cells to reaggregate only with their own type. To emphasize how these results cannot yet be put forth to support a general concept, different results are obtained with most other sponge species (see Curtis and van de Vyver, 1971). Whatever the exact nature of the mechanism of reaggregation in *Haliclona* and *Microciona*, it is not the same mechanism for all sponges. The reader is referred to some of the recent work by Lilien (1968), which suggests that for some vertebrate tissues there might also exist specific extracellular substances.

Using sponges, Curtis (1962) has performed experiments that show that species-specific reaggregation is not necessarily due to a specific factor but may be due to differing rates of reaggregation. There may occur quantitative, but not qualitative, differences in aggregation factors. If sponge A has a reaggregation time of 3 hours and sponge B has a reaggregation time of 8 hours, a mixture of the two sponge cell populations will eventually sort out. If they have equivalent reaggregation times, they will not sort out. These results were predicted and obtained, using the British sponges *Microciona sanguinea*, *Halichondria panicea*, *Hymeniacidon perleve*, and *Suberites suberites* (Curtis, 1962). The respective reaggregation times for these sponges were 3, 8, 15, and 15 hours. The only two species that formed chimeras and would not sort out were the two species having the same reaggregation time of 15 hours (*Hymeniacidon* and *Suberites*).

There is no unanimity in interpreting these rather simple experiments. Another hypothesis has been advanced by Jones (1966) positing the "contraction and relaxation of actomyosin-like proteins with ATPase activity located at the cell surface." According to Jones and Kemp (1970), there is contractile material, presumed to be actomyosin, at the surface of cells. Upon contraction there would be an increased viscosity within the cell, a change in membrane charge, and nonadhesion. Upon relaxation, these conditions would be reversed, and the cells would become more adhesive. According to Jones (1966), the aggregating factors of Moscona and Humphries do nothing other than act as "relaxation" factors. Upon chemical dissociation of the sponges, this hypothetical relaxation factor is removed, the contractile system contracts, and the cells do not adhere (i.e., reaggregate).

There is ample evidence that most, if not all, cells do contain actomyosin-like material (see Spooner's article in this book; Jones and Kemp, 1970). Thus there may be some validity to Jones's interpretation. Either way, it is obvious that different investigators, studying the same system, have come up with widely differing interpretations. Even though the problem of sponge reaggregation is still unsettled, it should be pointed out that the various interpretations of this most simple case are not mutually exclusive. In fact, Curtis's most recent work, in which two different strain types of the fresh water sponge *Ephydatia fluviatilis* (Curtis and van de Vyver, 1971) were used, attempts to reconcile the contradictory results. Each of the two different strain types of *Ephydatia* (alpha and delta) produces a soluble factor (as yet not identified) that is responsible for specific reaggregation. Using his method of measuring collision coefficients of a cell population undergoing adhesions, Curtis concludes that the alpha factor increases the adhesiveness of alpha cells and decreases the adhesiveness of delta cells, resulting in specific reaggregation. The delta factor behaves in a similar fashion. These experiments stress that the adhesion of cells is nonspecific but that factors can control adhesion and determine its quantitative value. Such an interpretation does tend to reconcile the various views.

Vertebrate cells have also been used extensively in studying problems of cell interactions during reaggregation, and conceptually the work has not been too different from the work on sponges. Holtfreter's pioneering work in 1947 showed that embryonic amphibian cells, if dissociated (in a manner very similar to the method of chemical dissociation of sponge cells), will reorganize to form recognizable tissue. Most of the work has been done on trypsin-dissociated embryonic tissues, and significant contributions to this area have been made by A. Moscona, P. Weiss, and M. Steinberg. Many of the tissues used by these workers are the same, but the interpretations are quite different. Moscona (1968) believes that specific "cell-ligands" are responsible for selective cell adhesions, whereas Steinberg (1970) invokes differential strengths of adhesion to account for sorting out.

With few exceptions, most of the work on vertebrate tissues has been performed using reasonably long-term cultures (2 to 3 days). Tissues used for these studies have included liver, cartilage, kidney, epithelium, neural and pigmented retina, heart, and muscle. The cells are disaggregated, usually by means of tryptic digestion (see Moscona, 1967), placed into a flask of culture medium on a gyratory shaker, and then observed days later to see how they have reaggregated.

Until recently, however, a very important aspect of this methodology was overlooked. Cells are, after all, living, metabolizing units with a constant turnover of cell membranes—regardless of whether the cells are dividing, moving, or still. If one disaggregates cells and allows them to reaggregate, and then observes the aggregate days later, it is very likely that the cell membrane will be different. This dif-ference is probably exaggerated by the trypsin damage created by the disaggregation process. Even if the membranes remain chemically the same, they have probably been broken down and replaced a number of times during the culture period, and they may even have been put back together slightly differently. Thus Roth (1968) looked at adhesions shortly after the cells had been disaggregated to see if they behaved in the same manner as in the longer-term experiments of Moscona, Steinberg, and others. He found that when he disaggregated different types of cells, they initially adhered to their own type more readily than to other cell types. There was no evidence of the differential adhesiveness of cells, as has been proposed by Steinberg. Rather, the hierarchy of adhesivity does not develop until later in the life of the culture, indicating that there are indeed membrane changes taking place during the culture period.

Recently, Roth et al. (1971a, b) have shown that the enzyme betagalactosidase renders the cells less specific in their reaggregation; the interpretation is that the complementary reaction between cell surfaces is between enzymes and substrates—specifically between carbohydrates and glycosyltransferases. That this approach—that is, the chemistry of the cell surface—may be a fruitful area of study is indicated by recent work in alterations of the cell surface components after viral infections. Inbar and Sachs (1969), Burger (1973), and Warren et al. (1973) have shown that cell surfaces of cell lines (i.e., altered cells, not primary embryonic cells) are changed after viral infections. The thrust of this work on cell lines is to be found in cancer research, but the techniques of studying the chemical aspects of the cell surface are beginning to be applied to embryonic cells (Roth et al., 1971a, b; Goldschneider and Moscona, 1972).

Growth Versus Differentiation: Antagonism or Compatibility?

Although the belief that there is an antagonism between mitosis and function, and between proliferation and cell differentiation, was a valid one when it was first proposed (Hertwig, 1920; Peter, 1930; Weiss, 1939), such a concept is no longer easily defended. In this section we shall present both sides of the question with an eye toward clarifying the difficulties that lie in the way of resolving the problem once and for all.

In the following discussion, a distinction will be made between "mitosis" and "proliferation." *Mitosis* will be used as a descriptive word for the division of a cell after chromosome duplication. This process of cell division (mitosis) is not, according to the previously stated concept, considered to be compatible with normal cell function. *Proliferation* will be used to mean the increase in number of a population of cells through mitotic divisions. It is this proliferative period that is considered by some to be incompatible with cell differentiation. This concept was originally based upon early descriptive embryology, at which time the observation was made that regions of the embryo that were undergoing rapid proliferation had no morphological manifestation of differentiation. Because the concept of differentiation was, when first developed, limited to what could be observed in histological sections under the light microscope, the conclusion that proliferation and differentiation were incompatible was a very natural one. Finally, we shall consider the term *differentiation* to mean the acquisition of tissue-specific, or characteristic, biosynthetic processes.

ANTAGONISM

The advent of tissue culture gave further impetus to the idea that proliferation and differentiation were antagonistic, and brought to light an additional aspect of this incompatibility—the distinction between *acquisition* and *maintenance* of the differentiated state. The early observations (see Weiss, 1939) had been based on the acquisition of the differentiated state, but in 1930 Doljanski observed that liver cells, in culture, began to proliferate, and, in so doing, lost their function. Many other experiments have since been performed with similar results, suggesting that when such cells proliferate, they become "dedifferentiated" (see

Weiss, 1939; 1968). Thus, whereas the stability of the differentiated state had seemed self-evident, the invention of tissue culture created the problem of whether or not proliferating cells underwent dedifferentiation. It is well known—and in fact is incontestable—that during embryonic development, some cells undergo repeated divisions before there is any manifest differentiation. After proliferation, many tissues undergo mitotic arrest, and thereafter remain mitotically quiescent. The term "postmitotic" was created for these mitotically quiescent cells to distinguish them from cells with a malignant potential (the "intermitotic" cells) (Cowdry, 1950). "Postmitotic" is probably an unwise designation when applied to embryonic development, however, because (1) it implies a distinction between a dividing and nondividing cell, which may not be warranted, and (2) all cells (even a zygote) are technically postmitotic except while undergoing the process of division.

The best evidence for the concept of antagonism between proliferation and differentiation, however, has been derived from studies on myogenesis in tissue culture. It has been agreed for many years that the nuclei in multi-nucleated muscle fibers do not divide except under rare and pathological circumstances. As single cells, myoblasts undergo proliferation before they fuse to form the multinucleated muscle fiber. Much evidence has been mustered to indicate that (*in vitro*) myoblasts have a programmed number of divisions that they must undergo before they are capable of fusion, and that the myoblasts do not differentiate until after they have undergone mitotic arrest (Holtzer, 1970). It must be emphasized that all these experiments were performed *in vitro*, and hasty conclusions should thus not be drawn about *in vivo* development.

Another instance in which there is an apparent conflict of interest between proliferation and differentiation is in mammary gland development (Topper and associates, Lockwood et al., 1967). The mammary gland requires the hormones insulin, prolactin, and hydrocortisone for the *in vitro* synthesis of its tissue-characteristic product casein. Insulin is considered to induce DNA synthesis (Stockdale and Topper, 1966), although the mechanism by which this occurs is not known.

Using colchicine to block mitosis, it was observed that casein synthesis was blocked, and from this it was concluded that mitosis was necessary for differentiation. The implication was that it is not mitosis *per se* that is required, but a *special* mitotic division. If blocking mitosis were the only effect that colchicine produced, this interpretation might be valid, but it is now understood that colchicine has more powerful effects than just the impairment of mitotic division (see Manasek, 1972). Therefore, the prerequisite for mitosis prior to the synthesis of casein is not firmly established. It is firmly established, however, that prolactin will stimulate casein synthesis in cells that have been induced by insulin to proliferate in the presence of hydrocortisone. As mentioned, the mechanism of action of these hormones upon casein synthesis is unknown, but the correlations are clear. Since hormones seem to be effective only upon dividing cells in this system (Turkington, 1971), it appears that hydrocortisone may gain entry only during the events of mitosis, thereby changing the potentiality of the cell to make it responsive to prolactin. In this case, the mystique of proliferation versus differentiation may be reduced to the mitotic event, permitting an inducing agent (i.e., hydrocortisone) to enter the cell (or nucleus) and modify the cell in such a way that it will now respond to prolactin. Hence, in this *in vitro* model, there is a requirement for proliferation (i.e., the mitotic event) before differentiation, which could be considered as evidence that the antagonism between proliferation and differentiation may be a real phenomenon. Couched in slightly different terms, however, such a requirement may show only that during certain events associated with mitosis (e.g., nuclear membrane breakdown), agents may enter the cell and effect changes. With this interpretation, the terms "postmitotic" or "quantal division" become less meaningful (see Turkington, 1971; Holtzer, 1970).

Yet another instance showing an antagonism between proliferation and differentiation, and an apparent requisite for mitosis, is the work of Wessells, using *in vitro* studies of pancreas differentiation (Wessells and Cohen, 1967). It was found that explanted embryonic pancreas had mitotic figures throughout the tissue early in culture—but that later the inner cells went into mitotic arrest, whereas the peripheral cells still underwent divisions. Correlated with the cessation of mitosis was the onset of pancreatic differentiation, as measured by the morphological appearance of pancreatic acini. Although there seems to be a clear difference between proliferation and differentiation in this instance, a word of caution should be inserted. Is the pattern of proliferation and differentiation as found in explants the same as that which occurs *in vivo*, where a blood supply is available? Or may these observations be the result of the *in vitro* environment, such as the depletion of nutrients in the avascular central region of the explant? Would chorioallantoic grafts of pancreatic anlagen show the same correlations between areas of proliferation and areas of differentiation?

As mentioned earlier, differentiation has been studied at various levels. When studying myogenesis, one can stain the preparation with iron hematoxylin and look for the formation of the typical striated fibers. Or, one can look at the ultrastructure for the formation of muscle proteins, the thick and thin myofilaments. With pancreas, one can study the differentiation of the glandular acini or, at another level, look for evidence of the synthesis of pancreas-specific products, Kallman and Grobstein (Grobstein, 1962) did just this, by looking for electron microscope evidence of zymogen granules. Using this level of assay for differentiation rather than that of gross morphology, it was found that the *in vivo* pancreas anlagen was synthesizing tissue-characteristic products (i.e., zymogen granules) at the time the tissues were dissected for culture (11-day-old mouse embryos). Comparable ultrastructural studies on cultured material (Kallman and Grobstein, 1964) have shown that zymogen granules do not appear until after 4 to 5 days of culture. It would appear, then, that the preparative procedures (e.g., trypsinization) had diminished the tissue's synthetic capabilities, which were then regained after a period of time in culture. We shall return later to this aspect of tissue interactions—that is, the time difference between *in vivo* and *in vitro* differentiation. For the present, the point is that in the case of pancreas, *in vitro* studies on the antagonism between proliferation and differentiation are inconclusive. *In vivo* differentiation and ultrastructural studies do not support the earlier contention that pancreatic tissue has to undergo a critical number of mitoses before it can differentiate. Whether the proliferation *in vitro*, which precedes overt differentiation, plays a significant role is unclear at this time.

COMPATIBILITY

Recently, new evidence has come to light that appears to show that for many tissues there is no antagonism between proliferation and differentiation and that many of the cases previously cited may have been due to artifacts of the culture situation. Cameron and Jeter (1971) have summarized the relation between proliferation and differentiation for various tissues and find that in most instances there is no real evidence of antagonism. Davis and Burnett (1964) have reported an interesting case in the hepatopancreas of the crayfish, where the same cell can go through multiple stages of differentiation without intervening periods of mitosis.

A major contributing factor to this controversy is the methodology of analysis. For reasons not known, when cells are explanted to tissue culture, they are frequently released from mitotic arrest and begin to proliferate. Similarly, when epithelial sheets are interrupted by injury, or explanted, they tend to spread and cover the available surface, then stop (Lash, 1955). This is an *in vivo* example of "contact inhibition" (see Abercrombie et al., 1957; Abercrombie, 1967). Why tissues and cells behave in such a characteristic manner when explanted is not yet known. Thus many experimenters have concluded—though tentatively—that there is some antagonism between proliferation and differentiation.

As noted previously, the studies that best support the case for antagonism are those done in *in vitro* myogenesis. Recently, however, Konigsberg (1971) performed experiments that call for a reevaluation of the myogenesis problem (see also the chapter by Konigsberg in this book). It had been shown previously that if embryonic muscle were dissociated with trypsin, and the resulting cells placed in culture, they would not fuse to form multinucleated myotubes until after having undergone critical, or "quantal," mitoses (Holtzer, 1970). The

original source of the mononucleated cells, whether from multinucleated myotubes or mononucleated myoblasts, is not known. This information is crucial in an attempt to give *in vivo* significance to the theory of quantal mitoses during myogenesis. If the cells are derived from multinucleated myotubes, they have presumably already finished the "critical" division. Konigsberg (1971) has obtained evidence for a diffusion-mediated control of *in vitro* myoblast fusion. A very simple experiment was performed, using different dishes with varying numbers of myoblasts. If the myoblasts had to undergo critical and programmed divisions before fusion, then myotubes should form in all dishes at the same time. It was found, however, that the dishes with the greatest number of cells had myotubes first and that the rate of fusion was correlated with cell concentration and the duration of the G_1 phase of the mitotic cycle—not the number of divisions. Similar experiments, in which the number of cells was kept constant but the amount of medium was varied, yielded similar results—that is, the cells of the culture with the least amount of medium were the first to exhibit fusion. Other experiments, in which the amount of DNA per culture was measured, showed that cells in a higher volume of medium go through another round of division before fusion, indicating that myoblast fusion is not related to critical divisions. Konigsberg (1971) has partially characterized a factor from embryo extract and has determined that "activity" resided in a molecular weight fraction $>300,000$. These results are similar to those of de la Haba and Amundsen (1972), who obtained two fractions from chick embryo extract, one promoting fusion, the other promoting myotube elongation. Whether Konigsberg's fraction is the same as that of de la Haba has yet to be determined. The result of these works is that it is not the mitotic event that is critical, but rather the production of a substance in the environment that promotes cell fusion, presumably by acting upon the cell membrane. Stockdale and O'Neill (1972) have come to the same conclusion—that myoblasts in culture proliferate or fuse into myotubes as the result of environmental factors, and not as the result of any intrinsic programs. This is reminiscent of the work mentioned previously with sponges, where an environmental factor (extracellular material) is held responsible for cell aggregation, again presumably working at the cell surface.

Needless to say, there are many other experiments on this subject. The proposed antagonism between proliferation and differentiation has to be particularized at various levels. What criterion of differentiation is to be used? How closely does the cell or tissue culture situation resemble what is taking place in the organism? *In vitro* methodology introduces artifacts. Proliferation is a prerequisite of normal development; otherwise there would be no multicellularity. Whether a function more basic than just proliferation can be ascribed to the mitotic event or not is a question as yet unanswered.

There is evidence that not all the gene material in cells is active—that is, some of the genetic machinery may be "masked." It has been proposed by Ebert (1968) that the act of chromosome duplication and cell division may actually "clean" inactive regions of the genome, thereby making them more active. Such a role for cell division would support the contention that proliferation plays a very important role in development over and above that of increasing cell numbers. That such a phenomenon—that is, a requirement for mitosis—occurs in viral-infected cells is incontestable (Burger, 1973). In primary cultures of embryonic tissues, however, there is no real evidence that anything like a "gene-cleaning" phenomenon is taking place. The present evidence is not strong enough to say that after division a cell can make product A but cannot so do before division. On the contrary, what little evidence there is indicates that some tissue-characteristic products are made much earlier than previously suspected (Marzullo and Lash, 1967: chondroitin sulfate; Klose and Flickinger, 1971: collagen; Kosher and Searls, 1973: chondroitin sulfate). Such early synthesis of tissue-characteristic products would be in agreement with the results of Daniel and Flickinger (1971), who report a decrease in the number of kinds of DNA-like RNA during early *Rana pipiens* development.

The level of assay is important in interpreting experiments. Morphological evidence of differentiation (i.e., phenotypic expression) may give an opposing interpretation. Although most embryologists lay claim to working in problems of cell differentiation, the most frequent analysis is upon tissue differentiation. Differentiation, in the strict sense, of systems such as blood cells or pigment cells, which differentiate as separate elements, may follow different rules with respect to division and differentiation than do tissue (i.e., multicellular and heterogeneous) systems (see the chapters in this book by Rifkind, Whittaker, and Wilde). If the ability to synthesize cell-characteristic (or tissue-characteristic) products is the criterion of differentiation, then few if any cells (or tissues) differentiate *after* proliferation (i.e., when in mitotic arrest). Additional evidence of this will be presented later.

With increasingly sensitive methods of assay, tissue-characteristic biosynthetic processes are being found at increasingly early stages of development (Marzullo and Lash, 1967; Klose and Flickinger, 1971; Kosher and Searls, 1973). This raises the subtle distinction between proliferation and cell division, as mentioned earlier. The act of division itself may play a more significant role during early cleavage stages, as the egg is being cut down into workable units (see Flickinger et al., 1973). In the proliferative stages, however, some characteristic synthetic patterns already seem to be established. To help sort out some of these problems, it may be necessary to amplify our terminology. "Differentiation" means many things to different people, and its use has created confusion when applied to different tissues or products. Hopefully it will not be necessary to redefine any terms but just to use more appropriate adjectives to define which level of differentiation is being discussed. Earlier we mentioned an operational distinction between "genotypic" and "phenotypic" levels of expression; similar terminology may be useful when applied to the concept of differentiation.

Tissue Mass and Differentiation

The question may be justly raised as to whether there is a minimum amount of tissue mass required for differentiation. Again, we must make the distinction between *in vivo* and *in vitro*. *In vitro*, there is no doubt that a single cell, or even a few cells, can become differentiated. Blood or pigment cells in culture become differentiated—even chondrocytes will synthesize their characteristic matrix. This is not, however, the level of differentiation implied in tissue-mass experiments. Rather, in these instances, differentiation refers to *tissue* differentiation, which implies the differentiation of a cellularly heterogeneous tissue, such as skeletal muscle, kidney, or thyroid, plus their supportive and vascular tissue elements.

In vivo, asexual budding in some of the colonial ascidians probably represents the smallest mass of tissue capable of independent differentiation. Sometimes the bud consists of just a few cells, which develop into a complete organism. In vertebrate tissues, very little has been done in his area. Probably the most significant work is that of Grobstein and Zwilling (1953), who cut chick blastoderms into different-sized pieces and followed subsequent differentiation *in vitro*. As a result of these experiments, it was concluded that the minimum size of tissue required for organ or tissue differentiation is no less than 0.1 to 0.2 mm in diameter. Similar *in vitro* experiments, utilizing embryonic chick somites, were reported by Holtzer (1964) in which an attempt was made to determine the relation between mass and chondrogenic differentiation. He, too, found a minimum size necessary for chondrogenesis, comparable in order of magnitude to that of Grobstein and Zwilling. Unfortunately, the differentiation of other tissue elements in the somite explants (muscle, dermis) was not assayed. These experiments, correlating tissue mass with differentiation are at present not interpretable in mechanistic terms. *In vivo*, this distinction does not hold up. Could this be another instance of tissue-culture artifacts?

Again working *in vitro*, Lash (1968a) was unable, using explants larger than 0.2 mm in diameter, to show a correlation between mass and chondrogenic differentiation, whereas Ellison et al. (1969), using smaller pieces, obtained evidence that there was a correlation. Whether there is a meaningful correlation between mass and differentiation is unclear. Clearly, in cellular systems, there is no correlation, whereas some tissue (multicellular) explants do show a correlation. One clue as to the mechanism behind some of these *in vitro* experiments is that the tissue explants used in all these experiments consisted of heterogeneous cell types. As concluded by Lash (1968), the increased differentiation in larger explants may be due solely to the chance inclusion of cells with a greater differentiative bias than others. The smaller the explant, the less likelihood of including such cells; the larger the explant, the greater the likelihood of including these cells. Another factor is the viability of these cells in culture. They may be perfectly capable of differentiation *in vivo* but, under the stresses of the culture environment, may not be able to function normally. For example, it has been shown that during the first 24 hours of culture, a number of chondrogenic cells of explanted somites will die (Gordon and Lash, 1974).

With embryonic somites, it is true that certain cells have a greater chondrogenic bias in culture than others (Lash, 1967), and differentiation depends upon the survival of these cells (O'Hare, 1972; Gordon and Lash, 1974). The correlation between mass and differentiation here is clearly created by the culture environment. More work will have to be done on the Grobstein–Zwilling system. It may be that these results are created also by the relatively hostile culture environment.

Tissue Interactions and Inductions

The earlier sections of this chapter dealt with various subsections of the field of tissue interactions—cell–cell interactions, proliferation, and the different levels of bioassay. An understanding of these areas is essential in the attempt to unravel the almost bewildering array of experiments and conclusions that have been published in recent years with respect to tissue interactions. It is worth repeating the statement made earlier about the advent of multicellularity. With multicellularity, it became necessary for cells to interact and communicate—not only for the development and maintenance of the organism but also for modulatory events. When interactions occur during embryonic development, and these interactions stimulate a particular line of differentiation, the term used to describe such events is "embryonic induction." The earliest examples of such interactions appeared in the literature almost simultaneously. Lewis (1903) and Spemann (1903) reported the induction of the amphibian lens by the wall of the forebrain (see Needham, 1950). In 1909 Browne (who later, as Ethel Browne Harvey, published extensively on echinoderms) demonstrated that a new bud in Hydra could be induced by the implantation of a piece of hypostome. These demonstrations were clear and unequivocal, bypassing even the need for microscopic examination—here was clearly an example of phenotypic expression alone being sufficient for analyzing the effects of interacting tissues.

A similar, and more widely known, example of tissue interactions resulting in a specific line of differentiation is that of Spemann and Mangold's (1924) demonstration that the amphibian chordamesoderm can induce the overlying tissue to undergo neurulation and that it in effect "organizes" the formation of the embryo. The early interpretation of such experiments was that one tissue instructed another to respond in a certain way. As the discipline developed, with various transplantation and recombination experiments, the problem of assay methods became more critical. If the end result of tissue interactions during embryonic development is a particular type of differentiation, how does one determine when a cell or tissue becomes "differentiated"? Is morphological (i.e., phenotypic) expression a valid criterion, or should metabolic activity be used as a more accurate assay for differentiation? Can, or should, one type of assay be dissociated from the other? It is becoming increasingly important to define morphological differentiation in ultrastructural or biochemical terms.

Although many different interacting systems have been analyzed (for the most part in the past 20 years*) the three systems that have been studied in the greatest detail are kidney tubulogenesis, pancreas development, and ver-

tebral chondrogenesis. With few exceptions, all interacting systems analyzed consist of epithelial–mesenchymal interactions. The mesenchymal tissue is usually the one that evokes a response from the epithelial tissue. For example, the pioneering work of Grobstein (1955, 1956) showed that the mesenchyme surrounding the epithelial ureteric bud in mouse embryos was necessary for kidney tubulogenesis. Although the mesenchyme does contribute to the metanephric tubular system (the proximal and distal convoluted tubules), the stimulus for tubulogenesis is the interaction of the epithelial ureteric bud and its surrounding mesenchyme. In most interacting systems, cellular contact was presumed to be unnecessary. Recent observations by Saxén (1972) indicate that, although the tissues are separated by a porous filter, cell contact does occur and may even be correlated with subsequent tubulogenesis. Although kidney development has been analyzed thoroughly (see Saxén et al., 1968, for a review of work to date), it has not been analyzed biochemically, because the only reliable assay for differentiation presently available is morphological.

Another complication in studying tissue interactions is that frequently a tissue that has absolutely no developmental significance to another tissue can promote the latter's differentiation. For example, the dorsal—not the ventral —half of the embryonic spinal cord will promote kidney tubulogenesis (Grobstein, 1956). Even though a great deal has been done to unravel the interactions involved in kidney tubulogenesis, most of these investigations are at the morphological level.

The development of the pancreas and vertebral cartilages have been studied at the level of genotypic response. Kallman and Grobstein (in Grobstein, 1962) had found that the pancreas anlagen had ultrastructural elements—indicating that they were genotypically active as pancreas cells—before any morphological aspects of "pancreas" were evident. Since the exocrine portion of the pancreas manufactures such readily identifiable products as amylase, lipase, DNAase, RNAase, trypsinogen, and chymotrypsinogen, and the endocrine portion synthesizes glucagon and insulin, it was possible to assay the pancreas for tissue-characteristic products without relying solely upon morphological characteristics. When such analyses

*Following is a partial list of interacting systems that have been studied in recent years; in all instances, epithelial–mesenchymal interactions are involved (see also Needham, 1950; Saxén and Toivonen, 1962): kidney, salivary, thymus, pancreas, thyroid, mammary, lung, taste buds, erythrocytes, liver, skin, feathers, limb buds, vertebral cartilage, otic capsule, and teeth.

were performed, it was found that at a very early stage of development, many days before morphological manifestations, the pancreas was making detectable amounts of tissue-characteristic products (Rutter et al., 1968). In this instance it was possible to dissociate metabolic function from phenotypic expression, and it was clear that morphology could be a misleading criterion of differentiation.

Another system analyzed in this manner is vertebral chondrogenesis. It has long been known that *in vitro* somite chondrogenesis can be induced by recombining isolated somites with either notochord or ventral spinal cord (Grobstein and Parker, 1954; Grobstein and Holtzer, 1955; Avery et al., 1956; Lash et al., 1957). The criterion used in these early experiments was the appearance of cartilage nodules, visible in culture and stainable with appropriate histological stains. As with the earlier work with the pancreas, morphological indications of differentiation did not appear in culture until after approximately 4 days (Lash et al., 1957; 1960). Over the years, as culture techniques improved and biochemical assays were perfected, it became obvious that the somite cells could be characterized as genotypically differentiated well before morphologically identifiable cartilage appeared in the explants (Lash, 1968b; Ellison and Lash, 1971; Minor, 1973). Primarily because of these two interacting systems—pancreas and vertebral cartilage—the concept of tissue interaction (or induction) as an instructive event has been reevaluated (Grobstein, 1967; Rutter et al., 1968; Zwilling, 1968; Ellison and Lash, 1971). It is now clear that most, if not all, systems of *in vitro* tissue interactions are permissive events. There is no convincing evidence yet of embryonic induction being an instructive event.

The artifacts of tissue culture misled early workers to presume that the onset of differentiation seen *in vitro* mimicked the *in vivo* development. In the case of pancreas and vertebral cartilage development, this is clearly not the case. The tissue-culture environment just reestablishes milieux for permissive differentiation. *In vivo* interactions are undoubtedly necessary and may constitute a central control mechanism during normal embryogenesis. Somehow embryos do develop into a harmonious organism with definite forms and functions

rather than into heterogeneous populations of cells with genotypic markers.

With appropriate manipulation of the environment, tissue culture shows that the determining factors in tissue differentiation occur at an unknown, but extremely early, stage of development (Ellison and Lash, 1971). This cannot be stated as a general rule for *all* developing systems, since most interacting systems have not yet been amenable to exacting biochemical analysis. Nor can it be stated that the interaction is always *solely* permissive. As with other areas of developmental biology, the presently available information precludes the construction of all-encompassing general concepts. One possible effect of *in vitro* interactions that is indicated both in pancreas development (Ronzio and Rutter, 1973) and vertebral chondrogenesis (Minor, 1973; Gordon and Lash, 1974) is the fact that inductive interactions may result in proliferation and less cell death in the culture environment. There is no evidence as yet that this proliferation does more than provide a suitable mass of tissue to become functionally differentiated. One problem that has plagued most investigations of *in vitro* interactions is the fact that after being transferred to the culture environment, most tissues exhibit varying degrees of abnormal cell death. In the case of vertebral chondrogenesis, the chondrogenic cells are particularly prone to culture-related death (Lash, 1967; Minor, 1973; Gordon and Lash, 1974).

Thus, at present, the conclusions on tissue interactions can only be tentatively stated. Interactions are, both phylogenetically and ontogenetically, older than the early workers had fully appreciated. The quest for an "inducing" molecule and the advent of *in vitro* studies have added invaluable insights to the general problem. Unfortunately, the complication introduced by the experimental manipulations frequently obscured the real nature of the developmental event under study. Current investigations utilizing the powerful tools of contemporary science are rapidly bringing to light details hitherto unsuspected. Every chapter in this book relates significant information acquired in just the past few years. Frequently, such information requires a serious reevaluation of previously accepted concepts, and, again, practically every chapter in this book could be cited as evidence for this statement.

A great deal of work done previous to 1950 in tissue interactions resulted from the stimulus of the work done in the 1920s and 1930s. In the 1950s the field expanded greatly, many rules were devised, and many experiments were designed to prove these rules. Now it appears that many such constructs hold true only for the artificial circumstances of cell or tissue culture and may have no (or little) bearing on what happens in the embryo. Only recently have increasing efforts been made to relate tissue-culture experiments to what happens in the embryo.

It would be difficult, if not presumptuous, to attempt to predict future work on interacting systems. Nonetheless, it would be reasonably safe to say that future studies will undoubtedly be similar in intent to the previous quest for the inducing molecule. With current methodology, however, the quest will be directed toward the *mechanism* of development, and involve two of the most rapidly expanding fields of modern biology: DNA and RNA biochemistry, and membrane structure and function.

References

ABERCROMBIE, M. (1967). Contact inhibition: the phenomenon and its biological implications. *Nat. Cancer Inst. Monogr. 26*, 249–264.

ABERCROMBIE, M., HEAYSMAN, J. E. M., AND KARTHAUSER, H. M. (1957). Social behavior of cells in tissue culture. *Exp. Cell Res. 13*, 276–304.

AVERY, G., CHOW, M., AND HOLTZER, H. (1956). An experimental analysis of the development of the spinal column. V. Reactivity of chick somites. *J. Exp. Zool. 132*, 409–426.

BROWNE, E. (1909). The production of new hydranths by the insertion of small grafts. *J. Exp. Zool. 7*, 1–23.

BURGER, M. (1973). Surface changes in transformed cells detected by lectins. *Fed. Proc. 32*, 91–101.

BURNETT, A. L. (1971). Cell association. *In* "Topics in the Study of Life," pp. 88–94. Harper & Row, New York.

CAMERON, I. L., AND JETER, J. R., JR. (1971). Relationship between cell proliferation and cytodifferentiation in embryonic chick tissues. *In* "Developmental Aspects of the Cell Cycle" (I. L. Cameron, G. M. Padilla, and A. M. Zimmerman, eds.), pp. 315–355. Academic Press, New York.

CHABRY, L. (1887). Contribution à l'embryologie normale et tératologique des ascidies simples. *J. Anat. Physiol. 23*, 167.

COWDRY, E. V. (1950). "Cancer Cells." Saunders, Philadelphia.

CURTIS, A. S. G. (1962). Pattern and mechanism in the reaggregation of sponges. *Nature (London) 196*, 245–248.

CURTIS, A. S. G. (1970). Problems and some solutions in the study of cellular aggregation. *Symp. Zool. Soc. Lond. 25*, 335–352.

CURTIS, A. S. G., AND VAN DE VYVER, G. (1971). The control of cell adhesion in a morphogenetic system. *J. Embryol. Exp. Morphol. 26*, 295–312.

DANIEL, J. C., AND FLICKINGER, R. A. (1971). Nuclear DNA-like RNA in developing frog embryos. *Exp. Cell Res. 64*, 285–290.

DAVIS, L. E., AND BURNETT, A. L. (1964). A study of growth and cell differentiation in the hepatopancreas of the crayfish. *Develop. Biol. 10*, 122–153.

DE LA HABA, G., AND AMUNDSEN, R. (1972). The contribution of embryo extract to myogenesis of avian striated muscle *in vitro*. *Proc. Nat. Acad. Sci. U.S.A. 69*, 1131–1135.

DOLJANSKI, L. (1930). Le glycogène dans les cultures de foie. *C. R. Soc. Biol. 105*, 504–506.

EBERT, J. D. (1968). Levels of control: a useful frame of perception. *Curr. Top. Develop. Biol. 3*, xv–xxv.

ELLISON, M. L., AMBROSE, E. J., AND EASTY, G. C. (1969). Chondrogenesis in chick embryo somites *in vitro*. *J. Embryol. Exp. Morphol. 21*, 331–340.

ELLISON, M. L., AND LASH, J. (1971). Environmental enhancement of *in vitro* chondrogenesis. *Develop. Biol. 26*, 486–496.

FLICKINGER, R. A., DANIEL, J. C., AND MITCHELL, R. A. (1973). Relative transcription from DNA of different degrees of redundancy in developing frog embryos. *Exp. Cell Res. 76*, 289–296.

GOLDSCHNEIDER, I., AND MOSCONA, A. A. (1972). Tissue-specific cell-surface antigens in embryonic cells. *J. Cell Biol. 53*, 435–449.

GORDON, J. S., AND LASH, J. W. (1974). *In vitro* chondrogenesis and differential cell viability. *Develop. Biol. 36*, in press.

GROBSTEIN, C. (1955). Inductive interaction in the development of the mouse metanephros. *J. Exp. Zool. 130*, 319–340.

GROBSTEIN, C. (1956). Transfilter induction of tubules in mouse metanephrogenic mesenchyme. *Exp. Cell Res. 10*, 427–440.

GROBSTEIN, C. (1962). Interactive processes in cytodifferentiation. *J. Cell Physiol. 60*, 35–48.

GROBSTEIN, C. (1967). Mechanisms or organogenetic tissue interaction. *Nat. Cancer Inst. Monogr.* 26, 279–294.

GROBSTEIN, C., AND HOLTZER, H. (1955). *In vitro* studies of cartilage induction in mouse somite mesoderm. *J. Exp. Zool.* 128, 333–359.

GROBSTEIN, C., AND PARKER, G. (1954). *In vitro* induction of cartilage in mouse somite mesoderm, by embryonic spinal cord. *Proc. Soc. Exp. Biol. Med.* 85, 477–481.

GROBSTEIN, C., AND ZWILLING, E. (1953). Modification of growth and differentiation of chorioallantoic grafts. *J. Exp. Zool.* 122, 259–284.

HERTWIG, O. (1920). "Allgemeine Biologie." Gustav Fischer. Jena, Germany.

HOLTFRETER, J. (1947). Observations on the migration, aggregation, and phagocytosis of embryonic cells. *J. Morphol.* 80, 25–56.

HOLTZER, H. (1964). Control of chondrogenesis in the embryo. *Biophys. J.* 4, Suppl. 239–256.

HOLTZER, H. (1970). Proliferative and quantal cell cycles in the differentiation of muscle, cartilage, and red blood cells. *Symp. Int. Soc. Cell Biol.* 9, 69–88.

HUMPHREYS, T. (1963). Chemical dissolution and *in vitro* reconstruction of sponge cell adhesions. I. Isolation and functional demonstration of the components involved. *Develop. Biol.* 8, 27–47.

HUMPHREYS, T. (1965). Aggregation of chemically dissociated sponge cells in the absence of protein synthesis. *J. Exp. Zool.* 160, 235–240.

HUMPHREYS, T. (1970). Biochemical analysis of sponge cell aggregation. *Symp. Zool. Soc. Lond.* 25, 325–334.

INBAR, M., AND SACHS, L. (1969). Interaction of the carbohydrate-binding protein concanavalin A with normal and transformed cells. *Proc. Nat. Acad. Sci. U.S.A.* 63, 1418–1425.

JONES, B. M. (1966). A unifying hypothesis of cell adhesion. *Nature (London)* 212, 362–365.

JONES, B. M., AND KEMP, R. B. (1970). Inhibition of cell aggregation by antibodies directed against actomyosin. *Nature (London)* 226, 262–262.

KALLMAN, F., AND GROBSTEIN, C. (1964). Fine structure of differentiating mouse pancreatic exocrine cells in transfilter culture. *J. Cell Biol.* 20, 399–413.

KLOSE, J., AND FLICKINGER, R. A. (1971). Collagen synthesis in frog embryo endoderm cells. *Biochim. Biophys. Acta* 232, 207–211.

KONIGSBERG, I. (1971). Diffusion-mediated control of cell fusion. *Develop. Biol.* 26, 133–152.

KOSHER, R. A., AND SEARLS, R. L. (1973). Sulfated mucopolysaccharide synthesis during the development of *Rana pipiens. Develop. Biol.* 32, 50–68.

LASH, J. W. (1955). Studies on wound closure in urodeles. *J. Exp. Zool.* 128, 13–27.

LASH, J. W. (1967). Differential behavior of anterior and posterior embryonic chick somites *in vitro. J. Exp. Zool.* 165, 47–56.

LASH, J. W. (1968a). Somitic mesenchyme and its response to cartilage induction. *In* "Epithelial-Mesenchymal Interactions" (R. Fleischmajer and R. E. Billingham, eds.), pp. 165–172. Williams & Wilkins, Baltimore.

LASH, J. W. (1968b). Chondrogenesis: genotypic and phenotypic expression. *J. Cell Physiol.* 72, 35–47.

LASH, J. W., HOLTZER, S., AND HOLTZER, H. (1957). Aspects of cartilage induction. *Exp. Cell Res.* 13, 292–303.

LASH, J. W., HOLTZER, H., AND WHITEHOUSE, M. (1960). *In vitro* studies on chondrogenesis: the uptake of radioactive sulfate during cartilage induction. *Develop. Biol.* 2, 77–89.

LILLIEN, J. E. (1968). Specific enhancement of cell aggregation *in vitro. Develop. Biol.* 17, 657–678.

LOCKWOOD, D. H., STOCKDALE, F. E., AND TOPPER, Y. J. (1967). Hormone dependent differentiation of mammary gland: sequence of action of hormones in relation to cell cycle. *Science* 156, 945–946.

MACLENNAN, A. P. (1970). Polysaccharides from sponges and their possible significance in cellular aggregation. *Symp. Zool. Soc. Lond.* 25, 299–324.

MANASEK, F. J. (1972). Interpretation of experiments using colchicine. *Develop. Biol.* 29, f–5.

MARZULLO, G., AND LASH, J. W. (1967). Acquisition of the chondrocytic phenotype. *Exp. Biol. Med.* 1, 213–219.

MINOR, R. R. (1973). Somite chondrogenesis: a structural analysis. *J. Cell Biol.* 56, 27–50.

MINTZ, B. (1965). Genetic mosaicism in adult mice of quadriparental lineage. *Science* 148, 1232–1233.

MOSCONA, A. A. (1963). Studies on cell aggregation: demonstration of materials with cell binding activity. *Proc. Nat. Acad. Sci. U.S.A.* 49, 742–747.

MOSCONA, A. A. (1967). *Nat. Cancer Inst. Monogr.* 26, 265–273, 295–299.

MOSCONA, A. A. (1968). Cell aggregation: properties of specific cell-ligands and their role in the formation of multicellular systems. *Develop. Biol.* 18, 250–277.

MOSCONA, M. H., AND MOSCONA, A. A. (1963). Inhibition of adhesiveness and aggregation of dissociated cells by inhibitors of protein and RNA synthesis. *Science 142*, 1070–1071.

NEEDHAM, J. (1950). "Biochemistry and Morphogenesis." Harvard University Press, Cambridge, Massachusetts.

O'HARE, M. J. (1972). Differentiation of chick embryo somites in chorioallantoic membrane culture. *J. Embryol. Exp. Morphol. 27*, 245–260.

PETER, K. (1930). Die Beziehungen zwischen Zellteilung und Zelltätigkeit, Darstellung and Versuch einer kausalen Betrachtung. *Protoplasma 10*, 613–625.

RONZIO, R. A., AND RUTTER, W. J. (1973). Effects of a partially purified factor from chick embryos on macromolecular synthesis of embryonic pancreatic epithelia. *Develop. Biol. 30*, 307–320.

ROTH, S. (1968). Studies on intercellular adhesive selectivity. *Develop. Biol. 16*, 602–613.

ROTH, S., MCGUIRE, E. J., AND ROSEMAN, S. (1971a). An assay for intercellular adhesive specificity. *J. Cell Biol. 51*, 525–535.

ROTH, S., MCGUIRE, E. J., AND ROSEMAN, S. (1971b). Evidence for cell-surface glycosyltransferases: Their potential role in cellular recognition. *J. Cell Biol. 51*, 536–547.

RUTTER, W. J., KEMP, J. D., BRADSHAW, W. S., CLARK, W. R., RONZIO, R. A., AND SANDERS, T. G. (1968). Regulation of specific protein synthesis in cytodifferentiation. *J. Cell Physiol. 72*, 1–18.

SAXÉN, L. (1972). Interactive mechanisms in morphogenesis. *In* "Tissue Interactions in Carcinogenesis" (D. Tarin, ed.), pp. 49–80. Academic Press, New York.

SAXÉN, L., AND KOHONEN, J. (1969). Inductive tissue interactions in vertebrate morphogenesis. *Int. Rev. Exp. Pathol. 1*, 57–128.

SAXÉN, L., KOSKIMIES, O., LAHTI, A., MIETTINEN, H., RAPOLA, J., AND WARTIOVAARA, J. (1968). Differentiation of kidney mesenchyme in an experimental model system. *Advan. Morphogen. 7*, 251–293.

SAXÉN, L., AND TOIVONEN, S. (1962). "Primary Embryonic Induction." Logos Press, London.

SPEMANN, H., AND MANGOLD, H. (1924). Über Induktion von Embryonalanlagen durch Implantation artfremder Organisatoren. *Wilhelm Roux Arch. Entwicklungmech. Organismem 100*, 599–638.

STEINBERG, M. (1970). Does differential adhesion govern self-assembly processes in histogenesis? Equilibrium configurations and the emergence of a hierarchy among populations of embryonic cells. *J. Exp. Zool. 173*, 395–434.

STOCKDALE, F. E., AND O'NEILL, M. C. (1972). Deoxyribonucleic acid synthesis, mitosis, and skeletal muscle differentiation. *In Vitro 8*, 212–225.

STOCKDALE, F., AND TOPPER, Y. J. (1966). The role of DNA synthesis and mitosis in hormone-dependent differentiation. *Proc. Nat. Acad. Sci. U.S.A. 56*, 1283–1289.

TARKOWSKI, A. K. (1961). Mouse chimaeras developed from fused eggs. *Nature (London) 190*, 857–860.

TURKINGTON, R. W. (1971). Hormonal regulation of cell proliferation and differentiation. *In* "Developmental Aspects of the Cell Cycle" (I. L. Cameron, G. M. Padilla, and A. M. Zimmerman, eds.), pp. 315–355. Academic Press, New York.

WARREN, L., FUHRER, J. P., AND BUCK, C. A. (1973). Surface glycoproteins of cells before and after transformation by oncogenic viruses. *Fed. Proc. 32*, 80–85.

WEISS, P. (1939). "Principles of Development." Holt, New York.

WEISS, P. (1968). "Dynamics of Development: Experiments and Inferences." Selected papers on developmental biology. Academic Press, New York.

WESSELLS, N. K., AND COHEN, J. H. (1967). Early pancreas organogenesis: morphogenesis, tissue interactions and mass effects. *Develop. Biol. 15*, 237–270.

WILSON, H. V. (1907). On some phenomena of coalescence and regeneration in sponges. *J. Exp. Zool. 5*, 245–258.

ZWILLING, E. (1968). Morphogenetic phases in development. *In* "27th Symposium of the Society for Developmental Biology" (M. Locke, ed.), pp. 184–207. Academic Press, New York.

CHAPTER ELEVEN

Morphogenesis of Vertebrate Organs

BRIAN S. SPOONER

During the course of embryonic development, cells that are initially alike gradually come to express different functional phenotypes. This phenotypic divergence or *differentiation* constitutes the core of embryonic development. The processes of morphogenesis and cytodifferentiation are the components of differentiation and the two major overt events that occur during development. *Cytodifferentiation* can be thought of as the acquisition of the cytological and biochemical features of the differentiated state by individual cells. *Morphogenesis* is the change in shape and position of cells and cell populations that results in the appearance of various mature tissues and organ systems. Both processes must occur if normal development is to proceed. This point becomes readily apparent in the case of organ development. Thus the ability of pancreas cells to produce secretory enzymes (cytodifferentiation) would be useless to the organism without the development of a secretory delivery system of acini, ductules, and ducts (morphogenesis). An understanding of morphogenesis, of cytodifferentiation, and of the relationship between them will ultimately lead to an understanding of much of development.

This chapter will concentrate on the process of vertebrate organ morphogenesis. The general approach will be to examine (1) the relationship between morphognesis and cytodifferentiation, (2) the role of tissue interactions and extracellular materials in morphogenesis, and (3) the role of intracellular organelles in morphogenetic movements. Finally, an attempt will be made to incorporate the current state of our knowledge into a cohesive picture of the forces controlling morphogenesis during embryonic development.

Relationship Between Morphogenesis and Cytodifferentiation

Since differentiation is in part a result of gene activity and since meaningful organ development requires both cytodifferentiation and morphogenesis, it is essential to understand the relationship between these events. Specifically, it must be determined if they are independently regulated or coregulated. That is, are the genes for morphogenesis and those for cytodifferentiation functionally coupled? A genetic analysis of this question would be quite profitable, but, in the case of organ development in birds and mammals, genetic tools have not yet been adequately applied. However, a number of observational and experimental correlations are available, and we shall consider several of these.

One example is embryonic pancreas development. The pancreatic diverticula of rodent embryos first appear at a specific developmental time and in a specific place along the primitive gut. This initial pancreatic morphogenesis begins at the same time that the first indications of cytodifferentiation are detectable (Rutter et al., 1968; Wessells and Evans, 1968). Furthermore, the rapid transition from the low, barely detectable levels of secretory enzymes that characterize the "protodifferentiated" pancreas to the high levels in the differentiated pancreas (Rutter et al., 1968) correlates temporarily with the morphogenetic formation of acini, ductules, and ducts (Grobstein, 1964). Similar correlations have been described for the development of the chick embryo thyroid gland (Shain et al., 1972). Here, too, morphogenetic phases can be correlated with the levels of cytodifferentiation, by analysis for thyroxine. The accumulation curve for thyroxine in the developing thyroid gland is consistent with the "concerted" model of differentiation described by Rutter et al. (1968) for the pancreas. The relevant point here is that thyroxine can first be detected at the time of initial thyroid morphogenesis and that its distribution is restricted to the early thyroid cells. These observations are consistent with the hypothesis that the genes for morphogenesis and for cytodifferentiation are "turned on" at the same time. But, the physicochemical basis for this temporal coupling has not been explained. A final case suggesting a regulatory coupling between morphogenesis and cytodifferentiation involves development of tubular glands in the chicken oviduct (Wrenn, 1971). In that system, administration of a single compound, estrogen, elicits *both* morphogenesis (initial gland formation) and cytodifferentiation (initial appearance of secretory granules).

The phenomenon of Wolffian lens regeneration provides another kind of correlation suggesting a positive relationship between morphogenesis and cytodifferentiation. If the lens of a *Urodele* eye is experimentally removed, the pigmented epithelial cells of the dorsal iris will form a new lens. However, the cells *first* lose their pigmented phenotype and divide; daughter cells then form a lens and begin to synthesize lens-specific proteins (see Yamada, 1967, for review). The noteworthy feature is that *pigmented* cells do not form the new lens. Lens morphogenesis correlates with lens cytodifferentiation, apparently excluding expression of the pigmented phenotype.

It would be informative to ask analogous questions of a system in which (1) the cells are already determined, (2) the cells have not yet acquired a differentiated phenotype, and (3) the morphogenetic pattern of the cells can still be changed experimentally. Potentially useful systems in this regard are the mammary epithelium, which will undergo salivary-like morphogenesis in the presence of salivary mesoderm (Kratochwil, 1969), and the bronchial epithelium, which will also branch in a salivary-like manner (Fig. 1) in a small percentage of cases when combined with salivary mesoderm (Spooner and Wessells, 1970a). In both these systems, it must be determined if cytodifferentiation is altered concomitantly with morphogenesis. That is, do the cells differentiate in the original mammary or lung mode, or do they differentiate as salivary tissue?

The use of cell culture procedures has also provided some insight into the relationship be-

a   b   c

FIGURE 1. *Tracings of embryonic mouse tissues in organ culture. (a) Dichotomous branching pattern of bronchial epithelium in response to bronchial mesoderm. (b) Salivary-like shape assumed by bronchial epithelium in response to salivary mesoderm. (c) Salivary epithelium isolated from a 13-day embryo. Note that the bronchial epithelium has begun to branch in a salivary-like manner in response to salivary mesoderm.*

tween morphogenesis and cytodifferentiation. The monolayer studies of Holtzer and his colleagues (Stockdale and Holtzer, 1961; Okazaki and Holtzer, 1966) and the clonal studies of Konigsberg (1963) and Hauschka and Konigsberg (1966) have shown that the morphogenetic event of myoblast fusion must occur if the cells are to differentiate as contractile striated muscle. Although fusion is required, it is not sufficient for myotube formation (de la Haba and Amundsen, 1972). However, since muscle differentiation also involves cessation of DNA synthesis and since mononucleated myoblasts *can* synthesize actin and myosin, the muscle case does not provide clear evidence on obligatory coupling between morphogenesis and cytodifferentiation. On the other hand, chondrocytes (Coon, 1966), retinal pigment cells (Cahn and Cahn, 1966), and cardiac cells (Cahn, 1964) all express differentiated phenotypes in clonal culture in the absence of tissue-level organization. In each of these cases, however, the cells were already differentiated when they were removed from the embryo. Analogous results have also been obtained with glandular epithelial cells. In studies with thyroid epithelium (Spooner, 1970; Spooner and Hilfer, 1971), the cells retained a differentiated cytostructure and continue to synthesize thyroxine when released from intact glands and grown in monolayer or clonal cell culture under appropriate nutrient conditions (Fig. 2). Thus the expression of differentiation (the result of cytodifferentiation) is independent of three-dimensional tissue architecture (the product of morphogenesis).

This kind of analysis does not reveal if "potential" glandular epithelial cells can *undergo*

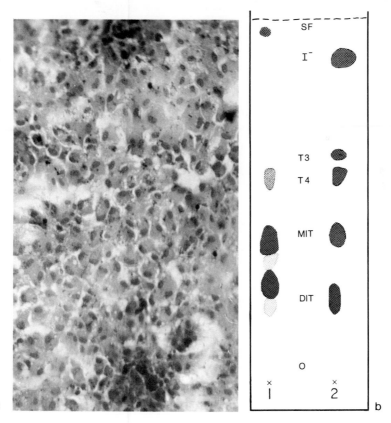

FIGURE 2. *Expression of differentiated function by 16-day chick embryo thyroid in cell culture. (a) Epithelial cells in a 32-day primary clone. (b) Chromatogram tracing showing thyroxine in a clonal culture: (1) iodamino acids in a 32-day clonal culture; (2) separation of a mixed standard of iodamino acids. O, origin; DIT, diiodotyrosine; MIT, moniodotyrosine; T4, thyroxine; T3, triiodothyronine; I⁻, iodide; SF, solvent front. (a), phase contrast, × 365.*

cytodifferentiation in the absence of morphogenesis. This kind of question has been approached by Shain (1971), again using the embryonic chick thyroid system. Cells were isolated at various stages during thyroid morphogenesis, grown in clonal culture, and then analyzed for thyroxine. The results showed that cells isolated after mesenchymal invasion began *in vivo* would continue to increase their thyroxine levels *in vitro* in the absence of further morphogenesis. Cells isolated from early stages would *maintain* the *in vivo* level of cytodifferentiation but would not *continue* differentiation. The experiments imply that prior to a key regulatory step, thyroid morphogenesis and cytodifferentiation are coupled. Although this kind of approach is suggestive, it is not clear whether the cells fail to continue cytodifferentiation because of the absence of morphogenesis or because the tissue culture environment is in some way inadequate. Nevertheless, the approach should be applied to the earliest cells of other organs, such as the pancreas. This technique, however, will provide definitive evidence only if the two events are not coupled—that is, if cytodifferentiation will occur in the absence of morphogenesis.

In summary, then, the information currently available implies, but does not prove, that cytodifferentiation and morphogenesis are coupled events in vertebrate organ differentiation. Definitive evidence to the contrary will require the demonstration that cytodifferentiation can occur in the absence of morphogenesis or that experimental alteration of a morphogenetic pathway does *not* result in a concomitant alteration of a cytodifferentiative pathway. Eventual understanding of the regulation of organ differentiation will require more detailed insight into the relationship between these two events.

Tissue Interactions and Extracellular Materials in Morphogenesis

The analysis of organ development has, during the past 20 years, yielded much information concerning epitheliomesodermal tissue interactions (see Grobstein, 1967, for review). Such interactions have been shown to be necessary for differentiation of salivary glands, kidney, pancreas, lung, lens, liver, mammary gland (Kratochwil, 1969), limb, skin, thymus, and thyroid (Hilfer, 1962). The technique of interposing a membrane filter of known porosity between the interacting tissues suggests that direct tissue contact is not required in many systems and has allowed definition of the distances over which the interaction can occur. Further, it has been shown that various epithelia possess different degrees of specificity with respect to the source of the mesoderm that will elicit a morphogenetic response. Thus salivary (Grobstein, 1967) or bronchial (Taderera, 1967) epithelia will undergo branching morphogenesis to form a salivary or bronchial "tree" only in response to specific salivary or bronchial mesoderm. The pancreatic epithelium, on the other hand, will respond to any embryonic mesoderm or even to an extract of whole chick embryos (Rutter et al., 1968). In addition to influencing morphogenesis of organ primordia, interactions have been shown to be required for the *initial* formation of liver (LeDouarin, 1964; Sherer, 1971), pancreas (Wessells and Cohen, 1967; Spooner et al., 1970), and lung (Spooner and Wessells, 1970a).

Some epithelia exhibit changing mesodermal requirements during the course of organ development. As described earlier, the pancreatic epithelium will differentiate in organ culture in the presence of a preparation of embryo extract (Rutter et al., 1968). However, embryo extract will not allow initial formation of the pancreatic rudiment from the gut endoderm (Wessells and Cohen, 1967). Initial morphogenesis requires the presence of mesoderm (pancreatic or salivary—Wessells and Cohen, 1967; gut mesoderm—Spooner et al., 1970). These results imply a change from a more to a less strict mesodermal requirement as development proceeds. Experimental analysis of interactions during mouse lung morphogenesis yields results that imply a change in the opposite direction (Spooner and Wessells, 1970a). A variety of embryonic mesoderms (gut, salivary, bronchial) will elicit bronchial bud formation from the primitive gut endoderm. However, branching morphogenesis of such bronchial buds is dependent upon interaction with bronchial mesoderm (Fig. 3). The possibility that this kind of result stems from generally inadequate culture conditions (see the chapter by Lash in this book) seems unlikely since other gut-derived organs develop normally in the same cultures. In lung development, then, bronchial bud formation (the first step in lung morphogenesis) occurs in the presence of "non-specific" mesoderm, while further morphogenesis requires "specific" bronchial mesoderm. The change is from a less to a more strict mesodermal requirement as development proceeds.

In the case of the pancreas, the experimental evidence has been obtained by enzymatic separation of epithelium and mesoderm, followed by recombination and organ culture *in vitro*. As pointed out by Wessells (1968), it is not yet clear whether the results reflect a real changing mesodermal requirement or if they reflect a difference in enzyme sensitivity between the younger and older endoderms. In the latter situation, the mesoderm might be required for "recovery" from the experimental manipulation. This is probably not the case with the lung system, since the organ culture results have been obtained both with enzymatic separation and microdissection techniques. In addition, experimental formation and branching morphogenesis of bronchial buds from the wall of the trachea fully corroborate these results (Wessells, 1970). That is, bronchial buds can be induced to form from the wall of the trachea by many embryonic mesoderms, but branching morphogenesis of such buds requires bronchial mesoderm. In sum, the restriction of effective mesodermal type during lung development suggests a differentiative change in the mesoderm that is critical for epithelial morphogenesis.

Although epitheliomesodermal interactions are required for epithelial morphogenesis, it is clear from transfilter organ culture experiments

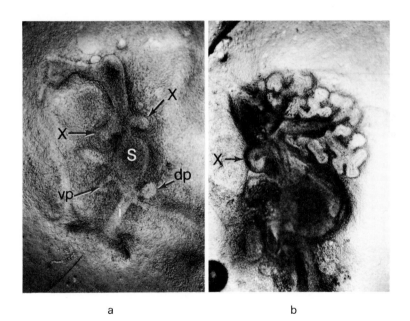

a   b

FIGURE 3. *Lung morphogenesis in organ culture. (a) Whole gut from a 12-somite-stage mouse embryo after 2 days of culture. The primitive stomach (s), intestine (i), and dorsal (dp) and ventral (vp) pancreas rudiments have formed. In addition, the two lung buds (x) have appeared. At this time, the gut mesoderm surrounding the right bud was removed and replaced with fresh 11-day bronchial mesoderm. (b) The same gut shown in (a), 4 days later. The treated lung bud has undergone extensive dichotomous branching in response to the bronchial mesoderm. The left bud has failed to branch in the presence of gut mesoderm. Living culture. × 119.*

that direct tissue contact is *not* required (see Grobstein, 1967, for review). This observation has led to investigation of the role of extracellular materials at the epitheliomesodermal interface in mediating morphogenetic interactions. Electron microscopy, histochemistry, and biochemistry have revealed collagen and mucopolysaccharides as major molecular species present at the junction between interacting tissues (see Bernfield and Wessells, 1970, and Cohen and Hay, 1971, for review). Isolated epithelial cells can synthesize collagen and are the source of early collagen during cornea development (Hay and Revel, 1969). In other developing organ systems, however, the mesoderm contributes the bulk of the collagen to the epitheliomesodermal interface (Kallman and Grobstein, 1965; Bernfield, 1970). Experiments employing the enzyme collagenase to remove extracellular collagen originally implicated that molecule in epithelial morphogenesis (Grobstein and Cohen, 1965; Wessells and Cohen, 1968). Thus collagenase treatment of salivary, lung, or ureteric bud epithelia, cultured transfilter to mesenchyme, causes a loss of the normal branched epithelial morphology and a temporary cessation of morphogenesis. It has been suggested (Grobstein and Cohen, 1965) that the epithelium is morphogenetically "quiescent" in regions of greatest collagen deposition and that differential stabilization of epithelia by collagen can result in specific branching patterns. Electron microscopic studies of developing rat and mouse lungs (Wessells, 1970) reveal that the collagen fibrils along the morphogenetically "quiescent" trachea are highly ordered and aligned parallel to the long axis of the trachea, while at the tips of the morpho-

genetically "active" bronchial buds, the collagen fibrils are randomly oriented (Fig. 4). These results suggest that highly ordered arrays of collagen may be correlated with inhibition of epithelial branching and that more random arrangements may permit branching morphogenesis. The way in which extracellular collagen influences epithelial morphogenesis probably differs from the way it influences skeletal muscle differentiation *in vitro*. Differentiation of myoblasts in clonal culture requires collagen (Hauschka and Konigsberg, 1966), but tropocollagen, purified α-1 and α-2 polypeptide chains, and two cyanogen bromide fragments from the α-1 chain will also support clonal differentiation (Hauschka, 1971).

In addition to collagen, the epitheliomesodermal interface contains mucopolysaccharide-protein complexes that are structural components of the basal lamina. In the case of the salivary gland, autoradiographic analyses have shown that both the mesoderm (Grobstein, 1967) and the epithelium (Bernfield and Wessells, 1970; Bernfield and Banerjee, 1972) make contributions to the extracellular mucopolysaccharide-protein found at the interface between the interacting tissues.

The role of surface-associated acid mucopolysaccharides in salivary morphogenesis has been explored (Bernfield and Wessells, 1970; Bernfield et al., 1972). The critical observation has been that commercial collagenase preparations contain appreciable mucopolysaccharidase activity (and protease activity), consistent with the earlier observation that treatment with collagenase removes associated materials as well as the basal lamina and collagen (Wessells and Cohen, 1968). If the epithelium is isolated from the mesoderm with low concentrations of collagenase and recombined in organ culture, there is no inhibition of morphogenesis. This continued morphogenesis correlates with autoradiographic observations that mucopolysaccharides are still present at the epithelial surface and electron microscopic observations that the basal lamina is diminished but still present. These results contrast with the effects of high collagenase treatment, in which mucopolysaccharides are removed, the basal lamina is absent, and morphogenesis is inhibited (Fig. 5). This same result can be obtained if low collagenase treatment is followed by brief

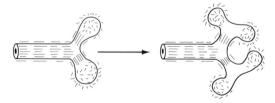

FIGURE 4. *Collagen distribution along the tracheal–bronchial epithelium of mammalian embryos. Collagen fibrils are randomly arrayed at the branching tips and highly oriented along the nonbranching trachea. See the text.*

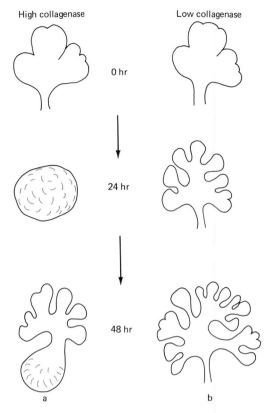

High collagenase   Low collagenase

0 hr

24 hr

48 hr

a   b

FIGURE 5. *Morphogenesis of salivary epithelia isolated by collagenase treatment (according to Bernfield et al., 1972) and cultured in combination with salivary mesoderm. (a) High collagenase treatment. Morphogenesis is inhibited and the epithelium has assumed a rounded "ball-like" shape by 24 hours of culture. By 48 hours, branching morphogenesis has resumed. (b) Low collagenase treatment. Epithelial shape is maintained and morphogenesis continues in culture. See the text for correlation with effects on surface-associated materials.*

treatment with hyaluronidase. Further, recovery from this inhibition due to removal of all surface-associated mucopolysaccharides correlates with a reappearance of those molecules at the epithelial surface.

These studies cast doubt on the conclusion that collagen is necessary for epithelial morphogenesis but do not disprove the possibility. They clearly demonstrate that the susceptibility of morphogenesis to collagenase treat-

ment does not *necessarily* mean that collagen is required for that process. Furthermore, they provide evidence that surface-associated mucopolysaccharides are involved in branching morphogenesis, and suggest the possibility that a specific combination of macromolecules, including mucopolysaccharides, collagen, and other extracellular glycoproteins, influences the pattern of morphogenesis.

It seems clear, then, from the evidence currently available, that epitheliomesodermal interactions are required for organ morphogenesis and that extracellular materials produced by the interacting tissues influence the resulting behavior of the epithelium. In systems such as the salivary or lung epithelia, where homologous mesoderm is required, it is likely that branching morphogenesis is dependent upon a qualitatively or quantitatively unique array of interfacial materials. The change from a less to a more strict mesodermal requirement during lung development may reflect a difference between the extracellular macromolecular requirements for evagination of bronchial buds from the gut wall and those for specific branching morphogenesis of the bronchial buds. It seems quite probable, however, that the specificity of morphogenesis in response to the interaction is a result of mesodermal activity, since salivary mesoderm can cause mammary epithelium and bronchial epithelium to undergo a salivary-like pattern of branching morphogenesis (see above) and since interspecific combinations of lung epithelium and mesoderm initially branch in a pattern characteristic of the mesoderm donor (Taderera, 1967).

Intracellular Organelles and Motive Forces in Morphogenetic Movements

Although extracellular materials and other tissues are intimately involved in organogenesis, it is the epithelial cells themselves that actively undergo "morphogenetic movements" to form definitive and unique organ systems. These movements seem to be accomplished by coordinating individual cell shape changes in discrete groups of cells within a population. The forces responsible for these movements must exist within the "moving" cells, since branching morphogenesis will still occur in transfilter culture (see above) and since isolated neural plate cells will continue to change shape (Holtfreter, 1947). Microtubules and microfilaments are the two kinds of intracellular organelles that have been implicated in generating cell shape changes during morphogenesis. Microtubules are thought to create changes in cell shape by their orientation and distribution in the cytoplasm, while microfilaments are thought to change cell shape by virtue of contractile activity.

Microtubules are long cylindrical structures, averaging 250 Å in diameter, that are found in the cytoplasm of cells, as well as in the mitotic spindle apparatus, cilia, and flagella. Developmentally, microtubules have been implicated in lens placode formation (Byers and Porter, 1964), sea urchin primary mesenchyme formation (Gibbins et al., 1969), neurulation (Schroeder, 1970a; Burnside, 1971), pollen tube elongation (Sanger and Jackson, 1971), and axon elongation of spinal ganglion neurons (Yamada et al., 1970, 1971). The drug colchicine and its analogue colcemid, which specifically disrupt microtubules (Borisy and Taylor, 1967; Tilney and Gibbins, 1969), have proved to be effective tools for analyzing microtubule function in systems undergoing morphogenesis (Tilney and Gibbins, 1969; Pearce and Zwaan, 1970; Yamada et al., 1970; Sanger and Jackson, 1971; Karfunkel, 1971; Handel and Roth, 1971; Spooner and Wessells, 1972).

Microfilaments are a class of cytoplasmic structures, averaging 50 Å in diameter, that are distinguished by their cross-sectional dimension from tonofilaments and neurofilaments (averaging 100 Å in diameter) as well as from microtubules. Microfilaments, of presumed contractile nature, appear to be essential for amphibian gastrulation (Baker, 1965), tail resorption during ascidian metamorphosis (Cloney, 1966, 1969), amphibian neurulation (Baker and Schroeder, 1967; Schroeder, 1970a; Burnside, 1971), elevation of the mammalian pancreatic diverticulum (Wessells and Evans, 1968), mouse lens invagination (Wrenn and Wessells, 1969), and morphogenesis of salivary epithelium (Spooner and Wessells 1970b, 1972). The drug cytochalasin B, a fungal metabolite independently isolated and identified by Aldridge et al. (1967) and by Rotweiler and Tamm (1966), inhibits cytokinesis and single cell locomotion (Carter, 1967). Since one action of cytochalasin appears to be interference with microfilament function (Schroeder, 1969, 1970b; Wessells et al., 1971), it has served as a useful tool in probing microfilament involvement in morphogenesis. The drug has been shown to inhibit salivary gland morphogenesis (Spooner and Wessells, 1970b, 1972), oviduct tubular gland formation (Wrenn, 1971), axon elongation in embryonic nerve cells (Yamada et al., 1970, 1971), tail resorption during ascidian metamorphosis (Cloney, 1972; Lash et al., 1973), and amphibian neurulation (Karfunkel, 1971). In each case, the inhibition of morphogenesis has been correlated with a concomitant and specific alteration in microfilament structure when analyzed with the electron microscope.

The drug also inhibits certain secretory processes (Schofield, 1971) and endocytosis (Davis et al., 1971). However, it is not yet known how these phenomena relate to the drug's effect on filament function.

MULTICELLULAR MORPHOGENESIS

The kinds of evidence implicating microtubules and microfilaments in morphogenesis are primarily electron microscopic observations and the use of specific drugs as discussed above. We shall now consider several morphogenetic systems in which microtubules, microfilaments, or both are thought to be operational.

The processes of neurulation and lens formation are analogous morphogenetic events and can be considered together. They can both be thought of as occurring in essentially three steps (Fig. 6). First, the ectodermal cells undergo a process of elongation resulting in the formation of a flat neural plate or lens placode. Second, the cells become constricted at their apical ends, imposing a curvature on the previously flat surface. This curvature represents the initial invagination of the epithelium that

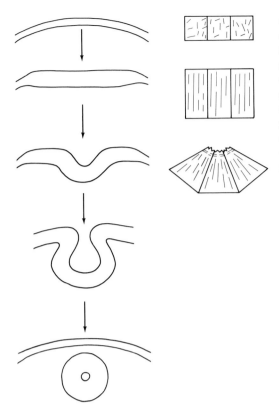

FIGURE 6. *Steps involved in neurulation or lens vesicle formation. See the text for description.*

of microtubules to push the cell out, sliding of tubules away from each other to push the cell out, and microtubule-dependent transport of materials causing elongation from one end of the cell (Burnside, 1971). The concept that microtubules slide from conditions of greater to lesser overlap is based on the model of microtubule sliding proposed to account for chromosome movements at anaphase (McIntosh et al., 1969). The transport mechanism is favored by Burnside (1971) for microtubule involvement in cell elongation during neural plate formation, although that mechanism could, in fact, generate directional polymerization by transport of microtubule subunit protein for assembly at the growing end of the tubule.

The invagination step of morphogenesis appears *not* to be dependent upon intact microtubules, since colchicine or colcemid disruption of microtubule integrity does not inhibit chick lens invagination (Pearce and Zwaan, 1970), cause flattening of the amphibian neural groove (Karfunkel, 1971), or cause opening up of the chick neural tube (Handel and Roth, 1971). Microfilaments appear to be responsible for the cell shape changes that cause invagination of the amphibian neural plate (Baker and Schroeder, 1967; Schroeder, 1970a; Burnside 1971). Baker and Schroeder showed that microfilaments appear in a discrete subpopulation of the neural plate cells at a specific developmental time. The microfilaments are found only at the cell apices, where they encircle that end of the cell. Their proposal is that contraction of the microfilaments, via a "pursestring" action, narrows the apical end of the cells and results in invagination of the neural groove. This idea is further supported by the electron microscope observations of Schroeder (1970a) and Burnside (1971), and by the observation that cytochalasin B causes opening up and flattening of the neural tube (Karfunkel, 1971). Similarly, microfilaments encircle the apical end of the lens placode cells and are thought to generate invagination by contractile activity (Wrenn and Wessells, 1969). Studies on estrogen-induced tubular gland formation in the chicken oviduct imply an analogous role for microfilaments in that system (Wrenn, 1971). The epithelial cells of the oviduct wall in an immature chicken contain no microfilaments. Within 24

establishes the early neural groove or lens vesicle. Finally, closure of the lateral folds over the top of the depression produces the neural tube or detached lens vesicle. The first two steps of these processes have been analyzed and are thought to result from the activities of microtubules and microfilaments respectively. Electron miscroscopic observations show that cell elongation during chick lens placode (Byers and Porter, 1964) and amphibian neural plate (Burnside, 1971) formation correlates with the presence of microtubules oriented with the elongating axis of the cell. In cells not undergoing elongation, microtubules are randomly oriented in the cytoplasm. The correlation between microtubule disposition and cell elongation appears to be a sound one, although, mechanistically, it is not clear how microtubules produce this change in shape. Possibilities include directional polymerization hours after estrogen injection, microfilaments are present at the apical ends of the cells, and by 36 hours cell shape changes (apical constriction) have caused numerous invaginations that represent early tubular glands. If such oviducts are treated with cytochalasin B, the early glands flatten back into the oviduct wall *and* the microfilaments are found to be disrupted. Thus there is good correlation between the presence of microfilaments, cell shape changes, and tubular gland morphogenesis. The cytochalasin experiments further imply that microfilament integrity is required to maintain the early invaginations, consistent with the idea that they maintain the constricted cell shape by virtue of contractile tension.

Thus microtubules seem to be responsible for cell elongation in those systems where a flat plate or placode forms as the primary morphogenetic event. It would strengthen the case to know if colchicine treatment would inhibit that event. The invagination step of morphogenesis,

on the other hand, seems to occur as a result of microfilament action to narrow the apical end of the cells.

What is the situation in the case of the repeated branching morphogenesis of glandular epithelia such as the lung and salivary? Are those "morphogenetic movements" also accomplished by microtubule and/or microfilament activities? The roles of microfilaments and microtubules in branching morphogenesis of the submandibular salivary epithelium have been explored *in vitro* by electron microscopy and by the use of cytochalasin B and colchicine (Spooner and Wessells, 1970b; 1972). The studies implicate microfilaments as the motive force for the cell shape changes required to form branch points (clefts) in the epithelium. Salivary epithelial cells contain bands of microfilaments in both the apical and the basal cytoplasm. The basal microfilaments are particularly interesting because cleft formation begins at the basal end of the epithelium, and could

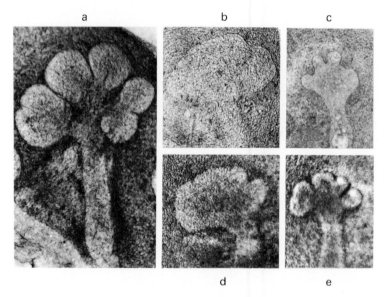

FIGURE 7. *Effects of cytochalasin B on salivary morphogenesis. (a) Control gland after 24 hours of organ culture. Note the numerous clefts and branches in the epithelium. (b) A gland treated for 24 hours with cytochalasin B (10 μg/ml). Morphogenesis has ceased, the epithelium has flattened, and the clefts and branches have disappeared. (c) The same gland shown in (b), 24 hours after withdrawal of cytochalasin. Morphogenesis has resumed, and clefts and branches have reappeared. (d) Another gland that has ceased morphogenesis in cytochalasin. (e) The same gland as in (d) after 23 hours of recovery from cytochalasin B. Cleft formation and morphogenesis have resumed. With continued culture, such recovered glands continue normal branching morphogenesis.*

be accomplished by coordinated constriction of the basal ends of discrete subpopulations of cells. Cultured glands undergo normal repetitive branching events (Fig. 7). If cytochalasin B is included in the culture medium, morphogenesis completely ceases, the epithelium flattens, and clefts that were present at the time of drug application disappear (Fig. 7). When cytochalasin is removed, clefts reappear and morphogenesis resumes (Fig. 7). The same results apply when the experiments are done in transfilter culture, and, furthermore, clefts will still reappear after cytochalasin withdrawal if the mesoderm is removed from the transfilter position. This indicates that recovery of branching capability is a function of release from cytochalasin inhibition in the epithelial cells themselves.

Electron microscope observations reveal that linear arrays of microfilaments are no longer present in cytochalasin-treated cells (Fig. 8). Masses of fine granular and amorphous material are found in place of both the basal and the apical microfilaments. The effect on microfilaments appears to be specific at the electron microscopic level, since there is no detectable alteration in the appearance of microtubules or other cell organelles. Furthermore, protein synthesis is unperturbed in these cultures. The masses of cytochalasin-induced material (granular and amorphous) are absent in glands that have recovered from the drug, *and* linear arrays of microfilaments have reappeared in both the apical and basal cytoplasm of the cells (Fig. 8). Microfilament reappearance even occurs when protein synthesis has been blocked with cycloheximide, suggesting that recovery involves reassembly of microfilaments from preexisting components. Thus, the cytochalasin B experiments establish clear and positive correlations between microfilament integrity and salivary gland morphogenesis. In the presence of cytochalasin, microfilaments are disrupted *and* morphogenesis ceases. When cytochalasin is removed, microfilaments reappear *and* morphogenesis resumes.

Colchicine or colcemid treatment, on the other hand, disrupts microtubules but does not cause regression of existing clefts (Fig. 9). In fact, such clefts seem to continue to deepen. Continuing morphogenesis is arrested, however, and no new clefts form, presumably be-

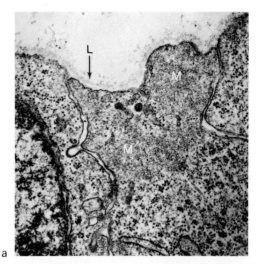

a

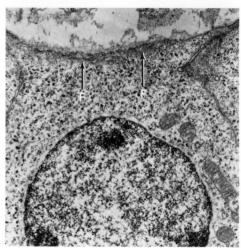

b

FIGURE 8. *(a) An example of cytochalasin-induced masses (M) of material found in place of filament bundles at the basal end of a salivary epithelial cell. L, basal lamina.* × *23,000. (b) Cell at the base of a recovery cleft showing basal microfilaments (F) and absence of masses at 4 hours of recovery.* × *9,500.*

cause mitosis is inhibited. Both microfilaments and clefts caused to disappear with cytochalasin will reform even if colchicine is included in the recovery medium (Fig. 10). This result shows that "recovery" clefts can form in the absence of microtubules. In sum, these experiments provide an exceedingly strong case for the conclusion that microfilament function is

a   b

FIGURE 9. *Salivary glands treated with (a) colchicine and (b) colcemid for 18 to 20 hours. Note the presence of normal-appearing clefts in the epithelia. Such glands do not continue morphogenesis, presumably because mitosis is blocked.*

indispensable for branching morphogenesis of salivary epithelium.

Since all the salivary epithelial cells contain both apical and basal microfilaments, the system differs from the lens placode, neural plate, and oviduct cases. Further, microfilament involvement in cleft formation is established only for the formation and maintenance of early clefts, since older, deeper clefts are not lost in the presence of cytochalasin. A working model to explain microfilament action in cleft formation is shown in Figure 11. The model presupposes microfilament contractility, an assumption to be discussed below. The essential idea is that the basal microfilaments in a discrete group of cells contract with a net force greater than the apical contractile forces in the same cells. This difference would result in constriction of the basal ends of the cells, thus generating a cleft. The rapid decrease in cross-sectional area accompanying such a contraction would cause folding of the basal cell surface and overlaying basal lamina. Such folding is observed at the basal ends of cells at the bottom of both normal and recovery clefts (Spooner and Wessells, 1970b, 1972; Bernfield and Wessells, 1970). A net increase in contraction by the apical microfilaments in neighboring groups of cells would shape the tips of the two branches associated with each cleft. As the cleft deepens, it becomes stabilized, and microfilaments are no longer required for its maintenance. This stabilization does not appear to be attributable to microtubules, since colchi-

cine treatment does not cause loss of clefts. The most plausible candidates for cleft stabilization seem to be extracellular collagen and mucopolysaccharides (see Bernfield and Wessells, 1970). This entire series of events is repeated as each new cleft is formed. Since new clefts begin at the tips of branches, this model requires a change in the net contractile force from the apical to the basal microfilaments at the time of new cleft formation.

It must be emphasized that this model alone is not intended to account for all the events in ongoing epithelial morphogenesis. It is intended to explain how microfilaments function in the single, but repetitive, event of cleft formation during branching morphogenesis. Morphogenesis also requires cell division. The salivary epithelium grows spectacularly during morphogenesis, both *in vivo* and *in vitro*, and contains numerous mitotic and ^{3}H-thymidine-incorporating cells. Since "recovery" clefts can form in glands recovering from cytochalasin in the presence of colchicine, mitosis appears not to be required for that event. Yet, *continuing* morphogenesis does not occur in the presence of colchicine, presumably because cell division is blocked. Thus a complete picture of morphogenesis must incorporate the feature of active cell division.

The most intriguing feature of neurulation and lens morphogenesis is the apparent two-step process that implies an active role for microtubules prior to microfilament-mediated invagination. A similar role for cytoplasmic microtubules does not, at first glance, seem to be required for cleft formation during branching morphogenesis. However, careful consideration of the data reveals that such a possibility has not been eliminated. Theoretically, a microtubule-dependent shape change could occur prior to the initial formation of the salivary evagination; palisading has been reported elsewhere in analogous epithelia. Moreover, a microtubule-dependent process could occur in the cells of an "incipient" cleft. Then microfilament contraction could form and maintain the early cleft, with no further requirement for microtubule integrity. Three pieces of evidence are consistent with this hypothesis. First, colchicine-treatment inhibits new cleft formation but existing clefts are not affected. Second, recovery from cytochalasin in the presence of

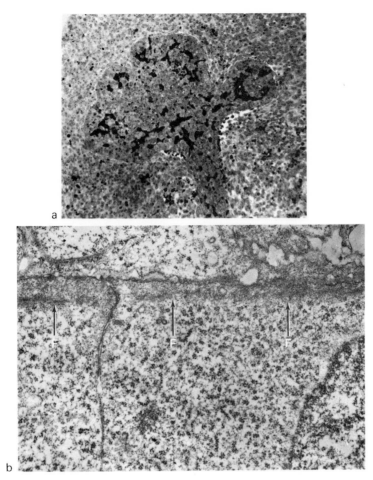

a

b

FIGURE 10. *Recovery from cytochalasin in the presence of colchicine. (a) A 1-μm section through a salivary gland that has undergone 4 hours of recovery from cytochalasin in the presence of colchicine. Despite some regions of necrosis, cleft reformation has begun. (b) Basal end of cells in a cleft that has reformed during 4 hours of recovery in the presence of colchicine. Linear arrays of basal microfilaments (F) have reappeared.*

colchicine results in reappearance of previous clefts, but no new ones form. Third, although microfilaments are present in the cells of the earliest clefts, they are not prominent structures in "incipient" clefts. These data certainly do not prove that a microtubule-requiring step precedes visible cleft formation during salivary morphogenesis. In fact, the colchicine experiments could instead reflect a mitotic requirement as discussed above. However, taken together, they are consistent with the idea. It is tempting, then, to suggest that cleft formation, like neurulation and lens formation, is a step-

wise series of events requiring the successive action of microtubules and microfilaments. The possibility that such a mechanism is a general feature of organogenesis is intriguing and deserving of experimental test.

SINGLE CELL MORPHOGENESIS

In addition to the morphogenetic behavior of cell populations during organ development, some kinds of development involve the activities of single cells. One of the best examples of single cell morphogenesis is the process of initiation and elongation of axons by embryonic neurons. The neuroblasts that populate the embryonic spinal ganglia are known to send out axons in response to a specific protein, nerve-growth factor (Levi-Montalcini, 1964), and it is clear that these events will still occur in individual neuroblasts that have been isolated and placed in cell culture (Scott et al., 1969).

The roles of microfilaments and microtubules in axon elongation have been studied during this single cell morphogenesis in culture (Yamada et al., 1970; 1971). As an axon elongates, its distal tip is extremely active, moving much like the undulating membrane to be desribed below. In addition to its undulatory movements, the distal tip or "growth cone" actively produces, moves, and withdraws fine microspikes as it advances over a substratum. Electron microscopic examination of the growth cone-microspike region reveals that the microspikes and peripheral growth cone are filled with a network of microfilaments. The body of the axon, on the other hand, contains numerous microtubules (neurotubules) and 100-Å-diameter neurofilaments, aligned with the long axis of the axon. Treatment with cytochalasin B causes a rapid cessation of elongation that correlates with withdrawal of microspikes and rounding up of the growth cone, *and* with a specific morphological alteration in the appearance of the microfilament network. Such an inhibited axon can retain a fixed length for many hours and still resume growth cone activity and active elongation when cytochalasin is removed (Fig. 12).

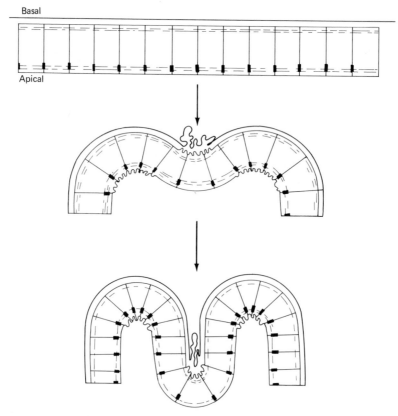

FIGURE 11. *Model to explain cleft formation during morphogenesis of salivary epithelium. A flat epithelium can be rapidly manipulated to form a cleft by microfilament-mediated shape changes in a small number of the cells. No increase in cell number is required for this initial morphogenetic movement. Further development of the cleft requires an increase in cell number and extracellular stabilization factors. See the text for further discussion.*

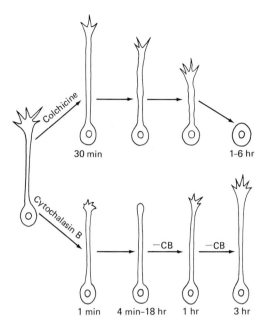

30 min   1-6 hr

1 min   4 min–18 hr   1 hr   3 hr

FIGURE 12. *Effects of cytochalasin B and colchicine on axon elongation in embryonic nerve cells in culture. Colchicine has its earliest effects on the axon, without affecting the distal growth cone-microspike area. Axonal collapse correlates ultrastructually with disruption of neurotubules in the axon. Cytochalasin B rapidly causes "wilting" and withdrawal of microspikes and "rounding up" of the growth cone. The axon does not collapse, but elongation is arrested. The microfilament network of the growth cone–microspike region is the only organelle morphologically altered. Removal of cytochalasin B results in reappearance of growth cone activity and resumption of axon elongation.*

Treatment with colchicine, on the other hand, produces entirely different results (Fig. 12). There is no immediate effect, but after approximately 30 minutes the axon begins to collapse back into the cell body. Axon collapse correlates ultrastructurally only with microtubule disruption. These experiments have led to the conclusion that both the microfilament network and microtubules are required for axon elongation. The microfilaments are thought to provide the motive force for elongation, possibly by contractile activity, and the microtubules are thought to function as cyto-

skeletal elements responsible for maintaining the rigid axon. Whether this is because they are in fact rigid skeletal elements, or because they support essential axonal transport processes, has not been resolved. Thus, just as is the case for cell elongation (as we noted above; see also Burnside, 1971), the mechanism of microtubule function has not been completely elucidated.

One further kind of single cell behavior is crucial for normal development, and that is single cell locomotion. The most dramatic example of this behavior is the migration of individual neural crest cells from their site of origin to diverse positions in the embryo (see Weston, 1970, for review). Depending on where they come to reside in the embryo, these cells can differentiate into either pigment, nerve, supportive, or adrenal cells.

Thus an understanding of development demands an understanding of the mechanism of single cell locomotion. By observing migratory vertebrate cells *in vitro*, Abercrombie (1961) and his colleagues (Abercrombie et al., 1970) have shown that cell movement is associated with undulating membrane activity at the leading edge of the cell. The undulating membrane exists as a broad thin expanse on one side of the cell periphery. It is not simply membrane but, in fact, an exceedingly thin region of the cell containing both cytoplasm and plasma membrane. This undulating membrane locomotor organ extends, retracts, and ruffles as it precedes a cell over a solid substratum. An undulating membrane can be immobilized, as when it contacts another cell, and the cell will stop moving. In such a case, a new undulating membrane can form elsewhere and lead the cell off in another direction. Although undulating membranes are always associated with cell movement, cells with undulating membranes do not necessarily move. Thus a cell can possess an undulating membrane that is actively ruffling and still not be translocating.

It is thought that active movement occurs as a result of three successive events (Ingram, 1969; Spooner et al., 1971): (1) extension of the cell's leading edge (undulating membrane); (2) attachment of the extended leading edge to the substratum; and (3) a contraction that pulls the cell forward. The extension step can be visualized by observing cells from the side

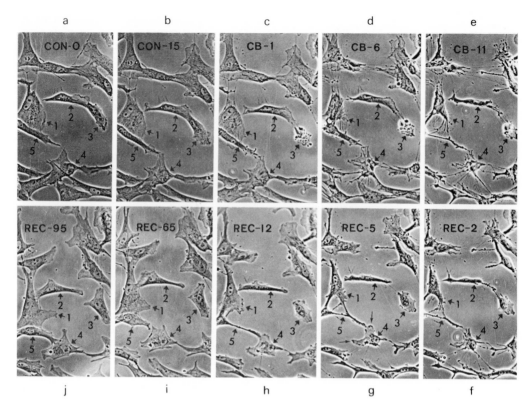

FIGURE 13. *Field of living heart fibroblasts. Five individual cells have been labeled so that they can be followed during cytochalasin B treatment and recovery. Phase contrast, × 160. (a) 0 time, control medium, (b) 15 minutes, control medium. Note that cell 3 has extended its undulating membrane. (c) 1 minute, 10 μg/ml cytochalasin B (CB). All undulating membrane activity has ceased. Cell 3 has undergone extensive blebing. (d) 6 minutes, CB. All cells are now morphologically affected. Note the appearance of numerous processes in undulating membrane regions. (e) 11 minutes, CB. Numerous processes are now the only indication of previous undulating membrane regions. (f) 2 minutes, recovery medium. The area between the processes is already "filling in." (g) 5 minutes, recovery medium. Most processes have disappeared because the area between them has "filled in." The recovered undulating membranes are already functional, as evidenced by the "ruffles" (arrows). (h) 12 minutes, recovery medium. All cells have fully recovered normal shape and ruffling activity. (i) 65 minutes, recovery medium. (j) 95 minutes, recovery medium.*

(Ingram, 1969). Current evidence suggests that extension involves the addition of new membrane at the leading edge (Abercrombie et al., 1971). Adhesion of the leading edge of the cell to the substratum is apparently stronger than posterior points of adhesion if movement is to result (Ingram, 1969). It is not known whether this reflects a quantitative difference due to the fact that a larger area of the cell contacts the substratum in the front or if some qualitative difference in adhesion is present.

The idea that a contraction occurs between the anterior point of adhesion and some more posterior position is consistent with the observation that migratory vertebrate cells possess contractile properties (Hoffman-Berling, 1959) and has prompted a search for the structural basis of such contraction. Electron microscopic analysis has revealed the presence of microfilaments, intermediate-sized (100 Å in diameter) filaments, and microtubules in the cytoplasm of migratory cells (Buckley and Porter, 1967;

Goldman and Follett, 1969, 1970; Spooner et al., 1971). In recent years, the use of cytochalasin B and colchicine has allowed some insight into the respective roles of microfilaments and microtubules in single cell migration.

The treatment of migratory vertebrate cells in culture with cytochalasin B results in an immediate cessation of undulating membrane activity and locomotion (Carter, 1967; Spooner et al., 1971; Goldman, 1972). In the case of embryonic fibroblast and glial cells, inhibition of movement is followed, within minutes, by morphological effects on the undulating membrane (Fig. 13). This response seems to be a collapse of the undulating membrane back toward the cell body. However, some regions remain fully extended, thus producing a number of long rigid processes in the area formerly occupied by the undulating membrane. At this time, the cells are still flattened on the substratum, but hours later they have assumed a "rounded-up" appearance with rigid processes still projecting from the "ball-like" cell body (Fig. 14). It is important to recognize that the early effects on undulating membranes and locomotion are temporally separated from the "rounding-up" phenomenon. Thus inhibition of locomotion by cytochalasin is not a secondary consequence of "rounding-up." All these effects of cytochalasin are fully reversible (Figs. 13 and 14). Within minutes after drug withdrawal, the area between rigid processes is "filled-in" to reform the undulating membrane, ruffling is evident, and movement resumes. The duration of drug treatment has no discernible effect on the speed of recovery. As in the salivary and nerve systems, cytochalasin does not inhibit protein synthesis and recovery from the drug does not require new protein synthesis.

In embryonic heart fibroblasts and glial cells, microfilaments are found to be organized either into linear, cross-linked arrays or into a network or meshwork. The linear array has been termed the microfilament sheath and the network is referred to as the microfilament network (Spooner et al., 1971). The microfilament sheath is found predominantly near the base of the cell, oriented with the long axis of the cell. The microfilament network is found just beneath the plasma membrane both at the leading edge and along the sides of the cell. In the presence of cytochalasin B, intermediate-sized filaments, microtubules, *and* the microfilament sheath (Fig. 15) are structurally intact and appear normal. The microfilament network, however, is no longer evident. Instead, large masses of fine amorphous material are present (Fig. 16), similar to those produced in epithelial cells treated with cytochalasin (Spooner and Wessells, 1972; Wrenn, 1971). These drug-induced masses have been demonstrated in cells as rapidly as 8 minutes after the application of cytochalasin B. The correlation, then, is that in the presence of cytochalasin, the microfilament network is disrupted and cell locomotion is inhibited. Furthermore, the resumption of motility after cytochalasin withdrawal correlates with reappearance of the microfilament network. The idea based on these experiments is that the microfilament network is essential for locomotion. The micro-

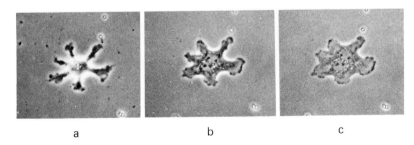

a   b   c

FIGURE 14. *Recovery from long-term cytochalasin B treatment. (a) A living heart fibroblast after 24 hours in cytochalasin. (b) The same cell after 9 minutes of recovery. (c) 15 minutes of recovery. The speed of recovery is not impeded by long-term exposure to the drug. Note that this cell has become binucleated, a typical effect of cytochalasin with long-term treatment. Phase contrast, × 320.*

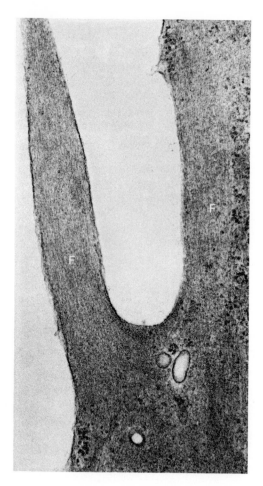

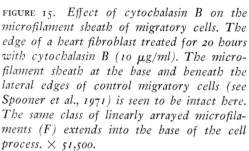

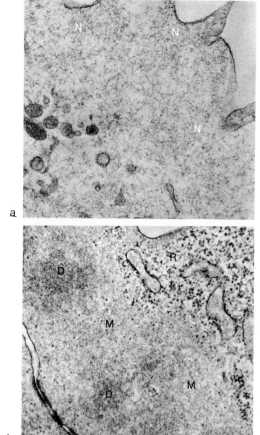

FIGURE 15. *Effect of cytochalasin B on the microfilament sheath of migratory cells. The edge of a heart fibroblast treated for 20 hours with cytochalasin B (10 µg/ml). The microfilament sheath at the base and beneath the lateral edges of control migratory cells (see Spooner et al., 1971) is seen to be intact here. The same class of linearly arrayed microfilaments (F) extends into the base of the cell process.* × 51,500.

filament sheath may also be required, but its structural integrity alone is not sufficient. However, this could simply mean that the sheath will not function without prior activity by the network. Thus the microfilament network could be responsible for extension of the leading edge, followed by contraction of the microfilament sheath that causes net forward displacement of the cell.

FIGURE 16. *Effect of cytochalasin B on the microfilament network. (a) The microfilament network (N) or meshwork found just beneath the plasma membrane at the leading edge of control migratory cells and filling the peripheral growth cone and microspike regions of elongating axons. (b) A cytochalasin B-treated migratory glial cell. The filament network is no longer resolvable. Instead, a mass (M) of fine filamentous and amorphous material is present. Dense regions (D) similar to those found in drug-treated nerve-growth cones (Yamada et al., 1971) are present. The masses are indistinguishable from those found in cytochalasin-treated epithelial cells (as in Fig. 8a). Note that ribosomes (R) approach very close to the plasma membrane, a condition not usually seen in control cells, presumably because the microfilament network forms a region of exclusion of other organelles. (a)* × 37,000; *(b)* × 30,000.

Are microtubules involved in cell locomotion? Treatment with colchicine or colcemid disrupts microtubules, but undulating membrane activity is not inhibited (Goldman, 1971; Rovensky et al. 1971; Spooner et al. 1971). Further, net cell movement is not affected (Fig. 17). However, there is an effect on directional translocation (Vasiliev et al. 1970). Thus, while control cells move in predominantly straight paths, colchicine and colcemid-treated cells wander in more random directions. These data imply that intact microtubules are required to add directionality to the microfilament-mediated movements of the cells. Vasiliev et al. (1970) have suggested that microtubules function to "stabilize" the cell surface. Such an action might, in some way, locally inhibit microfilament activity. Observation of cells that are plated into culture suggest a way in which microtubules could function (Fig. 18). After attaching, cells spreading on the substratum initially show ruffling around their entire periphery. As the cell assumes a migratory shape, ruffling activity becomes confined to the leading edge of the cell. Microtubules beneath the leading edge are randomly arrayed with respect to the long axis of the cell and are not located near the cell surface, where ruffling is occurring. Microtubules in the center and tail of the cell tend to be oriented parallel to the cell axis. Colchicine disruption of microtubules results in the loss of the migratory shape and the appearance of ruffling around the entire cell periphery. That is, colchicine-treatment results in a reversal of the events leading to an elongate shape and directional translocation. The implication is that local polymerization of microtubule subunits could produce and maintain the migratory shape.

Changing direction, as in contact inhibition, could be accounted for by local polymerization of microtubules at the leading edge with concomitant cessation of microfilament activity, followed by local depolymerization of microtubules and activation of microfilaments in another region of the cell. The idea inherent in this model is that the microfilament network, which is just under the plasma membrane everywhere in the cell, provides the motive force for ruffling (extension), and that microtubules function as a guidance system.

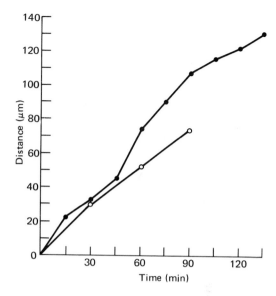

FIGURE 17. *Effect of colchicine on single-cell locomotion. The curves show the net movement of representative individual cells in the presence of colchicine. These cells were treated for 48 hours with colchicine (1 μg/ml) and then the measurements were made while the cells were still exposed to the drug. The solid circles show the net movement of an embryonic glial cell and the open circles show the net movement of an embryonic heart fibroblast. Colchicine does not inhibit the capability for net movement over a substratum. However, directional translocation (not shown in these curves) is impaired by the drug. See the text.*

The information currently available is consistent with the hypothesis that microfilaments and microtubules function in this way during locomotion, but they do not prove that hypothesis.

If microtubules limit the directions of cell movement as just outlined, there must be intracellular local control over where and when microtubule polymerization is to occur. One of the more exciting possibilities is that cyclic AMP is involved in this regulation. It is now known that dibutyryl cyclic AMP will cause a shape transformation in established cell lines to an elongate fibroblast-like condition, and that the transformation is blocked by colchicine (Hsie and Puck, 1971; Johnson et al.,

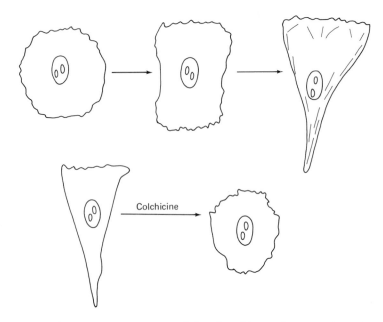

FIGURE 18. *Steps leading to restriction of peripheral ruffling and assumption of a migratory cell shape. Colchicine treatment leads to a reversal of these steps. See the text for further discussion.*

1971). When tumor cells are exposed to this cyclic nucleotide, contact inhibition is seen, a phenomenon normally absent in such cells. Furthermore, treatment with dibutyryl cyclic AMP causes a decrease in cell motility within a period of 20 minutes (Johnson et al., 1972). All these results would be expected if an elevated intracellular concentration of cyclic AMP either directly or indirectly resulted in microtubule polymerization. Normal directional movement and direction changes might then be regulated by local increases in intracellular cyclic AMP concentrations. Although such a regulatory role for cyclic AMP is appealing, it is highly speculative at this point. The idea would be strengthened if it were demonstrated by electron microscopy that dibutyryl cyclic AMP-treated cells possessed more microtubules than control cells.

To summarize briefly, it is quite clear that microfilaments and microtubules are intimately involved in the analogous events of axon elongation and single cell movement. Electron microscopy and cytochalasin B experiments strongly implicate the microfilament network as a source of motive force in both processes. For axon elongation, microtubules are also required, presumably functioning as cytoskeletal agents.

However, the demonstration that new surface is added at the distal tip of the axon (Bray, 1970) raises the possibility that microtubules are also involved in transport of surface materials. For single cell movement, microtubules are responsible for the elongate cell shape and for adding directionally to cell movements. The role of the microfilament sheath is not yet clear; however, it is worth noting that no equivalent structure has been found in embryonic nerves, but growth cone activity and elongation still take place (Ludueña and Wessells, 1973). This difference may mean that axon elongation is analogous only to the extension phase of cell movement.

MICROFILAMENT CONTRACTILITY AND CYTOCHALASIN B

A major assumption that has been made in the foregoing discussion is that the microfilaments in these nonmuscle embryonic cells are contractile. Although contractility is the simplest way to explain their apparent involvement in these various kinds of movements, it is imperative to examine the evidence supporting that assumption. In addition, since cytochalasin B is the tool that provides a correlation between microfilaments and movement, it is essential to consider the drug's effect on known contractile systems and to examine the ways in which it could act so specifically.

There is no longer any doubt that nonmuscle cells contain actin or actin and myosin. The technique of producing ATP-, calcium- and magnesium-dependent contractile models by glycerination (Szent-Gyorgyi, 1951) has been successfully applied to fibroblasts (Hoffman-Berling, 1959), amoebas (Simard-Duquesne and Gouillard, 1962), myxomycete plasmodia (Kamiya and Kuroda, 1965), and leucocytes (Norberg, 1970). Actin and myosin have been isolated from blood platelets (Pollard and Adelstein, 1971) and slime moulds (Nachmias, 1972) that will interact with each other or with skeletal muscle actin and myosin. In addition, embryonic neurons undergoing axonal growth contain substantial amounts of a protein that is similar to muscle actin in molecular weight, electrophoretic pattern, and peptide map (Fine and Bray, 1971). The microfilaments of nonmuscle cells seem to be analogous to muscle actin in both their size and in some of their properties. For example, an actin-like protein that can be isolated from many nonmuscle cell types can be shown to undergo polymerization from a globular (G) form to a fibrous (F) form (Behnke et al., 1971; Pollard and Adelstein, 1971). This "G-actin" to "F-actin" conversion has ionic requirements identical to those required for the G to F conversion of muscle actin. Furthermore, many of the microfilaments in nonmuscle cells have, in common with skeletal muscle actin, the ability to bind heavy meromyosin (HMM), a tryptic fragment of the myosin molecule (Ishikawa et al., 1969; Behnke et al., 1971). The ability to bind HMM is considered to be a property specific to actin, the "arrowhead" pattern being dictated by the right-handed, double-stranded, helical chain that forms the actin filament (Huxley, 1963). The microfilaments in some of the morphogenetic systems discussed above have been analyzed, and they too will bind heavy meromyosin (Ludueña and Wessells, 1973; Spooner et al., 1973). For example, both the apical and basal microfilaments in glycerinated salivary epithelial cells bind HMM (Fig.

19). The data are consistent with the observations that many nonmuscle cells contain muscle-like contractile proteins and support the idea that the microfilaments in morphogenetic systems are actin-like. In addition to possessing properties in common with muscle-derived contractile proteins, it is clear that the microfilaments are present in systems that do contract. Thus there is little doubt that animal cell

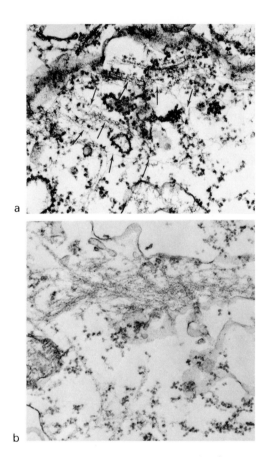

FIGURE 19. *Heavy meromyosin binding to microfilaments in salivary epithelial cells. (a) Glycerinated control cell. The microfilaments (arrows) are thin, and their surfaces are clean and smooth. × 44,000. (b) Heavy meromyosin-treated epithelial cell. HMM binding to the microfilaments produces an increased electron density. Individual filaments appear thicker, their surfaces have short "hair-like" projections, and a rough binding periodicity is suggested in favorable areas. Other organelles do not exhibit binding. × 33,000.*

cytokinesis involves the generation of contractile tension (see the chapter by Rappaport in this book) or that tail resorption during ascidian metamorphosis is the result of nonmuscle contractions (Cloney, 1966, 1969). In sum, the evidence is quite strong that nonmuscle cells possess contractile systems and that microfilaments are similar to muscle actin. Nevertheless, microfilament contractility during morphogenetic movements is, although attractive, still an *unproven* hypothesis. Proof of that hypothesis will require direct demonstration of contractility in isolated microfilaments.

The relationship between microfilaments, contractility, and cytochalasin B also deserves some comment. Following the initial observation by Schroeder (1969, 1970b) that inhibition of cytokinesis by cytochalasin B correlated with a disappearance of the "contractile ring" microfilaments, the drug has been used to probe microfilament involvement in many cell activities (see Wessells et al., 1971; Carter, 1972; and above). In many cases, this same correlation has been demonstrated; that is, cytochalasin B inhibition of a biological process thought to be microfilament-mediated concomitantly results in alteration of microfilament morphology. The interpretation has been that the drug interferes with microfilament *function*. Although the morphological alteration of microfilaments has provided useful correlations, there is no reason to assume that drug interference with microfilament function will necessarily derange the filaments. It has been pointed out that the mechanism by which cytochalasin acts is not known (Spooner et al., 1971). Possible sites of action include both the plasma membrane and the microfilament systems. If filament contractility moves plasma membrane and changes cell shape, the filaments must insert either directly or indirectly on the inner face of the plasma membrane (Wessells et al., 1973). The evidence from electron microscopy is consistent with the idea that microfilaments *do* insert on the plasma membrane (e.g., Wrenn, 1971; Cloney, 1972; Yamada et al. 1971). Thus cytochalasin could interfere with filament function by disturbing (breaking?) insertion points of microfilaments on membranes. Alternatively, effects on membrane permeability could inhibit passage of molecules required for filament function or

integrity. The idea that cytochalasin primarily affects membranes is favored by Bluemink (1971a, b) and by Estensen et al., (1971) to account for inhibition of cleavage in amphibian eggs. Schroeder (1972) has recently discussed the differences between cleavage in amphibian eggs and cleavage in cells where cytochalasin inhibition correlates with effects on microfilaments. In addition to its effects on filament function, cytochalasin strongly inhibits sugar transport (Kletzien et al., 1972; Mizel and Wilson, 1972; Cohn et al., 1972; Yamada and Wessells, 1973) and causes an apparent depression of mucopolysaccharide synthesis (Sanger and Holtzer, 1972). In the case of mucopolysaccharides, synthesis per se is *not* inhibited; the result reflects the inhibition of precursor sugar transport into the cells (Cohn et al., 1972). However, attribution of *all* cytochalasin effects directly to alterations in sugar transport and incorporation into macromolecules is not an entirely satisfactory explanation. For example, treatment with glucose-free medium (thus eliminating glucose transport) does not mimic the effects of cytochalasin on membrane ruffling, growth cone–microspike activity, salivary morphogenesis, or microfilament integrity (Yamada and Wessells, 1973; Taylor and Wessells, 1973). However, the effects of the drug on membrane-bound transport systems *may* be a visible monitor of a more basic action at the level of the plasma membrane that does interrupt filament function (e.g., filament-membrane attachment sites).

Direct action of cytochalasin on contractile filaments is an extremely controversial possibility. Experimentation with known contractile systems both support and refute the idea. The contraction of embryonic cardiac and smooth muscle in culture was reported to be inhibited by cytochalasin (Wessells et al., 1971), but similar experiments reveal no inhibition of spontaneous contraction in cardiac, smooth, or skeletal muscle cultures (Sanger and Holtzer, 1972; Sanger et al., 1971). Manasek et al. (1972) reported that chick embryo heartbeat is reversibly stopped by cytochalasin and thin filaments are disrupted. Their results furthermore imply an age and dose dependency that might explain the lack of effect in other systems. However, Forer et al. (1972) find that cytochalasin does not cause the breakdown of F-actin, does not inhibit the transformation of G-actin to F-actin, does not inhibit the binding of heavy meromyosin to F-actin, and does not inhibit the ATP-induced release of heavy meromyosin from F-actin. Spudich and Lin (1972), also working with skeletal muscle actin, have obtained quite different results. They find that cytochalasin B interacts with highly purified actomyosin (or acto-HMM) causing a decrease in viscosity of the actomyosin complex. The drug does not affect myosin alone, but it *does interact* with *purified* actin. Forer et al. (1972) used electron microscopy to monitor the effects of cytochalasin on actin, while Spudich and Lin (1972) used viscometry as the analytical tool. However, the difference between the two sets of results may well reflect the purity of the actin preparation. Spudich (1973) has begun to test this possibility. Thus cytochalasin B causes an alteration in morphology of purified F-actin, obtained either from skeletal muscle or from blood platelets. These preparations do not contain troponin–tropomyosin, a routine contaminant of isolated actin. Electron microscopy of negatively stained material shows the presence of "collapsed" aggregates of short filamentous and amorphous material resulting from drug treatment. Furthermore, addition of troponin–tropomyosin to the F-actin prevents the cytochalasin effect. The apparent inhibition of the cytochalasin effect on actin morphology by troponin–tropomyosin may account for the contradiction between the results of Forer et al. (1972) and those of Spudich (Spudich and Lin, 1972; Spudich, 1973).

In summary, there is overwhelming circumstantial evidence that nonmuscle cells contain muscle-like contractile proteins. The microfilaments of these cells appear to be analogous to muscle actin by several criteria. The drug cytochalasin B has been useful in relating microfilaments to morphogenetic events, but final interpretation of those results must await discovery of the primary mode of drug action.

Controls in Organ Morphogenesis

The current state of our knowledge clearly demonstrates that both intracellular and extra-cellular forces are involved in regulating the pattern of organ morphogenesis. The interrelationships between these forces are not yet clear. However, it is of value to attempt to speculate on the possible interplay that controls morphogenesis. The discussion that follows will concentrate on branching morphogenesis of organs like the lung and salivary gland.

The evidence described earlier is consistent with the idea that changes in epithelial shape are produced by coordinated shape changes in individual cells and that microfilament contraction is the immediate causal agent. In the salivary system, epithelial branching occurs *only* in the presence of salivary mesoderm, yet microfilaments are still intact in the presence of bronchial mesoderm (Fig. 20). Thus, specific morphogenetic patterns imply differential *regulation* of microfilament action. This regulation requires an interaction between the epithelium and the homologous mesoderm. As a result of the interaction, extracellular materials are deposited at the interface between the two tissues. These materials include collagen and mucopolysaccharide–protein complexes, and their presence is necessary for branching to occur. Such surface-associated materials probably function in stabilizing epithelial shape and may also be involved in stimulating mitosis in the epithelium. In addition, they may also control where and when microfilament contraction occurs. Calcium ions are required for muscle contraction, and they are also required for the cortical contractions in *Xenopus* eggs (Gingell, 1970). The salivary microfilament contractile system could require an external supply of calcium analogous to smooth muscle. There is as yet no direct evidence for this, but preliminary experiments are suggestive. The drug papaverine inhibits contraction and causes relaxation of smooth muscle (von Hattingberg et al., 1966). Although its mode of action at the molecular level is unknown, papaverine is thought to compete with calcium for sites on the cell surface and thus prevent calcium entry (Imai and Takeda, 1967). Treatment with papaverine will reversibly inhibit salivary gland morphogenesis in culture. Interestingly, the time course and pattern of inhibition differs from that of cytochalasin B, and microfilaments are morphologically still intact (Ash, Spooner, and Wessells, 1973). These experiments (and ones using cal-

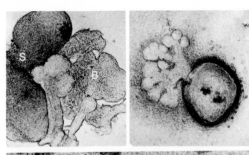
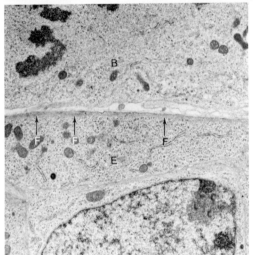

the calcium involvement, the surface-associated materials could function in regulating calcium availability. Acid mucopolysaccharides, in particular, could bind calcium at the cell surface. A unique array and distribution of the interface materials might then provide differential calcium availability along the epithelium. Such a system could account for both the specific mesodermal requirement and the resultant specificity in branching patterns. At this point, such control is highly speculative; however, it is consistent with the available information and provides a possible coupling between what is known about extracellular control factors and about intracellular controls.

In summary, we are beginning to construct a reasonably concrete picture of morphogenetic processes in epithelia. As this task advances, the goal must be to link such diverse elements as morphogenetic shape changes, single cell movements, and filament construction and function—with cell adhesion, mitotic activity, and cell communication. In so doing, possible relationship to underlying genetic bases for cell activity in development must be considered. Eventual integrated understanding of these processes and relationships is essential to a clear understanding of organ morphogenesis.

FIGURE 20. *Effect of bronchial mesoderm on microfilaments and morphogenesis of salivary epithelium. (a) Isolated salivary epithelium combined with salivary mesoderm (S) on one side and bronchial mesoderm (B) on the other side. (b) 3 days of culture. The epithelium undergoes morphogenetic branching where it contacts salivary mesoderm but is morphogenetically quiescent where it contacts bronchial mesoderm (B). (c) Region of morphogenetic quiescence where salivary epithelium (E) contacts bronchial mesoderm (B). Microfilaments (F) are still present, shown here at the basal end of the epithelium. A mesodermal cell is seen in mitosis. The persistence of microfilaments in the presence of foreign mesoderm shows that homologous mesoderm is not required for filament stability, and suggests that it functions to regulate filament activity in morphogenetic branching.*

cium-free medium) suggest that external calcium is involved in the salivary system but do not provide definite proof. However, assuming

References

ABERCROMBIE, M. (1961). The bases of locomotory behavior of fibroblasts. *Exp. Cell Res.*, Suppl. 9, 188–198.

ABERCROMBIE, M., HEAYSMANN, J. E. M., AND PEGRUM, S. M. (1970). The locomotion of fibroblasts in culture. I. Movements of the leading edge. *Exp. Cell Res.* 59, 393–398.

ABERCROMBIE, M., HEAYSMANN, J. E. M., AND PEGRUM, S. M. (1971). The locomotion of fibroblasts in culture. IV. Electron microscopy of the leading lamella. *Exp. Cell Res.* 67, 359–367.

ALDRIDGE, D. C., ARMSTRONG, J. J., SPEAKE, R. N., AND TURNER, W. B. (1967). The structures of cytochalasins A and B. *J. Chem. Soc. (C) 17*, 1667–1676.

ASH, J. F., SPOONER, B. S., AND WESSELLS, N. K. (1973). Effects of papaverine and calcium-free medium on salivary gland morphogenesis. *Develop. Biol. 33*, 463–469.

BAKER, P. C. (1965). Fine structure and morphogenetic movements in the gastrula of the tree frog, *Hyla regilla. J. Cell Biol. 24*, 95-116.

BAKER, P. C., AND SCHROEDER, T. E. (1967). Cytoplasmic filaments and morphogenetic movements in the amphibian neural tube. *Develop. Biol. 15*, 432-450.

BEHNKE, O., KRISTENSEN, B. I., AND NIELSEN, L. E. (1971). Electron microscopical observations on actinoid and myosinoid filaments in blood platelets. *J. Ultrastruct. Res. 37*, 351-369.

BERNFIELD, M. R. (1970). Collagen synthesis during epitheliomesenchymal interactions. *Develop. Biol. 22*, 213-231.

BERNFIELD, M. R., AND BANERJEE, S. D. (1972). Acid mucopolysaccharide (glycosaminoglycan) at the epithelial-mesenchymal interface of mouse embryo salivary glands. *J. Cell Biol. 52*, 664-673.

BERNFIELD, M. R., BANERJEE, S. D., AND COHN, R. H. (1972). Dependence of salivary epithelial morphology and branching morphogenesis upon acid mucopolysaccharide-protein (proteoglycan) at the epithelial surface. *J. Cell Biol. 52*, 674-689.

BERNFIELD, M. R., AND WESSELLS, N. K. (1970). Intra- and extracellular control of epithelial morphogenesis. *Develop. Biol.*, Suppl. 4, 195-249.

BLUEMINK, J. G. (1971a). Effects of cytochalasin B on surface contractility and cell junction formation during egg cleavage in *Xenopus laevis. Cytobiologie 3*, 176-187.

BLUEMINK, J. G. (1971b). Cytokinesis and cytochalasin-induced furrow regression in the first-cleavage zygote of *Xenopus laevis. Z. Zellforsch. 121*, 102-126.

BORISY, G. C., AND TAYLOR, E. W. (1967). The mechanism of action of colchicine. Binding of colchicine-^{3}H to cellular protein. *J. Cell Biol. 34*, 525-533.

BRAY, D. (1970). Surface movements during the growth of a single explanted neuron. *Proc. Nat. Acad. Sci. U.S.A. 65*, 905-910.

BUCKLEY, I. K., AND PORTER, K. R. (1967). Cytoplasmic fibrils in living cultured cells. A light and electron microscope study. *Protoplasma 64*, 349-380.

BURNSIDE, B. (1971). Microtubules and microfilaments in newt neurulation.. *Develop. Biol. 26*, 416-441.

BYERS, B., AND PORTER, K. R. (1964). Oriented microtubules in elongating cells of the developing lens rudiment after induction. *Proc. Nat. Acad. Sci. U.S.A. 52*, 1091-1099.

CAHN, R. D. (1964). Maintenance of beating and dissociation of biochemical and functional differ-

entiation in clones of chicken embryo heart cells. *J. Cell Biol. 23*, 17A.

CAHN, R. D., AND CAHN, M. B. (1966). Heritability of cellular differentiation: Clonal growth and expression of differentiation in retinal pigment cells *in vitro. Proc. Nat. Acad. Sci. U.S.A. 55*, 106-114.

CARTER, S. B. (1967). Effects of cytochalasins on mammalian cells. *Nature (London) 213*, 261-264.

CARTER, S. B. (1972). The cytochalasins as research tools in cytology. *Endeavor 31*, 77-82.

CLONEY, R. A. (1966). Cytoplasmic filaments and cell movements: epidermal cells during ascidian metamorphosis. *J. Ultrastruct. Res. 14*, 300-328.

CLONEY, R. A. (1969). Cytoplasmic filaments and morphogenesis: the role of the notocord in ascidian metamorphosis. *Z. Zellforsch. 100*, 31-53.

CLONEY, R. A. (1972). Cytoplasmic filaments and morphogenesis: effects of cytochalasin B on contractile epidermal cells. *Z. Zellforsch. 132*, 167-192.

COHN, R. H., BANERJEE, S. D., SHELTON, E. R., AND BERNFIELD, M. R. (1972). Cytochalasin B: lack of effect on mucopolysaccharide synthesis and selective alterations in precursor uptake. *Proc. Nat. Acad. Sci. U.S.A. 69*, 2865-2869.

COON, H. G. (1966). Clonal stability and phenotypic expression of chick cartilage cells *in vitro. Proc. Nat. Acad. Sci. U.S.A. 55*, 66-73.

DAVIS, A. T., ESTENSEN, R. D., AND QUIE, P. B. (1971). Cytochalasin B. III. Inhibition of human polymorphonuclear leucocyte phagacytosis. *Proc. Soc. Exp. Biol. Med. 137*, 161-164.

DE LA HABA, G., AND AMUNDSEN, R. (1972). The contribution of embryo extract to myogenesis of avian striated muscle *in vitro. Proc. Nat. Acad. Sci. U.S.A. 69*, 1131-1133.

ESTENSEN, R. D., ROSENBERG, M., AND SHERIDAN, J. D. (1971). Microfilaments and "contractile" processes. *Science 173*, 356-358.

FINE, R. E., AND BRAY, D. (1971). Actin in growing nerve cells. *Nature New Biol. 234*, 115-118.

FORER, A., EMMERSON, J., AND BEHNKE, O. (1972). Cytochalasin B: does it affect actin-like filaments? *Science 175*, 774-776.

GIBBINS, J. R., TILNEY, L. G., AND PORTER, K. R. (1969). Microtubules in the formation and development of the primary mesenchyme in *Arbacia punctulata*. I. The distribution of microtubules. *J. Cell Biol. 41*, 201-226.

GINGELL, D. (1970). Contractile responses at the surface of an amphibian egg. *J. Embryol. Exp. Morphol. 23*, 583-609.

GOLDMAN, R. D. (1971). The role of three cytoplasmic fibers in BHK-21 cell motility. I. Microtubules and the effects of colchicine. *J. Cell Biol.* 51, 752–762.

GOLDMAN, R. D. (1972). The effects of cytochalasin B on the microfilaments of baby hamster kidney (BHK-21) cells. *J. Cell Biol.* 52, 246–254.

GOLDMAN, R. D., AND FOLLETT, E. A. C. (1969). The structure of the major cell processes of isolated BHK-21 fibroblasts. *Exp. Cell Res.* 57, 263–276.

GOLDMAN, R. D., AND FOLLETT, E. A. C. (1970). Birefrigent filamentous organelle in BHK-21 cells and its possible role in cell spreading and cell motility. *Science* 169, 286–288.

GROBSTEIN, C. (1964). Cytodifferentiation and its controls. *Science* 143, 643–650.

GROBSTEIN, C. (1967). Mechanism of organogenetic tissue interaction. *Nat. Cancer Inst. Monogr. 26,* 279–299.

GROBSTEIN, C., AND COHEN, J. (1965). Collagenase: effect on the morphogenesis of embryonic salivary epithelium *in vitro. Science* 150, 626–628.

HANDEL, M. A., AND ROTH, L. E. (1971). Cell shape and morphology of the neural tube: implications for microtubule function. *Develop. Biol. 25,* 78–95.

HAUSCHKA, S. D. (1971). Myogenesis: molecular specificity of the myoblast-collagen interaction for clonal differentiation and cell attachment. American Society Cell Biology, eleventh annual meeting, New Orleans, abstract 235, p. 122.

HAUSCHKA, S. D., AND KONIGSBERG, I. R. (1966). The influence of collagen on the development of muscle clones. *Proc. Nat. Acad. Sci. U.S.A. 55,* 119–126.

HAY, E. D., AND REVEL, J. P. (1969). Fine structure of the developing avian cornea. *In* "Monographs in Developmental Biology" (A. Wolsky and P. S. Chen, eds.), pp. 16–46. Karger, New York.

HILFER, S. R. (1962). The stability of embryonic chick thyroid cells *in vitro* as judged by morphological and physiological criteria. *Develop. Biol. 4,* 1–21.

HOFFMAN-BERLING, H. (1959). The role of cell structures in cell movement. *In* "Cell, Organism, and Milieu" (D. Rudnick, ed.), p. 45. Ronald Press, New York.

HOLTFRETER, J. (1947). Observations on the migration, aggregation, and phagocytosis of embryonic cells. *J. Morphol. 80,* 25–56.

HSIE, A. W., AND PUCK, T. T. (1971). Morphological transformation of chinese hamster cells by dibutryl adenosine cyclic 3′:5′-monophosphate

and testosterone. *Proc. Nat. Acad. Sci. U.S.A. 68,* 358–361.

HUXLEY, H. E. (1963). Electron microscope studies on the structure of natural and synthetic protein filaments from striated muscle. *J. Mol. Biol.* 7, 281–308.

IMAI, S., AND TAKEDA, K. (1967). Effects of vasodilators on the isolated taenia coli of the guineapig. *Nature (London) 213,* 509–511.

INGRAM, V. M. (1969). A side view of moving fibroblasts. *Nature (London) 222,* 641–644.

ISHIKAWA, H., BISCHOFF, R., AND HOLTZER, H. (1969). Formation of arrowhead complexes with heavy meromyosin in a variety of cell types. *J. Cell. Biol. 43,* 312–328.

JOHNSON, G. S., FRIEDMAN, R. M., AND PASTAN, I. (1971). Restoration of several morphological characteristics of normal fibroblasts in sarcoma cells treated with adenosine 3′:5′-cyclic monophosphate and its derivatives. *Proc. Nat. Acad. Sci. U.S.A. 68,* 425–429.

JOHNSON, G. S., MORGAN, W. D., AND PASTAN, I. (1972). Regulation of cell motility by cyclic AMP. *Nature (London) 235,* 54–56.

KALLMAN, F., AND GROBSTEIN, C. (1965). Source of collagen at the epitheliomesenchymal interface during inductive interaction. *Develop. Biol. 11,* 169–183.

KAMIYA, N., AND KURODA, K. (1965). Movement of the myxomycete plasmodium. I. A study of glycerinated models. *Proc. Japan Acad. 41,* 837–841.

KARFUNKEL, P. (1971). The role of microtubules and microfilaments in neurulation in *Xenopus. Develop. Biol. 25,* 30–56.

KLETZIEN, R. F., PERDUE, J. F., AND SPRINGER, A. (1972). Cytochalasin A and B. Inhibition of sugar uptake in cultured cells. *J. Biol. Chem. 247,* 2964–2966.

KONIGSBERG, I. R. (1963). Clonal analysis of myogenesis. *Science 140,* 1273–1284.

KRATOCHWIL, K. (1969). Organ specificity in mesenchymal induction demonstrated in the embryonic development of the mammary gland of the mouse. *Develop. Biol. 20,* 46–71.

LASH, J., CLONEY, R. A., AND MINOR, R. R. (1973). The effect of cytochalasin B upon tail resorption and metamorphosis in ten species of ascidians. *Biol. Bull. 145,* 360–372.

LEDOUARIN, N. (1964). Induction de l'endoderme pré-hépatique par le mésoderme de l'aire cardiaque chez l'embryon de poulet. *J. Embryol. Exp. Morphol. 12,* 651–664.

LEVI-MONTALCINI, R. (1964). Growth control of nerve cells by a protein factor and its antiserum. *Science 143,* 105–110.

LUDUEÑA, M. A., AND WESSELLS, N. K. (1973). Cell locomotion, nerve elongation, and microfilaments. *Develop. Biol. 30*, 427–440.

MANASEK, F. J., BURNSIDE, B., AND STROMAN, J. (1972). The sensitivity of developing cardiac myofibrils to cytochalasin B. *Proc. Nat. Acad. Sci. U.S.A. 69*, 308–312.

McINTOSH, J. R., HEPLER, P. K., AND VAN WIE, D. G. (1969). Model for mitosis. *Nature (London) 224*, 659–663.

MIZEL, S. B., AND WILSON, L. (1972). Inhibition of hexose transport in mammalian cells by cytochalasin B. *J. Biol. Chem. 247*, 4102–4105.

NACHMIAS, V. T. (1972). Electron microscope studies on myosin from *Physarum polycephalum*. *J. Cell Biol. 52*, 648–663.

NORBERG, B. (1970). Amoeboid movements and cytoplasmic fragmentation of glycerinated leucocytes induced by ATP. *Exp. Cell Res. 59*, 11–21.

OKAZAKI, K., AND HOLTZER, H. (1966). Myogenesis: fusion, myosin synthesis, and the mitotic cycle. *Proc. Nat. Acad. Sci. U.S.A. 56*, 1484–1490.

PEARCE, T. L., AND ZWAAN, J. (1970). A light and electron microscopic study of cell behavior and microtubules in the embryonic chicken lens using colcemid. *J. Embryol. Exp. Morphol. 23*, 491–507.

POLLARD, T. D., AND ADELSTEIN, R. S. (1971). The ultrastructure of purified platelet contractile proteins. *American Society Cell Biology*, eleventh annual meeting, New Orleans, abstract 454, p. 231.

ROTWEILER, W., AND TAMM, C. (1966). Isolation and structure of phomin. *Experentia 22*, 750–752.

ROVENSKY, Y. A., SLAVAJA, I. L., AND VASILIEV, J. M. (1971). Behavior of fibroblast-like cells on grooved surfaces. *Exp. Cell Res. 65*, 193–201.

RUTTER, W. J., KEMP, J. D., BRADSHAW, W. S., CLARK, W. R., RONZIO, R. A., AND SANDERS, T. G. (1968). Regulation of specific protein synthesis in cytodifferentiation. *J. Cell Physiol. 72*, 1–18.

SANGER, J. M., AND JACKSON, W. T. (1971). Fine structure study of pollen development in Haemanthus Katherine Baker. II. Microtubules and elongation of the generative cells. *J. Cell Sci. 8*, 303–315.

SANGER, J. W., AND HOLTZER, H. (1972). Cytochalasin B: effects on cell morphology, cell adhesion, and mucopolysaccharide synthesis. *Proc. Nat. Acad. Sci. U.S.A. 69*, 253–257.

SANGER, J. W., HOLTZER, S., AND HOLTZER, H. (1971). Effects of cytochalasin B on muscle cells in tissue culture. *Nature New Biol. 229*, 121–123.

SCHOFIELD, J. G. (1971). Cytochalasin B and release of growth hormone. *Nature New Biol. 234*, 215–216.

SCHROEDER, T. E. (1969). The role of "contractile ring" filaments in dividing *Arbacia* eggs. *Biol. Bull. 137*, 413–414.

SCHROEDER, T. E. (1970a). Neurulation in *Xenopus laevis*. An analysis and model based upon light and electron microscopy. *J. Embryol. Exp. Morphol. 23*, 427–462.

SCHROEDER, T. E. (1970b). The contractile ring. I. Fine structure of dividing mammalian (Hela) cells and the effects of cytochalasin B. *Z. Zellforsch. 109*, 431–449.

SCHROEDER, T. E. (1972). The contractile ring. II. Determining its brief existence, volumetric changes, and vital role in cleaving *Arbacia* eggs. *J. Cell Biol. 53*, 419–434.

SCOTT, B. S., ENGLEBERT, V. E., AND FISHER, K. C. (1969). Morphological and electrophysiological characteristics of dissociated chick embryonic spinal ganglion cells in culture. *Exp. Neurol. 23*, 230–248.

SHAIN, W. G. (1972). Development of the embryonic chick thyroid: an analysis of thyroxine synthesis in the developing gland *in vivo* and in clonal cell culture. *Ph.D.* thesis, Temple University.

SHAIN, W. G., HILFER, S. R., AND FONTE, V. G. (1972). Early organogenesis of the embryonic chick thyroid. I. Morphology and biochemistry. *Develop. Biol. 28*, 202–218.

SHERER, G. K. (1971). Tissue interaction in chick liver development: a reevaluation. *American Society Cell Biology*, eleventh annual meeting, New Orleans, abstract 533, p. 271.

SIMARD-DUQUESNE, N., AND GOUILLARD, P. (1962). Ameboid movement. I. Reactivation of glycerinated models of *Amoeba proteus* with adenosinetriphosphate. *Exp. Cell Res. 28*, 85–91.

SPOONER, B. S. (1970). The expression of differentiation by chick embryo thyroid in cell culture. I. Functional and fine structural stability in mass and clonal culture. *J. Cell Physiol. 75*, 33–48.

SPOONER, B. S., ASH, J. F., WRENN, J. T., FRATER, R. B., AND WESSELS, N. K. (1973). Heavy meromyosin binding to microfilaments involved in cell and morphogenetic movements. *Tissue and Cell 5*, 37–46.

SPOONER, B. S., AND HILFER, S. R. (1971). The expression of differentiation by chick embryo thyroid in cell structure. II. Modification of phenotype in monolayer culture by different media. *J. Cell Biol. 48*, 225–234.

Spooner, B. S., Walther, B. T., and Rutter, W. J. (1970). The development of the dorsal and ventral mammalian pancreas *in vivo* and *in vitro*. *J. Cell Biol. 47*, 235–246.

Spooner, B. S., and Wessells, N. K. (1970a). Mammalian lung development: interactions in primordium formation and bronchial morphogenesis. *J. Exp. Zool. 175*, 445–454.

Spooner, B. S., and Wessells, N. K. (1970b). Effects of cytochalasin B upon microfilaments involved in morphogenesis of salivary epithelium. *Proc. Nat. Acad. Sci. U.S.A. 66*, 360–364.

Spooner, B. S., and Wessells, N. K. (1972). An analysis of salivary gland morphogenesis: role of cytoplasmic microfilaments and microtubules. *Develop. Biol. 27*, 38–54.

Spooner, B. S., Yamada, K. M., and Wessells, N. K. (1971). Microfilaments and cell locomotion. *J. Cell Biol. 49*, 595–613.

Spudich, J. A. (1973). On the effects of cytochalasin B on actin filaments. *Cold Spring Harbor Symp. 37*, 585–593.

Spudich, J. A., and Lin, S. (1972). Cytochalasin B, its interaction with actin and actomyosin from muscle. *Proc. Nat. Acad. Sci. U.S.A. 69*, 442–446.

Stockdale, F. E., and Holtzer, H. (1961). DNA synthesis and myogenesis. *Exp. Cell Res. 24*, 508–520.

Szent-Gyorgyi, A. (1951). "Chemistry of Muscular Contraction." Academic Press, New York.

Taderera, J. V. (1967). Control of lung differentiation *in vitro*. *Develop. Biol. 16*, 489–912.

Taylor, E. L., and Wessells, N. K. (1973). Cytochalasin B: alterations in salivary gland morphogenesis not due to glucose depletion. *Develop. Biol. 31*, 421–425.

Tilney, L. G., and Gibbins, J. R. (1969). Microtubules in the formation and development of the primary mesenchyme in *Arbacia punctulata*. II. An experimental analysis of their role in development and maintenance of cell shape. *J. Cell Biol. 41*, 227–250.

Vasiliev, J. M., Gelfand, I. M., Domnina, L. V., Ivanova, O. Y., Komm, S. G., and Olshevskaja, L. V. (1970). Effect of colcemid on the locomotory behavior of fibroblasts. *J. Embryol. Exp. Morphol. 24*, 625–640.

von Hattingberg, M., Kuschinsky, G., and Rahn, K. H. (1966). Der Einfluss von pharmaka auf calcium-gehalt und ^{45}-calciumaustausch der glatten Muskulatur der Taenia coli von Meerschweinchen. *Naunym-Schmiedelberg Arch. Exp. Path. Pharmak. 253*, 438–443.

Wessells, N. K. (1968). Problems in the analysis of determination, mitosis, and differentiation. *In* "Epithelial-Mesenchymal Interactions" (R. Fleischmajer and R. Billingham, eds.), pp. 131–151. Williams & Wilkins, Baltimore.

Wessells, N. K. (1970). Mammalian lung development: interactions in formation and morphogenesis of tracheal buds. *J. Exp. Zool. 175*, 455–466.

Wessells, N. K., and Cohen, J. H. (1967). Early pancreas organogenesis: morphologenesis, tissue interactions and mass effects. *Develop. Biol. 15*, 237–270.

Wessells, N. K., and Cohen, J. H. (1968). Effects of collagenase on developing epithelia *in vitro*: lung, ureteric bud, and pancreas. *Develop. Biol. 18*, 294–309.

Wessells, N. K., and Evans, J. (1968). Ultrastructural studies of early morphogenesis and cytodifferentiation in the embryonic mammalian pancreas. *Develop. Biol. 17*, 413–446.

Wessells, N. K., Spooner, B. S., Ash, J. F., Bradley, M. O., Ludueña, M. A., Taylor, E. L., Wrenn, J. T., and Yamada, K. M. (1971). Microfilaments in cellular and developmental processes. *Science 171*, 135–143.

Wessells, N. K., Spooner, B. S., and Ludueña, M. A. (1973). Surface movements, microfilaments, and cell locomotion. *In* "Locomation of Tissue Cells" (Ciba Symposium). *14*, 53–77. Associated Scientific Publishers, Amsterdam.

Weston, J. A. (1970). The migration and differentiation of neural crest cells. *Advan. Morphogen. 8*, 41–114.

Wrenn, J. T. (1971). An analysis of tubular gland morphogenesis in chick oviduct. *Develop. Biol. 26*, 400–415.

Wrenn, J. T., and Wessells, N. K. (1969). An ultrastructural study of lens invagination in the mouse. *J. Exp. Zool. 171*, 359–368.

Yamada, K. M., Spooner, B. S., and Wessells, N. K. (1970). Axon growth: roles of microfilaments and microtubules. *Proc. Nat. Acad. Sci. U.S.A. 66*, 1206–1212.

Yamada, K. M., Spooner, B. S., and Wessells, N. K. (1971). Ultrastructure and function of growth cones and axons of cultured nerve cells. *J. Cell Biol. 49*, 614–635.

Yamada, K. M., and Wessells, N. K. (1973). Cytochalasin B: effects on membrane ruffling, growth cone and microspike activity, and microfilament structure not due to altered glucose transport. *Develop. Biol. 31*, 413–420.

Yamada, T. (1967). Cellular and subcellular events in Wolffian lens regeneration *Curr. Top. Develop. Biol. 2*, 247–283.

CHAPTER TWELVE

Time Flow in Differentiation and Morphogenesis

CHARLES E. WILDE, JR.

Recent advances in the understanding of the behavior of macromolecules have had profound effects on developmental biology. New terminologies, techniques, and interpretations have led to reexamination of the classic problems of development. Recent long reviews testify to the vitality of current inquiry (Davidson, 1968; Schjeide and deVellis, 1970).

Yet the problems remain. It is the purpose of this chapter to restate some of the major problems of developmental biology, to assess briefly the progress in their solution, to analyze critically the "state of the art," and hopefully to suggest paths for further advances. Many of the areas to which I shall make brief, or perhaps superficial, reference are discussed in detail and with high competence by my colleagues elsewhere in this book. Embryologists, like the field of development itself, are haunted by "relics of their demonology" (Harrison, 1933), and so the problems discussed here inevitably reflect the present author's emphases. Other problems in the field, of deep interest and high merit, obviously exist. Their omission

from this chapter should not be interpreted as demeaning them but only as the fallibility of the writer.

Analysis Versus Synthesis

If one were to take one's temerity firmly in hand and venture an overall criticism of the impact of molecular biology upon developmental biology, it would be that the analytical methods involving extraction, purification, degradation, and identification of the macromolecular constituents of living matter are intrinsically destructive. These methods, although often exquisitely quantitative in nature, yield no portal for the observation of *development*, which is an ongoing life process requiring the integrity of cellular function. The degradative essentials of analysis provide static "time slices," leaving the investigator to undertake the fallacy-prone process of reconstruction.

This is not to decry the contemporary analytical process; it is an absolute necessity. Yet it would appear that analysis, at the moment, predominates, while the development of conceptual, interpretive ideas of the coordinated means of development is singularly lacking.

The lack of interpretive syntheses, at present, creates a void, since many classical integrating concepts are under serious attack or have been shown to be epiphenomena, with alternative interpretations. For example, consider the problems in classical embryonic induction (Spemann and Mangold, 1924), where Curtis (1962) has developed, through experiment, the possibility of alternative explanations. This follows from the demonstration by Curtis that, by transplantation of fragments of the cortex of *Xenopus* zygotes from the gray crescent of the two-cell stage into another site, a second embryo will be formed. The experiment advanced into the very early embryo the dynamic capabilities found in the cells of the dorsal lip of the blastopore of the gastrula by Spemann and Mangold (1924) and seats these capabilities in a portion of the cortex, a subcellular portion of the first blastomeres. The possibility that the phenomena are an expression of genetically preprogrammed segregants must be raised.

New or integrating concepts have been historically rare and are often immediately challenged. This is healthy, provided the challenger and the challengee are thereby motivated to devise more penetrating experiments, which will then be subjected to the same test processes.

Time Flow

It is implicit in development that a zygote passes from fertilization to mature organism through time. The flow of time is customarily monitored by the observer by reference to fixed external standards in terms of seconds, minutes, hours, days, months, or years. In biological studies it is considered to be undirectional (Gal-Or, 1972). So also is it commonplace to understand that the embryo is not the same between any two time points, no matter how small. This implies a flux in the constituents of the organism reflected in those covert and overt changes that are usually considered under the broader categories of development, morphogenesis, and differentiation.

The student of development can never afford to neglect the flow of time. Since some of the changes are observable and, at various levels, codifiable, there underlies these processes a regularity and a governance of extreme precision.

Not only, therefore, are we concerned with *what happens* but also *when it happens*. The embryologist must be concerned with dynamics at all levels of analysis and the order of expression of these phenomena, always and continuously correlating these with changes in the physical state of the developing embryo as a whole, the appearance of its parts, and their coordinate function.

Intrinsic temporal control is clearly seen in the following experiments, chosen as representative from many available examples. The differentiation of pigment cells and other cellular derivatives of the neural crest in amphibia is dependent, in part, upon the availability in the microenvironment of phenylalanine. The expression of this dependence can be demonstrated throughout gastrulation and neurulation (Wilde, 1955, 1956). When phenylalanine is presented, in low concentration, to explanted blastula cells of stage 8 (Harrison), although a precocious increment in melanin is produced, they remain ungainly, large and rounded, typical blastula cells in form. At stage 9, this treatment leads not only to normal melanization but also to the development of dendritic processes, and the cells adopt to a form much more closely approaching that of normal melanocytes. Treatment at stage 10 (onset of gastrulation) leads to the appearance of normal melanocytes. Clearly the biosynthetic machinery for the fabrication of melanosomes is present at stage 8 and can be brought into function by the presentation of phenylalanine. Macromolecular processes such as yolk degradation and microfilament protein synthesis are now out of synchrony with melanin production, apparently undisturbed by the environmental change wrought by the experimenter. The details of macromolecular chemistry in this experiment are not directly known. However, we may presume that at least an increment in activity or an activation of tyrosinases has been induced. Since, like all proteins, tyrosinase must be genomically encoded, we must also presume that the ability for both transcription and the translation of this protein on polyribosomes is currently available. The gene is available, although under normal conditions its function would not be appropriate in normal embryogenesis until several days later.

The experimental alteration of temporal control is clearly demonstrable in the well-documented case of glutamine synthetase in retinal differentiation (Rudnick and Waelsch, 1955; Moscona and Hubby, 1963; Kirk and Moscona, 1963) in chick late embryogenesis. Here normal control is mediated by synthesis of the 11 β-hydroxyl corticosteroids, a result of the differentiation of the appropriate cells of the adrenal gland. The target of the circulating hormone, in this case, is the retina. Its presence in the microenvironment leads to the onset of transcription and translation processes and through these to the rapid increase of glutamine synthetase in the cells of the neural retina. The ability for the enzyme to be induced by the steroid 4 or 5 days in advance of its normal rapid rise means that the system is already available for function. Surprisingly, as Schwartz (1972) has shown, the induction by steroid involves only a subtle change, which is revealed by the appearance of a single protein peak of 27,000 molecular weight, appropriate for the polypeptide subunit of the enzyme within the spectrum of retinal proteins present. This is accompanied by a single major change in the RNA, and the new component is of the proper molecular weight (450,000), which puts it in the proper class of polyribosomes to contain the message for the translation of the enzyme subunit. There are essentially no other changes in either the protein or RNA (polyribosomal) profiles between enzyme-producing and non-induced neural retina cells. The response to the hormonal "call," even though advanced in time, is specific, at least at the level of a single protein subunit.

We are faced with the fact that specific biosynthetic mechanisms are present and capable of function well before their normal time of appearance or are rapidly inducible at similar earlier periods (Crawford and Batchelder, 1967; Epel et al., 1969; King and Mykolajewycz, 1973). Indeed, as the Gurdon group has recently shown (Gurdon et al., 1971), specific biosynthesis (as of hemoglobin) characteristic of late embryogenesis can be elicited in the activated mature oocyte of the African clawed toad, Xenopus, by artificial provision of the appropriate messenger RNA (mRNA). It is legitimate, therefore, to inquire how it is that embryogenesis is so precise in the face of these individual variants brought about by the experimenter with such relative ease.

The subtlety and apparent ease of manipulation of temporal control contrasts vividly with the extreme fidelity of normal embryogenesis when it is undisturbed. For instance, in the teleost fish, Fundulus heteroclitus, thousands of isochronously fertilized eggs under carefully controlled conditions of cleanliness and temperature will develop to hatching with less than 1 percent observable anomalies and essentially no mortality (Wilde and Crawford, 1963). Similar data can be obtained for many other developing systems. It is this very regularity of temporal control that permits the experimental analysis of development.

If it is established that mechanisms of temporal control of development constitute a primary (if often neglected) problem, what are the others? Broadly, they can be catalogued under the headings "cellular differentiation" and "morphogenesis." These traditional words have remained continuously valuable in the broader classifications of problems of development, as a sharp distinction can be made between them.

There would appear to be little doubt that cellular differentiation is under genomic control. Still, phenotypic expression in many mature cells obviously does not involve the overt expression of the total genome, (however, see King and Mykolajewycz, 1973, for a stimulating new idea*). There is an overwhelming

---

*They present data from the morphogenesis of bacteriophage tails that indicate that the assembly of proteins in the morphogenesis of tails may depend upon the presence of a single finished base plate. If any one of the base plate proteins is missing, tail cores do not form. Introduction of the missing protein allows rapid morphogenesis of the complete tail. Thus any *one* of the missing proteins could exert morphogenic control. King and Mykolajewycz (1973) have foreseen that rate-controlling mechanisms could control differentiation and morphogenesis and correlate them in time.

expression of the genes for various globin polypeptides in differentiating reticulocytes or, similarly, for actin and myosin in muscle. Yet, with the exception of such terminally differentiated cells as erythrocytes or moribund keratin squames of the skin, which are enucleated, the presence of a functional nucleus indicates presumptively the presence of the total genome. Where this can be or has been tested, as in the transplantation of nuclei in certain amphibia (Briggs and King, 1952; King and Briggs, 1956; King and McKinnel, 1960; Gurdon, 1962), the resultant normal or almost normal embryos indicate the presence of the total genome whose whole spectrum of function is recoverable by experiment.

Yet in a functioning nucleated and differentiated cell, in addition to the function of those genes of major expression by which the *type* of differentiation is made clear to the microscopist and chemist through the appearance of myosin, melanoprotein, crystallin, and the like, there must function, less obviously, genes for the cytochromes, the dehydrogenases, proteases, RNAases, DNAases, and many other proteins. These proteins are concerned with the maintenance of the life functions of the cell and must also be genomically encoded. They are called upon to function in the process of differentiation, if only to supply energy source systems. We have little data concerning the rate of synthesis and turnover of these essential macromolecules that represent apparently continuously functioning units whose aggregate bulk must make up a very large segment of a cell's protein family. Attention tends to be directed to those specific biosyntheses that are hallmarks of a cell's phenotypic expression. These constitute a relatively small group among the much larger catalogue of macromolecules active in the cell, whatever its specialized state.

In spite of an increasing mass of elegant studies on cellular differentiation, much remains obscure. Two primary problems must be faced here. The first is: In what manner are specific genes selected for expression? The second is: What mechanism controls the choice of expression in time and topography in the embryo? These problems increase in complexity where serial or multiple expression in terms of biosyntheses is required in a differentiative

process. To this is added the further complexity of dependence, whereby one cellular family makes a differentiative step that is a prerequisite to the taking of a different differentiative step by a second cellular family, which may either be in contact with the first or some distance from it.

The evidence in developmental biology is overwhelming that "information" is exchanged between cells. This is nowhere more clearly seen than in the experiments of Grobstein (1953) and many others who have adopted his techniques, where, in the now classical demonstration, salivary gland capsular mesenchyme contains and transmits, at the appropriate time, signals that determine the form of the response of the epithelial bud. These signals appear to be capable of transmission over small distances and through porous membranes, which would exclude whole cells but, perhaps, not cellular processes. In other instances, possible or probable informational material (steroid hormones) is passed over long distances (Lillie, 1916; Ohno, 1971) in a humoral fashion.

The chemical configurations of the exchanged materials are heterogeneous. They range from amino acids (Wilde, 1955, 1956), vitamins (Fell and Mellanby, 1953; Levenson, 1969), and steroids (Moscona and Hubby, 1963; Kirk and Moscona, 1963; Rudnick and Waelsch, 1955; Schwartz, 1972) to nucleic acids (Saxén and Toivonen, 1962) and, perhaps, proteins (Tiedemann, 1967) and particulates of higher order (e.g., the whole field of virus information transfer).

We need to know much more concerning the nature of information exchange, the forms that it takes, and the order in which it is transmitted. We need to understand the means of reception and perception of the input information and the mechanism of response whether optional or obligate, and whether secondary ordering or "noise reduction" of the information is required.

Interactions of molecular exchange take place between cells and through the cell's microenvironment. Attention must be turned to the functions of surfaces, for these constitute the first physical delimitations of the exchange sites. I shall not go into stringent detail of surface configuration or biochemical constituents here, as this is an area richly under analysis today (Rothfield, 1971). For the present, the membrane can be treated almost topographically as a basis for the following questions: Is the membrane of an embryonal, information-exchanging cell structurally and chemically homogeneous over all its parts, or are there topographical differences? If there are differences, what role do they play in information exchange? Are there generalized receptors? Are there step-up transformers in an energetics sense (Na^+- and K^+-activated ATPases) that mediate passage of appropriately "energized" molecules? Or, alternatively, are there mosaics of specific information receptors? If so, do the membranes themselves contain informational macromolecules? There are beginnings of answers in the current elegant work on immunologically reactive surfaces of certain lymphocytes and thymocytes and the currently interesting studies with more generalized receptor sites (Burger and Noonan, 1970; Moscona, 1971) in response to externally applied macromolecules (e.g., concanavalin A).

We recognize differentiated cells by rather simple methods: they look different from their neighbors; they function differently from their neighbors; or they fabricate a recognizable intra- or extracellular product, usually a macromolecule that is not obviously present in their neighbors. It is only natural that we emphasize in studies of differentiation the onset and completion of these spectacular arrays. While it is legitimate to use these dramatic changes as criteria in experiments, we often forget several important logical points in our enthusiasm.

Differentiation is the act of becoming different. To become different implies a comparative: Different from what? We noted above that cells are or become different from their neighbors. Cells may become different from their predecessors in time and also, perhaps more subtly, in space. Levenson (1970) has shown that the behavioral characteristics of chondroblasts of neural crest origin (Meckel's cartilage) are strikingly different from those of mesodermal origin (long bones). At the tissue level this is equally well shown for cartilage of differing embryonic source by the earlier work of Weiss and Moscona (1958). There are many ways in which cells become different from each other. They appear to follow lines of par-

ticular synthesis. It is obvious, therefore, that we have no present yardsticks to compare the compositions of differences. For example, there is no chondrocyte that is more or less differentiated than a fibrocyte.

What is the *function* of the differentiated state? There is a point here to be made that is beyond the obvious. Surely an erythrocyte in form and content is a more effective oxygen transporter than is a lymphocyte; yet beneath their specialties lies a vast array of similarities —for example, their cytochromes, dehydrogenases, and the like—in which areas they are perhaps quite identical in content and efficiency of function. The experimenter comparing these internal functions would perhaps find no differences between the two cell types and therefore to that test he could not differentiate between them. Differentiation, then, is some form of functional overload or specialization. If we restrict ourselves, for the moment, to multicellular organisms, the fact of multicellularity may be a requirement for differences to be functional. Indeed, single early embryonic cells (gastrula) in total isolation do not differentiate, whereas of two cells, one at least will (Wilde, 1960). This is evidence of the requirement for some exchange process to occur before the process of differentiation can take place.

If one were to discuss this point for a moment in another sense, the student of histones, on examination of the widest possible variety of eukaryote organisms, would conclude that all were indeed alike, the range of histone variability being so small as to be negligible when one considers the range of animal and plant life (Bonner et al., 1968).

All vertebrates have a liver. The anatomist, the morphologist would all find it in similar location throughout the class, whatever species was chosen. The histologist would recognize the proper cellular associations and cell types to identify liver, while the biochemist could extract from these cells, whatever the source, liver enzymes and other proteins more similar than not. Probably in a "blindfold" test the histologist and the biochemist would find it impossible, in some cases, to identify from which species a liver specimen came. In this sense all liver parenchymal cells that exist in whatever vertebrate are less different than liver parenchyma is different from the pancreas of the same organism.

Having now discussed the lines along which differences progress, we may return to the temporal nature of the process. Thus, in a zygote, there is a time when blastomeres are more alike than different, both in structure and in function. Yet it must be emphasized that such blastomeres are different from the fertilized ovum and are appropriately specialized (differentiated) for this particular time in development and this particular function.

Within a developing organism the options available to cells in the choice of pathways of differentiation are limited not as much by the contained genome as by time and place. The classical blastomere separation studies (see (Hörstadius, 1939, for review) indicate complete *genomic* encodement and function in each early blastomere (e.g., two-cell stage) for the development of a whole organism. Yet *positionally* each early blastomere is cued to express with its progeny cells only a certain portion of the resultant embryo. Positional information is required in morphogenesis (Harrison, 1921; Wolpert, 1971).

While it is most appropriate to discuss both positional and temporal elements together, it is difficult. Since there is more to be developed in the context of *cellular* differentiation, we are forced to forego this for the moment, bearing in mind that much data in the field of cellular differentiation have been derived by the experimental abrogation of positional control, as in tissue culture.

As cleavage continues after zygote formation, there is a progressive restriction in the options available to the blastomeres, and as the classic transplantation experiments have shown, there is, at length, a firm and almost complete delimitation to one (we must, however, within this context, recognize a list of blurrings expressed by the pathologist's term *metaplasia* or the *modulation* of Weiss, 1939).

There is a temporal expression or control of this process that varies among different organisms and, within a single embryotype, among the different "track options" as they become restricted. Nowhere is this more dramatically shown than by review of the classical transplantation experiments of Hans Spemann and his many students and of W. H. Lewis, where temporal control of a graft's differentiation was expressed either by its intrinsically preselected path (self-differentiating—usually a graft taken later in time) or extrinsically responding to signals of pathway direction from the neighboring cells of the graft (site-dependency—usually a graft taken earlier in time). This dichotomy as presented here is far too simple a statement since many intergrades have been demonstrated (Saxén and Toivonen, 1962).

If there is a restriction of options to cells with time (and place) in embryonic development, an important series of questions requiring answers lies in the general area of determining what factors or agencies control these decisions. It must be admitted that we know relatively little concerning this. But what in fact do we know? Let us examine the differentiation of two cell types that become specialized in the embryo at approximately the same time and that have highly divergent functional attributes and "hallmarks"—namely, striated muscle cells and erythrocytes.

Striated muscle cells of vertebrates differentiate in large part from the myotomes of somites and from the lateral mesoderm. Erythrocytes have variant areas of genesis changing through time: yolk sac, liver, and bone marrow. Erythrocytes are recognized by their shape—in mammals usually a biconcave disc, their very smooth surface contours, their lack of cytologically evident cytoplasmic organelles, and their very high content of the protein hemoglobin. Hemoglobin is a complex molecule, which itself varies through time in normal development. Its component parts are sufficiently complex to indicate that it must be controlled by a multiple-gene system (α chain, β chain, heme synthesis, and the enzymes for it) (Hunt et al., 1972).

Striated muscle cells in maturity are syncytia (although this is not obligatory—Wilde, 1959); as such they are multinucleate. However, their nuclei in *normality* no longer serve replicatory functions (Yaffe, 1969; Holtzer, 1970), having apparently lost the ability to synthesize DNA. The cytoplasm is filled with orthogonal arrays of myofibrils, themselves orthogonal assemblies of two proteins, actin and myosin, which are the functional contractile substances of these cells. Certain other specific proteins (e.g., T-system proteins) characteristic of striated muscle are present.

Erythrocytes normally exist in a liquid medium, the blood plasma. In formative stages they are found in very loose reticular tissue, fragile, tender and hidden in other organs or parts. In their formative areas are responsive stem cells that serve the progenitor function lost in maturation by the individual erythrocyte. The whole system is continuously in a proliferative phase and is sensitively responsive to a variety of control or homeostatic mechanisms (e.g., erythropoetin).

Striated muscle cells lie in a perimysium—a tough, collagen fiber-rich, fibroblastic membrane of some complexity. In intimate subperimysial relation lie occasional satellite cells next to the cell membrane of the muscle cell itself. These may serve as reserve function for new myoblast formation. The system is not in an obvious manner continuously proliferative. In fact, regeneration is difficult in homiothermic vertebrates (Godman, 1957) and somewhat less so in poikilotherms.

If we consider the embryogenesis of these two systems, which are temporally very close in development (and positionally not too distant!), several important points should be made. First, it would probably be very difficult indeed for the morphologist or the chemist to distinguish between the promyoblast and the proerythroblast. Both would be similar to many other progenitor mesenchymal cells. Chemically they would be expected to contain a predominantly similar, if not identical, roster of enzymes. Indeed, until the very terminal phases of differentiation, the similarity between the rosters would probably be far greater than the differences. As an example, most mitochondrial (sarcosomal) proteins would be identical between the two until the degeneration of normoblast mitochondria in the terminal phase of erythrocyte differentiation. This is to state that the similarities between two cells, each differentiating along a different pathway, may be much greater in sum than are the differences of which we make so much.

The point has recently been developed by Ohno (1971) that the control of differentiative steps in eukaryotes is apt to be surprisingly simple. In defense of this somewhat heretical statement, he cites the simplicity to be found in the control of sex organ differentiation as seen in the gene-controlled testicular feminization syndrome. This idea is further supported by data from other laboratories. These deal with extrinsic control of RNA synthesis and of dependent, specific protein synthesis elicited by environmental presence of particular steroid hormones in very low titer. While the events treated in these experiments occur relatively late in embryogenesis or in young adults, still much of import can be developed therefrom. Palmiter et al. (1972) and O'Malley et al. (1972) have dealt with the specific role of two steroid hormones, each of which elicits, in the appropriate oviductal cells, the synthesis and outpouring of two specific proteins, ovalbumin and avidin.

We shall now discuss the work of Schwartz (1972), as it is very recent and does not deal with a sex-determined and -determining system. In very brief outline the findings are as follows.

Glutamine synthetase (GS) is an enzyme that occurs in high concentration in the neuroretina and in the visual tracts of the brain (Piddington, 1971). The enzyme mediates the transfer of ammonium to glutamate and thus acts in a detoxifying sense. There is some probability that GS also functions in neurotransmission. The enzyme normally increases rapidly in the chick retina after day 16 (of incubation). This rapid phase is preceded by the functional differentiation of the adrenals and the appearance in the circulation of 11 β-hydroxyl corticosteroids. Hydrocortisone (an 11 β-hydroxyl corticosteroid) can precociously induce the enzyme in organ culture (Piddington and Moscona, 1967) by several days. The question to be answered is: What does the hormone do? Schwartz has examined the polyribosomal profiles of hormone-treated and control retinas of the same age. He has examined the proteins synthesized on isolated polyribosomes from hormone-treated and control retinas. He has further examined the mRNA extracted from polyribosomes of hormone-treated and control retinas. His findings of import are as follows. The polyribosomal profiles, the mRNAs, and the proteins synthesized by his experimental (hormone-treated) cases and his controls are predominantly similar—indicating that whatever the induction function of the hormone is, the cellular response is subtle. The response in terms of hormone treatment lies in the production of a new mRNA of the appropriate molecular weight to code for glutamine synthetase. This mRNA is extractible only from the appropriate polysome group whose size would accommodate it. It is not a degradation product of ribosomal RNA being unmethylated by experimental test. A single new protein appears among the welter of proteins synthesized but only in the hormone-treated experiments. This protein is of the correct molecular weight to be the subunit of glutamine synthetase and when assayed immunologically is serologically identical to purified glutamine synthetase of commercial origin. This rapid rise is remarkable since its presence cannot be demonstrated in the retina by current immunological techniques of high precision at 12 days (Schwartz, 1972).

We thus have a situation in terms of this particular step in the differentiation of the cells of the eye where the advent of circulating corticosteroid (reflecting the maturation of the differentiation of the adrenal cortex) brings about a single mRNA synthesis, which in its own turn is located on a particularly sized polysome complex. One new specific protein, the enzyme glutamine synthetase, appears in the catalogue of proteins being contemporaneously synthesized in these cells. The normal appearance of enzyme immediately precedes temporally certain advances in the histological and functional differentiation of the retina. Visual signals can be received and transmitted by the chick embryo at day 19.

This work, while meritorious and stimulating to the student of development, would constitute just another addition to the catalogue of macromolecular reactions associated with the multiplicity of differentiative processes were it left where it stands. O'Malley et al. (1972a) and others have recently begun to obtain data as to how the process of steroid induction works and some insight into its specificity. Of particular importance is the finding that steroid receptor proteins of high specificity exist. One, a cytosol protein, appears to pick up steroid at the cell membrane and transport it to the nucleus, where it is transferred to a nucleus limited receptor whose function presumably is to interact with the genome (Jensen et al., 1967).

Two things are immediately apparent. First,

it may be that all steroid function is mediated intranuclearly and, therefore, perhaps at the transcriptional level. Also there may be a single class of steroid receptor proteins. Such a datum is of significance in itself, yet within such a class of proteins may lie still further specificities, one to the particular steroid and another positionally in the genome or more crudely the chromatin. Here perhaps is an advance in understanding how a differentiative step takes place and what the particular function of some of the molecules concerned is. Can such double specificities (Ohno, 1971) explain, in part, why, for example, one cell becomes a chondroblast and its immediate neighbor a fibroblast, or how separate pathways of differentiation are elected and followed with the precision with which this is done *in time* and *in space*?

I would add to this segment a final cautionary portion. Normal cellular differentiation takes place in coordinate harmony with morphogenesis in developing organisms. Much of the data, informative and elegant as they are, are derived from systems of tissue culture that remove cells from their normal conditions for some other sets of conditions, perhaps quite bizarre. One often observes overt differentiation under these situations (and more rarely, partial morphogenesis). Sometimes the differentiation is extreme (overdone) as in striated muscle myotubes (Yaffe, 1969; Holtzer, 1970); sometimes it is incomplete or inapparent. Sometimes cells will travel two pathways of overt differentiation at once (Wilde, 1960), or differentiation will be abnormal (transformation). Since the cells with which we deal are alive, they must be expected to obey the laws of natural selection as will any living population. With the passage of time, there will inevitably be selective processes at work, some observable, some covert, that will result in a major change in the artificial population. The student of differentiation should be continually wary of this and in the referral of his findings from such systems back to the organismal state.

Many data are derived from cellular families in tissue culture that have had a preliminary treatment to dissociate them. The techniques are variant: trypsin, hyaluronidase, collagenase, salt exclusion, and the like. What is the nature of the insults applied to the cells *even before the experiment begins?* If one observes cells immediately after dissociation, they are often in no way similar in morphology to the cells of the tissue of their origin. In an area of research where we all emphasize surfaces of contact, membrane transfer, membrane interaction, and transmembranal information exchange, what is known about the surface of our donor cells prior to enzyme treatment, or after? Recently Maslow (1970) has shown a change in surface charge in cells immediately following treatment with trypsin. What is the relation of these posttreatment surfaces to those functioning in normal morphogenesis?

Morphogenesis— The Great Problem

Morphogenesis represents the quintessential problem to the developmental biologist. Thus it is understandable that discussion of, or research directly involved with, morphogenesis has been historically avoided except in a descriptive sense. It is difficult indeed to get a "handle" on the process. We presume, of course, that morphogenesis, as a multitude of processes, and morphology, as the result, are controlled in the genome. We presume also that the portion of the genome involved is very conservative in an evolutionary sense and protected from drastic evolutionary change in some manner. The primary function of the genome lies in the control of types of protein synthesized. The types and classes are very conservative, apparently changing little through long evolutionary periods.

This fact becomes manifest by observation both of the similarities of proteins with similar functions among many phyla and the very minor changes—often found among proteins within a species of somewhat differing but related functions (Dayhoff and Eck, 1968).

Many of the early preparations for future morphogenesis are the property of the maternal genome and are carried out during oogenesis. It is probable that morphogenetic information, processing, and control of morphogenesis up to the beginning of gastrulation

are maternally regulated in all multicellular animals, to judge from positive data from a broad spectrum of phyla (for a modern review see Tyler and Tyler, 1970). Such control may extend further, being gradually abandoned to the zygote genome function. Zygote genome function appears, in most of the cases studied, to begin during cleavage and to control morphogenesis at the beginning of gastrulation.

If this is true, then several questions of detail can be asked: What is the nature of the maternal control? How is the input conserved? Of what does the input information consist? How is the expression of input regulated? How is control transferred to the zygote? What accounts for the extreme regularity of these processes such that the result, though variant to a degree, falls within such restricted boundaries as to permit organisms to be classified as to species, subspecies, race, and so on?

In manufacturing, precision in reproducing hard goods and control of variance of product is greatly benefited by the previous fabrication of molds and templates. Precision is also furthered by control of subassembly input through time control such that the pieces and parts are economically placed in function when temporally advantageous. I have previously emphasized temporal control in cellular differentiation. It is equally important in morphogenesis.

We have some understanding of some of the templates of morphogenesis, their evolutionary stability, and their functional efficiency. They are genes—their products enzymes and other proteins whose structure and function are evolutionarily very ancient, as noted above. A large catalogue of macromolecules continues to be developed, many of which are identical or only slightly variant among a wide range of organisms and over a wide range of functions within a single species.

We have little or no understanding of important morphogenetic mechanisms that lead to the regulation of symmetry, of morphogenetic movements, of location of organs, of independent or dependent development of organs and parts, of regulation of size, and many other questions of a similar nature.

We have little information concerning the temporal regulation of morphogenetic processes. How is an organ ready when its function is required? What at first seems a silly question leads directly into questions of greater merit. How is the operations schedule of the genome so regulated that the heart is formed and functional as the blood vascular system is readied, as the embryo requires more efficient metabolite exchange? What events previous to visible morphogenesis are required? How are they scheduled and fed into the synchrony?

What evidence we have with regard to the genomic load of somatic or embryonic cells is that every cell contains the whole genome. Cells that do not, become transformed or moribund. Yet such genetic information is manifestly redundant, unless each somatic cell should be afforded the opportunity to recreate the whole organism.

Yet the very redundancy of the total genome in a somatic cell of an organism provides one of the fundamental requirements for the solution of many problems of morphogenesis.

It is a requirement for the organismic solution of the "French Flag Problem" of Wolpert (1971). This theoretical situation whereby "cells" within a planar French tricouleur are able to generate the whole, and the information with regard to location in space that they would require for this process, have been elegantly treated in two essays by Wolpert (1969, 1971) as a model of morphogenesis.

Redundancy of genetic information in a very similar sense is required for the most general solution of a more simple situation that may be called the "last man on the right" problem. The analogy is to a rank of private soldiers, all similarly instructed, drilled and equipped (Fig. 1). When ordered into formation, the last man on the right has duties somewhat different from those of his comrades. All privates are instructed what to do in case they find themselves to be the last man on the right. In any particular case this information is redundant and in reserve for all but one individual. Yet he knows what to do through training and recall from his redundant or excess information. Another solution to this problem is to inform only one individual and require that he always be in the proper position at the proper time. If he is not present, then the formation cannot be made since no other man has the required information.

There are many organogenetic situations when a cell finds itself "the last man on the

FIGURE 1. *A rank of soldiers, all identically trained and equipped, responding to the command "Dress right. Dress!" Any soldier could be the last man on the right and will behave appropriately by recall and use of information redundant except when he is the last man on the right. A similar argument can be made for the last man on the left.*

right" and thus must have recall of preinstruction in order to function correctly. In spiralian (e.g., molluscan, annelid) development, preinstruction of the single unit (cell) system seems to apply, and the A cell cannot substitute for the D cell. In many, perhaps most, morphogenetic systems, the redundant information and recall system appears to function.

There are obvious theoretical intergrades between the two systems of information utilization, and *in vivo* examples can be found in morphogenesis and organogenesis. An epithelial primordium (as that of the ureteric bud) grows into the metanephrogenic mesenchyme. As these cells proliferate and advance and nephron units begin to form, there comes a time when one tubule cell along a particular vector finds itself the "last man on the right." How does it receive the information that it has a glomerulus associated fibroblast for a neighbor rather than sister parenchymal cells, and from what reservoir of information does it draw the appropriate response—namely: quiescence, adaptation to the glomerular contents, reception of a micromilieu to a degree different from other parenchymal cells, and so on? Since any epithelial cell could find itself in this position, all must be equipped with the appropriate information, redundant except in the specific case.

In the resolution of temporal and positional problems that are inherent in morphogenesis, it would appear that the genome (again we assume its primacy) must have a managerial function. Such a managerial function would have the ability to control a rather large number of chemical reactions coordinately in time and in place, dictating the start of some and fading out or turning off of others.

There are no simple cases to discuss and all analogies are dangerous, but some are, perhaps, occasionally instructive. A horde of army ants in many ways acts as an organism and can show certain aspects of morphogenesis in its behavior. One form of the horde is that of a series of interlocking streams exhibiting approximate directional "flow" of marching individual ants, workers in the centers of the streams, soldiers forming outliers, and guards at the sides of the streams. As this order of form marches through the forest, night falls and a bivouac is required. Explorer ants diverge to sense an appropriate location. The column is perhaps diverted, but still there is an imperative in the horde for bivouac. Under these circumstances, an interlocking mass of ants is formed that serves a protective function for the night. New cues initiate a return to the stream. The change from interweaving streams to a condensed mass is a morphogenesis involving response of external cues, perhaps of a very small number, whereby the assembly and interlocking of the ant bodies is called for and to which essentially all respond. The cues could be, for example, alteration in the angle of polarization of light or degree of hydration, and would require a genomic behavioral response (which would be complex in itself). Any group of soldiers, returning from scout and receiving these cues, begins a response (i.e., forming the mass) to which now all other ants respond, the informative cues being transferred either behaviorally or by means of pheromones (Schnierla, 1971). The possibility that small cues or information "chips" are functional in morphogenesis is treated in detail at the end of this chapter.

This is to say that a managerial function of the genome (or grouped individual cellular genomes) does not require an intelligence or an "entelechy" (Driesch, 1929)—only evolutionarily long-ingrained, appropriate responses to appropriate clues, however complex they may have become, now stored and in most circumstances redundant. What evidence can be adduced for a managerial complex? Perhaps the following experiments can help.

The demonstration of a managerial function in the genome with regard to morphogenesis has been indirect and requires the union of the techniques of molecular biology and those of classic descriptive morphogenesis and teratogenesis. The application of this combination tends to restrict the types of eggs and embryos that are available for experiment. Certain experiments with the eggs, zygotes, and embryos of *Fundulus heteroclitus*, a teleost fish, have been undertaken that make it probable that a managerial function exists. This organism is readily available and has been much studied. Its normal stages are established (Oppenheimer, 1937; Armstrong and Child, 1965; New, 1966; and the distributed but unpublished color photographs of Wilde), and synchronous development of very large numbers of zygotes and embryos can be established by timed external fertilization controlled by the experimenter. One can therefore observe with ease the external events of development and make continuous observations of abnormal development. At the same time, since the numbers of synchronously developing embryos are large, adequate samples can be taken at any desired time point for chemical analysis and macromolecular assay.

The stiff and rapidly developed chorion of the egg permits the entrance of a variety of molecular species of varying molecular weight. In many instances both inflow and outflow can be measured. For example, actinomycin D passes readily in both directions; indeed its recovery can be almost stoichiometric (Wilde and Crawford, 1966). Therefore, dose response curves can be easily established with many inhibitors, and the specific effects on morphogenesis brought on by the inhibitors can be studied by observation of the details of morphogenesis of the survivors. One can then, by knowing the time at which an inhibitor has been acting and with some evidence as to what its mode of action is (e.g., for pactamycin, inhibition of protein synthesis) and by understanding how a particular defect develops and what parts are affected, tentatively assign a direct causal relationship between the event and the effect of the drug through its mode and time of action. We are quite accustomed to doing this in a variety of living systems where our depth of analysis into a system is not now, or, for some reason, cannot be complete to the molecular or genetic level. It can only be hoped for that later more sophisticated experiments will give us more penetrating insights.

We must, however distantly, work with the catalogue of molecules initially to be found in egg and sperm. We can hope, ultimately, to record through time, accessions and deletions, new associations and reassociations, as morphogenesis continues. Where a defect is demonstrated in a survivor and the time of the insult is known, then one may justifiably correlate the time of insult, the form of the insulting agent, and the long-term result. Should the insult be chemical, then we may associate the resultant morphogenetic defect with the mode of action of the chemical. We may also associate the time of action of the chemical with the morphogenetic defect since this is also known to the experimenter.

Should the defects be highly repeatable and the time of insult be as precise as possible, then a catalogue of defects can be made and correlated with the time of the insult as well as correlated with the general mechanism of action of the drug.

In *Fundulus* we have developed such a catalogue and such correlates (Wilde and Crawford, 1968; Crawford, et al., 1973). Two drugs

concern us here, actinomycin and pactamycin (Fig. 2). Actinomycin D has been widely used. Its primary mode of action is to prevent transcription of RNAs from DNA by binding guanine residues of the latter (Reich et al., 1962). The drug has many side effects and care must be taken in its use (Schwartz, 1973). However, in *Fundulus*, with very precise experiments and with controls for respiratory effect and mitotic effect, depression of approximately 55 percent in the incorporation of ^{14}C uridine into RNA can be demonstrated (Wilde and Crawford, 1966). The incorporation is restored to normal following the removal of the drug, but the survivors, and their number is very large, retain a common defect, a morphological abnormality throughout their embryonic period until hatching of the controls (i.e., the length of time they have been studied prior to fixation, 40 ± 1 days). The defects form a catalogue that reflects the serial order of the events of morphogenesis. The time of onset of the insult correlates precisely with the catalogue (Wilde and Crawford, 1968; Crawford, et al., 1973).

Actinomycin D, at the dose used ($20 \mu g/ml$), acts as a general inhibitor of RNA synthesis as measured by the decrease in incorporation of isotopically labeled precursor (uridine, uracil) into the macromolecule. In *Fundulus*, the drug, over the periods used, does not inhibit protein

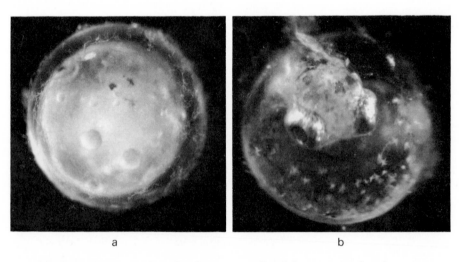

a   b

FIGURE 2. *(A) Defective, nonaxiate development of* Fundulus *zygote placed in actinomycin D immediately upon fertilization for 1 hour and then reared until hatching of controls. (B) Control normal embryo of the same age (almost hatching, 38 to 40 days) as A.*

TABLE 1

TERATOLOGICAL EFFECTS OBSERVED IN EMBRYOS OF FUNDULUS
HETEROCLITUS IN RELATION TO AGENTS USED AND TIME
(INITIATION TIME) INHIBITORY PULSE WAS APPLIED[a]

Pulse initiation time after fertilization (min)	Effect on morphogenesis observed in survivors	Inhibition of RNA synthesis (actinomycin D)	of Protein synthesis (pactamycin)
0–1	Prevention of early cleavage	−	+
1–2	Defective blastula	+	+
2–3	Defective gastrulation	+	+
3–4	Defective axiation	+	+
4–5	Anencephaly	+	+
5–10	Less severe anencephaly	+	+
10–30	Microcephaly	+	±
30–60	Normal development	±	±

[a]All drugs are at 20 µg/ml. The data are reported in C. E. Wilde, Jr., and R. B. Crawford, *Exp. Cell Res. 44*, 471–488, 1966; C. E. Wilde, Jr., and R. B. Crawford, Epithelial-mesenchymal interactions in the lower vertebrates, *in* "Epithelial–Mesenchymal Interactions" (R. Fleishmajer and R. Billingham, eds.), Chap. 8, pp. 98–113, Williams & Wilkins, Baltimore, 1968; and R. B. Crawford, C. E. Wilde, Jr., M. H. Heinemann, and F. J. Hendler, *J. Embryol. Exp. Morphol. 29*, 363–382, 1973.

synthesis nor respiration (Crawford and Wilde, 1966b). The effect appears, in *Fundulus*, to be concentrated primarily upon RNA specifically associated with morphogenesis, since cyanide, at low concentrations (Crawford and Wilde, 1966a, b), is an absolute and reversible inhibitor of all RNA synthesis and, acting through another mode, the inhibition of oxidative phosphorylation. Maintenance of this inhibition forces the arrest of morphogenesis at the high blastula; relief of the inhibition permits both renewed morphogenesis and renewed RNA synthesis that is *without defect* (Crawford and Wilde, 1966a, b; Wilde and Crawford, 1966). Experiments with a nitrogen environment are precisely the same. Therefore, we are probably not dealing with nonrespiratory effect.

The contrast here lies with the retention of the morphogenetic defect, the teratum, following actinomycin D, and its absence following cyanide or N_2 (Wilde and Crawford, 1966), in the embryos so treated. This suggests rather

forcefully that the effect of the drug lies within the mRNA fraction. The catalogue of defects appears in Table 1.

The antibiotic pactamycin acts to inhibit protein synthesis by affecting the binding of aminoacyl transfer RNA to ribosomal subunits (Cohen et al., 1969) and is completely reversible, protein synthesis being restored to normal 4 hours following removal of the agent (Crawford et al., 1973). Happily, at the concentrations used, pactamycin does not affect RNA synthesis in *Fundulus* (Crawford et al., 1973). Therefore, experiments in parallel with those mentioned above can correlate the effects of precisely timed inhibitions of protein synthesis upon morphogenesis of the survivors with those that were observed in experiments using actinomycin D. When this is done, the terata observed reflect the same time dependence and the same series of types of abnormality seen with actinomycin D (Fig. 1, Crawford et al., 1973).

An additional series of observations with pactamycin is of interest. The drug reduced incorporation of amino acid into protein by between 75 and 99 percent and a teratogenic pulse of 15 minutes duration leads to gross abnormalities (Crawford et al., 1973). When the inhibitor is applied immediately upon fertilization or very shortly thereafter (seconds and minutes), cleavage is inhibited. Usually the inhibition of cleavage is manifested after a short delay; thus the next cleavage occurs, but the following one tends to be aborted. This indicates that there is a required protein synthesis for normal cleavage inhibitable by pactamycin and that a small reserve of synthesized proteins functioning in the cleavage process is available (usually enough for one division). Since the cleavage period is unaffected by treatment with actinomycin D, there is strong presumption here that the proteins transcribed during this period are fabricated upon maternal templates. Indeed this has been specifically shown by identification of certain proteins in recent experiments (Raff et al., 1972).

Is there a *change* in the catalogue of proteins as an early embryo develops? Certain preliminary experiments dealing with the proteins of *Ilyanassa* embryos would indicate that this is so (Teitelman, 1973). Recent data in *Fundulus* have identified distinctive changes beginning shortly following gastrulation (Schwartz and Wilde, 1973).

The value of these experiments with regard to the understanding of morphogenesis lies in the ability through this method to study the deviant morphogenesis of the survivors presumably caused by strictly time-controlled insults to their genetically controlled macromolecular chemistry. In part at least, which events in cellular chemistry are being inhibited are known. This then gives us a roster of the events in morphogenesis; it tells us what morphogenetic activity is begun at what time. It gives us a set of antecedent–succedent steps, so that, at least in early morphogenesis in *Fundulus*, it is possible to say which precede others and thus to speculate on the order in which the genome is activated for morphogenesis.

Thus it would appear that the genomic control of the steps of gastrulation is begun at the second minute after fertilization and is fully confirmed by the fifth minute. So also the anteroposterior axis of the body is probably expressed and decided by macromolecular synthesis at the third minute and is complete by the sixth minute. The whole "package" of mechanisms concerned with the formation of the head is set shortly thereafter (Wilde and Crawford, 1968), and indeed within this "package" there is evidence of a subserial order determining the individual organs of the head.

One extremely important facet of morphogenesis, of which more will be said later, is the fact that, with the control of gastrulation and axiation, a confirmatory decision of absolute value is made with regard to the symmetry of the resultant embryo and adult. The egg of *Fundulus* is equipped with a micropile and thus, along with the stratification of heavy entities such as yolk, an animal–vegetal axis is expressed in the zygote. However, this is overridden by inhibition of transcription or translation at the second minute after fertilization, for the embryos resulting after recovery from this insult have no axes at all but spherical masses of cells colonizing the yolk surface in which cells are differentiated at random (Wilde and Crawford, 1966).

One of the remarkable things about the analysis of control of morphogenesis in *Fundulus* is the extremely early period (during the one-celled stage prior to the onset of cleavage), when inhibitions are effective. One advantage for such precocity may lie in the possibility that those genes that are open and functioning in a family of genes (which perhaps are predominantly shut off) will pass equally to the blastomeres at cleavage with an immediate prehistory of function. The importance of this remains to be assessed. Such a system requires the presence of a mechanism that will sequester the transcribed products intact or in proper shape for function only after a considerable period of time, which is, in *Fundulus*, at least 40 hours. No such system is presently known, although Spirin (1966) and Tyler and Tyler (1970) among others have made proposals for mechanisms. Recently, however, analyses and findings in the sea urchin, reported by Slater et al. (1972) on the biosynthesis of polyadenylic acid in relation to protein synthesis, may give a preliminary insight into mechanisms of early control.

Of equal importance are those aspects of morphogenesis that escape from or recover from these insults.

If the analysis and/or the understanding of morphogenesis is indeed the essential problem of the student of development, one somewhat antecedent problem is: "Where does one begin?" There are many answers to this question varying from the choice of an apparently simple problem—for example, tubulin synthesis (Raff et al., 1972)—to attempts to find the *gestalt* of embryogenesis. We have chosen to view the embryo as a whole as it develops, rationalizing as follows: If every identifiable process is genomic and expressed in relation to the microenvironment, so also must be the whole of embryogenesis; and, further, it is possible to see the whole in phenogenetic terms. We have shown that there must be an obligate order of synthesis of morphogenetically meaningful macromolecules (nucleic acids and proteins) whose early biosynthesis probably in strict temporal order is essential for normal morphogenesis. This obligate order cannot be disturbed without quite specific morphogenetic failure (teratogenesis is the inverse notion). Such morphogenetic failure is permanent and, although protein and other macromolecular syntheses are undertaken anew following the experimental insult, some sequence, probably in a temporal sense, is missing and, having been missed, *cannot* be restored. The result is a *permanent* defect in the embryo that is expressed later in its morphogenesis.

In morphogenesis, however, it is easier to think about the process while it is going on—that is, contemporaneously—and not in terms of antecedent control and decision. In order to understand morphogenesis, one must consider the causal events that occurred long before the visible change. The gap in time may be extensive. Consider the following experiment conducted by Dr. Robert J. Schwartz in my laboratory.

Inhibition of RNA synthesis in *Fundulus* by actinomycin D within minutes of fertilization (2 to 3 minutes, Wilde and Crawford, 1968) leads to the abortion of normal gastrulation after stages 9 and 10 (Oppenheimer, 1937) have been reached normally as compared with isochronous controls. The treated embryos therefore do not axiate. Schwartz compared the protein profiles by sodium dodecyl sulfate acrylamide gel electrophoresis in actinomycin-treated and control embryos. The assays were made when controls were at stages 15 and 19, that is,

during late gastrulation (15) and after the beginning of axiation (19).

There is an early "pattern" of protein synthesis exhibited by the early controls (gastrulae, stages 10 through 12). The experimental 15s and 19s, which, of course, did *not* axiate or gastrulate properly, show a "pattern" similar to these. However, the pattern of normal or control 15 is different, indicating new protein synthesis in normally gastrulating embryos over the primitive pattern. The pattern of new synthesis is different again at stage 19, indicating that the process or period of axiation requires new protein synthesis associated with these processes. The pattern at 15 is different from that at 19. We were able also to identify a "yolk protein" configuration quite distinct from these proteins, which appear to be more directly related to morphogenesis (Schwartz and Wilde, 1973).

Such data only tantalize and require much more specific answers to such questions as follow:

1. What is the identity of these morphogenetically associated proteins?
2. If they are indeed functional in morphogenesis: (a) How do they act? (b) What energy systems are required? (c) What other cognates are required? (d) How are these proteins different from "nonmorphogenetically active" proteins (if such exist)? (e) How do they behave in causal morphogenesis? (f) Do such series of syntheses have an order that would or could lead to left–right or anterior–posterior asymmetry?

These last questions lead to the heart of morphogenesis. How is symmetry determined?

Symmetry

Discussions of symmetry must always require reference to the fascinating, always new and refreshing "On Growth and Form" of Sir D'Arcy W. Thompson (1948). We shall consider only certain pertinent questions relating to the symmetries of vertebrate morphogenesis.

I paraphrase here a question I once wrote in the discussion of a paper (Wilde and Crawford, 1968): If a cell exists in the midline of an

A–P axis of an embryo, how does it "know" it so lies? And further, how do its neighbors topographically left and right know, recognize, and react to the position in which they find themselves? In normal morphogenesis such positional sensing must exist at least within 1 or a few cell diameters, else the lateral borders of organs (with differently differentiating cells) could not exist with the precision we know (the reader will recall the last man on the right problem).

Consider for a moment the problems of regeneration of amphibian limbs. Stocum (1968a,b) has clearly shown that the blastema of regeneration in Urodeles is a mosaic of "determined" organogenetic parts. Thus a mechanism of genetic encodement must exist in the genome of the cells of the amputation site that "tells" the cells in the wound area several things. They are demonstrably equipped with "knowledge" of their position in the limb and with the information as to what has to be made peripherally to complete the regeneration. Stocum has indicated this to be so in elegant transplantation experiments (1968b). Why is there no centripetal regeneration? What geometric parameters lead to this being forbidden? Even the experiments of E. G. Butler (1955) on regeneration of limbs with an artificially reversed polarity indicate the denial of centripetal regeneration and the obviously awkward assumption of peripheral regeneration by these reversed limbs.

Finally, in this context, given permission for only peripheral regeneration in the limb, how is it that the peripheral regeneration is anatomically precise, in that no extra organogenesis is undertaken? A more intriguing situation dealing with the possible reversal of polarity of certain organogenetic cells is to be found in the case of the dental lamina. In tooth and jaw regeneration it may, under certain circumstances, have its polarity reversed (Graver, 1973).

Concluding Remarks

We yearn to know how the genome regulates morphogenesis. Those systems with which we can play games of synthesis (e.g., globin) give us no insight into the mechanism of formation of an erythrocyte. We recognize that there is, in morphogenesis, a conservatism and regularity that is extreme. This is, after all, the basis of the *ability to classify* animals and plants, the discipline of the taxonomists upon which we all rely daily. Such genes must be exceedingly well conserved, protected from extensive crossing over, deletion, and mutation. Where can they be and what is their nature? An enticing possibility is that they are protected and lie close to the centromeres of their chromosomes. The latter position would ensure the greatest conservatism in terms of crossing-over. An intriguing possibility lies in the fact that heterochromatin in many species is near or about the centromeres. Is the DNA of morphogenes contained in heterochromatin or protected by it? Do morphogenes exist and how would they function?

Dame Honor Fell, during the period of her active studies on the differentiation and morphogenesis of long bones in organ culture, demonstrated femur morphogenesis in a remarkable film made by time-lapse photomicrography. One observed the growth and, particularly, the morphogenesis of the isolated organ rudiment. I retain a vivid image of the gradual appearance of the three-dimensional shape of the rudiment within a sea of single cells in motion that "jittered" flame-like over the organ. Cells moved in and out of the forming organ in an apparent random manner; yet out of this "noise" the morphogenesis of the rudiment was carried out. Obviously more and more cells were becoming positioned in the developing rudiment, while others not yet positionally fixed or programmed exhibited migratory activity.

Morphogenesis was going on through time, and one must presume that earlier migratory cells or their progeny were those that contributed to the establishment of organ form and responded appropriately to positional information. We return to the two problems—time and positional information among the units of the morphogenetic process. The importance of these two elements of the morphogenetic process requires this final discussion. What could positional information be? Of what would it consist?

Positional information in developing systems

could consist of a large series of comparative "bits" of information of the following types:

RIGHT OF	ATTACHED TO
LEFT OF	OVERLAPPING
ABOVE	(OCCLUDING)
BELOW	ISOLATED (ONE)
INSIDE	GROUPED (MANY)
OUTSIDE	LARGER THAN
AT THE CENTER OF	SMALLER THAN
SURROUNDED BY	LONGER THAN
NEAR	SHORTER THAN
FAR	MORE THAN
NEXT TO	LESS THAN

These items of comparative bits have been taken directly (and therefore out of the author's context) from the stimulating article by Barlow et al. (1973), entitled "Visual Pattern Analysis in Machines and Animals." I was struck forcefully by the application of comparative "bits" of this nature to the whole dynamic of positional information and its usage by cells in morphogenesis. To this list I would add:

ANTERIOR TO	VENTRAL TO
POSTERIOR TO	MEDIAL TO
DORSAL TO	LATERAL TO

given that positional information and its utilization by cells in morphogenesis require in some manner the cells' adherence to axes of organismic symmetry. Finally, in a dynamic sense, in terms of movement, I would also add:

SLIDING BY LEFT	SLIDING BY DORSAL
SLIDING BY RIGHT	SLIDING BY VENTRAL

and with respect to time:

PRECEDING

FOLLOWING

IN TIME WITH

and lastly:

SITE RECOGNITION GENERAL

SITE RECOGNITION PARTIAL

SITE RECOGNITION SPECIFIC

We have previously discussed the morphogenetic and differentiation advantage of the retention of redundant information provided a means of recall is available. Positional informa-

tion in the frame of the current discussion could be made up of such comparatives as in the lists above, that is, "bits."

In avian gastrulation, for instance, the delaminatory phase to form the hypoblast could be driven if a slight preference to the signal "below" would occur amid the noise of other signals. Such a slight preference we will call an *imperative*. Along the appropriate time frame this imperative is expressed, and as a result there would be an increasing number of cells dividing with planes parallel to the blastula surface. The lower or inner daughters would ineluctably find themselves positionally in the hypoblast area where and whence other imperatives would be responded to.

However, here, and most importantly, we must ask: Of what would such an imperative consist? The only "language" function we know of is in individual cells and is expressed in protein synthesis. In these terms, certain possibilities present themselves. There could be expressed in the appropriate time flow of genomic expression—a change in the topography of membrane proteins such that the rest configuration in terms of energetics would tend to place mitotic spindles 90° from a previous mean parallel position. This might be expressed by the temporally controlled biosynthesis of a single or a few membrane associated proteins. Thus the imperative "morphogene" concerned with a delamination could consist of an operon translating for a single protein of differing surface character such that spindle verticality tended to increase for a certain period. My example is manifestly too simple but does indicate how, hypothetically, "morphogenetically meaningful information" or "positional information" may be directly linked to the protein synthetic "language" of the cell.

I now return to the "shimmer" seen so clearly in Dame Honor's motion picture of femur morphogenesis *in vitro*. The "shimmer" that reflects the apparent random movements of individual cells can be seen as an expression of the "noise," while the slow appearance of form and growth represents the function of the appropriate imperatives expressed in positional information.

It would follow, therefore, that the "noise" is an integral and necessary part of the morphogenetic process, inevitably so—owing to the amount of redundant information present in each unit of the system.

Morphogenesis, then, joins the ranks as an extremely difficult problem in protein synthesis, given the serial order of genetic expression in development. Cellular differentiation, complex as it is, can be treated as a homologous but simpler case.

Some of our difficulties may perhaps disappear if we consider that in the interpretation of morphogenesis we must work amid the "noise." Only rarely can the process become clear, such as in the computerized "clean up" of fuzzy photographs.

Finally, however, we may see the beginnings of a morphogenetic "language" of positional information that must be translated back to the classical language of genome-controlled, polyribosome-translated protein synthesis.

References

ARMSTRONG, P. B., AND CHILD, J. S. (1965). Stages in the normal development of *Fundulus heteroclitus. Biol. Bull. 128*, 143–168.

BARLOW, H. B., NARASUNHAR, R., AND ROSENFELD, A. (1972). Visual pattern analysis in machines and animals. *Science 177*, 567–575.

BATESON, WILLIAM (1894). Materials for the study of variation. Macmillan, London.

BONNER, J., DAHMUS, M. E., FAMBROUGH, D., HUANG, R. C., MARUSHIGE, K., AND TUAN, D. Y. A. (1968). The biology of isolated chromatin. *Science 159*, 47–55.

BRIGGS, R., AND KING, T. J. (1952). Transplantation of living nuclei from blastula cells into enucleated frog's eggs. *Proc. Nat. Acad. Sci. U.S.A. 38*, 455.

BURGER, M. M.. AND NOONAN, K. D. (1970). Restoration of normal growth by covering of agglutinin sites on tumour cell surface. *Nature (London) 228*, 512–515.

BUTLER, E. G. (1955). Regeneration of the urodele forelimb after reversal of its proximodistal axis. *J. Morphol. 96*, 165–182.

COHEN, L. B., GOLDBERG, I. H., AND HERNER, A. C. (1969). Inhibition by pactamycin of the initiation of protein synthesis. Effect on the 30S ribosomal subunit. *Biochemistry 8*, 1327–1344.

CRAWFORD, R. B., AND BATCHELDER, M. L. (1967). Hexokinase in the developing embryo of *Ambystoma maculatum. Amer. Zool. 7*, abstract.

CRAWFORD, R. B., AND WILDE, C. E., JR. (1966a). Cellular differentiation in the anamniota. II. Oxygen dependency and energetics requirements during early development of teleosts and urodeles. *Exp. Cell Res. 44*, 453–470.

CRAWFORD, R. B., AND WILDE, C. E., JR., (1966b). Cellular differentiation in the anamniota. IV. Relationship between RNA synthesis and aerobic metabolism in *Fundulus* heteroclitus embryos. *Exp. Cell. Res. 44*, 489–497.

CRAWFORD, R. B., WILDE, C. E., JR., HEINEMANN, M.-H., AND HENDLER, F. J. (1973). Morphogenetic disturbances from timed inhibitions of protein synthesis in *Fundulus. J. Embryol. Exp. Morphol. 29*, 363–382.

CURTIS, A. S. G. (1962). Morphogenetic interactions before gastrulation in the amphibian *Xenopus laevis*—the cortical field. *J. Embryol. Exp. Morphol. 10*, 410.

DAVIDSON, E. H. (1968). "Gene Activity in Early Development." Academic Press, New York.

DAYHOFF, M. D., AND ECK, R. V. (1968). "Atlas of Protein Sequence and Structure." National Biomedical Research Foundation, Bethesda, Maryland.

DRIESCH, H. (1929). "The Science and Philosophy of the Organism." Black, London.

EPEL, D., PRESSMAN, B. C., ELSAESSER, S., AND WEAVER, A. M. (1969). The program of structural and metabolic changes following fertilization of sea urchin eggs. *In* "The Cell Cycle" (G. M. Padilla, I. L. Cameron, and G. L. Whitson, eds.) Chap. 13. Academic Press, New York.

FELL, H. B., AND MELLANBY, E. (1953). Metaplasia produced in cultures of chick ectoderm by high vitamin A. *J. Physiol. 119*, 470–488.

GAL-OR, BENJAMIN. (1972). The crisis about the origin of irreversibility and time anisotrophy. *Science 176*, 11–17.

GODMAN, G. C. (1957). On the regeneration and redifferentiation of mammalian striated muscle. *J. Morphol. 100*, 27–82.

GRAVER, H. T. (1973). The polarity of the dental lamina in the regenerating salamander jaw. *J. Embryol. Exp. Morphol.* (in press).

GROBSTEIN, C. (1953). Morphogenetic interaction between embryonic mouse tissues separated by a membrane filter. *Nature (London) 172*, 869.

GURDON, J. B. (1962). Adult frogs derived from the nuclei of single somatic cells. *Develop. Biol. 4*, 256.

GURDON, J. B., LANE, C. D., WOODLAND, H. R., AND MARBAIX, G. (1971). The use of frog eggs and oocytes for the study of messenger RNA and its translation in living cells. *Nature (London) 233*, 177–182.

HARRISON, ROSS G. (1921). On relations of symmetry in transplanted limbs. *J. Expl. Zool. 32*, 1–136.

HARRISON, ROSS G. (1933). Some difficulties with the determination problem. *Amer. Natur. 67*, 306–317.

HOLTZER, H. (1970). "Myogenesis." *In* "Cell Differentiation" (O. A. Schjeide and J. deVellis, eds.), Chap. 27. Van Nostrand Reinhold, New York.

HÖRSTADIUS, S. (1939). The mechanism of sea urchin development, studied by operative methods. *Biol. Rev. Camb. Phil. Soc. 14*, 132.

HUNT, T., VANDERHOFF, G., AND LONDON, I. M. (1972). Control of globin synthesis: the role of heme. *J. Mol. Biol. 66*, 471–481.

JENSEN, E. V., SUZUKI, T., KAWASHIMA, T., STUMPF, W. E., JUNGHENT, P. W., AND DE-SOMBRE, E. R. (1967). A two step mechanism for the interaction of estradiol with rat uterus. *Proc. Nat. Acad. Sci. U.S.A. 59*, 632–638.

KING, J., AND MYKOLAJEWYCZ, N. (1973). Bacteriophage T4 tail assembly: proteins of the sheath, core and baseplate. *J. Mol. Biol. 75*, 339–358.

KING, T. J., AND BRIGGS, R. (1956). Serial transplantation of embryonic nuclei. *Cold Spring Harbor Symp. Quant. Biol. 21*, 271.

KING, T. J., AND MCKINNELL, R. G. (1960). An attempt to determine the developmental potentialities of the cancer cell nucleus by means of transplantation. *In* "Cell Physiology of Neoplasia." University of Texas Press, Austin, Texas.

KIRK, D. L., AND MOSCONA, A. A. (1963). Synthesis of experimentally induced glutamine synthetase (glutamotransferase activity) in embryonic chick retina *in vitro*. *Develop. Biol. 8*, 341–357.

LEVENSON, G. E. (1969). The effect of ascorbic acid on monolayer cultures of three types of chondrocytes. *Exp. Cell Res. 55*, 433–435.

LILLIE, F. R. (1916). The theory of the free-martin. *Science 43*, 611.

MASLOW, D. E. (1970). Electrokinetic surfaces of trypsin-dissociated embryonic liver cells. *Exp. Cell Res. 61*, 266–270.

MOSCONA, A. A. (1971). Embryonic and neoplastic cell surfaces: availability of receptors for concanavalin A and wheat germ agglutinin. *Science, 171*, 905–907.

MOSCONA, A. A. AND HUBBY, J. L., (1963). Experimentally induced changes in glutamotransferase activity in embryonic tissue. *Develop. Biol. 7*, 192–206.

NEW, D. A. T. (1966). "The Culture of Vertebrate Embryos." Academic Press, New York.

OHNO, S. (1971). Simplicity of mammalian regulatory systems inferred by single gene determination of sex phenotypes. *Nature (London) 234*, 134–137.

O'MALLEY, B. W., ROSENFELD, G. C., COMSTOCK, J. P., AND MEANS, A. R. (1972a). Steroid hormone induction of a specific translatable messenge RNA. *Nature, New Biol. 240*, 45–48.

O'MALLEY, B. W., SPELSBERG, T. C., SCHRADER, W. T., CHYTIL, F., AND STEGGLES, A. W. (1972 b) Mechanisms of interaction of a hormone-receptor complex with the genome of a eukaryotic target cell. *Nature (London) 235*, 141–144.

OPPENHEIMER, J. M. (1937). The normal stages of *Fundulus heteroclitus*. *Anat. Rec. 68*, 1–15.

PALMITER, R. D., PALASCIOS, R., AND SCHIMKE, R. T. (1972). Identification and isolation of ovalbumin synthesizing polysomes. *J. Biol. Chem. 247*, 3296.

PIDDINGTON, R. (1971). Distribution and development of glutamine synthetase in the embryonic cerebral hemisphere. *J. Exp. Zool. 177*, 219–227.

PIDDINGTON, R., AND MOSCONA, A. A. (1967). Precocious induction of retinal glutamine synthetase by hydrocortisone in the embryo and in culture. *Biochim. Biophys. Acta 170*, 263–270.

RAFF, R. A., COLOT, H. V., SELVIG, S. E., AND GROSS, P. R. (1972). Oogenetic origin of messenger RNA for embryonic synthesis of microtubule proteins. *Nature (London) 235*, 211–214.

REICH, E., GOLDBERG, I. H., AND RABINOWITZ, M. (1962). Structure activity correlations of actinomycins and their derivatives. *Nature (London) 196*, 743–748.

ROTHFIELD, L. E., ed. (1971). "Structure and Function of Biological Membranes." Academic Press, New York.

RUDNICK, D., AND WAELSCH, H. (1955). The development of glutamotransferase and glutamine synthetase in the nervous system of the chick. *J. Exp. Zool. 129*, 309–326.

SAXÉN, L., AND TOIVONEN, S. (1962). "Primary Embryonic Induction." Logos Press, London.

SCHJEIDE, O. A., AND deVELLIS, JEAN. (1970). "Cell Differentiation." Van Nostrand Reinhold, New York.

SCHNIERLA, T. C. (1971). "Army Ants, a Study in Social Organization." W. H. Freeman, San Francisco.

SCHWARTZ, R. J. (1972). Steroid control of genomic expression in embryonic chick retina. *Nature, New Biol. 237*, 121–125.

SCHWARTZ, R. J. (1973). Control of glutamine synthetase synthesis in the embryonic chick neural retina: a caution in the use of actinomycin D. *J. Biol. Chem. 248*, 6426–6435.

SCHWARTZ, R. J., AND WILDE, C. E., JR. (1973). Proteins associated with early morphogenesis in *Fundulus. Nature (London)* (in press).

SHERMAN, M. R., CORVAL, P. L., AND O'MALLEY, B. W. (1970). Progesterone binding components of chick oviduct. *J. Biol. Chem. 245*, 6085.

SLATER, D. W., SLATER, I., AND GILLESPIE, D. (1972). Post-fertilization synthesis of polyadenylic acid in sea urchin embryos. *Nature (London) 240*, 333–337.

SPEMANN, H., AND MANGOLD, H. (1924). Über Induktion von Embryonalanlagen durch Implantation artfremder Organisatoren. *Wilhelm Roux Arch. Entwicklungmech. Organismen 100*, 599–638.

SPIRIN, A. S. (1966). On "masked" forms of messenger RNA in early embryogenesis and in other differentiating systems. *Curr. Top. Develop. Biol. 1*, 1–38.

STOCUM, D. L. (1968a). The urodele limb regeneration blastema: a self-organizing system. I. Differentiation *in vitro. Develop. Biol. 18*, 441–456.

STOCUM, D. L. (1968b). The urodele limb regeneration blastema: a self-organizing system. II. Morphogenesis and differentiation of autografted whole and fractional blastemas. *Develop. Biol. 18*, 457–480.

TEITELMAN, G. (1973). Protein synthesis during Ilyanassa development: effect of the polar lobe. *J. Embryol. Exp. Morphol. 29*, 267–281.

THOMPSON, D. W. (1948). "On Growth and Form." Macmillan, New York.

TIEDEMANN, H. (1967). Biochemical aspects of primary induction and determination. *In* "The Biochemistry of Animal Development" (R. Weber, ed.), pp. 3–55. Academic Press, New York.

TYLER, A., AND TYLER, B. S. (1970). Informational macromolecules and differentiation. *In* "Cell Differentiation" (O. A. Schjeide and J. deVellis, eds.), Chap. 4, pp. 42–118.. Van Nostrand Reinhold, New York.

WEISS, P. (1939). "Principles of Development." Holt, New York.

WEISS, P., AND MOSCONA, A. (1958). Type-specific morphogenesis of cartilages developed from dissociated limb and scleral mesenchyme *in vitro. J. Embryol. Exp. Morphol. 6*, 238–246.

WILDE, C. E., JR. (1955). The urodele neuroepithelium. II. The relationship between phenylalanine metabolism and the differentiation of neural crest cells. *J. Morphol. 97*, 313–344.

WILDE, C. E., JR. (1956). The urodele neuroepithelium. III. The presentation of phenylalanine to the neural crest by archenteron roof mesoderm. *J. Exp. Zool. 133*, 409–440.

WILDE, C. E., JR. (1959). Differentiation in response to the biochemical environment. "Cell, Organism and Milieu" (D. Rudnick, ed.), Chap. 1. Ronald Press, New York.

WILDE, C. E., JR. (1960). Factors concerning the degree of cellular differentiation in organotypic and disaggregated tissue cultures. *In* "La Culture organotypique: associations et dissociations d'organes en culture *in vitro*." pp. 183–201. Editions du CNRS, Paris.

WILDE, C. E., JR., AND CRAWFORD, R. B. (1963). Cellular differentiation in the anamniota. I. Initial studies on the aerobic metabolic pathways of differentiation and morphogenesis in teleosts and urodeles. *Develop. Biol. 7*, 578–594.

WILDE, C. E., JR., AND CRAWFORD, R. B. (1966). Cellular differentiation in the anamniota. III. Effects of actinomycin D and cyanide on the morphogenesis of *Fundulus. Exp. Cell Res. 44*, 471–488.

WILDE, C. E., JR., AND CRAWFORD, R. B. (1968). Epithelial-mesenchymal interactions in the lower vertebrates. *In* "Epithelial–Mesenchymal Interactions" (R. Fleischmajer and R. Billingham, eds.), Chap. 6, pp. 98–113. Williams & Wilkins, Baltimore.

WOLPERT, L. (1969). Positional information and the spatial pattern of cellular differentiation. *J. Theoret. Biol. 25*, 1–47.

WOLPERT, L. (1971). Positional information and pattern formation. *Curr. Top. Develop. Biol. 6*, 183–224.

YAFFE, D. (1969). Cellular aspects of muscle differentiation *in vitro. Curr. Top. Develop. Biol. 4*, 37–77.

CHAPTER THIRTEEN

Development of Immunity

ROBERT AUERBACH

Discussion of the development of immunity can be focused on a number of aspects, rather different in approach, all of which together constitute the ontogeny of the capacity to respond specifically to foreign material. At one end of the spectrum is the morphology of the immune system, which must consider the development of the effector organs such as the spleen, lymph node, and the lymphoid system in general; at the other end is the origin of immunological diversity, the molecular basis for specific recognition of an almost infinite variety of molecular entities. Central to the entire spectrum is the concept of differentiation, and for this reason alone the development of immunity represents one of the most challenging areas in the field of developmental biology.

Nature of Immune Reactions

This chapter cannot provide even a superficial summary of immunological concepts, and the reader is therefore referred to the excellent overview of immunology presented by a number of recent publications (CIBA Foundation, 1972; Humphrey and White, 1970; Amos, 1971; Nossal and Ada, 1971; Abramoff and LaVia, 1970). For our purposes, we will restrict ourselves to a few general statements that are intended to serve only as an orientation for the developmental aspects of immunity that will be covered in this chapter.

Immune reactions are those in which a foreign material, be that a protein, virus, or cell, induces a specific response leading to the inactivation or destruction of that material. In general, the response is not only highly specific but leads to relatively long-lasting memory; that is, a second challenge by the same foreign material elicits a faster and stronger reaction. Although numerous cells function in immune reactions, including phagocytes and a variety of blood-borne leukocytes, the primary effector cells belong to the lymphoid series. Immune reactions may be subdivided into two major groups: cell-mediated or *cellular immunity*, and secretory or *humoral immunity*. For humoral immunity we measure the production of immunoglobulins, usually in serum; these immunoglobulins have highly unique properties and have been well characterized in terms of the amino acid sequence of light and heavy chains, bonding between chains, and the nature of combining properties with antigens and complement. Cells secreting antibody have been shown to have immunoglobulin-like material on the cell surface that is capable of combining with antigen; and synthesis, assembly, and secretion of immunoglobulins appear to follow the classical rules for protein synthesis.

Less is known about the nature of cellular immunity. Classical examples of cell-mediated immunity are delayed hypersensitivity, homograft rejections, and the graft-versus-host (gvh) reaction. Operationally, specific lymphocytes can attack and destroy foreign cells, without the obvious elaboration of secretory immunoglobulins. Yet the absence of demonstrable immunoglobulin does not preclude the implication of immunoglobulin-like material in cell-mediated immunity as well, and an as yet uncharacterized immunoglobulin has been postulated to account for the striking similarities between cellular and humoral immunity. Difficulties in identification of the immunoglobulin presumed important for cellular immunity arise from the fact that both cellular and humoral immunity are evoked by foreign material, so that the occasional presence of immunoglobulins in cell-mediated reactions does not warrant the conclusion that these immunoglobulins must play an integral role in cellular immunity. It does seem clear, at any rate, that if immunoglobulins are involved in cell-mediated immunity, they are apt to be nonsecretory, surface-bound molecules, only partially analogous to the typical immunoglobulins found in serum and body fluids; and it is likely that they function more to bring together target and effector cells than to play a direct killing or destroying role in the cellular type of immune reaction.

"T" and "B" Cells and the Concept of Cell Interactions

A major recent finding in immunology has been the recognition that there are two main types of lymphoid cells that are active in immune reactions: one, the "T" cells, represents thymus-derived cells, and these are clearly implicated in cellular immunity and are essential for facilitation of some humoral immune reactions as well. The other, "B" cells, are cells that in birds are initially associated with the bursa of Fabricius (hence "B"). This organ is a lymphoid structure dorsal to the cloaca, large at birth, and degenerate in adult animals; its removal during embryogenesis or shortly after birth had been shown to interfere with the capacity of birds to produce antibodies. Only birds possess a bursa of Fabricius; yet clearly an equivalent must exist in mammals, and bursal function is believed to be associated first with the embryonic liver and then with the bone marrow. Thus embryonic liver-derived or bone marrow-derived cells include precursors of cells that actively synthesize humoral antibodies. The role of B and T cells in tolerance is discussed in Chapter 14.

That immune reactions require cell interactions has been known for some time but a series of recent experiments has led to a clearer delineation of cell functions. A complete review of these experiments is beyond the scope of the present chapter, and the reader is referred to several recent symposia for detailed information (Amos, 1971; Cohen et al., 1971; Makela et al., 1971). We will include here only a brief summary of these studies.

Cell interactions in humoral immunity were clearly defined by the work of Claman and collaborators (Claman and Chaperon, 1969), working with the humoral response to sheep red blood cells (SRBC), and the nature of these interactions was further described by the studies of Miller and his colleagues (Miller et al., 1971). Subsequent work with protein antigens and with defined carrier-hapten systems have led to additional insight into the nature of cell interactions in humoral immunity (Mitchison, 1971). When lethally irradiated animals were injected either with bone marrow cells, with thymus cells or both, and then immunized with SRBC, it was found that the response to SRBC was greatly enhanced when both thymus and marrow cells were injected; that is, there was synergism between T- and B-cell populations in irradiated host animals (Claman and Chaperon, 1969). Moreover, cells from the thoracic duct, known to be rich in thymus-derived cells, were even more effective in synergistic interaction with bone marrow cells than were thymus cells themselves (Miller et al., 1971). The sequence of events was demonstrated to include specific antigen-stimulation of T cells first, followed by activation of B cells (Miller et al., 1971).

It had long been known that some antigenic moieties were too small to be, by themselves, immunogenic (haptens), and that to immunize against such haptens a larger carrier molecule was necessary. Taking advantage of carrier-hapten systems to further delineate the nature of T- and B-cell collaboration, Mitchison and coworkers demonstrated that T-cell function was directed against carrier components, while B-cell function was associated with antihapten antibody production (Mitchison, 1971).

More recently, techniques of filter-mediated cell interactions, long used in embryological studies (Grobstein, 1956; Auerbach, 1968) were extended to T- and B-cell interactions by Feldman and coworkers (Feldman and Basten, 1972). It appears that antigen-specific T-cell stimulation leads to the production of a factor (immunoglobulin-like?) that can cross a thin, 0.1-μm-porosity nuclepore filter but that is not capable of passing through dialysis membranes. This antigen-mediated, T-cell-derived factor is adsorbed onto macrophage-like cells, presumably because of its cytophyllic nature, and in this condition is capable of enhancing the B-cell response to antigen.

The similarity of this type of cell interaction to that seen in nonimmunological developing systems is, of course, striking (see Auerbach, 1971, for discussion), and it may well be, indeed, that the immunologist has finally found the ideal system for study of what has so long been the domain of the developmental biologist, the area of inductive tissue interactions.

Cell interactions are also known to be involved in cell-mediated immunity, such as the gvh reaction. This reaction, which serves as an excellent model for cellular immune systems in general, represents a complex series of processes that, although initiated by grafted lymphoid cells, ultimately includes a whole array of sequential responses by the immunologically impaired host. A detailed description of this immunological system cannot be included in this chapter, and the reader is referred to the excellent review by Elkins (1971) (see also Umiel and Auerbach, 1973, and Chapter 14).

Historically the gvh reaction was a developmental observation of host spleen enlargement after transplantation of adult chicken spleen cells into embryonic chick hosts. The immunological basis of this reaction was recognized by Simonsen (1962) and was further clarified by analysis of genetic factors such as transplantation antigens, immunological tolerance, and the effect of immunosuppressive agents. On the basis of studies with thymus cells and thoracic duct lymphocytes, as well as by the use of surface antigenic markers directed at T-cell populations, it was demonstrated that the effectors of cell-mediated immunity such as the gvh reaction were T cells.

That two subpopulations of T cells were actually involved in this reaction was shown by recent work of Cantor and Asofsky (Asofsky et al., 1971; Cantor and Asofsky, 1972). These investigators have presented evidence that just as there is B–T synergism in the humoral response to SRBC, there is T–T synergism in the eliciting of a gvh response. Their evidence is based on the finding that T cells obtained from the thoracic duct or thymus are different from T cells obtained from lymph nodes, and that appropriate mixtures of these two types of T cells (termed T_1 and T_2) show a synergism when measured in the gvh reaction. Their suggestion that the two subpopulations of T cells represent different stages in the differentiation of T cells is interesting but speculative, and the possibility of a variety of populations of thymic or thymus-derived cells differing in embryological history as well as functional capacity needs to be seriously considered. That, indeed, there are many subpopulations of T cells, although unpopular from a reductionist standpoint, may in fact best explain a variety of distinctions seen in analysis of T-cell functions. For example, embryonic thymus cells can induce a gvh reaction as effectively as can adult thymus cells (see below); yet their ability to respond in the mixed leukocyte interaction system, another parameter of cellular immunity, arises somewhat later, and the ability to respond to phytohemagglutinin (PHA) matures at yet another age (Stites et al., 1973).

It should not be assumed that cellular interactions in immunity are restricted to T and B cells. Indeed in our discussion of T–B cell interactions in the response to SRBC, we have already alluded to a possible function of macrophages, that of adsorbing thymus-mediated, antigen-specific factors. A number of other studies point to important functions of macrophages. In the SRBC system, for example, it has been shown that the immune response *in vitro* is dependent on either the presence of adherent cells (macrophages), or on factors produced by such cells. This *in vitro* function of macrophages may be largely nutritional in nature, but the possible function of macrophages in holding antigen or in permitting appropriate physical cell contact between B and T cells should not be overlooked. Indeed the macrophage in studies of both humoral immunity and in cellular immunity appears operationally to resemble the mesenchyme in such cell–cell interactions as seen in the development of the pituitary gland or the eye lens, permitting the appropriate contact and transfer of "inductive" information between two cell types fundamentally incapable of adhering with each other.

Ontogeny of Immunocompetence

There is no sudden moment during development when animals become immunocompetent. Not only is there variation between species, or between strains of a given species, but even for a single individual, responsiveness to a variety of antigens becomes demonstrable at different points in ontogeny (Silverstein, et al., 1966). Much of the variation in response between species does disappear when one compares animals not on the basis of chronological age but on the basis of developmental equivalence (see Sterzl and Riha, 1971). Similarly, at least some of the discrepancies between responsiveness to different antigens may be based not on differentiation of immunocompetence in general but on ancillary processes such as antigen handling, or on efficiency of detection of reactivity (e.g., antibodies to virus are more readily detected than antibodies to polypeptide antigens).

Certainly, to the extent that several distinct cell populations (such as B cells and T cells and their subclasses) are involved in immune capacity, one would expect some broad differences to be apparent between the rise of cell-mediated and of humoral immunocompetence. In dissecting out the processes required for immune reactions, moreover, one would expect to find that the property of recognition of foreignness may precede the ability to respond, and, certainly, recognition functions play a critical role in all aspects of development, beginning with the very event of fertilization.

We may consider the maturation of responsiveness of young mice to SRBC as a model of ontogeny of humoral immune reactions. When mice are immunized with SRBC, they normally fail to respond if they are less than 3 to 4 days of age, although there are strain variations in responsiveness (Auerbach, 1972; Shalaby and Auerbach, 1973). Using BDF_1 mice, which normally give detectable responses only when immunized as 3-day-olds or later, it was possible for us to show that a single injection of SRBC given on the first or second day of life leads to an accelerated response to a second injection of that antigen given 2 days later. This finding is most readily interpreted as due to the differential maturation of T and B cells; the first injection activates T cells, which mature first, while the second injection permits the B-cell response to proceed. A number of other observations support this interpretation. For example, immunological memory leading to 7S antibody formation is believed to be carried by T cells, and a single injection of SRBC at birth permits an increased, secondary response to occur 6 weeks later. Similarly, transfer studies carried out with young spleen cells indicate that T-cell function is demonstrable in spleens prior to B-cell function.

An alternative interpretation may be placed on the results of double injection of antigen, however. It is conceivable that T-cell migration occurs in response to antigen, so that the first injection of SRBC leads to an increase in the number of SRBC-reactive T cells in the spleen, not only by stimulating cell division but by enhancing cell seeding. It should be emphasized, however, that these two alternative interpretations are not mutually exclusive, and that the priming effect of SRBC may involve both stimulation of splenic lymphocytes and of differential cell migration.

The ontogeny of cellular immune capacity has been studied most extensively in mice by use of the gvh reaction (Umiel and Auerbach, 1973). Simonsen (1962) showed that immunocompetence in the mouse arises within 3 to 4 days of birth, and his findings were confirmed by *in vitro* studies of Auerbach and Globerson (1966). There does not appear to be a sudden rise in immunocompetence, however, and the first few days of life may simply represent a time when the relative number of cells in the spleen capable of evoking a gvh reaction increases.

Evidence that immunocompetent cells exist in newborn mice has been shown by examination of thymus cell suspensions. Although the proportion of thymus cells capable of evoking a gvh reaction is low, the newborn thymus is already as capable as the adult thymus of inducing a gvh reaction (Cohen et al., 1963). *In vitro* studies, moreover, indicate that even embryonic thymus cells are immunocompetent as measured in this system (Chakravarty et al., 1973).

This finding of relative equivalence of newborn and adult thymus cells, but not of newborn and adult spleen cells, raises some interesting developmental questions. Can it be, for example, that immunocompetent cells exist in other places and that it is only the migration and proliferation of these cells that is delayed? Even for the humoral immune system, immunocompetent cells have been seen in the bone marrow prior to their appearance in the spleen (Saunders and Swartzendruber, 1970), and it has been suggested that the failure of immune response is due to the fact that these cells do not respond to antigen by cell division and hence normally escape detection. Similarly, functional T cells may exist in the thymus, thoracic duct, bone marrow, or embryonic liver long before they are detected in the peripheral lymphoid organs, such as the spleen and lymph nodes (Umiel and Auerbach, 1973). Recent *in vitro* studies indeed indicate that T-cell function can be measured by the gvh reaction of liver cells obtained from 12-day mouse embryos (Umiel et al., 1968) or from human embryos of 5 weeks of gestation (Stites et al., 1973).

Phylogenetic Aspects of Immunological Differentiation

As has been so frequently the case for other developing systems, a phylogenetic survey frequently reveals underlying features of ontogenic significance. The immune system is no exception. Thus, for example, the clear delineation of T- and B-cell functions, of cellular and humoral immunity, still rests primarily on work carried out on developing chicken embryos and hatchlings, where the removal of the bursa of Fabricius (B) was found to impair humoral immunity, while the removal of the thymus (T) was associated with deficiencies in cellular-immune mechanisms.

The ontogeny of immune responsiveness in the chicken has been well studied and carefully reviewed by Solomon (1971). Chicken embryos already show some competence in humoral and cellular immunity, according to Solomon, although other investigators place immunological maturation at posthatching stages. It does seem clear that prior to 16 to 18 days of chick development there is little manifestation of any significant immune reactivity for most antigens, as measured by humoral antibody production, and that prior to 14 days (but later for most workers) there is no significant immunocompetence as seen in gvh reactivity during cell transfer experiments.

Of the many studies in lower vertebrates, perhaps the most provocative is that of Du Pasquier (1970) working with frog embryos. He reported that even early tadpole stages of anurans are competent to respond specifically to antigens such as SRBC, which suggests that specific immunocompetence in humoral immunity may actually occur much earlier in ontogeny than had previously been suspected. If indeed this is the case, the possibility is raised that immunological maturation may be blocked or masked by inhibitors and that development of the basic genetic machinery for antibody formation precedes observable immunocompetence in the higher vertebrates.

Ontogeny of Lymphoid Cells, Tissues, and Organs

The thymus is perhaps the best studied of the many organs that play a significant role in the ontogeny of immune systems. In the mouse the thymus can be readily recognized and dissected at about 12 days of development, at which time it appears as an epitheliomesenchymal rudiment. This early thymus rudiment can "self-differentiate" *in vitro* to produce within a few days a complex organ, rich in lymphoid cells, appropriately subdivided into medullary and cortical regions (Auerbach, 1960, 1961, 1964, 1965, 1967). As might be expected from analogy with other developing systems, separation of epithelial from mesenchymal portions of the rudiment abrogates differentiation (Auerbach, 1960, 1961). Recombination and transfilter studies indicate that the epithelial portion of the rudiment gives rise to lymphoid cells and that the mesenchymal requirement for differentiation is relatively nonspecific (Auerbach, 1961).

More recently, however, it was demonstrated (Moore and Owen, 1967) that, even in the early rudiment, a few cells can be recognized as distinct by Giemsa staining, and these have been classified as stem cells of the lymphoid system. Their existence, moreover, has been confirmed by electron microscope analysis (Mandel and Russell, 1971). These stem cells are currently believed to be migratory in origin, entering the early, developing rudiment from the circulation, and originating in the yolk sac (Moore and Owen, 1967; Metcalf and Moore, 1971; Auerbach, 1970) (Fig. 1).

Whether, indeed, all thymic lymphoid cells arise from this population of stem cells, however, is once again in doubt, partially because of the demonstration that there are many subpopulations of T cells, as mentioned earlier, partially because of data obtained from a variety of tissue-culture studies involving radiation or cortisone treatment, and partially because of

transplantation experiments carried out in amphibian embryos (workshop 9 in Amos, 1971).

Currently, much new information is becoming available based on the ontogeny of thymus-specific cell surface antigens (Owen, 1972). Several T-cell specific markers now exist: TL, the thymus leukemia antigen, is limited to cells in the thymus, with approximately 90 to 95 percent of thymic lymphocytes demonstrably TL+ in appropriate strains of mice. This antigen can be recognized as early as 16 days after fertilization—that is, as soon as recognizable medium-sized lymphocytes can be seen. Similarly, the θ antigen appears at this time. θ antigens are carried on all thymus and thymus-derived lymphoid cells. It appears that the maturation of lymphoid cells in the thymus includes a progression from cells that are strongly θ positive TL+ to cells that are TL− and only weakly θ positive. The loss of the TL antigen is correlated with an increase in H-2 antigenicity. Other lymphocyte antigens associated with thymus-derived cells, such as LyA and LyB, have not been characterized developmentally (see Mitchison, 1971).

In contrast to studies on the development of thymic cells, the developmental history of B cells is less well documented. While, as with T cells, B-cell origins seem to lie in the embryonic yolk sac (Moore and Owen, 1967; see Metcalf and Moore, 1971), the major organs of B-cell maturation appear to be the embryonic liver and the bone marrow in mammals, and the bursa of Fabricius in birds. It is noteworthy that the embryonic liver of birds appears to have no hematopoietic function, while the bursa of Fabricius has no obvious analogue in mammals. Moreover, since the yolk sac persists as a source of erythrocytes in birds, the possible analogy between the bursa of Fabricius and the lymphoid component of the embryonic liver should not be overlooked.

Little is known concerning the controlling events of early liver morphogenesis, but it seems reasonable to suggest that the initial outpocketing of the gut and the structuring of the hepatic rudiment involves typical inductive interactions. Tissue-culture studies of Umiel (Umiel et al., 1968; Umiel, 1971; see Umiel and Auerbach, 1973) indicate that the subsequent development of liver lymphoid cell precursors is influenced by the thymus, but the precise nature of that thymic influence has not been determined. A major difficulty in analysis of liver development is the fact that the cell population of immunological interest is a transient one and that many of the developmental events in the liver are directed toward differentiation of hepatic cords and functional liver cells rather than of functional hematopoietic cells. Even among the latter, the lymphoid cell line represents only a minority population.

The chicken bursa of Fabricius arises as a structure on the seventh day, but lymphoid cell precursors enter only at a later time, being derived from the yolk sac (Moore and Owen, 1967). The initial development of the bursa involves inductive interaction between epithelium and mesenchyme; disturbance of that differentiation, as effected by injection of 5-to-9-day-old embryos with 19-nortestosterone or testosterone propionate, prevents entirely the subsequent differentiation of bursal lympho-

cytes. It is presumed that the bursal environment plays a critical role, therefore, in the maturation of B cells in the chicken.

The development of bone marrow is even less well documented. Bone marrow stem cells appear capable of giving rise to B cells, as seen in restitution experiments. At the same time they are capable of entering the thymus, there becoming functional T cells through division and maturation. Precisely what pathway of differentiation a given stem cell of the bone marrow will take must in part be dependent on the intramarrow microenvironment, for the marrow contains many subpopulations of cells in varying stages of definitive differentiation into various blood cell elements. At the same time, within the marrow, the machinery appears to exist for the retention and replication of stem cells capable of being induced into further differentiation by extrinsic environments, furnished, for example, by the thymus.

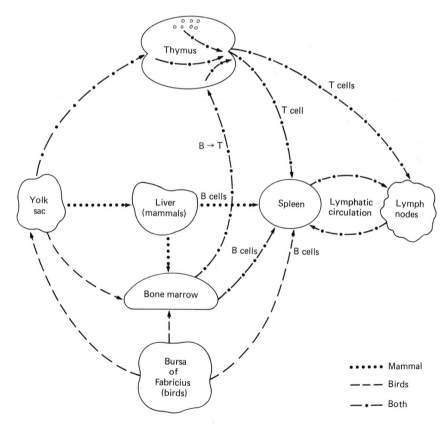

FIGURE I. *Cell migration patterns in the developing immune system. For detailed description see text.*

In contrast to differentiation of marrow, thymus, bursa, or liver, the differentiation of the spleen appears to be almost entirely due to the entry of definitive cell types. Thus, in the spleen, the small lymphocyte—representing a differentiated cell—makes its appearance prior to the presence of immature lymphoid cells, and a variety of studies indicate that the spleen is a reflection of, rather than a cause of, immunological maturation. On the other hand, within the spleen, a number of areas exist that reflect unique microenvironments for the differentiation or proliferation of thymus-derived lymphoid cells, B-type cells, and granuloid or erythroid cells. It has been suggested that the spleen plays an essential role in the maturation of immunocompetence (Battisto et al., 1971); recent studies, however, tend to minimize the influence of the spleen in that maturation (Chakravarty et al., 1973).

Finally, it should be pointed out that the lymphoid system involves a dynamic rather than a static one, in that there is a continuous and massive migration of cells through the blood and lymphatic system, involving lymph nodes, spleen, thymus, marrow, Peyer's patches, and all the other organs harboring lymphoid cells. Thus, in the animal, we are always faced with the problem of continuing influx and exit of cells that may be in varying stages of differentiation. In tissue-culture systems, on the other hand, the problems of cell survival and the absence of normal migrational events may lead to results that at best must be interpreted with great caution.

Theoretical Considerations

Two major classes of antibody theories have evolved over the past 35 years: instructive and selective ones. The *instructive* hypotheses all were based on the concept that antigen played a role in determining the structure of antibodies, while the *selective* ones all took the central thesis that antibody variability arose in the absence of antigen. The parallel between these two theoretical approaches and the Lamarckian and Darwinian evolutionary principles seems obvious. Evidence of the last decade appears to favor almost exclusively the selective theories of antibody variability.

Among selective theories, however, there still exist many alternative solutions to the problem of generation of antibody diversity. There are those who believe in germ-line variability and those that favor somatic generation of diversity. Germ-line variability theories begin with the tenet that the genetic information in the zygote is sufficient to account for the total spectrum of immunological reactivities. Individual cells are thus restricted to synthesizing only one or a few of the variants of immunoglobulins or cell surface recognition factors by mechanisms similar to those determining other restrictions in differentiation—presumably by some mechanism of differential gene activation. Somatic theories of generation of antibody variability, on the other hand, assume that diversity is generated epigenetically through some mutational or recombinational series of steps, combined with a strong selective pressure for newly generated variants.

The increasing evidence that a single polypeptide chain of the antibody molecule may be coded for by at least two genes, a v gene for the variable (antigen-specific) portion and a c gene for the constant portion of the chain, implies that previous constraints on generation of diversity may not be applicable (Edelman and Gall, 1969; Hood, 1971; Jerne, 1971; Smithies, 1970). For example, several v genes may exist—only one of which would be functional in a given cell by appropriate hookup to the c portion. This would permit a relatively small amount of DNA to be sufficient to account for much variability in the expressed amino acid sequence of the total antibody molecule. Furthermore, there is no valid reason for assuming that epigenetic evolution of antibody variability (somatic generation of diversity) may not be superimposed on a primary germ line group of variants that serve as a starting point for antibody variability.

Considerable evidence exists that specificity of the T system is not as great as that of the B system. For example, cross-reactivity of erythrocyte antigens appears to be more evident among T cells than among B cells (Hartou and Argyris, 1972). If one assumes that any given cell can make only one antibody, then one would predict that no more than one cell per million would be capable of reacting against any specific antigen; yet in cellular immune reactions such as that seen in the graft-versus-host reaction or in the mixed lymphocyte reaction, as many as 1 per 100 or at least 1 per 1000 are capable of responding to specific stimulation (Bach et al., 1969; Auerbach and Globerson, 1966). We may suggest that T-cell variability is germ-line in nature, and by the same reasoning one would then propose that B-cell specificity is initially of the same order. Further epigenetic evolution, by somatic mutation or recombination, could then lead to additional elaboration of more precise specificity of appropriate antibody-forming cells that would subsequently be favored by selective cell division in response to further antigenic stimulation. The affinity of antibody for a given antigen has been observed to increase with increased time after immunization, a finding that would support this speculative suggestion.

Concluding Remarks

Progress in immunology over the last few years has been almost overwhelming. Only a decade ago knowledge of immunoglobulin structure was rudimentary, the thymus was only just emerging as an organ of immunological significance, surface receptors in immunity were as yet unstudied, and tissue-culture procedures were not available for the analysis of primary immune reactions; indeed, much of the terminology of present-day immunology, such as idiotypes, blocking antibodies, B and T cells, or allotype suppression, had not even been coined. But just as molecular biology made phenomenal strides in the mid-1950s to mid-1960s, so immunology has, in recent times, shown such remarkable accomplishment in laying a basic foundation that the work yet to be done in the future seems clearly mapped out.

It is not surprising, therefore, that the immunologist, as the molecular biologist before him, has become increasingly fascinated by the problems of cellular differentiation: the interrelation between genetic and epigenetic events, the nature of cellular communication and in-teractions, and the characteristics of stability of the differentiated state. This area of cellular differentiation is still an unknown continent, and it is here that the immunologist must meet the challenge of applying his exquisitely specific tools to the exploration of a new territory, where the terrain is different and the landmarks keep changing.

References

ABRAMOFF, P., AND LaVIA, M. (1970). "Biology of the Immune Response." McGraw-Hill, New York, 492 pp.

AMOS, B., ed. (1971). "Progress in Immunology." Academic Press, New York, 1554 pp.

ASOFSKY, R., CANTOR, H., AND TIGELAAR, R. (1971). Cell interactions in the graft-versus-host response. In "Progress in Immunology" (B. Amos, ed.), pp. 369–381. Academic Press, New York.

AUERBACH, R. (1960). Morphogenetic interactions in the development of the mouse thymus gland. Develop. Biol. 2, 271–284.

AUERBACH, R. (1961). Experimental analysis of the origin of cell types in the development of the mouse thymus. Develop. Biol. 3, 336–354.

AUERBACH, R. (1964). Experimental analysis of mouse thymus and spleen morphogenesis. In "The Thymus in Immunobiology" (R. A. Good and A. Gabrielsen, eds.), Chap. 5, pp. 95–113. Harper & Row, New York.

AUERBACH, R. (1965). Experimental analysis of lymphoid differentiation in the mammalian thymus and spleen. In "Organogenesis" (R. DeHaan and H. Ursprung, eds.), pp. 539–558. Holt, New York.

AUERBACH, R. (1967). The development of immunocompetent cells. Develop. Biol., Suppl. 1, 254–263.

AUERBACH, R. (1968). In "Epithelial–Mesenchymal Interactions" (R. Fleishmajer, ed.). Williams & Wilkins, Baltimore.

AUERBACH, R. (1970). Toward a developmental theory of antibody formation: the germinal theory of immunity. In "Developmental Aspects of Antibody Formation and Structure" (J. Sterzl, and I. Riha, eds.), Vol. 1, pp. 23–33. Academic Press, New York.

AUERBACH, R. (1971). Towards a developmental theory of immunity: cell interactions. In "Cell Interactions and Receptor Antibodies in Immune Responses" (O. Makela, A. Cross, and T. E.

Kosuhen, eds.), pp. 393–398. Academic Press, New York.

AUERBACH, R. (1972). Studies on the development of immunity: the response to sheep red blood cells. *Curr. Top. Develop. Biol.* 7, 257–280.

AUERBACH, R., AND GLOBERSON, A. (1966). *In vitro* induction of the graft-versus-host reaction. *Exp. Cell Res.* 42, 31–41.

BACH, F. H., BOCK, H., GRAUPNER, K., DAY, E., AND KLOSTERMANN, H. (1969). Cell kinetic studies in mixed leukocyte cultures: an *in vitro* model of homograft reactivity. *Proc. Nat. Acad. Sci. U.S.A.* 62, 377–384.

BATTISTO, J. R., BOREK, F., AND BUCSI, R. A. (1971). Splenic determination of immunocompetence: influence on other lymphoid organs. *Cell Immunol.* 2, 627–633.

CANTOR, H., AND ASOFSKY, R. (1972). Synergy among lymphoid cells mediating the graft-versus-host response. III. Evidence for interaction between two types of thymus derived cells. *J. Exp. Med.* 135, 764–779.

CHAKRAVARTY, A., KUBAI, L., LANDAHL, C., ROETHLE, J., SHALABY, M. R., AND AUERBACH, R. (1973). Current studies on the development of immunity in the mouse. *In* "Colloque de la Société Française d'Immunologie: Phylogenic and Ontogenic Study of the Immune Response and its Contribution to the Immunological Theory," pp. 269–278.

CIBA FOUNDATION SYMPOSIUM. (1972). "Ontogeny of Acquired Immunity." Elsevier, Amsterdam, 283 pp.

CLAMAN, H. N., AND CHAPERON, E. A. (1969). Immunological complementation between thymus and marrow cells—a model for the two-cell theory of immunocompetence. *Transpl. Rev. 1,* 92–113.

COHEN, M. W., THORBECKE, G. J., HOCHWALD, G. M., AND JACOBSEN, E. G. (1963). Induction of graft-versus-host reaction in newborn mice by injection of newborn or adult homologous thymus cells. *Proc. Soc. Exp. Biol. Med.* 114, 242–244.

COHEN, S., CUDKOWICZ, G., AND MCCLUSKEY, R. T., eds. (1971). "Cellular Interactions in the Immune Response." Karger, New York.

DU PASQUIER, L. (1970). Ontogeny of the immune response in animals having less than one million lymphocytes: the larvae of the toad Alytes obstetricians. *Immunology 19,* 353–362.

EDELMAN, G. M., AND GALL, W. E. (1969). The antibody problem. *Ann. Rev. Biochem.* 38, 415–466.

ELKINS, W. L. (1971). Cellular immunology and the pathogenesis of graft-versus-host reactions. *Progr. Allergy 15,* 78–187.

FELDMAN, M., AND BASTEN, A. (1972). Cell interactions in the immune response *in vitro*. III. Specific collaboration across a cell impermeable membrane. *J. Exp. Med. 136,* 49–67.

GROBSTEIN, C. (1956). Transfilter induction of tubules in mouse metanephrogenic mesenchyme. *Exp. Cell Res. 10,* 427–440.

HARTOU, H., AND ARGYRIS, B. F. (1972). Evidence for cross-reactivity of antigens at the level of thymus-derived cells. *Cell. Immunol.* 4, 179–181.

HOOD, L. E. (1971). Two genes, one polypeptide chain—fact or fiction? *Fed. Proc. 31,* 177–187.

HUMPHREY, J. H., AND WHITE, R. G. (1970). "Immunology for Students of Medicine." 3rd ed., F. A. Davis, Philadelphia, 757 pp.

JERNE, N. (1971). The somatic generation of immune recognition. *Eur. J. Immunol. 1,* 1–9.

MAKELA, O., CROSS, A., AND KOSUNEN, T. U., eds. (1971). "Cell Interactions and Receptor Antibodies in Immune Responses." Academic Press, New York, 472 pp.

MANDEL, T., AND RUSSELL, P. J. (1971). Differentiation of foetal mouse thymus: ultrastructure of organ cultures and of subcapsular grafts. *Immunology 21,* 659–674.

METCALF, D., AND MOORE, M. A. S. (1971). "Haemopoietic Cells: Their Origin, Migration and Differentiation." Frontiers of Biology Vol. 24. North Holland, Amsterdam, 550 pp.

MILLER, J. F. A. P., BASTEN, A., SPRENT, J., AND CHEERS, C. (1971). Interaction between lymphocytes in immune responses. *Cell. Immunol. 2,* 469–495.

MITCHISON, N. A. (1971). The carrier effect in the secondary response to hapten-protein conjugates. II. Cellular cooperation. *Eur. J. Immunol. 1,* 18–27.

MOORE, M. A. S., AND OWEN, J. J. T. (1967). Experimental studies on the development of the thymus. *J. Exp. Med. 126,* 715–725.

NOSSAL, G. J. V., AND ADA, G. L. (1971). "Antigens, Lymphoid Cells, and the Immune Response." Academic Press, New York, 324 pp.

OWEN, J. J. T. (1972). The origins and development of lymphocyte populations. *In* "Ontogeny of Acquired Immunity" (CIBA Foundation Symposium), Elsevier, Amsterdam, pp. 35–52.

SAUNDERS, G. V., AND SWARTZENDRUBER, D. (1970). Maturation of hemolysin-producing cell clones. II. The appearance and localization of precursor units in lymphoid tissues of neonatal mice. *J. Exp. Med. 131*, 1261–1270.

SHALABY, M. R. S., AND AUERBACH, R. (1973). Studies on the maturation of immune responsiveness in the mouse. I. The *in vivo* response to sheep red blood cells. *Differentiation, 1*, 1–5.

SILVERSTEIN, A. M., PARSHALL, C. J., AND UHR, J. W. (1966). Immunological maturation in utero: kinetics of the primary antibody response in the fetal lamb. *Science 154*, 1675–1677.

SIMONSEN, M. (1962). Graft-versus-host reactions: their natural history and applicability as tools of research. *Progr. Allergy 6*, 349–467.

SMITHIES, O. (1970). Pathways through networks of branched DNA. *Science 169*, 882–886.

SOLOMON, J. B. (1971). "Foetal and Neonatal Immunology." Frontiers of Biology, Vol. 20. North-Holland/Elsevier, Amsterdam, 364 pp.

STERZL, J., AND RIHA, I., eds. (1971). "Developmental Aspects of Antibody Formation and Structure." 2 vol. Academic Press, New York, 1054 pp.

STITES, D. P., CARR, M. C., AND FUDENBERG, H. H. (1973). *In* "Colloque de la Société Française d'Immunologie: Phylogenic and Ontogenic Study of the Immune Response" (in press).

UMIEL, T. (1971). Thymus-influenced immunological maturation of embryonic liver cells. *Transplantation 11*, 531–535.

UMIEL, T., AND AUERBACH, R. (1973). Studies on the development of immunity: the graft-versus-host reaction. *Pathobiol. Ann. 3*, 27–45.

UMIEL, T., GLOBERSON, A., AND AUERBACH, R. (1968). Role of the thymus in the development of immunocompetence of embryonic liver cells *in vitro*. *Proc. Soc. Exp. Biol. Med. 129*, 598–600.

CHAPTER FOURTEEN

The Phenomenon of Immunological Tolerance and Its Possible Role in Development

R. E. BILLINGHAM

Early workers in the field of immunology were motivated by the thesis that the active immunological responses of animals and men against alien substances such as proteins and intruding pathogenic microorganisms reflected the activity of an important protective or homeostatic mechanism, and that such responses were always of advantage to the subject. Consequently, research activities were largely concerned with the production of antibodies (which at that time were believed to be the mediators of *all* immunological responses), the analysis of their properties, the augmentation of their production, and their transfer to normal, unimmunized subjects (i.e., passive immunization) for protective and other purposes.

Beginning about 1950, however, increasing attention was devoted to the means of preventing or eliminating a subject's ability to respond to an antigen. The impetus for this change in emphasis or bifurcation of activity was threefold:

1. The awareness that knowledge of how to terminate or prevent a particular sequence of events leading to an immunological response offers an approach to an analysis of the *modus operandi* of the response.

2. The recognition by surgeons that the ability to make use of tissue and organ grafts required the development of means of overcoming the destructive immunological response—that is, the so-called *homograft reaction* (see Billingham and Silvers, 1971)—that individuals normally develop against cells and tissues from unrelated donors of their own species.

3. The increasing suspicion that certain diseases might be due to specific reactivity of the patients against constituents of their *own* bodies—that is, *autoimmunity.*

One important agent that was known to interfere with a host's capacity to undertake immunological responses was whole body X-irradiation. Another agent was antigen itself, as evidenced by the work of Glenny and Hopkins, who, in 1924, had shown that, as a result of a prolonged course of injections of horse serum proteins, an adult rabbit's capacity to react against this antigen becomes attenuated. The significance of this observation remained unrecognized for many years.

The principal type of immunological unresponsiveness that can be induced by exposure to an antigen is known as *immunological tolerance* or *immunological paralysis*. This has been broadly defined as a state of specific nonreactivity to a normally effective antigenic challenge that is induced in an animal by prior exposure to the antigen concerned. The subject of this article is an account of the discovery of this phenomenon, of the ways in which we have had to modify our thoughts about it from time to time as the science of immunobiology advanced, and its possible role in normal development. Although the tolerance principle applies to a wide range of antigens, we shall deal mainly with the phenomenon as it applies to that special category of genetically determined cellular isoantigens that are responsible for the phenomenon of transplantation immunity. The latter includes, in addition to the conventional host-versus-graft, or homograft reaction, the graft-versus-host reactivity, which is responsible for homologous, runt, or transplantation disease (see Billingham and Silvers, 1971).

In 1951, in the course of an attempt to use skin grafting to differentiate between monozygotic and dizygotic twins in cattle, my colleagues and I (Anderson et al., 1951; Billingham et al., 1952) made the surprising discovery that, despite their genetic disparity, dizygotic cattle twins nearly always accepted skin grafts exchanged between them. This anomalous *tolerance* of grafts which, on genetic grounds, should have been rejected immunologically, was highly specific in that a "tolerant" calf would accept homografts from no other donor but its twin: grafts from its parents or from full siblings of separate birth were rejected with normal promptitude, rarely surviving longer than 9 or 10 days. The explanation of this anomalous acceptance of these skin grafts appeared to turn on two previous observations: first, that of F. R. Lillie (1916) that there are

nearly always anastomotic connections between the placental blood circulations of fetuses in multiple births in cattle, so that blood flowing in the circulation of the one twin has free access to the circulation of the other; and, second, Owen's discovery, in 1945, that at birth and throughout life most dizygotic cattle twins are chimeras, having two serologically distinguishable types of erythrocyte, one corresponding to the animal's *own* genotype and the other corresponding to that of its cotwin, and against which it might reasonably have been expected to react immunologically. Recognizing that red blood cells are "end cells" of finite life-span, Owen postulated that their presence must reflect the prenatal interchange and more or less permanent establishment of genetically alien hematopoietic stem cells.

In 1949 Burnet and Fenner propounded an important theory of antibody response that was greatly influenced by Owen's observation and reasoning, since it hinted at the existence of a mechanism whereby antibody-forming cells must "learn" to distinguish between substances that are proper to the individual and those that may later gain entry to his body from the outside—that is, to distinguish between "self and non-self." Although there is no need to go into details about this theory, it is important to note that it made the important prediction that if embryos or very young animals (which hitherto had been regarded as incapable of undertaking immunological responses) are exposed to antigenic substances, their ability to react against these *same* substances—when rechallenged after they had grown up and their immunological response machinery was functionally mature—would be abolished or impaired.

These three separate events—that is, Owen's discovery of red cell chimerism in cattle twins, Burnet and Fenner's speculations about "self and non-self," and Medawar and his associates' discovery that dizygotic cattle twins are mutually tolerant of each others' skin grafts—prompted laboratory experiments to reproduce the state of affairs that happens naturally in twin cattle. (Subsequently, it has been found that a similar situation also occurs naturally in marmosets, in which the incidence of dizygotic twinning is high, in the twin chicks that can sometimes be hatched from double-yolked eggs, and very rarely in sheep and man.)

Experimental Induction of Tolerance in Very Young Animals

Initially Medawar and his associates (see Billingham et al., 1953, 1956) inoculated CBA strain mouse fetuses of about 15 or 16 days' gestation with viable suspensions of cells derived from a variety of tissues obtained from adult mice of an unrelated donor strain, A. Then, about 8 weeks after their birth and when their capacity to undertake immunological responses was fully mature, the subjects were challenged with skin homografts from mice of the original donor strain, which they would normally be expected to reject within about 11 days. It was found that test grafts on some of the prenatally inoculated animals either survived permanently, or significantly outlived control grafts on untreated hosts (see Fig. 1). That the state of unresponsiveness induced was highly specific followed from the observation that CBA mice that had been rendered unresponsive to grafts from A-strain donors rejected homografts from donors of other unrelated strains with unaffected vigor.

Essentially similar results were obtained in experiments conducted in chickens, in which White Leghorn embryos of 11 days' incubation were inoculated intravenously with whole blood from Rhode Island Red donor embryos of comparable maturity.

At about the same time Milan Hašek, in Czechoslovakia (see Hašek, 1953, Hašek et al., 1961), developed an ingenious laboratory model that reproduced very faithfully the dizygotic twin-cattle situation. He procured

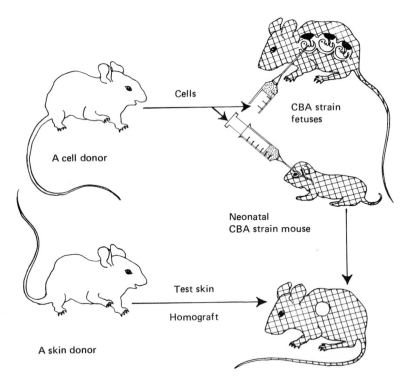

FIGURE 1. *Methods of inducing tolerance of A strain tissue antigens in CBA strain mice. A suspension of viable cells, usually prepared from the spleen, from an adult A strain donor is inoculated into either CBA strain fetal hosts in utero or intravenously into newborn CBA mice. When the treated mice have grown up, challenge with homografts of skin from an A strain donor reveals that their capacity to reject these grafts has been partially or completely abrogated; that is, they have been rendered immunologically tolerant of the alien transplantation antigens concerned.*

the development of vascular intercommunications between the circulations of paired genetically dissimilar avian embryos by maintaining in close approximation small, exposed areas of the highly vascular choriollantoic membranes of the parabionts. This allowed vascular interconnections to become established and an extensive interchange of blood to take place over a period of a week or more.

After hatching it was shown that such former parabionts were (1) red cell chimeras, (2) incapable of producing isoagglutinins when attempts were made to immunize them against their partners' erythrocytes, and (3) tolerant of grafts of their partner's skin.

These simple experiments confirmed Burnet and Fenner's prediction and initiated a whole, new, fast-developing field of immunology. It was soon discovered that tolerance of tissue homografts could be induced by intravenous injection of newly hatched chicks, and of newborn mice and rats, facilitating the production of tolerant animals for experimental study.

CONCEPT OF "TOLERANCE-RESPONSIVE PERIOD"

When tolerance was at an early stage of its experimental elucidation, it was believed that it could only be induced during a finite "critical period" of an animal's life-span, usually before or for a short time after birth, which varied from one species to another, depending upon the degree of maturity of the host's machinery of immunological response at the time of birth. This period was referred to as the *tolerance-responsive period*. In animals with relatively long gestation periods (e.g., man, sheep, cattle, and guinea pigs), it was believed to terminate well before birth. Evidence that fetal lambs of about 80 days' gestation could reject homografts as rapidly as adult sheep lent support to this concept, as did findings that in animals with relatively short gestation periods, such as rabbits, mice, chickens, and hamsters, the tolerance-responsive period appeared to terminate at or about the time of birth (see Solomon, 1971).

Since it was evident that the same antigenic stimulus that induced unresponsiveness in fetuses or young animals evoked an active immune response in older subjects, the logical conclusion was drawn that there must exist a gradual transition from one mode of response to another. The existence of a "neutral" or "null-period" was therefore postulated—a period in life during which exposure to a particular antigen confers neither tolerance nor immunity (see Billingham et al., 1953). In the light of subsequent findings, especially those which have established that animals of *any* age can be rendered tolerant, these terms have lost most of their theoretical significance. However, from a purely practical point of view, animals that are immunologically immature are much more susceptible to tolerance-induction. This is well illustrated by Brent and Gowland's (1961) finding that if the number of homologous cells injected into infant mice is kept constant in relation to the host's body weight, the proportion of animals rendered tolerant declines rapidly as the time of inoculation after birth is extended. To achieve approximately the same degree of unresponsiveness in 2-week-old mice as can be induced with a single, weight-adjusted cellular inoculum in neonatal hosts required injection of the older animals 3 times per week for up to 4 weeks. It is important to note that before these animals became tolerant, they passed through a phase when they behaved as if they had been weakly *immunized*.

Using the tolerance-responsive period purely as a convenient, operationally useful designation of the age range over which tolerance is easily inducible by a particular antigenic material, it has been found that, for a given species, the period is not constant even for antigens belonging to the same general class. Where transplantation antigens are concerned, the duration of the period depends upon the magnitude of the immunogenetic disparity between donor and recipient (or, in other words, upon the "strength" of the antigenic stimulus); the period is longer where differences with respect to only minor transplantation antigens are involved than where there are differences determined by alleles at the major histocompatibility locus for the species concerned (the H-2 locus for the mouse or the Ag-B locus for rats) (see Billingham and Silvers, 1971; Silvers and Billingham, 1969).

In mice, where strong, H-2 locus-determined

antigens are involved, tolerance of skin homografts can only be induced if the homologous cells are introduced by the intravenous route within a few hours of birth. There are, indeed, some H-2 locus-incompatible donor–host strain combinations in which, at best, only very feeble degrees of tolerance are inducible by neonatal inoculation of cells. However, this refractoriness can easily be overcome by exposure of the neonatal subjects to a sublethal dosage of irradiation before inoculation (see Billingham and Silvers, 1962 and page 280).

In the rat, by contrast, there appears to be no strain combination in which high degrees of tolerance of skin homografts cannot be induced by neonatal intravenous inoculation.

Where only weak transplantation antigens are concerned, tolerance is inducible by administration of the antigenic material by the intraperitoneal or even the subcutaneous routes after much greater delays postpartum. Furthermore, to induce tolerance of grafts in situations in which only "minor" transplantation antigens are involved requires lower doses of cellular inoculum than when major antigenic differences are involved (see Table 1).

CROSS-REACTIVITY IN TOLERANCE

Results of the early investigations on tolerance suggested that cellular inocula of similar genetic constitution but of widely different histological origins were equally effective in inducing unresponsiveness of test skin or other types of homograft from the same alien donor strain—that is, that no tissue specificity was involved. This premise was sustained by findings that chimeric twin cattle are tolerant of each others' kidneys as well as skin grafts, and substantiated by numerous empirical observations that rats or mice inoculated at birth with cell suspensions derived from various components of the lymphohematopoietic tissue system (which is comprised of spleen, bone marrow, lymph nodes, thymus, and blood)—either singly or in combination—were tolerant of subsequent test grafts of skin, adrenal cortex, ovary, kidney, various tumors, and so on.

However, as a result of subsequent comparative studies of the capacities of various types of cell suspension to induce tolerance with a variety of donor–host strain combinations in both mice and rats, we now know that cellular inocula of different histological origins are *not* equivalent with regard to their ability to confer tolerance of skin grafts. For example, in mice where major histocompatiblity antigens are involved, lymph node and spleen cells are more effective than marrow cells, whereas in rats tolerance of antigens determined by the Ag-B locus is most easily induced by bone marrow cells (Billingham and Silvers, 1962; Silvers and Billingham, 1969). In both species thymocytes afford a very inferior "tolerogenic" stimulus, unless only weak antigenic differences are involved.

SPLIT TOLERANCE

A partial explanation for some instances in which inoculation of infant hosts with cells of lymphohematopoietic tissue origin fails to induce tolerance of skin homografts of similar genetic makeup is afforded by observations that rats inoculated at birth with suspensions of lymph node cells from Ag-B locus-incompatible donors may become tolerant of these cells (as evidenced by the findings of chimera and other tests) yet display no tolerance of skin grafts of similar alien genetic origin (Silvers et al., 1970). Bone marrow cells, on the other hand, are highly effective in inducing tolerance of themselves, leading to chimerism, as well as tolerance of skin homografts. This superiority on the part of marrow cells as inducers of tolerance in respect of skin grafts has long been held to reflect a sharing of an antigenic determinant(s) with skin cells that is poorly expressed by lymph node cells. An alternative explanation for "split tolerance" will be considered on page 287.

There are other important observations that animals that are chimeric with regard to cells of one tissue type do not necessarily display tolerance towards genetically similar cells or tissues of other histological types (see Warner et al., 1965). For example, newborn A-strain mice inoculated at birth with (C57BL x CBA)F$_1$ hybrid spleen or marrow cells may, when they grow up, display only a moderate

TABLE 1

RESULTS OF EXPERIMENTS IN WHICH LEWIS RATS WERE RENDERED TOLERANT
OF SKIN HOMOGRAFTS BY MEANS OF NEONATAL INTRAVENOUS
INJECTIONS OF BONE MARROW CELLS[a]

	Donor–Recipient Rat Strain Combination			
	BN → Lewis		Fischer → Lewis	
Median survival times of skin homografs exchanged between adult rats:	7.0 ± 0.2 days		7.6 ± 0.4 days	
Whether strains are compatible at the Ag-B locus:	No; Ag-B³ → Ag-B¹		Yes; both strains Ag-B¹	
Percentage hosts rendered tolerant[b] and highly tolerant[c] of donor strain skin allografts following neonatal inoculation of *indicated* doses of bone marrow cells[d]:				
	Tolerant	*Highly Tolerant*	*Tolerant*	*Highly Tolerant*
0.1 × 10⁶	0	0	0	0
0.25	0	0	88	63
1.0	0	0	100	89
5.0	0	0	100	95
10.0	23	0		
20.0	52	19		
40	93	66		
60	94	76		
80	96	82		
120	100	96		

[a]Data from R. E. Billingham and W. K. Silvers, *J. Cell. Comp. Physiol.* 60, Suppl. 1, 183–200, 1962; W. K. Silvers and R. E. Billingham, *Transplantation* 8, 167–178, 1969.
[b]Tolerant rats are those whose test grafts significantly outlived similar grafts on untreated control hosts.
[c]Highly tolerant animals are those whose grafts lived for a minimum of 50 days.
[d]Note that dosage of the tolerogenic stimulus is an important variable. Tolerance was much easier to induce when donor and host were alike at the major Ag-B locus than when they differed.

degree of tolerance toward grafts from hybrid or C57 donors but a much higher degree of tolerance of subsequent CBA grafts (see Brent and Courtenay, 1962). More important, it has been shown that even after such animals have rejected hybrid or C57BL skin homografts, they may remain chimeric, containing in their lymphohematopoietic tissues surviving foreign cells that bear both C57 and CBA antigens on their surfaces. There is also evidence that dizygotic twin cattle that have proved to be incompletely tolerant of grafts of each others' skin usually remain tolerant of each others' hematopoietic cells, as evidenced by erythrocyte chimerism, long after rejection of their co-twins' skin (see Stone et al., 1965).

TYPE OF ANTIGENIC STIMULUS THAT WILL INDUCE TOLERANCE OF SKIN HOMOGRAFTS

It has been found that tolerance of skin grafts can only be conferred by means of inocula comprising cells of the erythropoietic, myeloid, or lymphoid series. With cells of other histological types success has been very fitful even when high dosage inocula have been employed. In rats, when Ag-B locus incompatibilities are not involved, neonatal inoculation of hosts with thymocytes will confer tolerance of skin homografts. However, suspensions of viable epidermal cells are totally ineffective as "tolerogens." Irradiation of lymphohematopoietic cells, though it does not affect their immunogenicity, totally abolishes their tolerogenicity, and in this light it is scarcely surprising that little success has attended attempts to render neonatal hosts tolerant by means of cell extracts.

On the basis of such findings as these, it became widely believed that the prerequisites for an effective tolerance-conferring inoculum are that the cells must be viable, and capable of migrating and proliferating—becoming permanently established of their own accord in anatomically appropriate sites (which include the host's seats of immunological response). All the available evidence indicates that, to establish tolerance initially, all the host's immunologically competent cells must be appropriately exposed to the tolerogen and the latter must be persistent. This latter requirement is met by survival and proliferation of the tolerance-conferring cells. In this regard the test skin graft that is used to record the tolerant status of the host also acts as a potential source of tolerance-maintaining antigen.

ANTIGENIC DOSAGE FACTOR

Apart from the *type* of cell comprising a putative tolerance-conferring inoculum, the actual number or dosage of cells is an important factor in determining whether tolerance will be induced and whether it will be complete or only partial—in other words, will a test skin homograft survive permanently or is the host's capacity to reject it only partially abrogated, allowing the graft to survive for only a few days or even many weeks before its eventual demise? The number of cells required to induce tolerance is much greater where strong transplantation antigens are involved than when only weak ones are involved (see Table 1). Indeed, in both mice and rats, where "strong" histoincompatibilities are involved, low dosage cellular inocula may elicit sensitivity, whereas higher dosage inocula may induce tolerance in neonatal recipients (Billingham and Silvers, 1962).

COMPLICATING FACTOR IN TOLERANCE-INDUCTION: GRAFT-VERSUS-HOST REACTIVITY

Of the various tissues of the body, it is those that comprise the lymphohematopoietic system that can most easily and efficiently, by purely mechanical means, be dissociated to yield viable suspensions of cells suitable for intravenous inoculation. Moreover, as we have seen, these are the only cells capable of inducing tolerance in neonatal subjects. A perplexing complication of many early attempts to induce tolerance was the development of an often fatal wasting disease, sometimes referred to as *runt disease*, in subjects inoculated with cells from adult donors (Fig. 2).

It was found that this disease was due to the presence of immunologically competent cells in the tolerance-conferring stimulus. On being confronted by the alien cellular antigens of the immature host, these donor cells were stimulated to react against them, generating a so-called *graft-versus-host (gvh) reaction*—effectively a homograft reaction in which the roles of graft and host have been reversed (see Billingham and Brent, 1959; Billingham, 1968; Elkins, 1971).

The risk of a tolerance-conferring inoculum causing clinically overt gvh reactivity can be reduced by using bone marrow or thymus cells rather than spleen or lymph node cells, since the former include much lower proportions of immunocompetent cells. This risk can also be reduced by employing cellular inocula from immunologically immature donors, in which

FIGURE 2. *Three 12-day-old A strain mice. The two on the right were injected at birth with a suspension of 5 million spleen cells from an adult donor of the C57 strain. These animals display retardation of growth, abnormalities of their skin, and are suffering from diarrhoea—symptoms of runt or graft-versus-host disease. The healthy mouse on the left is an uninjected litter mate.*

case the inoculated immunologically competent cells may become tolerant of the host's transplantation antigens. It can, of course, be totally eliminated by using appropriate F_1 hybrid animals of *any* age as the source of the tolerance-conferring stimulus. For example, to render an infant A-strain mouse tolerant of strain C57 skin, inoculate it with splenic cells from a $(C57 \times A)F_1$ hybrid donor. The hybrid donor, having inherited a complete set of transplantation antigens from each of its parents, is incapable of reacting against them—that is, it is *genetically tolerant* or unresponsive.

Recent work has established that a necessary condition for the development of gvh reactivity is that the host must confront the putative attacking cells with a strong transplantation antigen, determined by the major histocompatibility locus for the species concerned, unless the cellular inoculum is derived from a specifically presensitized donor. The host target cells against which the gvh reactivity is directed appear to be the ingredients of its lymphohematopoietic tissue system and

its skin; in other words, contrary to expectation all tissues are not at equal risk.

ABOLITION OF TOLERANCE

All the observations cited so far suggest that tolerance of a tissue homograft is the outcome of a specific central failure of the host's immunological response machinery—some kind of adaptive response to the tissue transplantation antigens involved. However, this does not entirely rule out the possibility that the cells that evoked the tolerant state initially (and their mitotic descendants), as well as subsequently transplanted homografts of skin or other tissues that have long been in residence on tolerant animals, may not also undergo some kind of antigenic change or adaptation. Cogent evidence that this does not take place, at least to any significant extent, was afforded by findings that if skin homografts that have long been in residence on specifically tolerant hosts are excised and returned to normal mice of the strain from which they were derived originally, they are permanently accepted, whereas if they are transplanted to normal mice of the tolerant mouse's strain, they are rejected.

A state of tolerance can be abolished quite rapidly, as evidenced by the destruction of a perfectly healthy skin homograft of long standing and the disappearance of cell chimerism in the host if the later is inoculated with a source of immunologically competent cells—for example, a suspension of lymph node cells, splenic cells, leucocytes, or thoracic duct cells from a normal, or unsensitized, donor of its *own* strain (Billingham et al., 1956; Billingham et al., 1963). The speed with which this takes place is dependent upon the actual number of lymphoid cells transferred. Besides affording independent proof that homografts borne by tolerant animals retain their antigenic autonomy, this observation indicates that there is no impediment in the efferent pathway of the immunologic reflex. Tolerance, it was concluded, is a specific *central* failure of the host's machinery of immunological response. The transferred normal isologous lymphoid cells reequip the tolerant animals with cells that can initiate an immunological response against the alien antigens associated with the tolerated grafts as well as with the widely distributed cells responsible for their chimerism.

As one might anticipate, tolerance is abolished with much greater promptitude if the transferred lymphoid cells are derived from donors that have already been sensitized to the antigens in respect of which the host is tolerant.

Repeated attempts to prejudice the wellbeing of healthy skin homografts of long standing on tolerant mice by passive transfer of specific isoantisera have been unsuccessful. However, Lubaroff and Silvers (1970) have shown that in rats repeated inoculation of relatively large doses of isoantiserum over a period of 7 days will abolish tolerance of Ag-B incompatible skin grafts. The finding that destruction of these grafts usually took several weeks, in conjunction with other observations, suggested that the transferred antibody itself was not the direct mediator of graft destruction. Rather, the antibody may have mediated the destruction of the alien lymphohematopoietic tissue cells responsible for the chimeric status

of the hosts. Deprived of the stimulus of this persisting antigen, the hosts probably regained immunological competence. Further evidence sustaining this conclusion was forthcoming from experiments in which the tolerant status of rats was abolished by adoptive immunization with lymphoid cells against which the hosts were capable of reacting (see Silvers and Billingham, 1970).

INTRINSIC IMMUNOLOGICAL TOLERANCE

Mintz (1962, 1965) has furnished us with the most dramatic of all models of prenatally induced immunological tolerance. If a single, "unified" mouse embryo is produced by fusion *in vitro* of the blastomeres from genetically dissimilar blastulas and it is then returned to the uterus of an "incubator" mother it will develop to maturity as a single "tetraparental" (or "allophenic") individual (Mintz, 1965). Indeed such mice may be comprised of two or more zygotes differing from each other at major and other histocompatibility loci. They are chimeric and may exhibit mosaicism in many different tissues, and they are usually fully tolerant of skin grafts from the strains that provided the blastocysts, though they can reject skin grafts from unrelated donors with normal vigor. Mintz and Silvers (1967) have termed this *intrinsic* immunological tolerance.

Induction of Tolerance in Adult Animals

A state of specific unresponsiveness to homografts, closely resembling if not indistinguishable from tolerance as procured by inoculating very young animals with homologous cells, can also be induced in adult, fully mature animals, though the strategy of its induction usually has to be more complex, and the development of the unresponsive state is frequently, if not always, preceded by a state of sensitivity.

Tolerance may be induced in adult subjects by inoculating them with a single massive dose of living cellular antigen (Guttmann and Aust, 1961; Wigzell, 1962) or, more effectively, by *repeated* inoculation with high dosages of cells (Brent and Gowland, 1962; Shapiro et al., 1961). An alternative procedure is to place the individual to be rendered tolerant in temporary surgical parabiotic union with a foreign partner, usually an F_1 hybrid, as a chronic and abundant source of the cellular antigens to which it is to be rendered unresponsive (Martinez et al., 1960). Since it is reasonable to assume that the efficacy of these various procedures to induce unresponsiveness is related to the number of cells in the host that are competent to react with the antigenic stimulus concerned, any means of reducing this number should facilitate tolerance induction. Treatment of adult animals with a number of so-called *immunosuppressive agents*, such as purine and pyrimidine antagonists, sublethal X-irradiation, heterologous antilymphocyte serum (ALS), or thoracic duct drainage, in some exact temporal relation to the administration of antigen, may result in tolerance under circumstances in which administration of the antigen alone would have evoked immunity (see Billingham and Silvers, 1971).

Antilymphocyte serum (ALS) is raised in animals of one species by injecting them with lymphoid cells from donors of the species in which it is to be used (see Medawar, 1969; Lance and Medawar, 1971). For example, ALS for use in mice is normally produced by injection of murine lymphocytes into rabbits. The genetic constitution of the donor is unimportant. In experimental animals this agent has a truly dramatic capacity to prolong the lives of homografts at dosage levels that are relatively harmless and nontoxic. Its effectiveness in abrogating homograft reactivity—it can even abolish a preexisting state of sensitivity against homografts—turns upon the fact that it acts principally, if not entirely, by a complement-dependent, selective killing of the thymus-derived population of long-lived, small lymphocytes (i.e., T cells—see page 286) in the recirculating pool.

Lance and Medawar (1969) have obtained very impressive prolongations of skin homograft survival in adult mice by means of a short course of ALS injections followed by a single, relatively low dose (25–100 × 10⁶) of donor

cells inoculated 8 to 12 days after grafting. The unresponsiveness obtained had all the characteristics of tolerance as induced in neonatal mice. In dogs impressive prolongations of survival of orthotopic limb homografts have been obtained by prior massive infusion of the hosts with donor cells and pretreatment with ALS and Azathioprine (see Lance, 1971).

Most attempts to induce tolerance by inoculation of antigenic extracts into both neonatal and adult animals have given very disappointing results. However, encouragement for continuation of this approach was afforded by Medawar's (1963) demonstration that both A-methopterin and X-irradiation act synergistically with crude, cell-free, semisoluble extracts of homologous lymphoid cells in inducing some degree of tolerance in adult mice. Brent and his colleagues (see Brent et al., 1971) have shown that ALS can be used in the mouse to facilitate the induction of specific unresponsiveness to skin homografts with extracts of liver or spleen. Where the antigens involved included strong ones determined by the H-2 locus, this unresponsiveness was most successfully induced by administration of a dose of extract equivalent to 250 mg of fresh tissue 16 to 26 days before skin grafting and ALS treatment. Zakarian et al. (1972) have shown that chronic treatment of adult hamsters with crude, cell-free antigenic material over a 90-day period induced a high degree of tolerance of skin homografts.

Phenomenon of Immunological Enhancement

Any consideration of immunological tolerance would be incomplete without saying something about another phenomenon—immunological enhancement, which antedates it and which, for a long time, seemed to be both operationally and causally distinguishable from it, despite the fact that it produced a similar end result—that is, specific unresponsiveness.

One of the earliest discovered means of promoting the growth of tumor grafts in genetically alien adult hosts involved pretreatment of the latter with various desiccates or extractives of normal or malignant tissues having the same genetic makeup as the future tumor graft. Although the phenomenon was discovered in 1907 by Flexner and Jobling, it was largely the work of Kaliss and Snell in the early 1950s that established its immunological basis and mediation by humoral antibodies (see Snell, 1970).

When adult animals are confronted by homografts, they normally respond in two distinct ways to the alien histocompatibility gene products on the cell surfaces. First, they respond by the generation of a population of "effector" lymphocytes that circulate in the blood stream and that, in some as yet incompletely understood manner, are capable of destroying the homografts that incited their formation by cellular immunity (Wilson and Billingham, 1967; see also Chapter 13). These effector cells are also capable of destroying target cells *in vitro* in the absence of humoral antibodies. Second, adult animals respond by the synthesis of humoral antibodies of a variety of types that are also capable of reacting with antigenic determinants on the surfaces of the target cells (Winn, 1970). In the presence of complement, some of these antibodies have a cytopathogenic action. This action varies in severity, depending upon the histological type of the target cell, lymphoid and myeloid cells being especially susceptible and sarcoma cells highly resistant.

In the case of certain kinds of tumor homograft, as well as some normal tissues and renal transplants, it has been shown that the presence of a high titer of serum antibodies may dramatically weaken the host's capacity to react against and destroy them. This immunosuppressive effect, which is highly specific, "enhances" or prolongs the survival of the graft. The requisite antibody levels may be obtained in two ways: (1) by *active* immunization of the host by repeated inoculation with homogenates, desiccates or other rather crude, nonliving preparations from the donor strain of the future living homograft (active enhancement), or (2) by *passive transfer* of these antibodies to a normal host. Passive enhancement can also occur naturally from an actively immunized (i.e., enhanced) mother to her perinatal offspring.

As in the case of tolerance, there are marked differences in the facility with which enhancement is procurable and its magnitude, depending upon the genetic relationship between donor and recipient. Where strong transplantation antigens are involved, the cellular immune responses provoked by many types of graft, including skin, can usually override any prior enhancing treatment. In many instances, seemingly minor variations in the enhancing protocol may be crucially important in determining whether an animal's capacity to reject a homograft is weakened or unaffected. The fact that animals that can mount an immune response against a homograft may, with further treatment, become unresponsive suggests that the difference between an enhanced and an immune state must be a very subtle one.

Interest in immunological enhancement has been greatly heightened by recent findings that treatment of rats with an intravenous injection of donor spleen cells, followed by transfer of specific isoantiserum, or by transfer of antiserum alone, will prolong the survival of Ag-B locus-incompatible renal transplants, but not skin grafts, for very long periods (Stuart et al., 1968; French and Batchelor, 1969).

Although there is no doubt that enhancement is an antibody-mediated phenomenon, final agreement as to how the antibodies mediate the unresponsiveness has yet to be achieved. Two contributory processes appear to be: (1) combination of antibodies with the antigenic determinant sites on the cells comprising the homograft, thus "masking" them, and so interfering with both the elicitation (following a "recognition" process) and the fulfillment of the host's cellular immune response—that is, "afferent" and "efferent" inhibition, and (2) by some kind of *central* inhibitory action on the host's immunologic response machinery by antibodies or antigen-antibody complexes, preventing the development of cellular immunity.

Recently the Hellströms and various associates, with the aid of a rather special *in vitro* test to detect "immune" or effector lymphocytes, have studied mice and dogs rendered tolerant and chimeric in adult life by wholebody irradiation followed by transfusion of alien bone marrow cells, as well as mice rendered tolerant by neonatal inoculation with alien cells, and intrinsically tolerant tetraparental mice (Hellström et al., 1970, 1971). They claim that in all these various types of tolerant subjects, "immune" lymphocytes are present that are capable of destroying appropriate "target" cells *in vitro*. However, serum factors, believed to be antibodies, are also present *in vivo* that can block this lymphocyte reactivity *in vitro* and that, Hellström et al. suggest, may be similarly active *in vivo*. These observations are in accord with previous findings of Voisin et al., 1968 (see also Voisin, 1971) that the serum of mice rendered tolerant by neonatal inoculation often contains hemagglutinins specific for donor strain cells and manifests "enhancing" activity on transfer to appropriate hosts. On the basis of these observations the Hellströms have suggested that some examples of tolerance of homologous cells may in fact be due to the presence of serum factors that can block lymphocyte reactivity, rather than a specific failure of response.

Tolerance of Heterografts

Unlike homografts, grafts exchanged between members of *different* species—heterografts (or zenografts)—rarely become established to the point where they begin to fulfill their normal function; that is, they scarcely have any meaningful survival time at all. The exceptions include grafts exchanged between members of species that are closely related taxonomically, such as birds of the order Galliformes, sheep and goats, or hamsters of the genus *Mesocricetus*.

In birds the technique of experimental parabiotic union in embryonic life, or even the injection of donor blood into newly hatched subjects, will accomplish transient erythrocyte chimerism and tolerance of skin grafts across certain species barriers. However, the results of most early attempts to induce tolerance of heterografts by inoculation of very young hosts with donor cells or to rehabilitate lethally irradiated adult hosts with heterologous bone marrow infusions have given very disappointing results. Nevertheless, rat–mouse radiation chimeras have been produced and shown to be tolerant of skin heterografts.

That the achievement of high degrees of interspecific tolerance is a feasible objective is indicated by the demonstration by Lance and Medawar (1969) that rat skin grafts can be enabled to survive on mice for hundreds of days. To accomplish this required a rather heroic procedure: thymectomy of the adult mice, followed by a short, intensive period of ALS treatment and a massive infusion of rat lymphoid cells. To eliminate the intervention of graft-versus-host reactions the lymphoid cell donors were treated with ALS beforehand. Tolerant mice produced in this way were chimeric with respect to their lymphoid cells and manufactured rat protein.

Tolerance to erythrocytes has also been accomplished both within species and across species barriers; for example, rats can be rendered tolerant of sheep red cells. Since these cells have a limited life-span and are unable to divide, they cannot induce permanent chimerism in their host, so that periodic reinoculation of this nonreplicating antigen is required to maintain tolerance.

Tolerance of Other Antigens

The phenomenon of tolerance has been found to apply to many different categories of antigen, and there is great diversity with regard to the conditions for its induction and manifestation (see Hraba, 1968; Weigle, 1971). It applies to serum proteins from alien donor species, which are particularly suitable for analysis of tolerance induced in adult subjects for a variety of reasons: (1) they have a relatively simple molecular structure, (2) they are relatively weak immunogens; (3) they are readily available in a pure form; (4) they are soluble and rapidly equilibrate between the intra- and extravascular compartments of the body; and (5) they persist in the circulation, which enables them to reach *all* the antigen-reactive lymphocytes in effective concentrations. These antigens are rapidly catabolized, and prolonged courses of injection with massive doses are usually followed by a relatively short period of complete tolerance. In addition to this well-established "high dosage" tolerance, there is evidence that with some antigens tolerance can be induced by extremely low subimmunogenic dosages of antigen ("low dose" tolerance) (see Mitchison, 1964, 1971a). The principle of tolerance also applies to synthetic polypeptides.

Very long-lasting tolerance can also be induced in respect of pneumococcal polysaccharide antigens in adult animals. This usually requires a single injection and is extremely persistent, probably a consequence of the recirculatory and almost indestructable properties of this antigen. Although taken up by the reticuloendothelial system, it is not catabolized and forms depots in the bodies of paralyzed mice that are subject to release. Extensive studies have established that, in part, this paralysis is due (1) to the continuous "treadmill" neutralization of antibody by the recirculating, nondegradable antigen, "pseudoparalysis," which masks the occurrence of an immune response, and (2) to central inhibition of response, the two phenomena coexisting (see Howard et al., 1971). In paralyzed mice the existence of antibody-producing cells in numbers comparable to those present in control normal mice has been demonstrated by an *in vitro* rosette technique.

Unresponsiveness cannot be induced if only *part* of the potential immunologically competent cell population in a recipient is exposed to antigen under conditions necessary for tolerance induction. This affords an explanation for the difficulty of inducing tolerance of microbial and viral antigens and heterologous erythrocytes since none of these persist in the circulation nor gain access to the extravascular spaces. In addition, since these antigens are highly complex, presenting many different determinants, tolerance may be induced to some but not all of these specificities. This consideration applies to homologous tissue cells since they bear surface antigens determined by histocompatibility genes at many different loci.

Induction of Tolerance *In Vitro*

It has been shown by Diener and his associates (Diener and Armstrong, 1968; Diener and Feldman, 1971) that if mouse spleen cells dispersed *in vitro* are exposed for 3 to 6 hours to high concentrations of polymerized flagellin antigen, derived from *Salmonella adelaide*, unresponsiveness results, as evidenced by the results of subsequent antigenic challenge of the treated cells *in vitro*, or *in vivo* after transfer of the cells to irradiated syngenic recipients. This tolerance is immunologically specific and fulfills all the criteria of tolerance.

This *in vitro* system has lent itself to the analysis of the mechanism involved in the induction of low-zone tolerance. It has been found that whereas high-zone tolerance to polymerized flagellin is inducible by exposure of the spleen cells to the antigen alone, tolerance induction by otherwise immunogenic concentrations of the same antigen requires the collaboration of specific antibody. Low-zone tolerance to flagellin or to a submolecular fraction of it (fragment "A") requires the presence of extremely small concentrations of specific antibody. Diener and his colleagues have suggested that a critical degree of interlinking of antigen-recognition sites on immunocompetent cells by antigen–antibody complexes may provide the stimulus that renders the cells unresponsive.

Mechanism of Immunological Tolerance

Attempts to elucidate the mechanism of immunological tolerance have been complicated by the wide diversity of the conditions for inducing it and the manner in which it manifests itself with regard to antigens of different classes. Early belief that a plurality of mechanisms was involved is evidenced by attempts to distinguish paralysis, tolerance, protein overloading, and unresponsiveness.

As already indicated, the most generally accepted theory of tolerance is still that it represents a highly specific *deletion* from an animal's repertoire of responses against possible antigenic determinants. This unresponsiveness was attributed to the selective, antigen-induced elimination or failure on the part of the particular clone of antigen-sensitive small lymphocytes that was potentially capable of mounting an immunologic response to the antigenic determinants concerned (Burnet, 1959).

Exactly how this functional inactivity is accomplished has been subject to a great deal of speculation (see De Weck and Frey, 1966; Hraba, 1968; Dresser and Mitchison, 1968). The *stem-cell theory* maintained that mature, immunologically competent cells are constantly being produced from newly differentiated stem cells throughout life. At some stage of their maturation, these differentiating cells respond to an antigen by becoming tolerant rather than immune. Thus, when an adult animal is exposed to antigenic material, the potentially reactive, mature, immunologically competent cells are caused to undergo transformation and mitotic activity that underlies the development of immunity, whereas exposure to antigen causes their functionally immature progenitors to become unresponsive or tolerant rather than immune. Obviously an immune response would mask any tolerance induced, though appropriate tactics of antigen administration could lead progressively to a change in the lymphocyte population of an animal from a sensitized to a tolerant one. The attraction of this theory lay in its capacity to account for such observations as the existence of incomplete tolerance and the dependence of the tolerant state on the persistence of antigen. A variant of this theory maintained that the target of the tolerance-conferring stimulus is the immunologically competent cell itself rather than one of its precursors. This would account for the facility with which very young animals can be rendered tolerant—they have very small numbers of immunologically competent cells.

As we have seen from our consideration of the phenomenon of immunological enhancement, there is now a case for consideration of the possibility that tolerance may be attributable to the production of humoral antibody, which blocks the access or availability of antigen to the specifically reactive antigen-sensitive cells. Several familiar examples of antigen-induced unresponsiveness, hitherto cited as examples of tolerance, have now been shown to be antibody mediated (see Mitchison, 1971a). For example, when newborn mice are inoculated with certain oncogenic viruses, such as Gross leukemia or polyoma, tumors usually develop. However, immunologically mature mice are usually refractory to similar inocula. These findings were formerly attributed to the induction of tolerance toward the virus-dependent, tumor-specific transplantation antigens. However, the recent finding that mice can indeed respond to neonatal inoculation of virus by the production of high titers of humoral antibody sustains the possibility that this antibody is responsible for the success of the neonatally inoculated virus in inducing tumors by enhancement.

So far as transplantation antigens are concerned, a variety of observations make it difficult to accept the premise that enhancement is responsible for neonatally induced tolerance (see Brent, 1971; Brent et al., 1972).

1. Neonatally induced tolerance can be abolished at will by the adoptive transfer of relatively small numbers of lymphocytes from normal syngenic donors or by parabiosis of tolerant animals to normal syngeneic partners. Prior irradiation of the host facilitates the activity of the transferred cells, which it should fail to do if antibodies are involved.
2. Procedures that "enhance" adult animals are ineffective in neonatal subjects.
3. Of all solid tissues, skin homografts are probably the most difficult to enhance; yet challenge with a skin homograft remains the standard test for tolerance where transplantation antigens are concerned.
4. Syrian hamsters, which can be rendered tolerant by neonatal inoculation of viable cells or by repeated inoculation of antigenic extracts in adult life (Zakarian et al., 1972), behave as if they are totally refractory to enhancing procedures. So far, all attempts to demonstrate that hamsters produce humoral antibodies in the course of the rejection of skin homografts have been unsuccessful.
5. Both *in vivo* and *in vitro* tests on lymphytes removed from mice and rats rendered tolerant by neonatal inoculation of homologous cells indicate that these cells retain their unresponsiveness *after* removal from the milieu of the tolerant animal. However, lymphoid cells removed from passively enhanced rats bearing long-standing renal homografts display normal immunological reactivity

against cells having the same alien genetic constitution as the renal homografts.

In the much-studied tolerance of bovine serum albumin (BSA) in the mouse, the results of critical adoptive transfers of genetically compatible normal lymphocytes to tolerant animals also sustain the view that the unresponsiveness is central in nature (see Mitchison, 1971a). More compelling evidence has come from highly sensitive *in vitro* tests, such as the Jerne plaque test, which have failed to establish the presence of antibody-producing cells among the lymphoid cell populations of tolerant animals.

A REEVALUATION OF THE TOLERANCE CONCEPT SINCE THE RECOGNITION OF T AND B LYMPHOCYTES

One of the most important recent discoveries in immunology is that immunologically competent lymphocytes fall into two distinct classes or populations (see Chapter 13): "T" lymphocytes, which are derived from or are "schooled" by the thymus, and "B" lymphocytes, which are derived from the bursa of Fabricius in birds, or its equivalent (probably bone marrow) in mammals. These cells migrate from the thymus and marrow respectively to the various lymphoid tissues of the body. B cells, and probably T cells, have immunoglobulin receptors for specific antigenic determinants on their surfaces, though T lymphocytes probably have fewer of these than their B homologues. Apart from being responsible for the cellular immunities, including the homograft reaction, T lymphocytes play an important and indispensible role in humoral immune responses to some heterologous antigens such as heterologous erythrocytes and human gamma globulin, interacting synergistically with B cells in the spleen and other lymphoid tissues.

Exactly how these two types of cell interact and why this is necessary for the initiation of antibody synthesis by B cells with respect to some antigens only is still open to speculation. Mitchison (1971a) has postulated that T lymphocytes serve as "helpers" that pick up an antigen via one of its many different determinants

(even a "simple" antigen such as a serum albumin may have as many as 30 determinants, each differing from all the others) and present other determinants to B lymphocytes. By presenting antigen to B cells in this way the helper cells constitute a sort of "antigen-focusing" or concentrating mechanism.

Weigle (1971) has suggested that in some immunological reactions macrophages also enter the picture, playing a nonspecific role, possibly by providing a surface on which antigen becomes attached and to which B and T cells subsequently attach in sufficient proximity to enable some kind of essential interaction to occur between them. This results in the proliferation and differentiation of the B cells into antibody-synthesizing plasma cells, which are end cells incapable of antigenic recognition.

It has been established that tolerance can be induced in both types of lymphocyte, though for some antigens there is great disparity in the ease with which this can be accomplished, B lymphocytes seeming to require exposure to much higher concentrations of antigen and for longer periods than T cells (see Mitchison, 1971b, Weigle, 1971). Moreover, unresponsiveness apparently persists longer in T-cell populations than in B-cell populations. However, the induction of tolerance of a particular antigen in one lymphocyte population only will still be reflected in unresponsiveness on the part of the individual.

At present the need for cooperation between B and T cells in transplantation immunity has not been established. However, from the viewpoint of the induction of tolerance of homologous cellular antigens, one can reconcile the occurrence of some antibody production by B cells in animals whose T-cell activity has been abolished by tolerance. As Brent (1971) has recently suggested, tolerance and enhancement may accompany one another, one supporting the other.

A central question that has yet to be answered in connection with tolerance and its spontaneous breakdown, usually as a consequence of the disappearance of "sustaining" antigen, is whether there are such things as "tolerant" cells. Although cell death or inactivation of potentially reactive clones have repeatedly been advocated as the basis of tolerance, decisive evidence is still awaited. One of the strongest arguments sustaining the view that immunological recovery in hitherto tolerant animals occurs through cell replacement from new stem cells is that thymectomy interferes, at least partially, with the disappearance of tolerance to protein antigens in mice. However, since, in mice, thymus-derived cells are not the precursors of antibody-producing cells, though they cooperate with bone marrow derived cells (i.e., B cells) in the humoral antibody response, such observations tell us nothing about the source of antibody-synthesizing cells in animals that lose their tolerant status.

The first among a variety of findings that pointed to the possible existence of reversibly inactivated or "tolerant" cells was that of McGregor et al. (1968) that thoracic duct lymphocytes (mainly T cells) from rats tolerant of sheep erythrocytes failed to transfer a tolerant state to irradiated hosts after overnight incubation *in vitro*.

Role of Tolerance in Development

Burnet and Fenner (1949) were aware of the need to explain why animals do not normally react immunologically against some of the highly specialized ingredients of their own bodies—that is, why autoantibodies are not formed. This important philosophical problem was appreciated by Ehrlich (Ehrlich and Morgenroth, 1900) at the beginning of this century, when he formulated his law of "horror autotoxicus," according to which autoantibodies are incompatible with normal life.

Until recently it was generally believed that all the cells of the body, including immunologically competent cells, express from an early stage of development all the transplantation antigens corresponding to an animal's genetic constitution, precluding "at source" as it were reactivity on the part of the animal against any of its own transplantation antigens; in other words, the body was believed to be genetically tolerant of its own transplantation antigens. According to this premise, the only conceivable way in which an animal might be attacked by its own immunocompetent cells is if,

through somatic mutation, antigen-sensitive cells lost the capacity to synthesize a particular transplantation antigen, or if through a similar process, or through viral infection, other cells of the body suddenly developed a new cellular antigen.

Lance and his associates (1971) have recently presented evidence that constitutes a very serious challenge to the concept that *all* cells of the body express exactly the same complement of transplantation antigens in the same way, and raises the possibility that becoming self-tolerant of certain tissue-specific, "differentiation" alloantigens may be an important developmental process. They found that adult C57BL/6 mice, lethally irradiated and rehabilitated by means of a transfusion of spleen or marrow cells from (C57BL/6 × A)F$_1$ hybrid donors, rapidly rejected skin grafts from A-strain or F$_1$ hybrid donors transplanted 11 weeks later, despite the fact that the host's lymphohematopoietic tissue system was ostensibly 100 percent chimeric with F$_1$ cells (see Fig. 3). On the basis of these and other findings these investigators postulate that rejection of these grafts is due to at least one A-strain, skin-specific alloantigen in respect of which the F$_1$ lymphoid cells lose their tolerance after removal from their native F$_1$ milieu and transfer into the irradiated C57BL/6 hosts, whose skin lacks the postulated A-strain skin-specific alloantigen. Evidence for the existence of a skin alloantigen was also forthcoming from experiments of similar design carried out with other strain combinations, some of which rules out the possibility that the histocompatibility locus involved is closely linked to the major H-2 locus of the mouse.

The existence of skin-specific alloantigens would certainly provide an explanation for the rejection of skin grafts in other animals tolerant and chimeric with respect to lymphoid cells from the same donor strain, as well as for the phenomenon of "split tolerance" (see page 276). It might also account in part for the peculiar susceptibility of skin homografts to transplantation immunity and for the vulnerability of the intact skin of animals undergoing gvh or homologous disease (Billingham, 1968; Streilein and Billingham, 1970). The possible existence of skin-specific "differentiation" antigens—phenotypic results of some kind of selective gene action—raises a variety of interesting questions, notably whether other tissues of the body may also express specific histocompatibility alloantigens.

The differentiation of many cells and tissues of the body is unequivocally associated with the appearance of specialized substances or ingredients that are obviously not shared in common with antigen-sensitive lymphocytes and that might therefore appear foreign or antigenic to the latter. The most plausible reason why an individual fails to react against such substances is that his lymphoid cells are exposed to them at an early stage in development and so become tolerant of them. Persistence of

the materials in his body maintains the tolerant state in being.

By far the most convincing evidence that the body's own components can indeed induce autotolerance comes from Triplett's (1962) ingenious experiments on tree frogs. He extirpated the hypophyses of embryos at an early stage of development, before the gland had differentiated and begun to produce adult-type proteins. These organ rudiments were then implanted temporarily beneath the skin in other immature animals (although genetically compatible hosts were not available, the immaturity of these hosts ensured that the "stored" organs would not be rejected). When the individuals

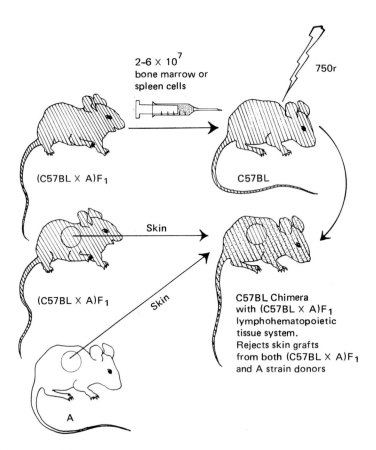

FIGURE 3. *Design of the important experiment carried out by Lance et al. (1971), which suggests that becoming self-tolerant of certain tissue-specific "differentiation" alloantigens may be an important development process. The lymphohematopoietic tissue system of C57BL mice was replaced by one from a (C57BL × A)F₁ hybrid donor by exposure to a high dose of X-irradiation followed by a transfusion of splenic or marrow cells. Such animals were found to be capable of rapidly rejecting skin grafts from either A strain or F₁ hybrid donors (see the text).*

whose organs had been extirpated had metamorphosed and attained immunological maturity, their *own* explanted hypophyses, which by this time had attained antigenic maturity, were returned to them. Most of these grafts underwent rejection. To exclude the possibility that the antigenic specificity of the "*in vivo* stored" glands might have undergone some change as a result of incorporation of host-type cells, for example, control experiments were performed in which only part of the hypophysis was removed, maintained in a host embryo, and then returned to its original donor. None of these returned grafts were rejected. So far no other direct experiments of this type have been reported.

It follows from the arguments pursued in this section that any specialized tissue or cell constituent that is not produced until relatively late in life, when tolerance induction becomes more difficult, should be at least potentially autoantigenic (see Weigle, 1971). Spermatozoa and the proteins of milk appear to belong to this category. Likewise, specialized tissue constituents that develop and remain in anatomically sequestered sites, so that the animal's antigen-sensitive cells do not get exposed to them, should also be potential autoantigens. The lens of the eye develops at a fairly early stage of development, but, being encapsulated and having no vascular supply, it is physiologically isolated from the reticuloendothelial system. It is scarcely surprising, therefore, that some of its specialized proteins are potential autoantigens. The protein thyroglobulin also appears to belong to this category. It develops in what appears to be a physiological quarantine—the follicles of the thyroid gland—from which it does not normally escape in significant amounts into the circulation. Consequently, this substance remains potentially autoantigenic as attested by the work of Ernest Witebsky and his associates (see Rose and Witebsky, 1971). They established that when thyroglobulin is administered to experimental animals under appropriate conditions, humoral antibodies appear in the blood stream, there is an infiltration of the gland with cells believed to be capable of forming antibodies, and progressive thyroiditis and impairment of glandular function occur.

Another anatomically sequestered tissue with which potentially autoantigenic material is associated is that comprising the brain, which has no lymphatic drainage or regional lymph nodes. It has long been known that various species of laboratory mammals develop allergic encephalomyelitis, with characteristic lesions that are confined to the central nervous system, a week or two after inoculation with an appropriate preparation of nervous tissue, usually guinea pig spinal cord homogenized in Freund's complete adjuvant (see Patterson, 1971). If newborn rats are injected with homogenized homologous spinal cord, their capacity to develop allergic encephalomyelitis, following reexposure to this antigen a month or two later, is markedly decreased. Various lines of evidence suggest that the protective influence of the neonatal injection turns upon its ability to induce tolerance of the distinctive nervous tissue antigens responsible for the paralytogenic response (see Kornblum, 1968).

Weigle (1971) has recently presented a cogent argument, with sustaining evidence, that the occurrence of certain autoimmune diseases, including thyroiditis in man, may be restricted to situations in which affected individuals have been living with only one type of lymphocyte (T cells) tolerant, so that, so far as a humoral antibody response is concerned, they remain unresponsive. As a result of some kind of alteration in the antigen concerned, tolerance at the level of the T cells is "broken," allowing T and B cells to respond to antigen and interact, leading to antibody formation.

References

ANDERSON, D., BILLINGHAM, R. E., LAMPKIN, G. H., AND MEDAWAR, P. B. (1951). The use of skin grafting to distinguish between monozygotic and dizygotic twins in cattle. *Heredity 5*, 379-397.

BILLINGHAM, R. E. (1968). The biology of graft-versus-host reactions. *In* "The Harvey Lectures," Series 62, pp. 21-78. Academic Press, New York.

BILLINGHAM, R. E., AND BRENT, L. (1959). Quantitative studies on tissue transplantation immunity. IV. Induction of tolerance in newborn mice and studies on the phenomenon of runt disease. *Phil. Trans. Roy. Soc. Lond. B 242*, 439-477.

BILLINGHAM, R. E., BRENT, L., AND MEDAWAR, P. B. (1953). Actively acquired tolerance of foreign cells. *Nature (London) 172*, 603-606.

BILLINGHAM, R. E., BRENT, L., AND MEDAWAR, P. B. (1956). Quantitative studies on tissue transplantation immunity. III. Actively acquired tolerance. *Phil. Trans. Roy. Soc. Lond. B 239,* 357–414.

BILLINGHAM, R. E., LAMPKIN, G. H., MEDAWAR, P. B., AND WILLIAMS, H. L. (1952). Tolerance to homografts, twin diagnosis, and the freemartin condition in cattle. *Heredity 6,* 201–212.

BILLINGHAM, R. E., AND SILVERS, W. K. (1962). Some factors that determine the ability of cellular inocula to induce tolerance of tissue homografts. *J. Cell. Comp. Physiol. 60,* Suppl. 1, 183–200.

BILLINGHAM, R. E., AND SILVERS, W. K. (1971). "The Immunobiology of Transplantation," pp. 207. Prentice-Hall, Englewood Cliffs, New Jersey.

BILLINGHAM, R. E., SILVERS, W. K., AND WILSON, D. B. (1963). Further studies on adoptive transfer of sensitivity to skin homografts. *J. Exp. Med. 118,* 397–420.

BRENT, L. (1971). Immunological tolerance 1951–71. *In* "Immunological Tolerance to Tissue Antigens" (N. W. Nisbet and M. W. Elves, eds.), pp. 49–66. Orthopaedic Hospital, Oswestry, England.

BRENT, L., BROOKS, C., LUBLING, N., AND THOMAS, A. V. (1972). Attempts to demonstrate an *in vivo* role for serum blocking factors. *Transplantation 14,* 382–387.

BRENT, L., AND COURTENAY, T. H. (1962). On the induction of split tolerance. *In* "Mechanisms of Immunological Tolerance" (M. Hašek, A. Lengerová, and M. Vojtíšková, eds.), pp. 113–121. Czechoslovak Academy of Sciences, Prague.

BRENT, L., AND GOWLAND, G. (1961). Cellular dose and age of host in the induction of tolerance. *Nature (London) 192,* 1265–1267.

BRENT, L., AND GOWLAND, G. (1962). Induction of tolerance of skin homografts in immunologically competent mice. *Nature (London) 196,* 1298–1301.

BRENT, L., HANSEN, J. A., AND KILSHAW, P. J. (1971). Unresponsiveness to skin allografts induced by tissue extracts and antilymphocytic serum. *Transpl. Proc. 3,* 684–687.

BURNET, F. M. (1959). "The Clonal Selection Theory of Acquired Immunity." Cambridge University Press, New York.

BURNET, F. M., AND FENNER, F. (1949). "The Production of Antibodies," pp. 142. Macmillan, Melbourne.

DE WECK, A. L., AND FREY, J. R. (1966). "Immunotolerance to Simple Chemicals." Monographs in Allergy Vol. 1. American Elsevier, New York.

DIENER, E., AND ARMSTRONG, W. D. (1968). Immunological tolerance *in vitro:* kinetic studies at the cellular level. *J. Exp. Med. 129,* 591–603.

DIENER, E., AND FELDMAN, M. (1971). Induction of low-zone tolerance *in vitro. Transpl. Proc. 3,* 663–665.

DRESSER, D. W., AND MITCHISON, N. A. (1968). The mechanism of immunologic paralysis. *Advan. Immunol. 8,* 129–181.

EHRLICH, P., AND MORGENROTH, J. (1900). *Ber. Klin. Wochschr. 21,* 453–462.

ELKINS, W. L. (1971). Cellular immunology and the pathogenesis of graft-versus-host reactions. *Progr. Allergy 15,* 78–187.

FLEXNER, S., AND JOBLING, J. W. (1907). On the promoting influence of heated tumor emulsions on tumor growth. *Proc. Soc. Exp. Biol. Med. 4,* 156–157.

FRENCH, M. E., AND BATCHELOR, J. R. (1969). Immunologic enhancement of rat kidney grafts. *Lancet ii,* 1103–1106.

GLENNY, A. T., AND HOPKINS, B. E. (1924). Duration of passive immunity. *J. Hygiene 22,* 208–221.

GUTTMANN, R. D., AND AUST, J. B. (1961). Acquired tolerance to homografts produced by homologous spleen cell injection in adult mice. *Nature (London) 192,* 564–565.

HAŠEK, M. (1953). Vegetative hybridization of animals by joining their blood circulations during embryonic development. *Cs. Biol. 2,* 265–280.

HAŠEK, M., LENEGEROVÁ, A., AND HRABA, T. (1961). Transplantation immunity and tolerance. *Advan. Immunol. 1,* 1–66.

HELLSTRÖM, I., HELLSTRÖM, K. E., AND ALLISON, A. C. (1971). Neonatally induced allograft tolerance may be mediated by serum-borne factors. *Nature (London) 230,* 49–50.

HELLSTRÖM, I., HELLSTRÖM, K. E., STORB, R., AND THOMAS, E. D. (1970). Colony inhibition of fibroblasts from chimeric dogs mediated by the dogs' own lymphocytes and specifically abbrogated by their serum. *Proc. Nat. Acad. Sci. U.S.A. 66,* 65–71.

HOWARD, J. G., CHRISTIE, G. H., AND COURTENAY, B. M. (1971). Studies on immunological paralysis. IV. The relative contribution of continuous antibody neutralization and central inhibition to paralysis with type III pneumococcal polysaccharide. *Proc. Roy. Soc. Lond. B 178,* 417–438.

HRABA, T. (1968). "Mechanism and Role of Immunological Tolerance," pp. 136. Karger, New York.

KORNBLUM, J. (1968). The application of the irradiated hamster test to the study of experimental allergic encephalomyelitis. *J. Immunol. 101,* 702–710.

LANCE, E. M. (1971). The induction of transplantation tolerance within and across species barriers. *In* "Immunological Tolerance to Tissue Antigens" (N. W. Nisbet and N. W. Elves, eds.), pp. 101–109. Orthopaedic Hospital, Oswestry, England.

LANCE, E. M., BOYSE, E. A., COOPER, S., AND CARSWELL, E. A. (1971). Rejection of skin allografts by irradiation chimeras: evidence for skin-specific transplantation barrier. *Transpl. Proc. 3,* 864–868.

LANCE, E. M., AND MEDAWAR, P. B. (1969). Quantitative studies on tissue transplantation immunity. IX. Induction of tolerance with antilymphocyte serum. *Proc. Roy. Soc. Lond. B 173,* 447–473.

LANCE, E. M., AND MEDAWAR, P. B. (1971). Antilymphocytic serum: its properties and potential. *In* "Immunobiology" (R. A. Good and D. W. Fisher, eds.), pp. 248–256. Sinauer Associates, Stamford, Connecticut.

LILLIE, F. R. (1916). The theory of the freemartin. *Science 43,* 611–613.

LUBAROFF, D. M., AND SILVERS, W. K. (1970). The abolition of tolerance of skin homografts in rats with isoantiserum. *J. Immunol. 104,* 1236–1241.

MARTINEZ, C., SHAPIRO, F., KELMAN, H., ONSTAD, T., AND GOOD, R. A. (1960). Tolerance of F_1 hybrid skin homografts in the parent strain induced by parabiosis. *Proc. Soc. Exp. Biol. Med. 103,* 266–269.

MCGREGOR, D. D., MCCULLAGH, P. J., AND GOWANS, J. L. (1967). The role of lymphocytes in antibody formation. I. Restoration of the haemolysis response in X-irradiated rats with lymphocytes from normal and immunologically tolerant donors. *Proc. Roy. Soc. Lond. B 168,* 229–243.

MEDAWAR, P. B. (1963). The use of antigenic tissue extracts to weaken the immunological reaction against skin homografts in mice. *Transplantation 1,* 21–38.

MEDAWAR, P. B. (1969). Immunosuppressive agents, with special reference to antilymphocyte serum. *Proc. Roy. Soc. Lond. B 174,* 155–172.

MINTZ, B. (1962). Formation of genotypically mosaic mouse embryos. *Amer. Zool. 2,* 432.

MINTZ, B. (1965). Genetic mosaicism in adult mice of quadriparental lineage. *Science 148,* 1232–1233.

MINTZ, B., AND SILVERS, W. K. (1967). "Intrinsic" immunological tolerance in allophenic mice. *Science 158,* 1484–1487.

MITCHISON, N. A. (1964). Induction of immuno-

logical paralysis in two zones of dosage. *Proc. Roy. Soc. B 161,* 275–292.

MITCHISON, N. A. (1971a). Perspectives of immunological tolerance in transplantation. *Transpl. Proc. 3,* 953–959.

MITCHISON, N. A. (1971b). Tolerance in T and B lymphocytes: evidence from hapten-specific tolerance. *In* "Immunological Tolerance to Tissue Antigens" (N. W. Nisbet and M. W. Elves, eds.), pp. 67–74. Orthopaedic Hospital, Oswestry, England.

OWEN, R. D. (1945). Immunogenetic consequences of vascular anastomoses between bovine twins. *Science 102,* 400–401.

PATTERSON, P. Y. (1971). The demyelinating diseases: clinical and experimental correlates. *In* "Immunological Diseases" (M. Samter, ed.), Vol. 2, pp. 1269–1299. Little, Brown, Boston.

ROSE, N. R., AND WITEBSKY, E. (1971). Experimental thyroiditis. *In* "Immunological Diseases" (M. Samter, ed.), Vol. 2, pp. 1179–1197. Little, Brown, Boston.

SHAPIRO, F., MARTINEZ, C., SMITH, J. M., AND GOOD, R. A. (1961). Tolerance of skin homografts induced in adult mice by multiple injections of homologous spleen cells. *Proc. Soc. Exp. Biol. Med. 106,* 472–475.

SILVERS, W. K., AND BILLINGHAM, R. E. (1969). Influence of the Ag-B locus on reactivity to skin homografts and tolerance responsiveness in rats. *Transplantation 8,* 167–178.

SILVERS, W. K., AND BILLINGHAM, R. E. (1970). Contributions of the rat to the immunobiology of tissue transplantation. *Transpl. Proc. 2,* 152–161.

SILVERS, W. K., LUBAROFF, D. M., WILSON, D. B., AND FOX, D. (1970). Mixed lymphocyte reactions and tissue transplantation tolerance. *Science 167,* 1264–1266.

SNELL, G. D. (1970). Immunological enhancement. *Surg. Gynecol. Obstet. 130,* 1109–1119.

SOLOMON, J. B. (1971). "Fetal and Neonatal Immunology," pp. 381. North-Holland, Amsterdam.

STONE, W. H., CRAGLE, R. G., SWANSON, E. W., AND BROWN, D. G. (1965). Skin grafts: delayed rejection between pairs of cattle twins showing erythrocyte chimerism. *Science 148,* 1335–1336.

STREILEIN, J. W., AND BILLINGHAM, R. E. (1970). An analysis of graft-versus-host disease in Syrian hamsters. I. The epidermolytic syndrome: description and studies on its procurement. *J. Exp. Med. 132,* 163–180.

STUART, F. P., SAITOH, T., FITCH, F. W., AND SPARGO,

B. (1968). Immunologic enhancement of renal allografts in the rat. *Surgery 64*, 17–24.

Triplett, E. L. (1962). On the mechanism of immunologic self-recognition. *J. Immunol. 89*, 505–510.

Voisin, G. A. (1971). Immunological facilitation, a broadening of the concept of the enhancement phenomenon. *Progr. Allergy 15*, 328–485.

Voisin, G. A., Kinsky, R. G., and Maillard, J. (1968). Démonstration d'une réactivité immunitaire spécifique chez des animaux tolérants aux homogreffes. Rôle possible dans le maintien de la tolérance. *In* "Advance in Transplantation" (J. Dausset, J. Hamburger, and G. Mathé, eds.), pp. 31–40. Munksgaard, Copenhagen.

Warner, N. L., Herzenberg, L. A., Cole, L. J., and Davis, W. E. (1965). Dissociation of skin homograft tolerance and donor type gammaglobulin synthesis in allogeneic mouse radiation chimaeras. *Nature (London) 205*, 1077–1079.

Weigle, W. O. (1971). Immunologic unresponsiveness. *In* "Immunobiology" (R. A. Good and D. W. Fisher, eds.), pp. 123–134. Sinauer Associates, Stamford, Connecticut.

Wigzell, H. (1962). Studies of prolonged survival of skin homografts in adult mice. *In* "Mechanisms of Immunological Tolerance" (M. Hašek, A. Lengerová, and M. Vojtíšková, eds.), pp. 267–272. Czechoslovak Academy of Sciences, Prague.

Wilson, D. B., and Billingham, R. E. (1967). Lymphocytes and transplantation immunity. *Advan. Immunol. 7*, 189–273.

Winn, H. J. (1970). Humoral antibody in allograft reactions. *Transpl. Proc. 2*, 83–103.

Zakarian, S., Streilein, J. W., and Billingham, R. E. (1972). Studies on transplantation antigen extracts in Syrian hamsters. *Proc. Roy. Soc. Lond. B 180*, 1–20.

IV

THE ORGANISM

CHAPTER FIFTEEN

Developmental Enzymology

IVAN T. OLIVER

The development of any higher organism from zygote to maturity can be characterized in chemical terms by the acquisition of new biochemical functions and the loss of some old ones.

Since enzymes are necessary for the chemical reactions of living tissue, the biochemical development of higher organisms can be elucidated at a fundamental level by a study of the development of enzymes and the mechanisms controlling such processes. Biochemical development cannot be described only in terms of enzymes, however. Protein hormones, structural and functional nonenzymatic proteins, are also an integral part of the picture; but since all enzymes are proteins, general principles that emerge may apply to development characterized by changes in nonenzymatic species of proteins. This chapter, although restricted to discussion of developmental enzymology, will attempt to present a picture of fundamental principles that may apply to the production of any specific protein.

A sophisticated understanding of developmental enzymology, apart from its intrinsic scientific value and contribution to an inte-grated view of developmental biology, should also be applicable in the diagnosis and therapy of human and animal diseases in the fields of teratogenesis and neonatal disorders.

Since the aim of this chapter is to emphasize general principles, the discussion is almost totally devoted to animal systems. The unity of biology appears to be a valid philosophy, and it is unlikely that fundamentally different principles apply to the developmental chemistry of plants and other organisms.

Biochemical development is a multiphase process in which new functions, and the enzymes that control them, are acquired by different precursor cells and tissues at various times during embryogenesis, organogenesis, perinatal life, and puberty. During interphasic periods, the quantitative levels of enzymes may show adaptive responses to alterations in the environment; however, a definition of development is adopted in this survey that restricts discussion of such enzyme modulation.

Cellular differentiation is an integral part of development; in fact, the terms themselves are nearly synonymous. A clear marker of cellular differentiation is the acquisition of a characteristic protein. Some studies using such markers have revealed a phased sequence in differentiation that may be of general importance.

Since developmental biology is currently in an intensive experimental phase, methodological problems and experimental philosophies are continually in need of critical appraisal and reappraisal. For this reason, some time is devoted to a general discussion of methodology.

When cells differentiate, they begin to synthesize previously undetectable enzymes in response to signals. Such signals can be called *inducers,* since they induce synthesis of a particular enzyme. However, recognition of physiological signals is a formidable task for the experimentalist.

Several other problems also arise in the study of developmental enzymology. Tissues are rarely composed of a single cell type, and the determination of the cellular location of particular enzymes may be important. Enzymes that are composed of subunits often exist in polymerized or hybrid forms. The synthesis of a new subunit against a background of active enzyme may be impossible to detect without isoenzyme analysis.

The role of the genome in development is of prime importance since the differential expression of genes is the underlying cause of cell differentiation. DNA replication may be a prerequisite to specific gene expression and the amplification of particular genes may precede the production of large amounts of the gene product.

Some of these aspects are illustrated by reference to particular enzyme inductive systems, but an encyclopedic account of enzyme development has not been attempted. Rather, the few examples discussed in detail have been chosen to illustrate principles and focus attention on problems of experiment and interpretation.

Molecular Biology of Gene Expression

It is now clear that cellular differentiation, and the specialized cell function that arises from it, is the result of differential gene expression. The elegant experiments of Gurdon and his associates based on earlier work of Briggs and King (see Gurdon, 1968) have shown that the nucleus of differentiated cells retains all the information necessary to specify the complete development of a eukaryotic organism from ovum to sexually mature adult. Cell nuclei from differentiated epithelial cells of the alimentary canal of the *Xenopus laevis* tadpole were transplanted by micropipet into enucleated eggs of the same species. Penetration of the egg membrane sets off cleavage and the embryo develops through the tadpole stage and metamorphosis to the adult, sexually mature frog. Although successful nuclear transplants are infrequent, further experiments have provided adequate explanation for this fact, and it seems clear that no essential information for development is lost or destroyed during cellular differentiation.

Hence the problem of development becomes one of differential expression of genes to produce products that distinguish one cell from another. In other words, in the transition to a more differentiated state, genes that were previously inactive are in some way activated, and some active genes may be deactivated. Part of the later discussion in this chapter will deal with the nature of the signals involved in the control of gene expression; it is thus unnecessary to elaborate on them here.

Knowledge of the essential features of gene expression involving the transcription of DNA to messenger RNA (mRNA) and the subsequent translation of mRNA into protein will be assumed.

Reference will be made to systems of gene control in bacteria. The essential features of the model of the *lac operon* in *Escherichia coli*, put forward by Jacob and Monod, are simple and pertinent to our discussion. A regulator gene codes for a product called the *repressor*. The repressor is continuously produced (constitutive gene expression) and is a protein. The repressor binds to a specific region in the bacterial DNA called the *operator*. The operator is linked to one or more structural genes that specify the amino acid sequences of their enzymatic products, and is spatially located between the first structural gene and a promotor region that is an attachment site for

RNA polymerase. The binding of the repressor to the operator prevents transcription of the structural genes, and the particular enzymes are absent from the cell. The *inducer*, or an intracellular metabolic product (called the *effector*) binds to the repressor and inactivates it by a conformational change. In the inactive state there is no affinity for the operator in the DNA, and RNA polymerase transcribes the structural genes to mRNA, which is then translated to proteins. Since the repressor is constitutively produced, once the inducer is removed, the active repressor concentration increases. The binding of the repressor to the operator terminates transcription and hence stops protein synthesis. Thus, in the presence of an inducer, the enzymes are synthesized and accumulate in the cells. When the inducer is removed, then enzymes gradually disappear from the cells. Other variations of the model are possible in which an aporepressor is produced that is activated by an effector molecule, thus terminating enzyme synthesis.

Eukaryotic systems for the control of gene expression may well be more complex than the above system, and some possible mechanisms are discussed in subsequent sections.

Definitions: Development Versus Modulation

At various times during the growth and development of higher organisms specific enzymes that were not previously detectable appear in specific tissues. At other times, or even coincidentally, the activities of preexisting enzymes may show variation. In the case of the initial appearance of an enzyme a number of examples show that the majority of such events are due to the *de novo* induction of specific enzyme synthesis as a result of the initiation of either the translation of preexisting mRNA or the initiation of specific gene transcription. There is little doubt that within the concepts of developmental biology such a process should be classed as a developmental change. The disappearance of a specific enzyme should also be classed as a developmental event, but little is

known of the mechanisms that control such processes. However, in the situation in which a variation in the amount of preexisting enzyme molecules occurs with age or other changed condition, there is some doubt about classification of this as a developmental event. Development is equated with cellular differentiation, and the most useful biochemical definition of cellular differentiation is "the production of a new cell specific protein." The second case cannot meet these criteria. In addition, the modulation devices that cells use to vary the amounts of particular proteins once synthesis has been initiated may not be related at all closely to the mechanisms of initial induction of synthesis of the same protein. To name but one example, the level of a specific enzyme in adult tissue could be modulated by variation solely in the rate of degradation without effects on the rate of synthesis, but it is unlikely in the extreme that initial production of specific enzyme will occur without an alteration in the rate of synthesis. The factors that are involved in controlling the rate of degradation, although largely unknown, are unlikely to be identical with those inducing synthesis. To avoid confusion, it is desirable to separate discussion of these two aspects of the control of enzyme levels, and in this chapter such a stratagem will be used.

Phases in Cellular Differentiation or Development

As a result of studies of development in the fetal pancreas, Wessells and Rutter (1969), have suggested a sequence of events in cellular differentiation that may be of general application to all developing systems. They define three stages in differentiation relating to the acquisition of cell specific proteins (see Fig. 1):

1. The predifferentiated state in which effectively none of the cell specific protein is present or, at the least, is below the limits of detection.
2. The protodifferentiated state in which some small amount of the cell specific protein is present.

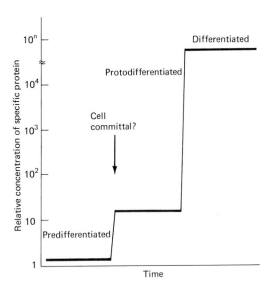

y-axis: Relative concentration of specific protein
10^n
10^4
10^3
10^2
10
1

Differentiated

Protodifferentiated

Cell committal?

Predifferentiated

Time

FIGURE I. *Theoretical phases of cellular differentiation. The time scale is unspecified since it will differ for different developmental events. The concentration scale is arbitrary but is calibrated in orders of magnitude.*

3. The differentiated state, the transition to which is characterized by a very large increase in the amount of cell specific protein. In this transition the quantitative increase is of *several orders of magnitude*.

In the differentiated state, modulation of the level of cell specific protein occurs, but the variation in amount is usually *not* of orders of magnitude.

In many cases, studies in development may detect only the final transition to the differentiated state since the amount of protein present in the protodifferentiated state may be below the limits of detection with current assay procedures. It is obviously desirable, then, to sensitize specific assays so that the generality or otherwise of the suggested pattern of differentiation may be revealed. The major experimental difficulty is to detect the transition from predifferentiated to protodifferentiated state, and only the most sensitive and specific assays will allow this detection. Cells that have progressed to the protodifferentiated state may be irreversibly committed to a specific line of

development (see later for discussion of cell committal).

Study of Developmental Enzymology

The first prerequisite to the study of developmental enzymology is to discover the developmental change. Despite this truism, much of the early work compared enzyme levels, for example, in newborn animals with levels in adults, near adults, or weanling animals. It soon became apparent that the more exciting developmental changes occurred relatively rapidly and that careful study at daily or even hourly intervals was necessary to accurately define the time at which such changes occur. This kind of study is of considerable importance because knowledge of the time scale of the event often gives a clue to the next level of analysis.

After completion of the essential mapping of the developmental event in time, the second problem is the identification of the physiological signal or signals. In this phase, accurate knowledge of the temporal scale of the event assumes importance since the timing may then be correlated with a known physiological change. Such physiological changes would include postnatal hypoglycemia in the placental mammal, a known hormonal flux, a condition such as hypoxia, and so on. Experiments may then be more rationally designed to identify the physiological signal.

The third problem is that of defining the molecular mechanism of the event. Analysis of the response to administration of inhibitors of DNA, RNA, and protein synthesis helps to distinguish between DNA transcription, mRNA translation, or enzyme activation as the fundamental initiation event that leads to cellular acquisition of biologically active enzyme. In addition, the involvement of intracellular effectors and their mechanisms of action have to be considered. Identification of such factors should lead to the discovery of DNA or RNA repressors, gene activators, or the mechanisms of activation of proenzymes or apoenzymes.

The three phases can be summarized as temporal mapping, physiological signal identification, and effector identification and mechanism.

General Methodology and Problems

CRITIQUE OF MODEL SYSTEMS

It was Joshua Lederberg, the Nobel Prize-winning geneticist, who remarked that the place to study embryology is in the embryo. The pragmatism of this remark can equally well be applied in the field of developmental enzymology. The place (or system) in which to study this process is in the developing intact organism. While it is true that studies of enzyme induction or modulation in cell cultures, organ explants in culture, microorganisms, or similar systems have often yielded valuable clues as to the nature of inductive events, it should not be forgotten that such systems are rather pale and incomplete shadows of the cellular complexity of a developing tissue or early embryo. They are simplified model systems that are easier to control, in the experimental sense, than are intact organisms, and they often allow the detection of errors of interpretation from results obtained on intact organisms. Nevertheless, cell and organ cultures often differ in significant properties from the tissues used to obtain the culture line and thus remain model systems. Some important differences lie, for instance, in the possession by cultured cells of "bizarre" chromosome numbers and by the occurrence of unexplained "refractory" periods in organ explants. There is, nevertheless, a distressing tendency for some investigators to allow the model to become the reality and the sole object of the inquiry, whereas in fact hypotheses derived from experimental models may be found wanting when they are tested in intact organisms. This is not to underemphasize the methodological and interpretative difficulties of working with organized tissues *in situ*, but theories derived from model systems must be tested in the physiological situation.

The advantages of the use of cells or tissue explants in culture and perfused organs in isolation is that conditions can be more rigorously controlled and the inductive effects of suspected agents can be tested for direct effects on cells or tissues. Consider the following fanciful example: a newly discovered hormone, called Ligondan PX_{22} by its discoverer, was injected into the footpads of the New Zealand white rabbit and resulted in the elevation of the activity of hexokinase type I in the kidney of the animal. Such a result does not exclude the possibility that Ligondan PX_{22} was metabolized to some derivative or even a dissimilar substance, or that the injection of the hormone resulted in the secretion of some other hormone that was the physiologically effective agent in the kidney. However, the elevation of enzyme activity after culture of rabbit kidney explants in the presence of Ligondon PX_{22} would at least identity the hormone as the true hormonal agent effecting the change. Such experiments can assist in the identification of *direct* tissue effects of suspected inductive or modulative agents.

ENZYME SYNTHESIS OR ENZYME ACTIVATION?

The accumulation of active enzyme in a tissue that previously had none can result from a variety of processes—including activation of preexisting molecules, *de novo* synthesis of the enzyme, or a sharp reduction in the enzyme's rate of degradation. The control may occur at one or more of several different loci, and the following summary attempts to present a variety of such loci that may be control points in any developmental event. The list is not intended to be exhaustive of all potential mechanisms.

1. Initiation of specific gene transcription to mRNA.
2. Release of specific mRNA from the DNA-RNA complex.
3. Transport of mRNA from the cell nucleus.
4. Initiation of translation of specific mRNA.
5. Release of the finished enzyme or its subunits from the polysome.
6. Activation of preexisting enzyme mole-

cules by proteolysis, phosphorylation, or some other process, such as the addition of a prosthetic group to an apoenzyme.

7. Reduction in the rate of degradation of the enzyme.

These possibilities have been separated out as monovalent control points, but some developmental events may have multiple control points, and multiple factors may be necessary to induce the accumulation of active enzyme.

The resolution of some of these various possibilities can be made by the use of certain inhibitors, even in the intact organism. Actinomycin D, which inhibits DNA-directed RNA synthesis, will prevent the accumulation of enzyme activity resulting from gene transcription and subsequent mRNA translation. When gene transcription is the initiating event, inhibitors of protein synthesis such as puromycin and cycloheximide will also block enzyme production, since protein synthesis is involved in the accumulation of the final gene product.

In systems where the control lies at the translational level (4 and above), actinomycin D is without effect, since specific mRNA is already transcribed, while inhibitors of protein synthesis blockade the ribosomal translation of mRNA into enzyme. There are some special characteristics of systems controlled at the release step of protein synthesis that require comment. In systems of type 4, inhibitors of protein synthesis at the correct concentration or dosage will prevent completely the accumulation of enzyme activity, since the control lies at initiation of translation and no protein synthesis will be achieved in presence of inhibitors. On the other hand, in systems of type 5, at least one ribosome on each mRNA molecule is loaded with the enzyme polypeptide, the synthesis of which is complete to the C-terminal amino acid. In the presence of inhibitors of protein synthesis, this polypeptide will still be released from the polysome, although the completion of polypeptide chains on the other ribosomes of the enzyme-specific polysome will be prevented. Thus the inhibitors will only partially prevent the accumulation of enzyme activity. It is also possible that an anomalous effect of puromycin might be obtained *in vivo*, since puromycin acts by the abortive release of polypeptides from ribo-

somes engaged in translation. Thus in a system consisting of polysomes preloaded with nascent enzyme chains and awaiting the release signal, the use of puromycin could result in elevation or appearance of enzyme activity, since the polypeptide complete to the C-terminal amino acid might have activity after its release, despite the covalent attachment of puromycin to the carboxy terminal.

Little is known of the mechanism of mRNA transport from the cell nucleus (type 3), but recent evidence suggests that mRNA has polyadenylic acid attached to the 3' end of the molecule before its extranuclear transport, and this process might also be a biologically advantageous control point.

The activation of inactive forms of enzymes and hormones is known to take place in cells, and where this is an enzymatic process, it is obvious that the first appearance of a particular enzyme activity might be due to the development of the activating enzyme. If this is so, then the systems should show the inhibition characteristics of control types 1, 2, 4, or 5, but if the usual inhibitors are without effects, then the accumulation of active enzyme probably results from activation of apoenzyme or proenzyme by a process that does not require the *de novo* production of any new protein.

The last system described (type 7) is a difficult one to investigate but is included for reasons of logic. This notion assumes that prior to the developmental event, specific enzyme synthesis is in progress but that the rate of degradation is so high that no active enzyme survives. The developmental event, then, consists of a sharp reduction in the rate of degradation, so that the steady state level of active enzyme rises to a detectable level. Since little is known of the mechanisms of enzyme degradation in tissues (see Schimke and Doyle, 1970, for review) it it rather profitless to discuss this potential control further.

All experiments using inhibitors of RNA and protein synthesis are open to criticism. All the commonly used agents are cytotoxic and have ill-defined side effects other than their major known reactions with cellular processes. A superior methodology for detecting *de novo* protein synthesis is to study the incorporation of labeled amino acids into the specific enzyme. This technique involves the isolation of the enzyme in homogeneous purified form in order to obtain an antibody to it. Immunological procedures can then be used to isolate the labeled enzyme protein from relatively crude cell extracts. In combination with the results of inhibitor experiments, the data obtained with the latter technique can yield definitive information about the characteristics of the developmental control mechanism.

PROBLEM OF SIGNAL RECOGNITION

If it is accepted that developmental changes occur as a result of differential gene expression, then the ultimate aim of developmental biology is to define in precise terms the mechanisms whereby each and every developmental process is effected. Reflecting the bias of a biochemist, the most precise definition of developmental mechanisms would appear to be in molecular terms. Once evidence is obtained that a previously absent gene product is now being produced in active form, it is necessary to consider the nature and origin of the physiological signal that activates the system. The formal logic is the same whether the system represents an ongoing or offgoing phenomenon. That is, the *disappearance* of a specific gene product is also part of developmental processes, and the detailed mechanisms require definition.

The word "signal" is deliberately used here as an all-embracing term to cover what may be a complex process involving a physiological change and the production of effector substances, which may be a multistage process. The nature of effector or inducer substances in developmental processes is largely unknown, although hormones and their secondary messengers in target cells have been clearly implicated in recent years.

This discussion might best be sharpened by reference to postnatal changes in placental mammals. The immediate postnatal synthesis of several enzymes in rabbit, guinea pig, and rat liver has been shown to be associated with birth. That is, premature delivery of fetuses by uterine section provokes the premature formation of enzyme, while in postmature fetuses produced, for example, by treatment of preg-

nant animals with anti utrin-S, the production of enzyme is delayed until birth of the animal.

This pioneer work was carried out by Nemeth, 1959). The physiological signals that appear to set up events for the induction of several different enzymes include a transient but intense phase of hypoglycemia during the first few hours after birth and the stress of birth itself, which appears to promote the release of corticosteroids from the adrenal gland. In response to hypoglycemia, newborn animals would be expected to release glucagon from the α cells of the pancreas, and adrenalin is known to be released from the adrenal medulla.

If these hormones are concerned with the induction of specific enzyme synthesis, then their administration to intact fetuses *in utero* might be expected to result in the premature synthesis of the specific enzyme. This experimental criterion can be used at any stage of development where the experiment is technically feasible. That is, a suspected inducer should in the first instance pass the above test —that of *premature* induction of enzyme synthesis when administered at a time prior to the normal appearance of the enzyme.

Unfortunately, these criteria are not always satisfied, even though the circumstantial evidence implicating a particular hormone or other agent may be quite strong. In such cases the system will probably be more complex than it appears. For instance, the target tissue may not respond to the inducer at a premature time. This could be due to the absence of receptors that develop at a later time in response to unknown factors.

Some hormones are known to give rise to secondary messengers inside target cells, while others require metabolism to intracellularly active forms. Adenosine $3',5'$-cyclic monophosphate (cyclic AMP) is a known intracellular messenger for glucagon and adrenalin in the liver and some other tissues. In this case, the intracellular effector can be tested and the normal response to the hormones should also be provoked by the effector given alone. Thus the "premature" test can also be mounted for the hormone derivative or messenger. This advances the investigation one step further to the identification of the intracellular effector.

The next step might be called "hunt the repressor." The identification of low-molecular-weight intracellular effectors for specific enzyme synthesis allows the identification of additional factors operating in individual systems. For example, the repressor protein for the *lac* operon in *E. coli* was identified by utilizing the fact that a synthetic nonmetabolizable effector molecule binds to the repressor protein (Gilbert and Müller-Hill, 1966). Hence the same approach can be used in eukaryote systems, in which proteins that bind intracellular effectors could be characterized. The major problem here is analytic, in that the concentration of repressor type proteins will in all probability, be vanishingly low. The same difficulty applies to activator substances. Nevertheless, the identification of such factors, whether they are activators or repressors of transcription or translation, remains a strong possibility once the identity of the effector molecule is known for a specific system and the characteristics of the inductive process are delineated.

CYTOLOGICAL PROBLEMS

A safe assumption to make about tissues and organs in eukaryote organisms is that they are nonhomogeneous with regard to cell type. This fact is often overlooked by biochemists, including those interested in developmental processes. Thus the development of a so-called *hepatic enzyme*, although it may be shown to be specific for the liver, may be a specific enzyme of parenchymal cells, bile canaliculi cells, Kupfer cells, or, in the perinatal liver, hematopoietic cell lines. The cytological origin of a particular developmental enzyme is thus a problem that needs more attention.

This difficulty can be approached with limited success by techniques of cell separation or by cytochemical techniques but only a relatively few enzymes can be detected using staining procedures. The fluorescent antibody technique can be applied in principle to any enzyme and in theory is the best method to resolve such cytological problems.

PROBLEM OF ISOZYMES

Many enzymes occur in more than one molecular form, and in the case of hybrid iso-

zymes of the lactate dehydrogenase type, consisting of tetramers comprised of two different subunits, the appearance or disappearance of a particular isozymic form may represent a developmental event as defined earlier. Failure to investigate the isozyme spectrum of an enzymic activity may lead to the classification of a developmental event as a modulation of preexisting enzyme levels. For example, where activity is already present, an abrupt increase in activity may be due to the initial *de novo* synthesis of a new subunit type, which would qualify as a developmental event.

Multiple molecular forms of enzymes may also arise in evolution from differential mutation of multiple copies of the structural gene specifying the primary structure of the enzyme, and a complex system for induction could arise from mutation of control genes in similar fashion. In at least two cases, multiple molecular forms of enzymes occur in mammalian liver, and the different forms appear to be induced during development by different factors. Some of these factors are hormones. Therefore, the analysis of enzymes for multiple molecular forms or isozymes is particularly relevant to studies in developmental enzymology, especially where the inductive situation appears to be multifactorial.

DNA Synthesis as a Prerequisite for Development

Cellular differentiation and development are indistinguishable one from the other when we consider the production of cell-characteristic protein as a necessary and sufficient criterion of the processes. In the organ culture system developed by Wessells and Rutter (and earlier by Grobstein) to study pancreas differentiation, the production of enzymes and proteins characteristic of the pancreas is preceded by DNA synthesis and cell division, and the specific enzymes are produced by non-dividing cells. The separation of mitotic activity and the synthesis of cell-specific proteins need not be obligatory for all systems, but one recent suggestion is that there is a critical mitosis (and thus a critical period of DNA synthesis) in cell differentiation that precedes the production of cell-characteristic protein and definitively commits the cell to such production. Such a mitosis need not be the terminal division, and the committed cell population may be increased by further cell division (Holtzer, 1972; but see also Chapter 9 of this book). In terms of the model of differentiation summarized in Fig. 1, such a critical mitosis may result in the protodifferentiated state.

In a cell as enzymologically complex as an hepatocyte, a number of critical mitoses may be required during development to elaborate the complex mature cell. On the other hand, the population of hepatocytes in the differentiated liver may be heterogeneous, consisting of a number of different cell lines that are committed to produce certain special enzymes at some stage in their (separate) development and that have experienced only a single critical or committing division. This question of heterogeneity might be resolved by cytological or fluorescent antibody staining of specific liver enzymes that are known to be induced at some stage of liver development. Such a study should reveal biochemical heterogeneity relevant to these notions.

The use of inhibitors of DNA synthesis at least in short-term experiments with suitable systems may shed light on the problem of an obligatory requirement for DNA synthesis, but other experiments not subject to criticism on the grounds of cytotoxicity would also need to be devised.

Gene Amplification as a Prerequisite for Development

If DNA synthesis is an obligatory prerequisite for cell differentiative processes in development, then the nature of the committal of cells may consist of specific gene amplification.

Such a process is known to occur during oocyte maturation in amphibia (Brown and Dawid, 1968). The developing oocyte of *Xenopus laevis* contains thousands of nucleoli instead of the four nucleoli expected from its tetraploid state. Each nucleolus appears to be the site of ribosomal RNA (rRNA) synthesis and quantitative molecular hybridization between rRNA and the DNA of the oocyte shows the presence of DNA coding for rRNA at thousands of times the normal somatic amount. There are normally about 900 copies of the genes coding for rRNA to be found in somatic cells of *X. laevis*, but these multiple copies are amplified 1000-fold in the oocyte. The oocyte is thus enabled to make very large numbers of ribosomes that carry the fertilized ovum through blastulation and gastrulation up to hatching of the larval tadpole. The extra copies of the rRNA genes laid down during oogenesis later become nonfunctional and are catabolized. Thus at some stage during oogenesis the rRNA genes are specifically replicated many thousandfold, and this process constitutes gene amplification. However, the amplification of rRNA genes alone seems unlikely to explain differentiation involving elevated synthesis of characteristic proteins. Recent findings in the silkworm *Bombyx mori* suggest that amplification of genes for specific proteins does not occur in differentiation (Suzuki et al., 1972). Molecular hybridization of silkworm DNA with silk fibroin mRNA was used to assess the relative abundance of fibroin genes in both the the body and silk glands of *B. mori*. The studies showed the same relative abundance in DNA isolated from both sources, even though the giant polyploid silk gland cells contain orders of magnitude more DNA than do cells from the animal's carcass. The number of fibroin genes per haploid amount of DNA was found to be between 1 and 3 in both gland and body cells. Thus the significance of DNA synthesis as a prerequisite for differentiation remains a controversial question.

Developmental Enzymology in Mammals

In this section the developmental formation of enzymes in some animal tissues will be discussed, and some examples that are illustrative of the points raised in the earlier section will be presented in detail.

Much work on developmental enzymology has been centered on the liver. This is not a result of personal bias of investigators. The liver has often been chosen since, even into adult life, it retains a high capacity to respond to environmental changes by quantitative alterations of enzyme levels. This capacity no doubt confers a distinct biological advantage to animals, since the liver (and the gut) is periodically exposed to fluctuations in the concentration of the various foodstuffs during the daily feeding cycle. The other internal tissues are of course maintained in an almost constant chemical environment. In addition, the liver, of all the tissues, has the broadest spectrum of metabolic capacity and therefore the broadest spectrum of enzymes.

Studies of enzyme development in the gut were initiated largely as the result of pediatric observations on dysfunctional conditions in the alimentary tract of neonatal human infants. These dysfunctional states were traced to the absence of certain enzymes, especially disaccharidases, from the gut tissue of affected individuals. Such enzymes normally appear in the gut cells or show large increases during postnatal life.

The first demonstration of a change in the level of an enzyme following hormone administration to the intact animal was made in rat

TABLE 1

ENZYME DEVELOPMENT IN ANIMALS: INDUCTION OF DE NOVO SYNTHESIS[a]

Tissue	Enzyme	Period	Inducer[b]	Intracellular Effector
Rat liver	Fructose-1-phosphate aldolase	Late fetal	—	—
	Arginase	Late fetal	Thyroxine (?)	—
	Phosphoenol-pyruvate carboxylase	Neonatal	Glucagon, epinephrine, norepinephrine	Cyclic AMP
	Tyrosine aminotransferase	Neonatal		
	Form A		Insulin	—
	Form B		Glucocorticoids	—
	Form C		Glucocorticoids and glucagon	Cyclic AMP (?)
			Epinephrine	Cyclic AMP (?)
	Serine dehydratase	Neonatal	Glucagon	Cyclic AMP (?)
	Glucokinase	Late postnatal	—	—
Mouse and rat jejunum	Alkaline phosphatase	Late neonatal	Hydrocortisone and other factors	—
Chick gut	Invertase	Late neonatal		
Rat jejunum	Galactosidase	Late fetal	Glucocorticoids and other factors	—
Tadpole liver	Urea cycle enzymes	Metamorphosis	Thyroxine	—

[a]These enzymes are presented as cases of *de novo* biosynthesis either based on the results of inhibitor experiments or because there is little evidence for significant preexisting enzyme before the time of initial change.
[b]Usually based on the capacity of the inducer to evoke premature synthesis.

TABLE 2

ENZYME DEVELOPMENT IN ANIMALS: INCREASED ENZYME FORMATION

Tissue	Enzyme	Period	Agents causing premature increase
Rat liver	UDPG-glycogen glucosyl transferase	Late fetal	Hydrocortisone
	Glucose-6-phosphatase	Late fetal	Glucagon, thyroxine, epinephrine, cyclic AMP
	Glucose-6-phosphatase	Neonatal	Glucagon, thyroxine, cyclic AMP
	Arginino succinate synthetase	Late fetal	Triamcinolone[a]
	Arginase	Late fetal	Thyroxine
	NADPH dehydrogenase	Late fetal	Thyroxine
	Drug metabolizing enzymes (microsomal)	Neonatal	Phenobarbitone (administered to pregnant dam)
	Ornithine aminotransferase	Late postnatal	Triamcinolone,[a] hydrocortisone
	Alanine aminotransferase	Late postnatal	Hydrocortisone
	Tryptophan oxygenase	Late postnatal	Hydrocortisone
	Malate dehydrogenase	Late postnatal	High glucose diet (weaning)
	ATP-citrate lyase	Late postnatal	High glucose diet (weaning)
	Pyruvate kinase	Late postnatal	High glucose diet (weaning)
Mouse and rat jejunum	Invertase	Late postnatal	Hydrocortisone

[a]Potent fluorinated glucocorticoid analogue.

liver by Lin and Knox (1957). This work stimulated investigations in the whole field of biochemical endocrinology, and there was inevitable fallout into the field of development. Additional impetus was derived from Karlson's pioneer studies on the importance of steroid hormones in insect metamorphosis and the identification of chromosomes as their locus of action (see Karlson and Sekeris, 1966, for review).

Information on enzyme development in mammalian tissues is, of course, still rudimentary, but it is convenient to summarize some of the information at this stage.

In the development of mammalian liver three distinct phases may be distinguished. There are late fetal, neonatal, and late postnatal phases of enzyme development. In some cases it is clear that de novo induction of enzyme synthesis occurs; in others the phenomenon may be one of modulation of the level of preexisting enzymes. In the tables that follow, some attempt has been made to distinguish between these phenomena, but some of the distinctions may be arbitrary and reflect the bias of the author.

The summarized data has been restricted to certain enzymes. This restriction is applied using two criteria. First, an exhaustive account of all the enzyme changes that have been charted during mammalian development adds nothing to the principle that such changes do occur. Second, the restriction is applied to those systems for which some clues are available as to the identity of inducing or modulating agents, (see Tables 1 and 2).

Some of these cases will be discussed in more detail to emphasize principles. References should be made to recent reviews for further detail and bibliography (Greengard, 1970;

Moog, 1970; see also Oliver and Yeung, 1972).

Several examples of enzyme development will now be discussed in which many of the principles described earlier can be illustrated.

NEONATAL INDUCTION OF PHOSPHOENOLPYRUVATE CARBOXYLASE

Phosphoenolpyruvate carboxylase catalyses the following reaction:

oxaloacetate + GTP (or ITP) $\rightleftarrows$ phosphoenolpyruvate + CO_2 + GDP (or IDP) The reaction is essential for the synthesis of glucose from lactate, pyruvate, and intermediates of the tricarboxylic acid cycle. The cytoplasmic enzyme occurs only in the liver and kidney. It is absent from the liver of the fetal rat and some other animals, probably including the human fetus. There is a mitochondrial form of the enzyme with unknown function, and the following discussion applies only to the cytoplasmic enzyme.

The fetal liver is incapable of gluconeogenesis from pyruvate, lactate, or amino acids, both *in vitro* and *in vivo*, and acquires this metabolic capacity only after the development of phosphoenolpyruvate carboxylase. Although all the other enzymes that are necessary for gluconeogenesis are, in fact, present in the late fetal liver, the postnatal acquisition of a single enzyme enables the liver to carry out a complex overall metabolic process that is important to the independent existence of the animal. Experiments in the author's laboratory attempted to elucidate the mechanisms involved in the development of this enzyme.

The postnatal development of some enzymes (see tyrosine aminotransferase below) is known to be dependent on adrenal steroids, and both experimental and clinical experience have shown that gluconeogenesis is elevated by the chronic or high-dose administration of adrenal glucocorticoids. Accordingly, a potent fluorinated synthetic steroid, triamcinolone, was administered to fetal animals *in utero* in order to test for inductive effects on phosphoenolpyruvate carboxylase. Some seven other enzymes were included in the study as various kinds of controls.

The results were surprisingly clear-cut: the carboxylase did not appear in the fetal liver, but the enzyme tyrosine aminotransferase did appear prematurely in the livers of animals treated with the steroid analogue. The latter finding was important since a negative result may have been due to the failure of the hormone to accumulate in the fetal liver. The results of these experiments seemed to eliminate glucocorticoids as factors involved in the induction of phosphoenolpyruvate carboxylase.

Since hepatic enzyme activity appears within a few hours after birth, further experiments along the lines of earlier work by Nemeth (1959) revealed that premature delivery of rat fetuses by uterine section resulted in the rapid appearance of enzyme activity (Yeung and Oliver, 1967). Using ether anesthesia of the dam, the rate of appearance of enzyme activity was identical in both surgically delivered and normal vaginally delivered animals. Hence the appearance of enzyme activity appeared to be related to birth and thus to some change in the environment or condition of the animal following its delivery (see Fig. 2).

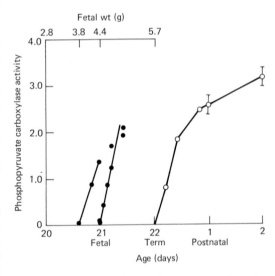

FIGURE 2. *Postnatal development of hepatic phosphopyruvate carboxylase and the effect of premature delivery. Fetal rats were delivered normally or by uterine section and kept in a Humidicrib at $37°C$ until they were killed for enzyme assay. Enzyme activities are μmoles phosphopyruvate produced/mg protein/hour. (From D. Yeung and I. T. Oliver, Biochem. J. 105, 1229, Fig. 2, 1967, by permission.)*

The surgical delivery of animals is extremely useful in the study of neonatal enzyme development since a whole litter of rats can be delivered in about 8 minutes. With a rapidly developing enzyme system, consistent results are thus easier to obtain with "surgical" animals than with normally delivered animals since normal delivery of a litter may take up to 2 or 3 hours. During this time, animals born early have already acquired the enzyme. There are few problems of animal husbandry, provided experiments occupy no more than about 12 hours of postnatal life.

Before the nature of the physiological signals are discussed, it should be stated that the accumulation of hepatic enzyme activity in this instance appears to be due to *de novo* synthesis. The administration of actinomycin D, puromycin, or amino acid analogues at delivery of the animals results in considerable reduction in the rate of enzyme accumulation when compared to controls. If actinomycin D is given to fetuses *in utero* at 30 minutes before their delivery, then no enzyme accumulates in the liver. This is because the drug takes about 30 minutes to accumulate to maximal levels in the liver following intraperitoneal injection. These data suggest that the enzyme is synthesized *de novo* as a result of structural gene transcription (Yeung and Oliver, 1968a).

The postnatal synthesis of the enzyme has recently been confirmed in F. J. Ballard's laboratory. Using an antibody prepared against the purified enzyme, it has been shown that there is negligible antigen present in the fetal liver, which eliminates the possibility of the activation of preexisting inactive enzyme. After birth, there is rapid incorporation of labeled amino acids into antibody-specific antigen (the newly synthesized enzyme), and there appears to be negligible degradation of the enzyme during the inductive phase (Ballard et al., 1972).

The physiological signal for enzyme induction arises from hypoglycemia at birth. In Fig. 3 it is shown that in the rat the pattern of hypoglycemia is the same in both surgically delivered and normally delivered animals.

This situation is readily subject to experimental manipulation. The injection of glucose at birth can be used to stabilize the blood glucose at normal levels and prevent the hypo-

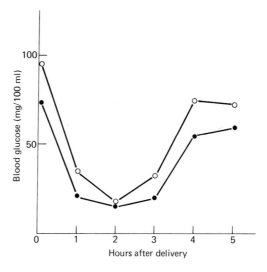

FIGURE 3. *Postnatal hypoglycemia in normal and surgically delivered premature rats.* O, *premature rats;* ●, *normal rats. (Data plotted from D. Yeung and I. T. Oliver,* Biochem. J. *108, 325, Table 2, 1968.)*

glycemia. When this is done, enzyme synthesis is considerably repressed (Fig. 4). It should be emphasized that manipulation of other obvious physiological parameters in the newborn, such as oxygen tension, are without effect on enzyme formation.

In response to low blood glucose levels, animals release glucagon from the α cells of the pancreas and epinephrine from the adrenal medulla. When glucagon is injected at delivery, the postnatal rate of enzyme accumulation increases. This effect is not prevented by glucose in doses sufficient to prevent hypoglycemia (Fig. 5). Glucagon, epinephrine, and norepinephrine also result in enzyme formation when administered *in utero* (Fig. 6). Glucagon at the low dose of 1 μg per fetus is effective for enzyme induction. This effect is prevented by simultaneous injection of actinomycin D (Yeung and Oliver, 1968a). Agents that result in adrenergic blockade reduce postnatal accumulation of the enzyme, indicating that epinephrine also plays a role in the postnatal induction of the enzyme (Yeung and Oliver, 1968b).

Catecholamines and glucagon exert some of their hormonal effects in liver through the secondary messenger cyclic AMP. In the experiments on the induction of phosphoenol-

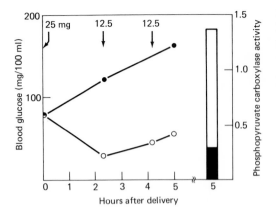

FIGURE 4. *Effect of glucose on postnatal synthesis of hepatic phosphopyruvate carboxylase. Fetal rats were delivered by uterine section and maintained in a Humidicrib at 37°C until killed for enzyme assay. Some were injected with glucose at the doses (in mg) and times indicated. The enzyme activity 5 hours after delivery is shown at the right for both groups of animals (μmoles oxaloacetate formed/mg protein/hour at 37°C). ●, Glucose-treated; ○, controls. (Data plotted from D. Yeung and I. T. Oliver, Biochem. J. 108, 325, Table 3, 1968.)*

pyruvate carboxylase, all the results with the hormones above can be replicated using cyclic AMP or its dibutyryl derivative (N^6-$O^{2'}$-dibutyryl adenosine 3',5'-cyclic monophosphate). Actinomycin D prevents the effects of cyclic AMP in both the fetal and postnatal animal. It should be noted that the injection of cyclic AMP to the newborn completely prevents hypoglycemia. No other cyclic nucleotides or related mononucleotides are effective (Yeung and Oliver, 1968b). Insulin injection, which results in prolonged hypoglycemia, represses postnatal enzyme formation, presumably because of the inhibition of adenyl cyclase (Fig. 7). The enzyme can also be induced in fetal liver explants maintained in tissue culture media *in vitro* by the same agents as described for the experiments *in vivo* (Wicks, 1969).

All these experimental data are consistent with the following interpretation of the mechanism of induction of phosphoenolpyruvate carboxylase. The reduction in the level of blood glucose that follows delivery of the placental

mammal causes release of glucagon and adrenalin from the respective endocrine glands. At the liver cell membrane, the hormones activate adenyl cyclase, and there follows an elevation in the intracellular concentration of cyclic AMP. This effector causes the specific transcription of the structural gene for phosphoenolpyruvate carboxylase, and translation of mRNA follows. The specific enzyme then accumulates in the liver cell cytoplasm. The locus of action of cyclic AMP appears to be at the level of gene transcription but no further details of the mechanism are known. Attempts in the author's laboratory to find factors in the cell nucleus that bind the effector molecule with high avidity have so far failed to yield significant results. Such factors, if they exist, may be gene activators or repressors.

In terms of the control possibilities listed earlier, this system seems to have a single locus

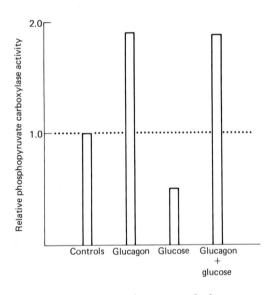

FIGURE 5. *Effect of glucagon and glucose on the postnatal synthesis of hepatic phosphopyruvate carboxylase. Animals were delivered by uterine section and maintained in a Humidicrib at 37°C. Animals were injected with 25 mg of glucose or with 25 μg of glucagon at delivery. Some animals received both glucose and glucagon. Enzyme activities were determined in the liver 5 hours after delivery. (Data plotted from D. Yeung and I. T. Oliver, Biochem. J. 108, 325, Table 6, 1968.)*

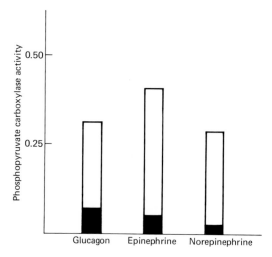

FIGURE 6. *Induction of hepatic phosphopyruvate carboxylase by hormones in fetal rats in utero. Fetal rats ranging in gestational age from 16 to 22 days were injected with glucagon (25 μg), epinephrine (5 or 10 μg) or norepinephrine (20 μg). Littermate animals in the opposite uterine horn were injected with saline. The dam was sutured and allowed to recover from the ether anaesthesia. Five hours later, all fetuses were removed for enzyme assay. The low activity that developed in the control animals is due to placental transfer of hormone since at a dose of 1 μg of glucagon no activity appears in the controls. Activities as in Fig. 4.* ■, *Controls;* □, *hormone-treated. (Data are means pooled from D. Yeung and I. T. Oliver,* Biochem. J. 108, 325, *Table 5, 1968.)*

of control at gene transcription. Postinductive modulation of enzyme level appears to occur by variation in the rate of enzyme degradation (Ballard et al. 1972).

In the rat kidney, the enzyme is present in late fetal life, and although injections of cyclic AMP *in utero* result in increased enzyme activity, the mechanisms and time of induction are not yet known.

NEONATAL INDUCTION OF SERINE DEHYDRATASE

Serine dehydratase catalyses a pyridoxal phosphate-dependent deamination of serine in which the amino acid is converted to pyruvate and ammonium ion.

The postnatal development of the enzyme is shown in Fig. 8. The late fetal liver may contain some enzyme, but if so the activity is very low and increases postnatally 20- to 30-fold. Thus system may therefore qualify as one in which *de novo* induction of synthesis occurs. In surgically delivered animals either actinomycin D or cycloheximide (an inhibitor of translocation reactions on the ribosome) prevents the normal accumulation of activity. Thus the system seems to involve both gene transcription and mRNA translation to form enzyme (Yeung and Oliver, 1971). Enzyme production in the postnatal animal is stimulated by glucagon and reduced by normalization of the blood glucose level by glucose administration. As with the carboxylase system, the effect of glucagon is not reduced by simultaneous glucose injection.

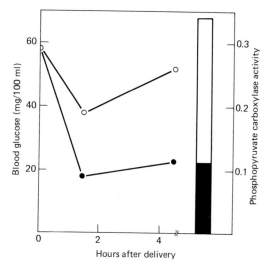

FIGURE 7. *Effect of insulin on postnatal synthesis of hepatic phosphopyruvate carboxylase. Animals were delivered by uterine section and kept in a Humidicrib at 37°C. Some animals were injected with 2 m units of insulin at delivery. Enzyme activity was assayed 4.5 hours later and blood was taken for glucose at the times indicated. The enzyme activities are given at the right. Units are as in Fig. 4.* ●, *insulin treated;* ○, *controls. (Data plotted from D. Yeung and I. T. Oliver,* Biochemistry 7, 3231, *Table 7, 1968.)*

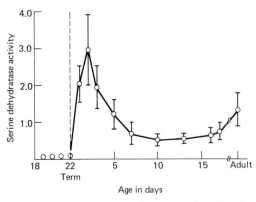

FIGURE 8. *Neonatal development of serine dehydratase in rat liver. Enzyme activity is in μmoles pyruvate formed/mg protein/hour. Each point represents the mean activity ± 1 SD (vertical bars) found in the livers of animals taken from different litters. (From D. Yeung and I. T. Oliver, Comp. Biochem. Physiol. 40A, 135, Fig. 1, 1971 with permission.)*

Cyclic AMP also stimulates enzyme production in the newborn animal when given at birth or some hours later. However, once synthesis has been initiated, the stimulatory effect obtained with cyclic AMP is completely prevented by simultaneous administration of actinomycin D (Fig. 9). This further indicates that cyclic AMP acts in this system at the level of DNA transcription and not at a subsequent level, at some stage of translation. It may also be of some interest to note that actinomycin D, when given 3 hours after birth, results in no significant fall in enzyme level over the next 3 hours. This result provisionally indicates that enzyme degradation is negligible in the early stages of induced synthesis of the enzyme. In these respects, further similarities with the phosphoenolpyruvate carboxylase system can be seen.

When the results of fetal experiments are considered, the picture is not so clear. First, epinephrine and cyclic AMP, when administered to rat fetuses *in utero* do not induce enzyme formation. Glucagon is effective in late fetal life (gestational days 20 to 22, birth occurring at the end of day 22 in the rat) but is ineffective in younger fetuses. Second, the dose of glucagon required to give an effect is 20 times that required for the induction of the carboxylase *in utero*. It is obvious that there

are completely unknown factors involved in this system. Although the postnatal enzyme development is partially blocked when hypoglycemia is prevented with glucose administration, the failure of epineprine and cyclic AMP to induce in the fetus and the high doses of glucagon required for induction suggest strongly that these hormones and cyclic AMP may not be the physiological factors involved in serine dehydratase induction.

Serine dehydratase consists of two forms of the enzyme that are separable by chromatography on anion-exchange cellulose. One form is induced (or increased) by glucagon in the adult rat liver, and it is this form that appears in early postnatal life. No inducers have been

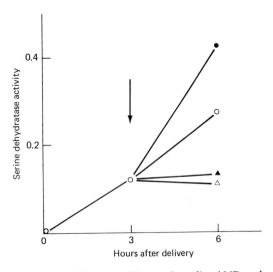

FIGURE 9. *Effect of dibutyryl cyclic AMP and actinomycin D on postnatal synthesis of serine dehydratase in neonatal rat liver. Fetal rats were delivered by uterine section and kept in a Humidicrib at 37°C. Three hours after delivery (arrow) animals received actinomycin D (4.4 μg), dibutyryl cyclic AMP (0.14 μmole), a combination of both, or saline. Three hours later all animals were killed for enzyme assay. Some control animals were assayed at 3 hours. Enzyme activities are in μmoles pyruvate formed/mg protein/hour. ○, Saline; △, actinomycin D (4.4 μg); ●, dibutyryl cyclic AMP (0.14 μmole); ▲, actinomycin D + dibutyryl cyclic AMP. (Data plotted from D. Yeung and I. T. Oliver, Comp. Biochem. Physiol. 40A, 135, Table 6, 1971.)*

identified for the second form, and this represents a further problem with this system.

POSTNATAL INDUCTION OF GLUCOKINASE

Glucokinase catalyses the first reaction of glucose metabolism as shown:

$$glucose + ATP \rightarrow glucose\text{-}6\text{-}phosphate + ADP$$

The enzyme is confined to the liver and is distinguished from the group of hexokinase isozymes by its high specificity for glucose but low affinity for the sugar. The concentration of glucose at which the enzymatic reaction achieves half-maximal velocity is 10 mM, which is in the middle of the range of portal blood glucose concentrations—which vary between 5 mM during carbohydrate fasting and 25 mM following carbohydrate feeding. The enzyme in the liver cell thus varies the rate of glucose phosphorylation according to the concentration of glucose presented to it, since the liver cell is freely permeable to glucose. The enzyme accounts for as much as 90 percent of the total rate of glucose phosphorylation by liver. Glucokinase thus functions as an important part of the control mechanisms for the regulation of the blood glucose concentration (Ballard and Oliver, 1964a,b).

The enzyme is not present in the liver of herbivores, carnivores, or marsupials but seems to be largely confined to omnivorous animals (Ballard 1965). In adult animals the level of activity of the glucokinase varies quite markedly with some conditions. For example, the activity falls to very low levels in fasting and may disappear altogether in experimental diabetes produced by alloxan destruction of pancreatic β cells. Insulin injection in diabetic animals and glucose refeeding in fasted animals restores enzyme activity, and the results of experiments using actinomycin D show that such restoration is due to enzyme synthesis.

All the evidence from adult animals indicates that the enzyme synthesis is dependent on insulin and that when blood insulin levels are low, as in fasting and diabetes, the degradation of the enzyme overtakes the rate of synthesis and, as a result, the activity falls.

The enzyme is absent from the fetal liver in the rat and appears for the first time at the fifteenth or sixteenth postnatal day, rising to maximal activity over the next 8 to 10 days of life. Walker and Holland (1965) have made a concentrated attempt to discover the inductive mechanism in the postnatal rat. Their experiments were based on findings in the adult animal and included the administration of glucose, insulin, or glucose and insulin in combination at all times between birth and the fifteenth postnatal day, but the enzyme was not induced prematurely by any of the various experimental regimes. Alloxan administration close to day 15 results in the death of many animals, but in those that survive the enzyme fails to appear until insulin is given (Walker and Holland, 1965). Apparently the liver does not respond to insulin until the fifteenth postnatal day, even when glucose is used to promote release of endogenous insulin.

In experiments in the author's laboratory, the plasma insulin level was measured in normal postnatal rats, and the levels remained unchanged throughout the first 20 days of postnatal life. Thus even the normal appearance of the enzyme is difficult to explain on any simplistic basis involving elevated pancreatic release of insulin at the time of initial enzyme appearance.

The molecular basis for the differential response of tissues to the same hormone seems best explained by the presence or absence of specific receptors. Specific binding proteins for steroid hormones and their metabolites have now been characterized in many tissues, and the existence of adrenergic receptors has been postulated for a long time. More recently still, insulin receptors have been demonstrated in the plasma membrane of liver cells using I^{125} labeled insulin (Freychet et al., 1971). The development of hormone responsiveness in any tissue may thus be due to a fundamental event such as the synthesis of specific receptors in response to uncharacterized inductive agents.

The failure of insulin to induce glucokinase synthesis in the early postnatal animal may well stem from the absence of insulin receptors, and a more incisive analysis should include investigation of this point, including the mechanism of receptor generation. However, in early postnatal liver there is another system that does respond to insulin. In 2-day-old rats, insulin injection induces one form of tyrosine aminotransferase that normally appears in the liver at postnatal ages of 3 to 4 days (Holt and Oliver, 1969, 1971). It can only be assumed that the liver cell interaction with insulin that results in synthesis of this enzyme is different from the interaction that promotes glucokinase synthesis, and there may exist a range of insulin receptors, specialized for different functions, that are generated by the liver at different times in response to different agents. This whole question will be discussed in a more general way at a later time.

Some recent work on glucokinase induction and the liver response to insulin has claimed the involvement of glucocorticoids in the process (Greengard, 1970). However, the hormone doses were unphysiological, and it is extremely difficult to equate the bizarre experimental protocols with anything known to happen in the intact postnatal animal.

The mechanism of glucokinase induction thus remains unsettled even at a tentative level, but it is hoped that the discussion presented here emphasizes some of the problems that developmental enzymologists will have to consider in the future. In particular, it is apparent that the problem of hormone receptors must figure largely in the thinking of enzymologists in this field.

NEONATAL INDUCTION OF TYROSINE AMINOTRANSFERASE

Tyrosine aminotransferase is one of the enzymes that initiate the metabolism of tyrosine and set in progress a number of subsequent reactions that lead to the production of many biologically important metabolites of the amino acid.

The reaction is a pyridoxal phosphate-dependent transamination between tyrosine and α-oxoglutarate:

L-tyrosine + α-oxoglutarate $\rightleftarrows$ p-hydroxy phenylpyruvate + glutamate

Kretchmer and his colleagues first showed that tyrosine aminotransferase was absent from the fetal rat liver and that activity developed within a few hours of birth (Sereni et al., 1959). Activity also rises sharply at birth in the rabbit and guinea pig and possibly in the human infant. The activity in adult rat liver is sharply increased by the administration of glucocorticoids, and the immunological isolation of the enzyme protein labeled with radioactive amino acids has shown this rise to be due to an increase in the rate of synthesis. Although early reports claimed substrate induction of the enzyme by tyrosine, later investigation showed more convincingly that this induction was brought about by endogenous glucocorticoid released as a result of stress provoked by the massive doses of tyrosine (See Oliver and Yeung, 1972, for review).

Surgical delivery of the fetal rat causes premature enzyme production that is blocked by either actinomycin D or puromycin. The technique of surgical delivery used in the author's laboratory has enabled the postnatal time course of enzyme development to be accurately charted, and there is a 2-hour lag before enzyme activity begins to accumulate (see Figure 10).

Adrenalectomy of naturally born or surgically delivered rats largely prevents the postnatal enzyme synthesis, while hormone replace-

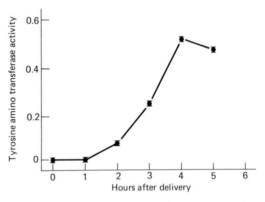

FIGURE 10. *Postnatal development of tyrosine aminotransferase in rat liver. Animals were delivered by uterine section and kept in the Humidicrib at 37°C until killed for enzyme assay. Animals from several litters were used for each point. Activities are in μmoles p-hydroxyphenyl pyruvate produced/mg protein/hour at 37°C. [Data from A. Armstrong and I. T. Oliver (unpublished experiments).]*

ment with hydrocortisone restores the enzyme production to normal. This hormone-stimulated enzyme production can be blocked with actinomycin D (Sereni et al., 1959; Holt and Oliver, 1969).

This evidence clearly implicates adrenal steroid hormones in the postnatal synthesis of tyrosine aminotransferase, but several anomalies exist in this system. It has been repeatedly observed in different laboratories over the last 12 to 13 years that hydrocortisone (or corticosterone) fails to induce the premature formation of enzyme in the fetal animal, even after prolonged exposure of the fetus to the steroid hormone. However, it was stated earlier in the discussion of phosphoenolpyruvate carboxylase that triamcinolone, a potent synthetic glucocorticoid analogue, will induce tyrosine aminotransferase in fetal rat liver when administered *in utero*. This effect takes about 10 hours to become apparent, but hydrocortisone in massive doses is ineffective even 10 hours after its injection into fetuses.

Further difficulties also become apparent upon examination of recent data. Insulin, glucagon, hydrocortisone, and cyclic AMP all stimulate synthesis of the enzyme in adult rat liver, both in the intact animal and in isolated perfused liver. The same agents and, in addition, epinephrine are also effective in fetal liver explants maintained in culture, although before hormonal responses are obtained, there is a refractory period lasting at least 24 hours after culture is initiated (Wicks, 1969).

Hydrocortisone, glucagon, epinephrine, and cyclic AMP all increase postnatal production of the enzyme in surgically delivered animals (Holt and Oliver 1969a), and insulin stimulates activity at the second postnatal day (Holt and Oliver 1969b).

In the intact fetal animal there are consistent negative findings for natural adrenal steroid hormones, but there is disagreement about other agents. For example, in the author's laboratory, dibutyryl cyclic AMP or epinephrine did not induce activity in the fetal liver, while glucagon, in high doses, was effective in Wistar albino rats—but only at a time when the fetal blood contained high concentrations of corticosterone (gestational days 18 to 20). During the last 2 days of gestation, when corticosterone levels are very low, glucagon was ineffec-

tive (Holt and Oliver 1969a). On the other hand, Greengard and her colleagues have reported enzyme induction by glucagon, cyclic AMP, and epinephrine in Sprague–Dawley rat fetuses during the last 2 days of gestation, although only dibutyryl cyclic AMP was effective at earlier ages (see review by Greengard, 1970). These contradictory findings are difficult to reconcile in the absence of information concerning blood steroid levels in Spague-Dawley rat fetuses.

In intact postnatal rats, high doses of glucagon approximately double the rate of enzyme accumulation, but in adrenalectomized newborn animals the hormone is about one third as effective as in the intact or sham-operated controls (Holt and Oliver, 1971). From these results, together with the fetal data, it could be argued that there is an interdependent effect of glucagon and adrenal steroids.

Before discussing this further, the last complicating facts about this system should be presented. It is difficult to integrate the various inductive phenomena and the wide range of inducing agents into any rational scheme for the control of tyrosine aminotransferase. Because the modes of action of insulin, adrenal steroid hormones, and the cyclic AMP mediated hormones are so different, and in some cases opposed (e.g., insulin inhibition of adenyl cyclase compared with glucagon and epinephrine stimulation), it is difficult to propose a satisfactory mechanism for stimulation of the synthesis of a single enzyme. Instead of considering multiple inductive mechanisms controlled by a range of physiologically unrelated hormones, it might be simpler to postulate multiplicity of the enzyme with independent and specific inducing mechanisms for each form.

Polyacrylamide gel electrophoresis of crude cytoplasmic extracts of liver followed by direct enzyme assay in gel sections show a distribution of enzyme activity that is suggestive of the existence of multiple forms. In the author's laboratory, experiments utilizing hormonal treatment of a 2-day postnatal rats, normal newborn rats at various times after delivery, and fetal rats have all indicated the occurrence of three forms of tyrosine aminotransferase that appear to be under independent hormonal control (Holt and Oliver, 1969b, 1971; see Figs. 11 and 12).

More recent data from Pitot's laboratory using hydroxylapatite chromatography shows the separation of the enzyme into four forms, three of which appear to be under independent hormonal control, while the fourth form is not inducible and appears to be a constitutive enzyme found in all tissues. The inducible forms are restricted to the liver (Iwasaki and Pitot, 1971).

The data are summarized in Table 3, in which the three forms separable in gel electrophoresis are named A, B, and C in order of their increasing electrophoretic mobility.

Some rationality now appears, but the situation is still complex in some respects. Insulin induces only form A of the enzyme, while epinephrine and cyclic AMP induce only form C. The duality of action of hydrocortisone and glucagon in induction of both forms B and C has not been explained. It is possible that glucagon has another action not mediated by cyclic AMP in this system but related to that of hydrocortisone.

Some further rationality emerges when the action of cyclic AMP is examined more closely. Cyclic AMP, presumably as the inter-

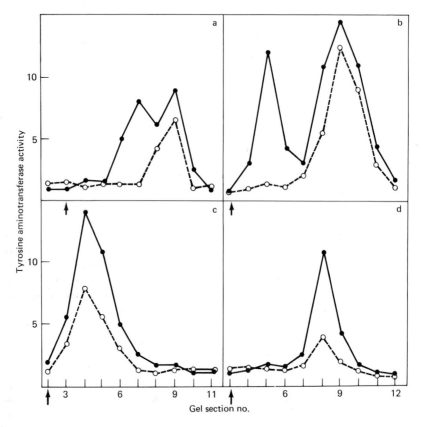

FIGURE 11. *Tyrosine aminotransferase activity in polyacrylamide gels following hormone treatment of 2-day postnatal rats. Liver extracts were prepared 5 hours after injection of animals with various hormones and then electrophoresed. The gel slabs were cut into 2-mm sections for assay of enzyme activity. (a) ●, Hydrocortisone + epinephrine; ○, epinephrine. (b) ●, Pyridoxin + epinephrine; ○, epinephrine. (c) ●, Pyridoxin + insulin; ○, insulin. (d) ●, Epinephrine + dibutyryl cyclic AMP; ○, dibutyryl cyclic AMP. Enzyme activity is in μmoles p-hydroxyphenylpruvate formed/gel section/hr. Hemoglobin is indicated by the arrow. The anode is at right. (From P. G. Holt and I. T. Oliver, FEBS Letters 5, 89, Fig. 1, 1969, with permission of North Holland Publishing Company Amsterdam.)*

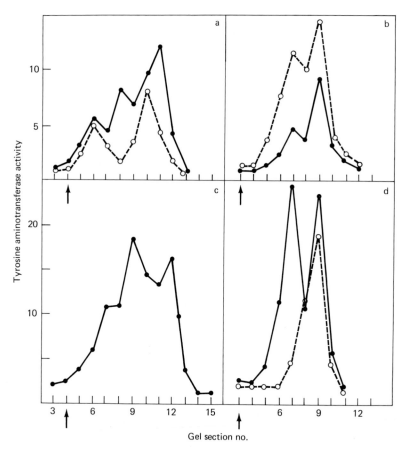

FIGURE 12. *Tyrosine aminotransferase activity in polyacrylamide gels following hormone treatment of 2-day postnatal rats. Liver extracts were prepared 5 hours after injection of animals with various hormones and then electrophoresed. The gel slabs were cut into 2-mm sections for assay of enzyme activity. (a)* ● *Pyridoxin + hydrocortisone + epinephrine;* ○ *pyridoxin + epinephrine. (b)* ● *Glucagon + actinomycin D;* ○ *glucagon. (c)* ● *Animals were normal, well-fed adults. (d)* ● *Hydrocortisone + epinephrine;* ○ *epinephrine. Enzyme activity is in μmoles p-hydroxyphenylpyruvate formed/gel section/hour. Hemoglobin is indicated by the arrows. The anode is at right. (Reproduced from P. G. Holt and I. T. Oliver, FEBS Letters 5, 89, Fig. 2, 1969, with permission of North Holland Publishing Company, Amsterdam.)*

mediary of epinephrine (and glucagon), appears to act in the production of form C of tyrosine aminotransferase at a posttranscriptional locus. This first became apparent from studies on the effect of cyclic AMP during postnatal synthesis of the enzyme. Injection of cyclic AMP at birth is followed by an increased rate of enzyme accumulation. Actinomycin D, given at birth, completely blockades all enzyme synthesis for at least 8 hours. The experiment was redesigned to employ a different stratagem. Enzyme synthesis was allowed to begin before injections were made at 4 hours after delivery. The amount of enzyme that accumulated in the controls over the next 3 hours was elevated twofold by cyclic AMP. The increased yield of enzyme was unchanged by simultaneous injection of actinomycin D at a (minimal) dose rate that completely blocks postnatal synthesis when given at birth. This effect occurs despite the fact that actinomycin D given alone in a similar regime, terminates enzyme accumulation at the 4-hour postnatal level (Holt and Oliver, 1969a). These results indicate that cy-

TABLE 3

HORMONAL INDUCTION OF MULTIPLE FORMS OF TYROSINE
AMINOTRANSFERASE

Enzyme form	Inducing agents	Order of normal postnatal appearance
A	Insulin	3rd (3–4 days)
B	Triamcinolone (fetal) hydrocortisone, glucagon	2nd (10 hr)
C	Triamcinolone (fetal) hydrocortisone, glucagon, epinephrine, cyclic AMP	1st (5 hr)

clic AMP promotes an event that is not dependent on DNA transcription and is located at a posttranscriptional level. The relevant data are illustrated in Fig. 13.

Studies on liver microsomes *in vitro* have added further evidence that locates the step at which cyclic AMP exerts its effect (Chuah and Oliver 1971). Microsomes isolated from epinephrine-inducible postnatal animals or from animals about 6 hours after delivery release active tyrosine aminotransferase to the supernatant when incubated with cyclic AMP *in vitro*. The effect is not blocked by cycloheximide and does not require the addition of amino acids, amino acyl synthetases, tRNA, ATP, or GTP. These properties mean that the mechanism operates only at the terminal step in protein synthesis, in which completed enzyme peptides are released from enzyme specific polysomes. The effect is obtained only with cyclic AMP, other cyclic nucleotides being without effect, but cyclic GMP and cyclic IMP are inhibitory to cyclic AMP. A protein that is extractable from rat liver microsomes at high ionic strength is also necessary for the effect, and the protein binds tritiated cyclic AMP with a high intrinsic association constant of about 10^{-9} M. The protein is present in fetal, postnatal, and adult liver microsomes.

This system is therefore of great complexity, showing multiple forms of the enzyme and dual control points at both transcription and translation (release) for one of its components.

The mechanisms of induction for forms A and B of the enzyme seem to involve only one control point, located at DNA transcription. The induction of form C appears to be controlled both at transcription of DNA and at the termination release step in polypeptide chain assembly. The two mechanisms appear to be independently controlled by different hormones and their effectors. There may be dualistic actions of glucagon by which the hormone induces transcription by one mechanism and controls release of the enzyme polypeptide by a reaction mediated by cyclic AMP. The apparent induction of enzymatic forms B and C by hydrocortisone and glucagon remains unexplained in any satisfactory way, as does the failure of hydrocortisone to induce enzyme synthesis in the fetus. One explanation might be that glucagon stimulates the production of steroid hormone receptor or the production of steroid metabolites that are normally not produced in the fetal liver. The normal postnatal release of glucagon in response to hypoglycemia may be the signal required for the liver to respond to hydrocortisone or corticosterone. Corticosterone is known to be released after birth, and glucagon appears to be more effective as an apparent inducer in the presence of endogenous adrenal steroid hormones than in their absence.

It is clearly not satisfactory to postulate the

hypoglycemia–glucagon mechanism as the sole signal required for postnatal induction of tyrosine aminotransferase. Even though form C of the enzyme is the first to appear, the polypeptide release mechanism mediated by cyclic AMP cannot alone account for this, because adrenalectomy considerably reduces the enzyme yield and actinomycin D blockades all enzyme synthesis. Hence adrenal glucocorticoids play some role in the initial induction, and a DNA-transcription step appears to be obligatory.

This system obviously requires much more investigation before the confusing array of results can be brought to reasonable order. In at least one respect, however, the system teaches a valuable lesson. Multifactor inductive situations may be clarified by a search for multienzymes.

The hormonal induction of tyrosine aminotransferase has also been studied in various cell culture systems. One such system consists of cells that are derived from a chemically induced liver hepatoma and grow well in culture conditions (HTC cells). The cells synthesize tyrosine aminotransferase in response to steroid hormones and synthetic analogues. However, they show no response to cyclic AMP-mediated hormones or to cyclic AMP itself and have been shown to possess no adenyl cyclase activity.

Tomkins and his associates have studied these cells intensively. A recent article reviews much of the work, and a suggestion emerges for the possibility of a third control point in eukaryote enzyme induction. Briefly, it is suggested that a control gene codes for a product that is a repressor of mRNA translation. The repressor gene product prevents the initiation of translation of enzyme message, but the induction of synthesis is brought about by steroid inactivation of the repressor. The details can be found in the paper by Tomkins et al. (1969).

However, the HTC cell is a model for tyrosine aminotransferase induction, and the complexities of the situation in the intact neonatal and adult liver have not been elucidated by the results from the culture system. The biochemistry of HTC cells shows that they are widely deviant from normal liver cells and that they have retained only some of the responses of the normal liver system. Perhaps these cells have even developed a unique system of control that is not present in normal liver.

The complexities of the tyrosine aminotransferase system in normal liver is perhaps best emphasized by the fact that since the original work of Lin and Knox (1957) on hydrocortisone elevation of enzyme activity, this system has received the most intensive study of any mammalian system in the field of hormonal regulation of enzymes.

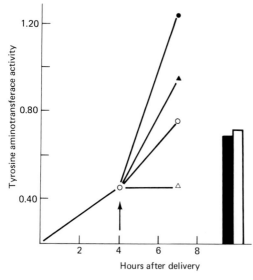

FIGURE 13. *Effect of cyclic AMP and actinomycin D on postnatal synthesis of tyrosine aminotransferase in rat liver. Late fetal rats were delivered by uterine section and kept in a Humidicrib at 37°C. Four hours after delivery (arrow) animals were injected with saline, dibutyryl cyclic AMP (0.14 μmole), actinomycin D (4.4 μg), or dibutyryl cyclic AMP plus actinomycin D. Three hours later all animals were killed and the livers assayed for tyrosine aminotransferase activity. Some animals were assayed at 4 hours after birth. The enzyme increments resulting from cyclic AMP injection alone (black) or in presence of actinomycin D (white) are shown at right. Enzyme activities are as in Fig. 10. O, Saline; △, actinomycin D (4.4 μg); ●, cyclic AMP (0.14 μmole); ▲, actinomycin D + dibutyryl cyclic AMP. (Data plotted from P. G. Holt and I. T. Oliver, Biochemistry 8, 1429, Table 5, 1969.)*

ENZYME MODULATION

The modulation of the level of a previously existing enzyme, especially where changes in level are less than one order of magnitude, appears to be less pertinent to development than the initial synthesis of a new enzyme. However, it might be an oversimplification to exclude modulation from the realms of development.

There are at least three factors that contribute to modulation effects. Changes in the rate of synthesis, changes in the rate of degradation of an enzyme, and alterations in the availability of essential cofactors will all alter the level of active enzyme. Dietary alterations often result in changes in the level of activity of enzymes in tissues, so that, among other things, weaning of young animals may result in such variations. However, it can be argued convincingly that dietary changes probably cause enzyme modulation through the agency of hormones, either as direct or indirect effectors. The interrelations of the endocrine system are not sufficiently well understood in detail, but indirect hormonal effects could well be modulated through interaction at target cells either through synergistic or inhibitory interplay. Various hormones have been shown to stimulate increased rates of synthesis of enzymes in adult tissues, but it is not always certain that modulating hormones are identical with inducing hormones.

Alteration of cofactor supply by dietary variation will also change enzymatic activity in a systematic way that will be determined by the various equilibria between different enzymes and a common cofactor. In addition, there will probably be other more complex relations. However, with the possible exception of hepatic tryptophane oxygenase, there are few cases where cofactor availability has been shown unequivocally to vary the rate of apoenzyme synthesis.

Quantitative data on enzyme degradation in animal tissues are being increasingly critically charted, but the mechanism of the process is so poorly understood that it would be unwarranted speculation to discuss the role of degradation in modulation. In principle, it can be said that modulative changes in enzyme levels in higher life forms have a causative component involving both synthesis and degradation.

Enzyme Development in Other Systems

In this chapter, little attention has been paid to nonanimal systems, but there are developmental enzymic changes in other life forms that would clearly repay study. The germination of plant seeds, bacterial sporulation, morphogenetic changes in slime moulds, fungi, and *Acetabularia* are but a few. However, it is probably reasonable to argue that the general principles of developmental enzymology will turn out to be similar for all living organisms. The mapping of systems, the identification of physiological signals, and their mechanism of action must still be elucidated in order to understand these phenomena at a molecular level, and although new or even special rules may emerge in time, the initial attack on the problems in diverse species should begin from the principle of biological unity.

A Final Word

The problems of specific enzyme induction in developing tissues generate further questions that begin to emerge more clearly only when the identification of specific inducers is achieved or even attempted. We have seen several cases in which a precondition of enzyme induction may be the development of specific tissue receptors for hormones or other inducers. Thus the mechanisms of receptor induction become an expansionist part of the problem but are integral with it. Techniques are now available for the detection of some hormonal receptors, and as more intracellular effectors are identified, similar techniques will allow the characterization of effector receptors and their role in inductive control systems. Such methodology will allow the next box in the Chinese puzzle of development to be opened.

The problem of cell committal also needs to be approached at the molecular level. The possible role of gene amplification in cell committal has already been briefly discussed. Another contributing process in cell committal could well be the generation of receptor substances in response to committal factors as yet unidentified. In this sense, a receptor cell is now committed to respond to specific inducers when they are presented, a response that is lacking in a progenitor cell from which it was derived. In this view, cell committal is a primary process creating the potentiality of a secondary response to inducing agents. The inducing agents, hormonal or otherwise, can then initiate the synthesis of a specialized array of enzymes and other proteins. Thus a particular cell becomes different from its neighbors, or, in a word, differentiates.

References

BALLARD, F. J. (1965). Glucose utilization in mammalian liver. *Comp. Biochem. Physiol. 14,* 437–443.

BALLARD, F. J., HOPGOOD, M. F., RESHEF, L., AND HANSON, R. W. (1972). Changes in protein synthesis and degradation involved in enzyme accumulation in differentiating liver. *Proc. Australian Biochem. Soc. 5,* 27.

BALLARD, F. J., AND OLIVER, I. T. (1964a). Keto-hexokinase, isoenzymes of glucokinase and glycogen synthesis from hexoses in neonatal rat liver. *Biochem. J. 90,* 261–268.

BALLARD, F. J., AND OLIVER, I. T. (1964b). The effect of concentration on glucose phosphorylation and incorporation into glycgen in the livers of fetal and adult rats and sheep. *Biochem. J. 92,* 131–136.

BROWN, D. D., AND DAWID, I. B. (1968). Specific gene amplification in oocytes. *Science 160,* 272–280.

CHUAH, C. C., AND OLIVER, I. T. (1971). Role of adenosine cyclic monophosphate in the synthesis of tyrosine aminotransferase in neonatal rat liver. Release of enzymes from membrane-bound polysomes *in vitro. Biochemistry 10,* 2990–3001.

FREYCHET, P., ROTH, J., AND NEVILLE, D. M., JR. (1971). Insulin receptors in the liver: specific binding of [^{125}I] insulin to the plasma membrane and its relation to insulin bioactivity. *Proc. Nat. Acad. Sci. U.S.A. 68,* 1833–1837.

GILBERT, W., AND MÜLLER-HILL, B. (1966). Isolation of the lac-repressor. *Proc. Nat. Acad. Sci. U.S.A. 56,* 1891–1898.

GREENGARD, O. (1970). The developmental formation of enzymes in rat liver. *In* "Biochemical Actions of Hormones" (G. Litwak, ed.), Vol. 1, pp. 53–87. Academic Press, New York.

GURDON, J. B. (1968). Transplanted nuclei and cell differentiation. *Sci. Amer. 219* No. 6, 24–35.

HOLT, P. G., AND OLIVER, I. T. (1969a). Studies on the mechanism of induction of tyrosine aminotransferase in neonatal rat liver. *Biochemistry 8,* 1429–1437.

HOLT, P. G., AND OLIVER, I. T. (1969b). Multiple forms of tyrosine aminotransferase in rat liver and their hormonal induction in the neonate. *FEBS Letters 5,* 89–91.

HOLT, P. G., AND OLIVER, I. T. (1971). Multiple forms of soluble rat liver tyrosine aminotransferase: premature induction and postnatal development. *Int. J. Biochem. 2,* 212–220.

HOLTZER, H. (1972). The obligatory requirement for DNA synthesis during myogenesis and erythrogenesis. *Proc. Australian Biochem. Soc. 5,* 14.

IWASAKI, Y., AND PITOT, H. C. (1971). The regulation of 4 forms of tyrosine aminotransferase in adult and developing rat liver. *Life Sci. 10,* 1071–1079.

KARLSON, P., AND SEKERIS, C. E. (1966). Biochemical mechanisms of hormone action. *Acta Endocrinol. 53,* 505–518.

LIN, E. C. C., AND KNOX, W. E. (1957). Adaptation of rat liver tyrosine transaminase. *Biochem. Biophys. Acta 26,* 85–88.

MOOG, F. (1970). Enzyme development and functional differentiation in the fetus. "Fetal Growth and Development" (H. A. Waisman and G. Kerr, eds.), pp. 29–48. McGraw-Hill, New York.

NEMETH, A. M. (1959). Mechanisms controlling changes in tryptophan peroxidase in developing mammalian liver. *J. Biol. Chem. 234,* 2921–2924.

OLIVER, I. T., AND YEUNG, D. (1972). Enzyme development in nutrition. *In Nutrition and Development* (M. Winick, ed.), pp. 27–48. John Wiley & Sons, New York.

SERENI, F., KENNEY, F. T., AND KRETCHMER, N. (1959). Factors influencing the development of tyrosine-α-keto glutarate transaminase activity in rat liver. *J. Biol. Chem. 234,* 609–612.

SCHIMKE, R. T., AND DOYLE, D. (1970). Control of

enzyme levels in animal tissues. *Ann. Rev. Biochem. 39,* 929–976.

SUZUKI, Y., GAGE, L. P., AND BROWN, D. D. (1972). The genes for silk fibroin in *Bombyx mori. J. Mol. Biol. 70,* 637–649.

TOMKINS, G. M., GELEHRTER, T. D., GRANNER, D., MARTIN, D., JR., SAMUELS, H. H., AND THOMPSON, E. B. (1969). Control of specific gene expression in higher organisms. *Science 166,* 1474–1480.

WALKER, D. G., AND HOLLAND, G. (1965). The development of hepatic glucokinase in the neonatal rat. *Biochem. J. 97,* 845–854.

WESSELLS, N. K., AND RUTTER, W. J. (1969). Phases in cell differentiation. *Sci. Amer. 220,* No. 3, 36–44.

WICKS, W. D. (1969). Induction of hepatic enzymes by adenosine 3',5'-monophosphate in organ culture. *J. Biol. Chem. 244,* 3941–3950.

YEUNG, D., AND OLIVER, I. T. (1967). Development of gluconeogenesis in neonatal rat liver: effect of premature delivery. *Biochem. J. 105,* 1229–1233.

YEUNG, D., AND OLIVER, I. T. (1968a). Factors affecting the premature induction of phosphopyruvate carboxylase in neonatal rat liver. *Biochem. J. 108,* 325–331.

YEUNG, D., AND OLIVER, I. T. (1968b). Induction of phosphopyruvate carboxylase in neonatal rat liver by adenosine 3',5'-cyclic monophosphate. *Biochemistry 7,* 3231–3239.

YEUNG, D., AND OLIVER, I. T. (1971). The postnatal induction of serine dehydratase in rat liver. *Comp. Biochem. Physiol. 40A,* 135–144.

Note: Original papers have been cited only where the substance of their content has not been recently reviewed.

CHAPTER SIXTEEN

Developmental Endocrinology

C. R. FILBURN AND
G. R. WYATT

The orderly process whereby an embryo develops into an adult organism capable of reproducing itself involves interactions between cells, mediated by chemical messengers. At given stages in this process, specific products of certain groups of cells have profound effects on the development of other adjacent cells, as with inducers, or on that of distant cells and tissues, as with hormones. These effects are dramatic and essential for the orderly development of the organism. For example, the action of inducers from the chordomesoderm of the vertebrate embryo on the overlying ectoderm is necessary for development of the entire nervous and sensory system. Increased thyroxine output from the thyroid gland of the tadpole is the signal that brings on metamorphosis into the adult amphibian; on the other hand, in larval insects it is decreased production of juvenile hormone, in the presence of ecdysone, that permits metamorphosis to the pupal or adult form.

Of these chemical messengers, the hormones have been most amenable to investigation, since pure compounds have been obtained that evoke specific responses. The general picture that has emerged is one in which specific hormones, individually or in combination, are necessary for the differentiation and maturation of specific embryonic cells and tissues, as well as for the function and replacement of certain differentiated cell types in the adult. The cells capable of responding to hormones are designated as *target cells*, and other cells remain unresponsive to the same substances. The action of the hormones usually involves stimulation of anabolic activities, particularly nucleic acid and protein synthesis, resulting in the production of new cellular constituents or products characteristic of the cell type. With some tissues, however, although there may be initial synthesis, the consequence is accelerated catabolic activity, resulting in the eventual destruction of the target cells themselves—for example, in the histolysis that occurs during metamorphosis.

In this chapter, we shall first consider some general aspects of endocrine organization, and then some selected systems that illustrate the salient features of hormonal regulation in development. We shall try to discover what exists in common and what is distinctive among them. For citation of the literature, we shall depend mainly on reviews, together with a few of the most significant and the most recent original papers. A readable and comprehensive textbook on endocrinology is that of Turner and Bagnara (1971).

Endocrine Organization, Hormones, and Hormone Action

ENDOCRINE SYSTEM OF VERTEBRATES

The vertebrate endocrine system is a complex communications system that uses chemical messengers. It is subject to regulation by the brain and also to internal feedback regulation, which together achieve a remarkable degree of coordination and stability in the control of body functions. A focal point in this system is the pituitary body, a structure made up of distinct neural and glandular subdivisions (the neurohypophysis and the adenohypophysis), attached by a neural stalk to the hypothalamic region of the base of the brain.

From the neurohypophysis, two hormones are released—vasopressin and oxytocin. These are true neurohormones, synthesized in neurosecretory cells in the hypothalamus and transported through axons into the neurohypophysis to be released into the blood. Since their principal functions—control of water reabsorption in the kidney (vasopressin) and milk release and uterine contraction (oxytocin)—are not essentially developmental, we shall dismiss them without further comment.

The adenohypophysis is the source of at least seven hormones, produced in different cell types in the gland. Four of these are *trophic hormones*, which act by stimulating the production of hormones by other endocrine glands, as indicated in Fig. 1. The other three act directly on nonendocrine tissues. Growth hormone (GH), or somatotrophin, stimulates body growth in a rather general sense, as will

be discussed shortly. Prolactin is named for its stimulation of milk production, but it also exerts diverse other effects in relation to metabolism, growth and reproduction, some of them overlapping those of growth hormone. Melanocyte-stimulating hormone (MSH) regulates pigment granule dispersal in chromatophores of poikilotherms, and it is not at all clear what role this hormone may have in mammals.

Secretion of these hormones from the adenohypophysis is regulated chiefly by yet another set of hormones, the *releasing factors*, which are produced in neurosecretory cells of the median eminence of the hypothalamus and transported directly to the adenohypophysis by a local system of portal blood vessels. According to present evidence, there are at least nine of these neurohormones, which serve either to stimulate or to inhibit the output of specific pituitary hormones. Release-stimulating factors have been shown for ACTH, TSH, LH, FSH, GH, MSH,* and prolactin, with strong indications that those for LH and FSH are identical, whereas release-inhibiting factors have been shown for GH, MSH, and prolactin (Burgus and Guillemin, 1970; Schally et al., 1973). The release of the releasing hormones themselves appears to be controlled in two ways: (1) by neural stimuli arising in response either to environmental signals such as light and temperature, to endogenous rhythms, or to changes in emotional state, particularly stress; and (2) by negative feedback mechanisms in which peripherally produced hormones (thyroid hormones and steroids) or pituitary hormones themselves (GH, MSH, and prolactin) act on the brain or hypothalamus to inhibit the release of the relevant releasing factors. Appropriately enough, the pituitary hormones for which hypothalamic inhibitors of release have been found are those whose target tissues do not produce hormones capable of feedback. In addition, output from the pituitary can be subject to direct inhibition by circulating hormones, such as that from the thyroid (Reichlin et al., 1972).

Thus neurosecretion represents the link be-

*ACTH, adrenocorticotrophic hormone; TSH, thyroid-stimulating hormone; LH, luteinizing hormone; FSH, follicle-stimulating hormone; GH, growth hormone; MSH, melanophore-stimulating hormone.

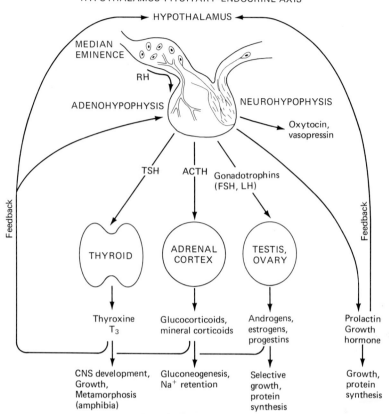

FIGURE 1. *Hypothalamus–pituitary–endocrine axis of the vertebrate endocrine system. Releasing hormones (RH) from neurosecretory cells of the hypothalamus are transported by hyphophysial portal veins to the adenohypophysis and regulate the synthesis and release of adenohypophysial hormones. The neurohypophysial hormones are neurosecretions, transported through axons from the hypothalamus, and are not under control by releasing hormones. The trophic hormones of the adenohypophysis stimulate peripheral endocrine glands to secrete hormones that modulate the activities of somatic target tissues and also, by negative feedback processes, act at different levels along the axis to control their own circulating levels.*

tween neural and endocrine systems of communication and control in the body, and the hypothalamus is the chief center for integration and transduction of signals. The pituitary provides for amplification, further control, and release of the message into the general circulation. These relationships are summarized in Fig. 1.

Some additional hormones are not under direct pituitary control, and their release is regulated by neural stimuli (for example, epinephrine and gastrointestinal hormones) or by feedback mechanisms based on blood levels of sugar (for insulin and glucagon) or calcium (for parathyroid hormone and calcitonin). The best understood hormones of vertebrates are listed in Table 1.

Chemically, the hormones fall into several distinctive groups (Table 1). The neurohormones are typically peptides; the structures of several hypothalamic releasing hormones have been determined recently, and all are small peptides of 3 to 12 amino acids. (Schally et al., 1973). All the adenohypophysial hormones and several of those of peripheral origin are either polypeptides, proteins, or glycoproteins. In

TABLE 1

STRUCTURES OF SOME MAMMALIAN HORMONES

Hypothalamus
    Thyrotropin: releasing hormone   L-(pyro) glu-his-pro-NH$_2$
    Gonadotropin: releasing hormone   L-(pryo) glu-his-trp-ser-tyr-
                                          gly-leu-arg-pro-gly-NH$_2$

    Growth hormone: releasing
      hormone $\leq$ decapeptide
Neurohypophysis
    Oxytocin, vasopressin   Peptides, 9 amino acids, cyclic
Adenohypophysis
    Thyrotropin, FSH, LH   Glycoproteins, MW 30,000
    ACTH   Peptide, 39 amino acids
    Growth hormone   Protein, 188 amino acids (human)
    Prolactin   Protein, MW 25,000 (sheep, ox)
Parathyroid
    Parathyroid hormone   Protein, 84 amino acids (bovine)
Thyroid
    Calcitonin   Peptide, 32 amino acids
Thymus
    Thymosin   Protein, 105 amino acids
Pancreas
    Glucagon   Peptide, 29 amino acids
    Insulin   Protein, 51 amino acids

Adrenal cortex

    Hydrocortisone   Aldosterone

    Testis   Ovary

      Testosterone   Estradiol   Progesterone

    Thyroid   Adrenal medulla
      Thyroxine   Epinephrine

some cases, the whole molecule is not required for biological activity: thus, up to 25 percent of certain somatotrophins may be digested away without loss of activity, and a synthetic peptide consisting of 19 of the 39 amino acids of ACTH is highly active (see Margoulies and Greenwood, 1971).

The vertebrate sex hormones and the hormones from the adrenal cortex are steroids, derived from cholesterol. The several classes of steroid hormones have distinctive structural features essential for their respective activities (Table 1). These relatively lipophilic compounds are only slightly soluble in water, and they are carried in the blood as complexes with specific transport proteins. The thyroid hormones are iodinated derivatives of tyrosine, which are released from the gland by cleavage from a protein, thyroglobulin. Both thyroxine (T_4) and the analogue 3,5,3'-triiodothyronine (T_3) occur naturally and are biologically active. Also derived from tyrosine (but having little else in common with the thyroid hormones) are the catecholamines epinephrine and norepinephrine. From this brief survey of hormone chemistry, one may justifiably conclude that substances of surprisingly diverse chemical structure can serve closely related functions.

ASPECTS OF HORMONE ACTION

In order to respond to a particular hormone, the cells of a target tissue must possess (1) receptors specific for the given molecule, and (2) the capacity to carry out, as a consequence of the interaction of hormone with receptor, a specific program of biochemical changes that yield the observed result. Hormone action has been investigated both by following the fate of hormones applied to target cells and by analysis of cell responses. Recently, these two paths have converged to yield some basic insight into subcellular mechanisms.

One such mechanism depends on the intracellular synthesis of cyclic 3,5'-adenosine monophosphate (cyclic AMP). This compound, discovered in 1958 through investigation of how epinephrine mediates the conversion of glycogen to glucose in mammalian liver, is now recognized as a "second messenger" in the action of many hormones, including the catecholamines as well as many of the peptide and protein hormones (Butcher et al., 1972; Robison et al., 1971). In this mechanism (Fig. 2), the receptors are located in the cell surface and the hormone can act on the cell without penetrating it, as has been shown by the activity of peptide hormones conjugated to cellulose or dextran. The immediate consequence is activation of adenyl cyclase, located in the inner surface of the cell membrane, which catalyzes the synthesis of cyclic AMP and causes an increase in its intracellular level. The cyclic AMP activates protein kinases that phosphorylate specific, key proteins ranging from particular enzymes to membrane proteins and chromatin constituents, and this results in changes in metabolism, permeability, and synthesis. The identification of the phosphorylated proteins, their roles, and their relations to particular protein kinases are currently active areas of research. These cyclic nucleotide-me-

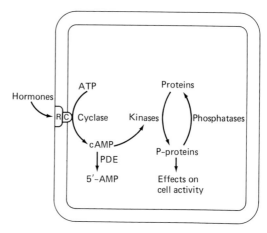

FIGURE 2. *Hormone action through cyclic AMP. Interaction of the hormone with cell surface receptors (R) modulates the activity of adenylate cyclase (C), producing a change in the intracellular concentration of cyclic AMP. Increased levels of the cyclic nucleotide activate protein kinases, then increasing the phosphorylation of various endogenous substrates (enzymes, membrane proteins, ribosomal proteins, and others), with resulting changes in cellular activities. Inhibition of cyclic nucleotide synthesis has opposite effects on these processes.*

diated responses usually occur with little or no lag time, and at least the initial steps are independent of protein and nucleic acid synthesis. At present, it remains unclear whether all cell surface receptors are linked to adenylate cyclase or whether there may exist additional mechanisms of primary hormone action via the cell surface.

A very different kind of hormone action is that of the steroids, which bind to cytoplasmic receptors forming a complex that migrates into the nucleus and apparently influences gene transcription. This mechanism will be discussed in some detail in the context of sex steroid action in a later section. Here, the overt response of the cell is usually observed only after a lag of hours or even days and is blocked by inhibitors of the synthesis of protein and RNA.

It would be wrong to assume that all the effects of vertebrate hormones are mediated by one or the other of the mechanisms just described. Insulin appears to inhibit adenylate cyclase in some tissues (Illiano and Cuatrecasas, 1972) but has other effects that may not involve cyclic AMP.* Thyroxine is known to activate myocardial adenylate cyclase (Levey and Epstein, 1969), but some of its actions in relation to growth and metamorphosis require protein and RNA synthesis and appear to involve intracellular receptors. In many instances of hormone action, the mechanisms are by no means clear.

Regardless of the mechanisms involved, some hormone actions are of far greater interest from the developmental viewpoint than others. Some effects are clearly metabolic—for example, the regulation of ion balance or blood sugar levels—while others are clearly developmental, as the morphogenetic effects of sex hormones. Some hormones, however, can produce both types, and some hormone actions are less easily categorized. Thyroid hormone, for example, regulates basal metabolic rate in adult tissue but is also required for nervous system development and amphibian metamorphosis. In the hope of

*Recently, a regulatory role for cyclic GMP (cyclic 3',5'-guanosine monophosphate) antagonistic and reciprocal to that of cyclic AMP has been proposed. Rapid elevation of cyclic GMP levels is produced in various tissues by acetyl choline, in cultured cells by mitogenic agents, and in fibroblasts and fat cells by insulin (Goldberg et al., 1973; Illiano et al., 1973).

gaining further understanding of the roles and mechanisms of action of hormones in animal development, we shall now consider some selected examples.

Endocrine Control of Vertebrate Growth and Development

HORMONAL REGULATION OF GROWTH

A basic process in development is growth. That growth in vertebrates is under pituitary regulation was known over 60 years ago from the marked reduction or cessation of somatic growth observed in animals adenohypophysectomized at an early age, but it was not until 1945 that a crystalline protein possessing growth-promoting activity was obtained (Li et al., 1945). It is now known that GH, given on a sustained basis to normal and hypophysectomized animals, causes skeletal and visceral enlargement, along with growth of muscular, connective, and lymphoid tissues. These effects are obtained, though to a lesser extent, in the absence of thyroid, adrenal, or sex hormones. Growth hormone apparently has little effect on the central nervous system and its derivatives, which grow at normal rates in its presence or absence.

The effects of GH are most apparent in conditions resulting from abnormal levels of the hormone. In humans, excessive production of GH before adolescence can result in giantism, with enlargement of all parts of the body in normal proportions. If excessive secretion (associated with tumors of the pituitary acidophilic cells or hypothalamic production of growth hormone-releasing hormone, GHRH) occurs after adolescence, there is no increase in height. However, disproportionate growth of skeletal tissues of the extremities (lower jaw, nose, feet, and hands) and of the tongue, liver, and various glands does ensue, producing the condition termed *acromegaly*. Deficient production of GH, on the other hand, results in pituitary dwarfism. Such dwarfs usually develop adult mental capacities but remain child-like in appearance and sexual development. During the past decade human growth hormone (HGH) has become available in sufficient quantities to permit successful treatment of such GH-deficient children (Tanner, 1972).

At the biochemical level, the regulation of growth by GH now appears to be the cumulative effect of changes in the protein, carbohydrate, lipid, and nucleic acid metabolism of various target tissues. In the intact animal, the hormone induces a positive nitrogen balance, lowering amino acid levels in the blood, and decreasing urea production, while increasing utilization of nitrogen in anabolic processes. In muscular, connective, and lymphoid tissues, GH increases DNA levels, indicating that cell or nuclear multiplication is also part of the growth response (Knobil and Hotchkiss, 1964; Snipes, 1968). The hormone alone has a slight stimulating effect on trophic hormone target tissues but acts synergistically to markedly increase the effectiveness of these adenohypophysial hormones.

The clearest responses to GH are shown by muscular and adipose tissue. Isolated rat diaphragms show increased amino acid uptake, protein synthesis, and glucose oxidation (Kostyo, 1968). Adipose tissue responds at first by increasing its rate of uptake and use of glucose, but within 2 to 4 hours these rates and that of fatty acid synthesis fall to subnormal values (Goodman, 1968). All these changes require either RNA or protein synthesis, usually both. At the level of transcription, GH *in vivo* causes increases in the liver in both nucleolar and nucleoplasmic RNA polymerase activities, synthesizing ribosomal precursor and other types of RNA (Salamon et al., 1972). An effect of GH not dependent on new RNA, however, is an early stimulation of protein synthesis that appears to be due to increased translational efficiency of membrane-bound ribosomes (MacDonald and Korner, 1971). This results from enhanced ribosomal capacity to promote peptide bond formation, with no change in the rate of chain initiation (Kostyo and Rillema, 1971).

While all of these observations illustrate the diversity of responses to GH, little is known about its primary action at the receptor level. The facts that a very brief *in vitro* exposure of target tissues to the hormone is sufficient to elicit a response (Goodman, 1968) and that the enzymes phospholipase C, trypsin, and collagenase destroy target tissue responsiveness (Rillema and Kostyo, 1971; Eastman and Goodman, 1971) suggest action through cell surface receptors. Since theophylline, an inhibitor of cyclic nucleotide phosphodiesterase, blocks the effects of GH on isolated rat diaphragms, it is possible that the hormone decreases the basal level of cyclic AMP, perhaps by inhibiting adenylate cyclase. The many similarities between the actions of GH and insulin, which can inhibit adenylate cyclase under some circumstances, support this possibility. However, since insulin has recently been shown to elevate levels of cyclic GMP in liver cells (Illiano et al., 1973), it is possible that some effects of GH may be mediated through altered levels of this cyclic nucleotide.

As mentioned earlier, one of the effects of GH in hypophysectomized animals is the stimulation of skeletal tissue growth, including the formation of cartilage. In 1957, however, Salmon and Daughaday found that GH *in vitro* fails to stimulate, as it does *in vivo*, the uptake of sulfate into cartilage of hypophysectomized rats. Uptake was stimulated by normal rat serum, while serum of hypophysectomized rats had no effect. Therefore, it was proposed that this action of GH might occur indirectly through a second substance operationally termed *sulfation factor*. This hypothesis has since been confirmed. Sulfation factor or somatomedin (SM) (Daughaday et al., 1972), a polypeptide found in normal serum, is deficient in hypopituitary rats and humans, and restored by treatment with GH (see Tanner, 1972). Thus, GH acts as a trophic hormone by stimulating the output of SM from a GH target tissue, possibly the liver.

Injected into hypophysectomized animals, SM causes widening of tibial epiphyseal cartilage, a response previously seen only with GH. Isolated cartilage responds to SM with increases in synthesis of chondromucoprotein, collagen, RNA, and DNA, in addition to the sulfation of chondroitin sulfate. Other GH-like effects *in vitro* include increased amino acid uptake and protein synthesis in rat diaphragms, as well as increased use of glucose by adipose tissue (Daughaday et al., 1972). Thus, while many of the observed responses to GH result from direct action on target tissues, some, including those affecting cartilage, are secondary to a stimulated output of SM.

In its primary action SM, like GH, acts at least partly through cell surface receptors. SM exerts metabolic effects on several tissues similar to those of insulin, and the sharing of receptors by these two hormones has, in fact, been demonstrated by competition for binding sites on isolated fat cells, liver membranes, and chrondrocytes (Hintz et al., 1972). Since SM inhibits the basal and hormone-stimulated adenylate cyclase activities of these and other tissues, its anabolic actions may be due in part to a reduction in intracellular cyclic AMP and cyclic AMP-stimulated catabolic activities. This possibility seems reasonable in view of the role that the cyclic nucleotide plays in regulating growth in cell cultures (Tell et al., 1973).

Much remains to be learned about the roles of SM and GH in the regulation of growth. Why, for example, does growth in normal animals cease when it does and why does the response to GH in GH-deficient children decrease as they approach their normal growth rates? This may result from a decreasing capacity of tissues to respond to SM, as suggested by the observation that costal cartilages of rats near the end of their normal growth are less responsive to SM than those from younger, rapidly growing animals. Alternatively, the GH regulation of SM may change, as indicated by the finding of low SM in children with diminishing "catch-up" growth rates under GH therapy (see Tanner, 1972). Obviously the roles of GH and SM in hypothalamic secretion of GHRH also have to be known for full understanding of the factors that determine the rate and duration of growth.

THYROID HORMONES AND CENTRAL NERVOUS SYSTEM

The role of the thyroid gland's secretions in the development of the central nervous system is amply demonstrated by the disorder of cretinism in humans. Cretins are reduced in physical stature, but unlike pituitary dwarfs they are also mentally deficient as a result of developmental abnormalities largely attributable to hypothyroidism. Since the demonstrations that T_4 (thyroxine) does, in fact, have direct effects on amphibian neural tissues (see Kollros, 1968), much effort has gone into determining when and how the hormone acts on developing nervous systems.

Most experimental work with mammals has been done on the rat. Its CNS undergoes most of its growth and maturation in the postnatal period, when both automatic and adaptive behavioral patterns develop. Thyroidectomy delays but does not prevent the emergence of automatic responses, while adaptive behavior, or the capacity for learning, is markedly impaired. This operation is ineffective, however, if delayed until at least 2 weeks after birth. Similarly, replacement therapy restores performance to normal, but only if begun within 10 days after birth. Impaired adaptive behavior may also result from an excess of T_4 but, again, only if hormone administration is started within this early critical period. In both cases, the degree of impairment correlates strongly with the time at which the hyper- or hypothyroid state begins (Eayrs, 1971).

At the structural level, hypothyroidism is characterized by reduced cerebral and cerebellar growth, decreased neuronal growth affecting the cell bodies as well as axons and dendrites, and altered vascularity resulting in fewer, larger blood vessels. A reduction in brain weight appears only after 14 days of postnatal development and results largely from a decrease in average cell size. Cell proliferation in the cerebellum is retarded to a small but significant extent, but the normal cell number is reached by 35 days. Postnatal hyperthyroidism, on the other hand, leads to reduction in brain size attributable not to cell size but to cell number, particularly in the cerebellum, where a reduced rate of formation (increase in total brain DNA) first appears at 12 to 14 days of age (Balazs et al., 1971).

At the subcellular level, thyroid deficiency appears to cause a reduction in nerve terminals, as shown by selective decreases in synaptosomal glutamate decarboxylase and succinic dehydrogenase. RNA synthesis is little affected, but the rate of protein synthesis in hypothyroid brains is somewhat decreased. Myelination, as judged by cerebroside content, is markedly impaired. Neurochemical maturation, considered in terms of the development of a characteristic compartmentation of glutamate and of conversion of glucose to amino acids associated with the tricarboxylic acid cycle, is advanced in hyperthyroids and retarded in thyroid deficiency (Balazs et al., 1971).

These changes in the developing brain are temporally—and probably causally—related to the development of behavioral impairment. But the mechanism(s) by which thyroid hormones influence nervous system development at present remain obscure. It has been suggested that T_4 pushes cells into the differentiative phase by inhibiting proliferation, thus permitting maturation to begin; if this transition occurs too soon or too late due to abnormal hormone levels, faulty coordination of maturational processes may result. The prolonged proliferation of "extragranular" cells in cerebella of hypothyroid rats supports this hypothesis (Hamburgh et al., 1971). The acceleration by T_4 of myelinogenesis in cultured rat cerebella demonstrates a direct action of the hormone on neural target tissues (Hamburgh, 1968). The situation *in vivo*, however, is complicated by the presence of other hormones that may affect some aspects of brain development (possibly GH—Gomez, 1971) and by hormonal imbalance likely to result from the effects of thyroidectomy on the pituitary (Schooley et al., 1966).

The actual biochemical mechanism of thyroid hormone action on neural tissues is poorly understood. While, in liver, the primary action appears to be at the genomic and possibly other levels (see discussion of amphibian metamorphosis), only recently have there been any data concerning thyroid hormone receptors in the brain. Luckock and Timiras (1972) found a greater uptake of ^{14}C-T_4 in the developing rat brain, particularly in the nuclear and mitochondrial fractions, at day 12 than at days 6 or 22 after birth. Since the twelfth day of development falls in the critical period for thyroid action on CNS development discussed earlier, this observation suggests nuclear and mitochondrial "receptors" involved in the maturational effects of the hormone. Further studies, perhaps comparable to those done with T_4 in amphibian and mammalian liver, are needed.

THYROID REGULATION OF AMPHIBIAN METAMORPHOSIS

In addition to an important role in relation to development of the vertebrate central nervous system, thyroid hormone is also required for maturation of the skeletal and reproductive systems. This capacity to influence the development of different tissues is common to a number of vertebrate hormones. Perhaps nowhere is this important fact of developmental endocrinology better illustrated in the vertebrates, however, than in the rapid and dramatic transition that thyroid hormones elicit in amphibian metamorphosis. In this process of developmental adaptation an aquatic larval form, with its associated anatomical and physiological features, is transformed into a terrestrial form. The herbivorous tadpole with its long tail, lidless eyes, gills, horny teeth, and lengthy intestine, turns into a four-legged carnivore possessing eyelids, a tongue, a shortened intestine, and many other adult-like features. Larval organs disappear or undergo considerable modification while new organs develop, all in an orderly sequence. Only after the legs are fully developed does the shrinking tail finally become nonfunctional. In many species this whole process is very rapid, occurring in only a few days.

Associated with this morphological transformation are biochemical changes that serve to equip the animal for its adult habitat. The liver begins producing the enzymes that catalyze the synthesis of urea, making possible the excretion of nitrogen in this form instead of the more toxic ammonia. Hemoglobin synthesis shifts from production of a larval to an adult species that is more suited to the breathing of gaseous oxygen. Along with these patently adaptive changes is a shift in serum protein patterns, the appearance of a secretory product of adult skin glands (serotonin), a shift in the type of intestinal proteolytic activity, and many other biochemical changes that together yield the characteristic makeup of the metamorphosed animal (Weber, 1967; Frieden and Just, 1970).

The various and specific responses of larval tissues result from the direct action of thyroid hormone, as demonstrated by topical application and organ culture experiments. Pellets of

T_4 implanted in the dorsal tail fin of the tadpole cause localized regression or, imbedded adjacent to the midbrain, stimulate only nearby cells to increase in size. Cultured tail tips survive quite well in a balanced salt solution but regress readily when T_4 or T_3 (triiodothyronine) is added.

The specificity and timing in the responses of the different larval tissues have important implications for the nature of the hormonal regulation. The amphibian larval tissues acquire sensitivity or "competence" to respond to experimentally administered hormone at different stages of development. While the tail fin becomes sensitive early in development, the cells of the mesencephalic nucleus are responsive much later, and the median eminence of the hypothalamus develops responsiveness only just prior to metamorphosis. Stage-dependent acquisition of hormonal sensitivity is known in other vertebrate systems as well—for example, sensitivity of rat liver to T_4-induced enzyme differentiation (Greengard, 1970), and chick hypothalamic sensitivity to androgenic modification of control of ovulation (Gorski, 1968). A basic question in all these systems is: Which of the two conditions for a hormonal response —(1) presence of receptors, or (2) the capacity to initiate changes following hormone-receptor interaction—is the last to be fulfilled during ontogeny. At present the only information on this question for an amphibian system comes from a study by Tata (1970a) on T_4 binding in whole *Xenopus laevis* larvae. Acquisition of metamorphic competence, defined as the capacity of whole embryos to alter their rates of RNA synthesis and phosphate uptake, at 36 to 60 hours after fertilization is accompanied by the appearance of sedimentable, temperature-sensitive molecules that strongly and specifically bind T_4 and its active analogue. Such evidence would be more informative if obtained for specific tissues, which may well differ from one another.

Some pertinent data do exist, however, with regard to the orderly sequence of metamorphic events. It is known that the degree of sensitivity to thyroid hormone varies over a wide range—the organs and structures that normally metmorphose first being more sensitive than those that begin changing later. This range is seen in the threshold levels of hormone needed to elicit a response—for example, minimal effective doses increase in the following order: hindlimbs < intestine < forelimbs < tail. A given organ may exhibit progressively increasing thresholds for successive stages of development; thus, increasing concentrations of hormone are needed for hindlimb development to proceed from the three-digit to the five-digit stage and on to the web-and-toe-pad stage (see Kaltenbach, 1968).

The existence of these different target tissue sensitivities can explain the sequence of onset of metamorphic changes if, in fact, the circulating levels of thyroid hormones gradually rise. During late larval development, the thyroid gland grows at a relatively high rate, reaches its maximum size at metamorphic climax, and has the structure characteristic of an actively secreting gland. Chemical, biochemical, and electrical studies do, in fact, lend qualitative support to the hypothesis of increasing thyroid activity prior to metamorphic climax. However, circulating levels of thyroid hormones (protein-bound iodine) appear to increase suddenly just prior to metamorphic climax and to a lesser total extent (Just, 1972) than the gradual increase predicted by Etkin (1968).

The actual increase in circulating hormone level apparently reflects secretion of hormone stored in the thyroid. By analogy with higher vertebrates, regulation by TSH from the pituitary, controlled in turn by thryotrophin-releasing hormone (TRH) from the hypothalamus, may be expected. In the tadpole, differentiation and development of the median eminence and portal vessels takes place just prior to metamorphic climax. This process does not occur, however, in thyroidectomized tadpoles and is accelerated by thyroid hormones only in animals that have reached a certain premetamorphic stage. Only when hypothalamic tissues become sensitive to the hormone would the positive feedback effect begin, leading to the observed sudden increase in circulating hormone. While an amphibian TRH remains to be directly demonstrated, the recent demonstration of stimulation by mammalian TRH of localized shrinkage in cultured pieces of tail fin containing contiguous pituitary and thyroid implants certainly supports this mechanism (Etkin and Kim, 1971).

These processes, however, may only partially account for the regulation of amphibian metamorphosis. Recent studies have shown that mammalian prolactin and pituitary growth hormones increase growth rates of normal and hypophysectomized amphibian larvae and at the same time inhibit tail reduction, hindlimb growth, and possibly thyroid gland function. While the effect on growth may result from some of the actions of GH previously discussed, the nature of the metamorphosis-inhibiting action remains obscure. A central antithyroid effect may occur, but a direct peripheral effect is demonstrated by inhibition of tail tissue regression by pituitary grafts (Derby, 1970; Frye et al., 1972). In either case, the onset and rate of metamorphosis may be governed by the blood levels of both prolactin (or GH) and thyroid hormones, and this presents an interesting analogy with the action of juvenile hormone in relation to insect metamorphosis.

Regardless of the number of hormones that may eventually prove to participate in regulating metamorphosis, the effects that are produced are impressively specific and generally independent of location in the body. While the hindlimb grows and develops, adjacent tissues regress; the muscular tissue of the limbs develops, while that of the tail undergoes histolysis leading to resorption. In a dramatic example of this position-independent specificity, Schwind (1933) transplanted an eyecup onto a tadpole tail, only to watch it remain healthy and eventually fuse to the body while surrounding tail tissues were completely resorbed.

Some understanding of the manner in which thyroid hormones stimulate and modify the biosynthetic activities of larval organs has been gained from studies on the liver. Unlike many of the developing tissues of the metamorphosing tadpole, the liver undergoes little or no cell division during this period, and thus the biochemical and structural modifications occur in a stable population of cells. Exposure to T_4 produces increases in the rates of synthesis of nuclear and cytoplasmic RNA, followed by the appearance of new enzymes and proteins. By the end of the normal metamorphosis the hepatocytes have a more elaborate ultrastructure, which is associated with a much increased capacity for RNA and protein biosynthesis.

Precocious induction with T_4 causes a shift in ribosomes from attachment to vesicular membranes to localization primarily on more complex double lamellar structures. Simultaneously, there is an increased synthesis of ribosomes and membrane phospholipid, along with increased breakdown of ribosomes and an enhanced rate of protein synthesis, especially in the heavy rough membranes of fractionated cells. Figure 3 shows a schematic representation of these events as they occur in the T_3-treated *Rana catesbeiana* tadpole (Tata, 1970b).

The prolonged lag periods shown in the diagram are not characteristic of all the effects of thyroid hormones on the liver. Eaton and Frieden (1968) found that within 3 hours after injection, T_3 produced an increase in incorporation of adenosine into RNA and in the specific radioactivity of nucleotide precursors. When measured by ^{32}P labeling, the RNA synthesized at 3 and 12 hours after hormone treatment had a DNA-like composition compared to the predominantly ribosomal RNA (rRNA) synthesized in controls.

The RNA synthesis observed after about 2 days of hormone treatment is marked by an increased production of species that range between transfer RNA (tRNA) and rRNA in size. This rapidly labeled RNA fraction is DNA-like in base composition, suggesting increased transcription of messenger RNA (mRNA). Recent work has shown that, in fact, both nucleolar and nucleoplasmic RNA polymerase activities are stimulated at this time (Griswold and Cohen, 1972), an effect that is due in part to an increase in chromatin template efficiency. Chromatin prepared from liver nuclei of T_4-treated tadpoles is 20 to 50 percent more efficient than that of untreated animals in supporting *Escherichia coli* RNA polymerase activity and, when partially melted by heating, is also more efficient in hybridizing pulse-labeled RNA (see Cohen, 1970).

This increase in chromatin template efficiency strongly suggests a primary effect of thyroid hormones on tadpole liver cells at the genomic level. A recent study by Griswold et al. (1972) of the intracellular distribution of ^{14}C-T_4 further supports this concept: the hormone accumulates only transiently in the supernatant, microsomal, and mitochondrial fractions of the cells but reaches maximal and

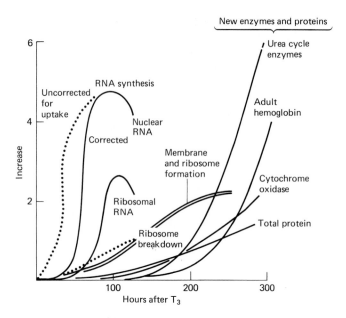

FIGURE 3. *Sequential stimulation of rates of RNA and phospholipid synthesis in relation to the increases in enzymes or protein synthesized upon the precocious induction of metamorphosis in* Rana catesbeiana *tadpoles with triiodothyronine. Curves show the rate of rapidly labeled nuclear RNA synthesis; specific activity of RNA in cytoplasmic ribosomes; breakdown of ribosomes labeled before induction; rate of microsomal phospholipid synthesis; urea cycle enzymes (carbamyl phosphate synthetase); cytochrome oxidase per milligram of mitochondrial protein; appearance of adult hemoglobin in the blood; total liver protein per milligram of wet weight. The values are expressed as percentage increases over those in the noninduced control tadpoles. The decreasing values in curves of RNA synthesis reflect the dilution of specific radioactivity in precursor molecules due to the onset of regression of tissues such as the tail and intestine. [From J. R. Tata, in "Biochemical Actions of Hormones" (G. Litwack, ed.), vol. 1, pp. 89–133, Academic Press, New York, 1970.]*

stable levels in the nuclei. Analysis of the nuclear components showed unmodified T_4 very tightly bound to the chromatin. In tadpoles maintained at $5°C$, however, the hormone accumulates in the mitochondrial and supernatant fractions, with negligible entry into the nuclei. Thus entry into the nucleus is temperature-dependent, a phenomenon observed with steroid hormones in mammalian target tissues, as discussed in a later section. This effect may well explain the ineffectiveness of thyroid hormones in inducing metamorphosis in amphibian larvae at $5°C$ (see Weber, 1967) and could be substantiated by comparable experiments with other larval tissues.

Other recent studies on the fate of thyroid hormones in vertebrate liver suggest that an action at the genomic level may be only one part of the picture. Dimino and Hoch (1972) found a considerable enrichment of iodine in liver mitochondria of rats injected with T_4. These mitochondria were more dense than those of untreated animals and appeared to contain iodine tightly bound to their inner membranes. Cohen (1970) also found that tadpole liver mitochondria have increased diameter and matrix density and are more frequently associated directly with endoplasmic reticulum following T_4 treatment. Direct effects of T_4 on isolated mitochondria have been known for some time, but they occur only at high, unphysiological concentrations and their significance is doubtful.

Important biochemical changes associated with the development of ureotelism occur in tadpole liver mitochondria after thyroid treatment. Two of the ornithine–urea cycle enzymes, carbamyl phosphate synthetase-I (CPS-I) and ornithine transcarbamylase, are localized in the mitochondria and increase in activity—CPS-I markedly so—in concert with the three nonmitochondrial enzymes of the cycle during induced metamorphosis (Wixom et al., 1972). The increase in CPS-I results from changes in the rates of extramitochondrial synthesis of enzyme precursor, conversion into immunoprecipitable enzyme, and degradation of active enzyme (Cohen, 1970). Since precursor conversion occurs in the mitochondria, enhanced transport into this organelle is implied. Thus the increase in this enzyme depends on hormonal stimulation at more than one site.

There is also evidence for an effect of thyroid hormones on the translation of RNA. Unsworth and Cohen (1968) found that ribosomes from T_4-treated tadpole livers were more efficient in incorporating amino acids than those of untreated controls. A similar effect of T_4 is known in mammalian liver, although in this case some controversy exists over the necessity for mitochondria for the stimulation (Sokoloff et al., 1968; Sokoloff and Roberts, 1972). While the nature of this enhanced efficiency of tadpole liver ribosomes remains to be elucidated, the similarity to effects on ribosomes obtained with insulin (Wool et al., 1968) and GH (Kostyo and Rillema, 1971) is noteworthy.

Among the larval organs that respond to T_4 by regression and eventual disappearance, the tail has received by far the most attention. Amputated tails survive for extended periods in a simple balanced salt solution but regress when T_4 is included, thus providing a model system for the study of hormonal regulation of tissue involution and the associated biochemical changes. As with many of the other metamorphic effects of T_4, the first visible signs of regression (tail fin retraction) appear only after a lag period of 3 to 4 days. At the enzyme level, resorption is marked by a selective retention of some acid hydrolases and actual increases in total activity per tail for cathespin, deoxyribonuclease, β-glucuronidase, and acid phosphatase (Frieden and Just, 1970). A shift of acid hydrolase activity from the bound to the soluble phase has been reported (see Weber, 1969; Robinson, 1972), but the small extent of these changes argues against an active role for lysosomal breakdown during this cytolysis. Inhibitors of RNA and protein synthesis prevent both regression and increased hydrolase activities, which has prompted the proposal that thyroid hormones selectively stimulate hydrolase synthesis and resultant degradation of tail tissues (Weber, 1969; Just and Frieden, 1970). Recent studies on isozymes of acid phosphatase have shown differential changes in activities that may be causally related to the initiation of cell death (Robinson, 1972; Filburn and Vanable, 1973). An early *in vitro* effect of T_3 is an increase, eventually 3- to 5-fold, in the total activity of an acid phosphatase specific for nucleoside di- and triphosphates (Filburn and Vanable, 1973), suggestive of a hormone-mediated disruption of energy metabolism. It remains to be demonstrated, however, whether the increased activity of this or other hydrolases represents a primary cause in autolysis and cell death or a consequence of these processes—for example, an activation by cellular debris of macrophages that increase their hydrolase content. Elucidation of the manner in which thyroid hormones affect these processes will add considerably to our understanding of the regulation of programmed cell death in developing systems.

MAMMARY GLAND DEVELOPMENT

In contrast to the multiple responses of amphibian tissues to stimulation by a single hormone, the development of mammalian mammary glands involves the responses of a single cell type to several hormones acting in concert and in sequence. Estrogens stimulate growth and development of the duct system of this exocrine gland, but only when progesterone is also present does full alveolar development occur. In hypophysectomized animals, however, these steroids fail to stimulate development. Experiments by ablation and replacement therapy pointed to roles for prolactin, growth hormone, and adrenal corticoids in normal lobuloalveolar growth (Lyons et al., 1958). But experimentation *in vivo* allows only limited control of hormonal milieu. Cultures of explanted mouse mammary tissue in media with controlled hormonal content have recently been used with great success for analyzing the roles of several hormones in the differentiation of the epithelial cells of the gland (Topper, 1970; Turkington, 1968, 1972).

The normal development of the gland during pregnancy encompasses a phase of proliferation, followed by differentiation for the production of the specific components of milk at parturition. Casein and lactose synthetase A protein (UDP-galactosyltransferase) are present at low levels during the first half of pregnancy and at much higher levels later, while the significant production of lactose synthetase B-protein (α-lactalbumin, which confers upon the transferase specificity for the synthesis of lactose) begins only at parturition. Lactose synthesis is thus initiated by the appearance of the B protein. These specialized protein products serve as biochemical indicators of the functional differentiation of the cells.

The epithelial cells of cultured mammary explants begin proliferation upon exposure to insulin, epithelial growth factor from mouse submaxillary glands, or to a lesser extent pituitary growth hormone, though it is unclear which is the effective mitogen *in vivo*. In tissue explanted from virgin mice, this response is delayed unless the animals are pretreated with prolactin (Oka and Topper, 1972), and since prolactin is present in the blood at rising levels during pregnancy, it appears that this hormone sensitizes the epithelial cells to mitogenic action.

For differentiation and the production of milk, both a corticosteroid and further exposure to prolactin are required. Cultured cells must undergo a round of cell division and be exposed to hydrocortisone (or another adrenal corticoid) before synthesis of the specific proteins can be induced. Prolactin induces the actual synthesis of these proteins, and thus this hormone has a second essential role, distinct from its earlier one of sensitizing the cells to insulin. In addition, the presence of insulin after cell division is necessary for the induction of protein synthesis by prolactin. In this synergistic role, the requirement for insulin is specific, and other mitogens cannot substitute, but prolactin can be replaced by placental lactogen, which closely resembles it structurally and functionally. These complex requirements are illustrated in Fig. 4.

Let us now consider in more detail the role

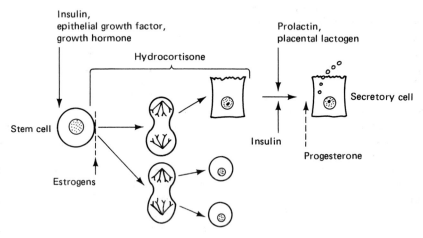

FIGURE 4. *Sequence of actions of hormones on mammary epithelial cells during alveolar (secretory) cell differentiation in organ culture. Prolactin in vivo prior to explantation sensitizes the stem cells so that the duplication process begins immediately upon exposure to mitogens. Hydrocortisone may be added after cell division is complete, but a longer culture period is then necessary for the cells to respond to prolactin with synthesis of secretory products. [From R. W. Turkington, in "Biochemical Actions of Hormones" (G. Litwack, ed.), vol. 2, pp 55–80, Academic Press, New York, 1972.]*

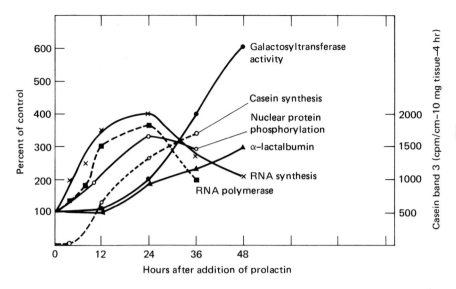

FIGURE 5. *Time course of the effect of prolactin and insulin on macromolecular synthesis in mammary epithelial cells pretreated with insulin and hydrocortisone for 96 hours. Rate of RNA synthesis (×); rate of phosphorylation of nuclear proteins (○); rate of synthesis of casein "band 3" (□). Each point represents the incorporation of isotopic precursor during the preceding 4-hour labeling period. Enzymatic activities of the galactosyltransferase of lactose synthetase (●) and of α-lactalbumin (▲); ■, RNA polymerase activity. [From R. W. Turkington, in "Biochemical Actions of Hormones" (G. Litwack, ed.), Vol. 2, pp. 55–80, Academic Press, New York, 1972.]*

of each class of hormone. The female sex steroids are known as potent stimulators of mammary gland growth and development *in vivo*. *In vitro*, estradiol at concentrations low in the physiological range (10^{-12} M) inhibits insulin-mediated cell proliferation, while at higher concentrations (10^{-10} M) this effect is reversed and maximal responses are permitted. Progesterone stimulates cell division selectively at sites of ductal budding, thus promoting morphogenesis of the gland, and also inhibits the induction of α-lactalbumin by prolactin. These effects of estradiol and progesterone are in accord with their increasing plasma concentrations during pregnancy and rapid decrease at parturition. However, since both hormones so strongly stimulate mammary growth *in vivo*, other possibly indirect actions may also be important in normal mammogenesis.

In preparing the mammary epithelial cells for maturation, the adrenal corticoids function at least in part by inducing changes in cytoplasmic structure, particularly a marked accumulation of rough endoplasmic reticulum (RER) characteristic of secretory cells. The accumulation of this RER, associated enzymes (e.g., NADH-cytochrome c reductase), and rRNA depends on treatment with both insulin and the steroid. Since prolactin induces casein synthesis only in cells containing newly accumulated RER, the entire sequence is essential for functional differentiation (Oka and Topper, 1971). Recently, a cytoplasmic glucocorticoid receptor protein has been demonstrated in lactating mouse mammary glands (Shyamala, 1973), which undergoes a change in sedimentation constant analogous to that observed with uterine estrogen receptors (described in a later section). It is thus likely that the primary action of the corticosteroids in mammary epithelial cell development is at the genomic level.

Of the polypeptide hormones, insulin stimulates RNA and protein synthesis, and RNA polymerase activity, prior to the increases in DNA and histone synthesis that mark the beginning of proliferation in the epithelial cells.

The addition of prolactin to the insulin and hydrocortisone pretreated cells also produces a rise in RNA synthesis, followed by production of the milk proteins (Fig. 5). In these stimulations of RNA synthesis, all three functional types—rRNA, tRNA, and mRNA—are affected (Green et al., 1971; Turkington, 1972). Competitive hybridization of the rapidly labeled nuclear RNA of virgin and lactational epithelial cells, however, showed that the latter contains molecular species lacking in the virgin tissue. This finding, together with the demonstration that prolactin stimulates polysome formation, suggests that some of the newly made RNA is messenger coding for the milk proteins.

Concurrently with the stimulation of RNA synthesis by insulin and prolactin, both histones and other nuclear proteins undergo increased phosphorylation, which is blocked by the presence of actinomycin. In this process, phosphate is esterified to serine and threonine residues of preexisting chromosomal proteins. Insulin alone acts to increase the phosphorylation of certain classes of histones. The stimulation by prolactin depends upon the simultaneous presence of insulin and the previous prolonged exposure of the cells to an adrenal corticoid; and in this case the phosphorylation of two histone fractions, designated as F_2a_2 and F_2b, is selectively stimulated. Upon investigation of the protein kinases responsible for phosphorylation, Majumder and Turkington (1971) found two cytoplasmic enzymes, one of them activated specifically by cyclic AMP. In cultured epithelial cells, these enzymes fall quickly to about 50 percent of initial levels, but the presence of insulin prevents this fall and causes a transient increase. Prolactin, in the presence of insulin, causes a more marked increase in kinase activities, preceding the rise in nuclear RNA synthesis. Both of the hormonally stimulated increases in protein kinase activity are blocked by inhibitors of RNA and protein synthesis. More recently, certain proteins of ribosomes and plasma membranes have been found to undergo increased phosphorylation when mammary cells are stimulated by prolactin, correlated with increases in activity of protein kinases (independent of cyclic AMP) found in these organelles (Majumder and Turkington, 1972). Although the role of this hormonally stimulated nonchromosomal protein phosphorylation is unclear, it seems likely that the nuclear protein phosphorylation, particularly that stimulated by prolactin, is causally linked with the changes in gene activity that occur during cell differentiation.

The nature of the primary actions of insulin and prolactin in this system remains to be determined. Insulin and prolactin bound to Sepharose elicit increased RNA synthesis, which shows that this response is initiated through cell surface receptors, but it has not been shown that cell surface interactions are sufficient to evoke all the responses to these hormones. It seems unlikely that insulin or prolactin is acting by stimulation of adenyl cyclase, since the increased levels of protein kinases observed with these hormones cannot be elicited by treatment of cells with cyclic AMP and its analogues, or with inhibitors of cyclic nucleotide hydrolysis (Majumder and Turkington, 1971). It appears that the levels of cyclic nucleotide-dependent and independent protein kinases may, in fact, be limiting elements in mammary gland development. We are thus left with the speculative alternatives of the existence of intracellular receptors for insulin and proclactin or an unidentified "second messenger" released as a consequence of hormone action at the cell surface that may mediate the effects that take place in the cell nucleus. In any case, the apparent sequence from increased protein kinase activity to nuclear protein phosphorylation, RNA synthesis, and eventual specific protein synthesis suggests a novel mechanism for hormonal regulation of a differentiative process.

Apart from questions of primary mechanism, the interdependence of hormone actions in this system is clear. This may require prior exposure to one hormone for the action of another, as with prolactin sensitization to the mitogenic action of insulin and adrenalcorticoid conversion to inducibility by prolactin; or it may necessitate simultaneous exposure to two hormones, as in the insulin requirement for induction of specific protein synthesis by prolactin. The outstanding feature of these studies is that, with a single cell type, they so clearly illustrate these interdependent relationships in hormone action, which undoubtedly pervade the growth and development of the entire organism.

DIFFERENTIATION OF REPRODUCTIVE SYSTEM

There is perhaps no better illustration of the essential role of hormones in vertebrate development than in the control of differentiation of the reproductive system. Although vertebrate embryos are genetically determined as to sex, during their development they pass through an ambisexual period wherein both sexes possess similar gonadal and accessory sex organ primordia. With the exception of cyclostomes and teleosts, the vertebrate primordial gonads consist of cortical and medullary regions that contain both somatic and germ cells. In males the medullary component continues to develop, giving rise to Sertoli and Leydig cells, with the gonocytes developing into sperm, while the cortical region regresses. In females the cortical region develops and gives rise to follicle cells, with the gonocytes developing into eggs, while the medullary region involutes. Similarly, the embryos possess both Wolffian and Mullerian ducts, but in males only the Wolffian duct persists to develop into the male genital ducts, while in females the Mullerian duct persists and gives rise to the female genital ducts.

Normally, the genetic constitution of the embryo determines which of these two pathways is followed. The phenotypic expression of this information, however, is not fixed. Intersexes occur occasionally in nature, and partial or complete sex reversal can be produced by experimental methods. In many amphibians the grafting of male primordial gonads into females or connecting oppositely sexed animals in parabiosis results in intersexuality or complete transformation of the female into a male. Similarly, castration of male toads allows a rudimentary ovary (Bidder's organ) to grow and eventually develop into a functional ovary. Thus, the presence of a male gonad inhibits the development of the female genital primordia.

A similar competitive interaction between primordial gonads, with one sex being dominant, occurs in higher vertebrates. In avian embryos the ovary inhibits male genital development, while in mammals the testis dominates. Upon explanation of rat genital tracts at the ambisexual stage and culture *in vitro*, only the Wolffian ducts persist in male tracts and only the Mullerian duct in female, as *in vivo*. Deletion of the testes from male tracts results in Wolffian duct regression, an effect that can be prevented by the addition of micropellets of testosterone. Removal of the ovary from female tracts, however, leaves Mullerian duct development unaffected. Normal male duct development thus requires the presence of a testis, while female duct development proceeds autonomously (Fig. 6; Price and Pannabecker, 1959; Price and Ortiz, 1965; Price, 1970).

Although production of hormones by relatively undifferentiated embryonic glands might not be expected, much evidence indicates that the differentiation of the genital tract is determined by androgens and estrogens from the fetal gonads. Guinea pig testes as early as 22 days of fetal age, before apparent differentiation, produce androgens active in bioassay on prostatic tissue; and with histochemical and isotopic methods, Haffen et al. (1971) found that chick gonads can synthesize sex steroids at stages when they appear indifferent or are just beginning to differentiate. The most dramatic effects of applied sex hormones have been obtained by administering estrogens to genetically male amphibian larvae (*Xenopus laevis*) and to young male fish (the medaka, *Oryzias latipes*): in both cases reproductively functional females were obtained (Yamamoto, 1959; Witschi, 1967). Curiously enough, in *Xenopus*, reversal can be achieved in either direction by appropriate treatment, so that it is possible to obtain functional males or females from any genotype. In avian and mammalian embryos, lesser degrees of sex change, without full functional reversal, have been achieved by steroid treatment (although gonadal reversal has been reported in the opossum). The control of differentiation of the gonads themselves is less clear than that of the associated genital structures; while some evidence indicates that the sex steroids can support early gonadal development and inhibit that of the opposite sex, other evidence obtained with antiandrogens and inhibitors of steroid synthesis indicates that the substances active in this process may not be the usual sex steroids and may, indeed, not be steroids at all (Josso, 1970; Elger et al., 1971).

During the period between embryonic differentiation of the genital system and puberty

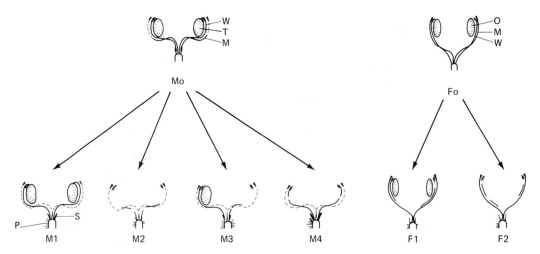

FIGURE 6. *Differentiation of rat genital tracts cultured* in vitro. *T, testis; O, ovary; W, Wolf- fian duct; M, Mullerian duct; S, seminal vesicle; P, prostate gland. Upon explanation at $17\frac{1}{2}$ days of development and then 4 days of culture, the Mullerian ducts of male tracts disappear while the Wolffian ducts persist and seminal vesicle and prostate gland rudiments appear (M1). Deletion of both testes leads to regression of both Wolffian ducts (M2); deletion of one testis plus wide separation of the two sides of the tract results in regression of the gonadless side (M3). Inclusion of testosterone micropellets with gonadless tracts maintains normal development (M4). In female tracts the Mullerian ducts persist (F1), even in the absence of ovaries (F2), and the Wolffian ducts regress (F1 and F2). (Adapted from D. Price and R. Pannebecker, Arch. Anat. Microsc. Morphol. Exp.* 48 bis, *223–244, 1959.)*

the adult pattern of pituitary gonadotrophin secretion is established. In a classic set of experiments, Pfeiffer (1936) showed that the cyclic output characteristic of female rats can be abolished by the presence of a testis during the neonatal period, and that castration of males shortly after birth will permit implanted ovaries to function cyclically. By transplanting male pituitaries under the median eminence of the hypothalamus of hypophysectomized females and obtaining normal female functions, Harris and Jacobson (1952) demonstrated that this testicular influence is at the level of the central nervous system (presumably the hypothalamus). It is now known that testosterone, and even estradiol, if administered to female rats during the critical postnatal period, will "masculinize" the brain. Since the responsiveness of target tissues to estrogens is diminished in androgenized females, it is likely that this treatment alters the normal elaboration of estrogen receptors, particularly those mediating the positive feedback of estrogens on hypo-

thalamic output of luteinizing hormone-releasing hormone (LHRH) (Van der Werff ten Bosch et al., 1971; Jost, 1970).

Human behavior can also be affected by abnormal steroidal exposure during the fetal stage (Money and Ehrhardt, 1972). Females may be androgenized either as a consequence of a genetically based adrenogenital syndrome, or when massive doses of progestin are given to the mother to prevent miscarriage. Girls belonging to both groups, in addition to showing variable degrees of masculinization of the external genitalia, tend to be "tomboys" with increased interest in male types of play and diminished interest in dolls and prematernal behavior, compared with controls. By contrast, individuals with testicular feminizing syndrome are genetic males whose body cells are unable to respond to androgens: despite the absence of ovaries and uterus, they are fully female in appearance and gender identity, and have made good mothers of adopted children. These observations illustrate the powerful influence of hormones, even during brief exposure at critical early stages, in determining life-long behavior tendencies.

MODES OF ACTION OF GONADAL STEROIDS

The postembryonic actions of the gonadal steroids ("sex hormones") are probably the most familiar developmental effects of hormones. Androgens released from the testis at puberty and thereafter facilitate spermatogenesis, stimulate development of the male sex organs and secondary sex characters (the comb of a rooster, the antlers of a deer, or the beard of a man), and have a general protein-anabolic effect in the body. Estrogens from the ovary stimulate growth of the female sex organs, and have selective effects on body growth, including the proliferation of mammary tissue just discussed. Progestins, produced chiefly in the corpus luteum, act on estrogen-stimulated tissues to produce various changes necessary for completion of the female reproductive cycle. Both male and female sex hormones, of course, also have profound effects upon behavior.

These various effects have been subject to a vast amount of research, both physiological and biochemical, and recently, progress has been rapid in building a picture of the steps by which, at the cellular level, steroid hormones exert an influence over gene expression (reviews: Williams-Ashman and Reddi, 1971; Mueller, 1971; Mueller et al., 1972; Jensen and De Sombre, 1972a, b). Two experimental systems have been especially productive: the action of estradiol-17β on the uterus of the castrated female rat, and the actions of estrogen and progesterone on the chick oviduct. We shall try to summarize the present state of biochemical understanding gained from these systems.

The uterus of the immature or ovariectomized female rat responds to a physiological dose of estradiol-17β or other estrogen by rapid initiation of growth. Within the first hour, there is imbibition of fluid and elevation of phospholipid content, and incorporation of RNA and protein precursors is stimulated at 1 to 2 hours, although net increases in these constituents are evident only later. Experiments with inhibitors (actinomycin, puromycin, cycloheximide) have shown that synthesis of both RNA and protein is a prerequisite for most other aspects of this hormone-stimulated growth. A limited range of early estrogen effects, however, notably the release of histamine and the uptake of water, appear to be independent of these macromolecular syntheses and may arise from a distinct primary mechanism, possibly at the level of the plasma membrane.

Much of the estrogen-induced RNA synthesis is ribosomal in character, and, in fact, a puromycin-sensitive increase in nucleolar RNA polymerase activity occurs at 1 to 4 hours. In addition, competitive RNA–DNA hybridization indicates the hormone-induced synthesis of new species of RNA, which, although quantitatively minor, may include essential, specific messengers. Prior to this increase in RNA polymerase activity and within 30 minutes of estrogen administration, there is synthesis of a new, specific protein, detected in acrylamide gel electrophoresis by a double labeling technique. The production of this protein, in turn, depends on RNA synthesis which does *not* require synthesis of protein (De Angelo and Gorski, 1970). Thus these findings point to the activation of a gene for a specific induced protein as an important early step that occurs well before the induced bulk synthesis of RNA and protein. The synthesis of this protein, unlike most of the physiological effects of estrogen on the uterus, has been induced by estradiol in uteri cultured *in vitro* (Katzenellenbogen and Gorski, 1972). The function of this protein is not yet known, but it may be related to, if not identical with, a key labile protein that appears to be necessary for RNA polymerase activation (Baulieu et al., 1972). Thus this "early" protein could serve in some way to activate transcription of rRNA and additional mRNAs (Fig. 7).

With the production of ³H-estradiol-17β of high specific activity, investigation of the intracellular fate of minute, physiological amounts of the hormone itself became possible. After injection of 0.1 μg or less of estradiol into immature female rats, tissues such as liver, kidney, and muscle took up radioactive hormone and then rapidly metabolized and eliminated it,

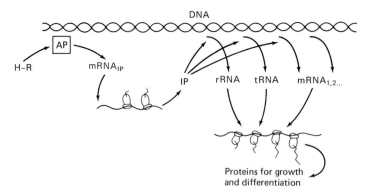

FIGURE 7. *Possible early steps in estrogen action. This model contains features from that of Baulieu et al. (1972) but differs in that IP is held to stimulate synthesis as well as function of mRNA. H-R, estrogen-receptor complex; AP, chromosomal acidic protein; IP, early induced protein.*

whereas uterus and vagina retained the hormone in unaltered form for a number of hours. This pioneering study (Jensen and Jacobson, 1962) opened a new era of research on steroid receptor proteins. Binding of estradiol can be shown in its known target tissues—pituitary, hypothalamus, mammary gland and estrogen-dependent mammary tumors, as well as uterus and vagina. The process appears to involve uptake into the cell by diffusion, followed by retention that is saturable, stereospecific, and subject to competitive inhibition. Upon fractionation of homogenates, the labeled hormone is found distributed between the high-speed supernatant and the nuclear fraction, but predominantly in the latter. Concentration of radioactive estradiol in uterine nuclei is also shown by autoradiography.

Steroid-receptor complexes in solution were readily demonstrated by gel filtration or sucrose gradient centrifugation (Gorski et al., 1968). From the cytoplasmic fraction an estrogen–protein complex is found at 8 to 9S in media of low ionic strength, but in 0.3 to 0.4 M KCl this dissociates to yield bound hormone at approximately 4S. From the nuclear fraction, 0.3 M salt is needed to release a soluble complex, and this is observed at approximately 5S but can aggregate to 8 to 9S if the salt is removed.

Transfer from the cytoplasm to the nucleus is a temperature-dependent, two-step process. The cytoplasmic complex accumulates in tissue incubated at 0 to 2°C, but upon warming to 37° the radioactive estradiol moves into the nucleus. The 8S cytoplasmic complex is readily formed *in vitro* simply by adding hormone to a cytosol fraction, and transfer to nuclei is then observed when the latter are added. Isolated nuclei will not take up the hormone in the free form but only when already bound to receptor protein.

The number of cytoplasmic estrogen receptors is estimated at about 100,000 per cell in the immature rat uterus, and depletion can be observed during transfer to the nucleus. Other, nontarget tissues apparently contain very much less binding protein, though some is present: therefore, one biochemical distinction between estrogen target and nontarget tissues is in the quantitative occurrence of hormone-receptor protein. The receptor protein exhibits very high affinity for the hormone: values of K_A, estimated by different techniques for cytosol complex from various animal species, range from 10^9 to 10^{12} M^{-1}. Binding may not be a simple reversible association, since release from the cytosol complex is extremely slow. Calcium-stabilized forms of the 8S and 4S proteins have been partially purified, and shown to have molecular weights of about 236,000 and 61,000, and isoelectric points of 6.2 and 6.6 to 6.8, respectively (Puca et al., 1971).

The 8 to 9S form of the uterine cytosol complex is not a simple dimer of the 4S form, as initially supposed, but a combination with other, nonsteroid-binding components (Mueller et al., 1972). Some insight into the role of the hormone may derive from the observation that when incubated at 25 to 37°C in low salt the cytoplasmic estrogen-receptor complex undergoes a transformation resulting in an altered sedimentation constant (Jensen et al., 1972). This appears to represent a conformational change in the protein dependent on the binding of estrogen. Once transformed, the complex is taken up by nuclei even in the cold. One view of these relationships is shown in Fig. 8.

Most interestingly, the hormone-protein complex is capable of stimulating nuclear RNA synthesis. Raynaud-Jammet and Baulieu (1969) first reported that nuclei prepared from heifer endometrium showed increased RNA polymerase activity when a mixture of estradiol and uterine cytosol was added, but not with either of these alone. With nuclei from immature rat uterus incubated in uterine cytosol, a threefold stimulation of RNA synthesis is evident with addition of estradiol. Alternatively, if the polymerase is assayed at 0°C to prevent transformation of the cytosol–estradiol complex, stimulation is produced only by addition of pretransformed complex (Jensen et al., 1972). These effects are observed only with nuclei from estrogen-dependent tissues; therefore, a second level of biochemical distinction in target tissues of this hormone must reside within the nucleus.

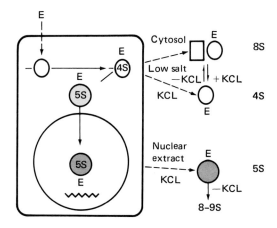

FIGURE 8. *Steps in the interaction of estradiol with receptors and transport into the nucleus in uterine cells. From E. V. Jensen, S. Mohla, T. Gorell, S. Tanaka, and E. R. de Sombre, J. Steroid Biochem. 3, 445–458, 1972.)*

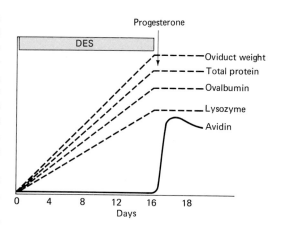

FIGURE 9. *Biochemical responses of chick oviduct to the administration of estrogen and progesterone. DES, diethylstilbestrol. From B. W. O'Malley, W. L. McGuire, P. O. Kohler, and S. G. Korenman, Rec. Progr. Horm. Res. 25, 105–160, 1969.*

For the further analysis of mechanisms of steroidal induction of biosynthesis, the uterus suffers from a disadvantage in the lack of identified protein products. In the chick oviduct, on the other hand, sex steroids induce the synthesis and secretion of several known proteins, and the study of this system has recently been highly fruitful (O'Malley et al., 1969, 1972). In the oviduct of the 5-day-old chick, the initial growth response to administration of the potent synthetic estrogen diethylstilbestrol (DES) appears within 24 hours (Fig. 9). After 4 days of continued daily treatment tubular glands begin to form, and after 6 days, active synthesis of ovalbumin, as well as of lysozyme and other proteins, begins. After 9 days, goblet cells, the sites of synthesis of the egg-white protein avidin, differentiate in the luminal epithelium. Production of avidin can then be induced within 12 hours by a single dose of progesterone. Avidin synthesis is also induced by progesterone without prior estrogen treatment, but the response is much weaker.

The biochemical consequences of DES stimulation of chick oviduct include elevation of RNA polymerases. After several days of estrogenic stimulation, the newly synthesized RNA differs from that of untreated tissue in base composition, nearest-neighbor frequency and in sequence pattern, as indicated by competition in RNA–DNA hybridization.

The synthesis of 4S tRNA appears to be disproportionately increased. That the new kinds of RNA include mRNA for ovalbumin has been demonstrated by the synthesis of this protein upon addition of oviduct RNA to a cell-free protein synthesizing system derived from rabbit reticulocytes (Means et al., 1972). The specific mRNA was not found in tissue from untreated chicks, but was present in hen oviduct and in chick oviduct after estrogen stimulation, increasing gradually during 17 days of DES treatment in parallel with the increasing production of ovalbumin (Comstock et al., 1972). The level of *total* mRNA in the tissue, indicated by the protein-synthesizing capacity of ribosomes, increased during 8 days of DES treatment and then dropped off. In addition, the content of peptide chain initiation factors increased during early estrogen stimulation and remained high for 15 days. These findings indicate that during hormonal stimulation the various elements in the protein-synthesizing system build up to support increased activity, but the production of a specific protein such as ovalbumin may be limited by the level of its specific mRNA.

The action of progesterone on the estrogen-treated oviduct is rapid and specific, synthesis of avidin being induced within 12 hours. The induction is blocked by actinomycin and cycloheximide, yet there is no measurable change in total oviduct RNA, little increase in RNA polymerase activity, and the rate of incorporation of amino acids into total protein actually decreases. A small qualitative change in the RNA made is shown by nearest-neighbor analysis and competitive hybridization. In the cell-free protein synthesis assay, mRNA for avidin is detectable at 6 and 18 hours after progesterone treatment; this activity is found to reside in an 8 to 9S RNA fraction that can be enriched by trapping on Millipore filters, which indicates a polyadenylate sequence. These experiments demonstrate the selective stimulation by two steroid hormones of synthesis of the specific mRNAs for two different proteins.

Further insight into mechanisms has been gained by studying the progesterone receptor protein of the oviduct and its interactions with the cell nucleus (O'Malley et al., 1972). Use of

^{3}H-progesterone has revealed the formation of a protein–hormone complex in the cytoplasm and its temperature-sensitive transfer into the nucleus, comparable to the binding and transfer of estradiol in the mammalian uterus. From the cytosol complex, two subunits, A and B, both sedimenting at 4S, have been separated by DEAE-cellulose chromatography. Unlike the estradiol-receptor subunits, both bind hormone very tightly ($K_D = 10^{-10} M$) (Schrader and O'Malley, 1972).

The oviduct progesterone-receptor complex is not only taken up by nuclei but is also bound directly by isolated chromatin. In both processes there is specificity for the target tissue, preparations from tissues such as liver, spleen, or lung being relatively inert. By treatment of chromatin with salt and urea at different pH values, histones and acidic proteins can be selectively extracted, and by dialysis they can be caused to recombine. By this means, "hybrid" chromatin can be prepared with components derived from different tissues, and it has been found that the tissue specificity for binding the progesterone-receptor complex resides not in the DNA nor in the histones but in the acidic protein fraction. The acceptor sites are located specifically in the fraction AP_3, extracted from chromatin with high salt and urea at pH 8.5 (Spelsberg et al., 1972). The purified A subunit of the soluble receptor complex binds to DNA and not to chromatin proteins, while the B component binds to chromatin and not to DNA (Schrader et al., 1972). This appears to provide a dual basis for the interaction between soluble hormone receptor and nuclear acceptor. At the time of writing, no influence of the hormone-receptor complex on RNA transcription in this system has been reported. However, the stimulation of RNA polymerase activity in uterine nuclei by added estradiol-receptor complex, described above, suggests that these interactions may have significance in relation to the regulation of gene expression.

Hormones in Insect Development

The development of insects, including their periodic molting and often-spectacular metamorphosis, is under endocrine control. Limitations of space prevent our dealing with this subject at length, but some of its more distinctive aspects may be worth pointing out (for particulars, see reviews by Wigglesworth, 1970; Wyatt, 1972; Doane, 1973; Gilbert and King, 1973).

When a molt is to be initiated, in response to various external and internal stimuli, a neurohormone (ecdysiotrophin) is released from the brain into the blood by way of a neurohemal organ, the corpus cardiacum, and this stimulates the production of the molting hormone, ecdysone, in a peripheral endocrine gland. This linkage between neural and endocrine communication systems is analogous to the hypothalamus–pituitary–endocrine axis of vertebrates but possesses one less stage of amplification. Ecdysone is a steroid that stimulates the epidermal events leading to molting—shedding of the old cuticle and synthesis of a new one—and also has some stimulatory effect upon growth, differentiation, and biosynthesis in many other insect tissues. Metamorphosis is controlled by the juvenile hormone, a modified terpenoid derivative produced by another endocrine gland, the corpus allatum. Its principal action in immature insects is to block the expression of adult characters, so that larval features are reexpressed during the syntheses that occur at each ecdysone-induced molt.

Neurosecretion is highly developed in the insects, and neurosecretory cells have been recognized in many parts of the central nervous system. Several neurohormones, in addition to the ecdysiotrophic hormone of the brain, have developmental actions. For example, a neurohormone is responsible for the induction of diapause in eggs of the mulberry silkworm (Isobe et al., 1973), and another for the timing of emergence of adult saturniid silkmoths (Truman and Riddiford, 1970); both these hormones are intermediaries in the expression of photoperiodic control over developmental events. Although no insect neurohormone has yet been purified to chemical homogeneity, considerable evidence indicates similarity in chemical nature and biochemical mode of action to the polypeptide hormones of vertebrates.

A major contribution of the study of insect development to endocrinology as a whole results from the possession by certain insects (the higher flies) of giant polytene chromosomes in which gene loci and the evidence of gene activity, in the form of chromosomal puffs, can be seen directly with the microscope. Clever and Karlson (1960) discovered that a developmentally specific puff was induced within 15 minutes after administration of ecdysone to appropriately staged larvae of the midge, *Chironomus*, followed by a regular pattern of other puffs at later times. The phenomenon has been reproduced and studied in detail in *Drosophila* species (review: Ashburner, 1970). This observation laid the basis for speculation about modes of hormone action (Karlson, 1963), which drew attention to the apparent relationship between certain hormones and the control of gene expression, and was influential in stimulating research on this problem in vertebrates as well as in insects.

The juvenile hormone is unusual in the extent to which, in the presence of ecdysone, it directs the expression of visible developmental features. It also belongs to a chemical group different from any hormone of vertebrates. Although a provocative theory to explain the action of juvenile hormone in regulating insect metamorphosis, involving master genes for stage-specific gene sets, repressors, and sigma factors, has been presented (Williams and Kafatos, 1971), there is, as yet, little relevant biochemical evidence. In adult insects of many species, juvenile hormone, in the absence of ecdysone, exercises a second, quite distinct function in the regulation of reproductive maturation. One aspect of this that is amenable to biochemical study is the induction of synthesis of specific yolk proteins (vitellogenins) in the fat body. The exercise of dual roles by juvenile hormone seems to represent an economy of nature in using the same hormone for two unconnected functions in different developmental stages of a species.

In insects, ecdysone and juvenile hormone are the only two known lipoidal hormones. This animal group lacks the range of functionally specific steroidal and other developmental hormones possessed by the vertebrates. It seems that direct genetic programming plays a greater role and that development is less plastic in insects than in the vertebrates. Thus, in insects, one cannot modify the development of secondary sexual characters, or induce the synthesis of yolk protein in a male individual (as can be done in amphibia or birds), by treatment with a sex hormone. The possession of mosaic eggs by insects, showing early regional determination, is perhaps another manifestation of the same tendency for more rigid programming. These differences between animal groups, however, are of degree and not of kind, and there is no reason to believe that growth and differentiation, and the endocrine control over them, do not depend upon similar molecular mechanisms throughout the animal kingdom.

Conclusion

From recent research on hormone action, several distinct biochemical mechanisms that may be associated with developmental effects are becoming evident: primary action at the gene level through intracellular protein receptors; action at the cell surface through modulation of adenylate cyclase and cyclic AMP-regulated processes; and action at the cell surface through mechanisms that are as yet poorly understood. While the range of hormonal stimuli and responses that occur through these mechanisms is both specific and varied, to a great extent hormones serve merely as triggers to evoke patterns of response that are already programmed into cells during their developmental history. Thus we are driven back to the perennially elusive problem of the control and nature of cell differentiation. Hormones may, perhaps, provide a key to this question through the analysis of the differences between target and nontarget cells and of the ontogeny of responsiveness.

References

ASHBURNER, M. (1970). Function and structure of polytone chromosomes during insect development. *Advan. Insect Physiol.* 7, 1–95.

BALAZS, R., COCKS, W. A., EAYRS, J. R., AND KOVACS, S. (1971). Biochemical effects of thyroid hormones on the developing brain. *In* "Hormones in Development" (M. Hamburgh and E. J. W. Barrington, eds.), pp. 357–379. Appleton-Century-Crofts, New York.

BAULIEU, E. E., ALBERGA, A., RAYNAUD-JAMMET, C., AND WIRA, C. R. (1972). New look at the very early steps of oestrogen action in uterus. *Nature New Biol.* 236, 236–239.

BURGUS, R., AND GUILLEMIN, R. (1970). Hypothalamic releasing factors. *Ann. Rev. Biochem.* 39, 499–526.

BUTCHER, R. W., ROBISON, G. A., AND SUTHERLAND, E. W. (1972). Cyclic AMP and hormone action. *In* "Biochemical Actions of Hormones" (G. Litwack, ed.), Vol. 2, pp. 21–54. Academic Press, New York.

CLEVER, U., AND KARLSON, P. (1960). Induktion von Puff-Veränderungen in den Speicheldrüsenchromosomen von *Chironomus tentans* durch Ecdyson. *Exp. Cell Res. 20*, 623–626.

COHEN, P. P. (1970). Biochemical differentiation during amphibian metamorphosis. *Science 168*, 533–543.

COMSTOCK, J. P., ROSENFELD, G. C., O'MALLEY, B. W., AND MEANS, A. R. (1972). Estrogen-induced changes in translation, and specific messenger RNA levels during oviduct differentiation. *Proc. Nat. Acad. Sci. U.S.A. 69*, 2377–2380.

DAUGHADAY, W. H., HALL, K., RABEN, M. S., SALMON, W. D., JR., VAN DEN BRANDE, J. L., AND VAN WYK, J. J. (1972). Somatomedin: proposed designation for sulphation factor. *Nature (London) 235*, 107.

DE ANGELO, A. B., AND GORSKI, J. (1970). Role of RNA synthesis in the estrogen induction of a specific uterine protein. *Proc. Nat. Acad. Sci. U.S.A. 66*, 693–700.

DERBY, A. (1970). Inhibition of thyroxine induced tail resorption *in vitro* by pituitary grafts from different stages of metamorphosis. *J. Exp. Zool. 173*, 319–327.

DIMINO, M. J., AND HOCH, F. L. (1972). Localization of endogenous and exogenous thyroid hormone in rat liver mitochondria. *Fed. Proc. 31*, 213, abstract.

DOANE, W. W. (1973). Role of hormones in development. *In* "Insects: Developmental Systems" (S. J. Counce and C. H. Waddington, eds.), Vol. 2, pp. 291–497. Academic Press, London.

EASTMAN, R. C., AND GOODMAN, H. M. (1971). Insensitivity of adipocytes of hypophysectomized rats to growth hormone and insulin. *Proc. Soc. Exp. Med. 137*, 168–174.

EATON, J. E., JR., AND FRIEDEN, E. (1968). Molecular changes during anuran metamorphosis; early effects of triiodothyronine on nucleotide and RNA metabolism in the bullfrog tadpole liver. *Gunma Symp. Endocrinol. 5*, 43–53.

EAYRS, J. J. (1971). Thyroid and developing brain: anatomical and behavioral effects. *In* "Hormones in Development" (M. Hamburgh and E. J. W. Barrington, eds.), pp. 345–355. Appleton-Century-Crofts, New York.

ELGER, W., NEUMANN, F., AND VON BERSWORDT-WALLRABE, R. (1971). The influence of androgen antagonists and progestogens on the sex differentiation of different mammalian species. *In* "Hormones in Development" (M. Hamburgh and E. J. W. Barrington, eds.), pp. 651–667. Appleton-Century-Crofts, New York.

ETKIN, W. (1968). Hormonal control of amphibian metamorphosis. *In* "Metamorphosis" (W. Etkin and L. I. Gilbert, eds.), pp. 313–348. Appleton-Century-Crofts, New York.

ETKIN, W., AND KIM, Y. (1971). Effects of TRH on tadpole tissue. *Amer. Zool. 11*, 654.

FILBURN, C. R., AND VANABLE, J. W., JR. (1973). Acid phosphatase isozymes of *Xenopus laevis* tadpole tails. II. Changes in activity during tail regression. *Arch. Biochem. Biophys.* (in press).

FRIEDEN, E., AND JUST, J. J. (1970). Hormonal responses in amphibian metamorphosis. *In* "Biochemical Actions of Hormones" (G. Litwack, ed.), Vol. 1, pp. 1–52. Academic Press, New York.

FRYE, B. E., BROWN, P. S., AND SNYDER, B. W. (1972). Effects of prolactin and somatotrophin on growth and metamorphosis of amphibians. *Gen. Comp. Endocrinol.*, Suppl. 3, 209–220.

GILBERT, L. I., AND KING, D. S. (1973). Physiology of growth and development: endocrine aspects. *In* "Physiology of Insects" (M. Rockstein, ed.). 2nd ed. Vol. 1, pp. 350–368. Academic Press, New York.

GOLDBERG, N. D., HADDOX, M. K., DUNHAM, E., LOPEZ, C., AND HADDEN, J. W. (1973). The Yin Yang hypothesis of biological control: opposing influences of cyclic GMP and cyclic AMP in the regulation of cell proliferation and other biological processes. Cold Spring Harbor Symp. Quant. Biol. (in press).

GOMEZ, C. J. (1971). Hormonal influences of the biochemical differentiation of the rat cerebral cortex. *In* "Hormones in Development" (M. Hamburgh and E. J. W. Barrington, eds.), pp. 417–435. Appleton-Century-Crofts, New York.

GOODMAN, H. M. (1968). Growth hormone and the metabolism of carbohydrate and lipid in adipose tissue. *Ann. N. Y. Acad. Sci. 148*, 419–440.

GORSKI, J., TOFT, D., SHYAMALA, G., SMITH, D., AND NOTIDES, A. (1968). Hormone receptors: studies on the interaction of estrogen with the uterus. *Rec. Progr. Hormone Res. 24*, 45–80.

GORSKI, R. A. (1968). Influence of age on the response to parental administration of a low dose of androgen. *Endocrinology 82*, 1001–1004.

GREEN, M. R., BUNTING, S. L., AND PEACOCK, A. C. (1971). Changes in labeling pattern of ribonucleic acid from mammary tissue as a result of hormone treatment. *Biochemistry 10*, 2366–2371.

GREENGARD, O. (1970). The developmental formation of enzymes in rat liver. In "Biochemical Actions of Hormones" (G. Litwack, ed.), Vol. 1, pp. 53–88. Academic Press, New York.

GRISWOLD, M. D., AND COHEN, P. P. (1972). Alteration of deoxyribonucleic acid-dependent ribonucleic acid polymerase activities in amphibian liver nuclei during thyroxine-induced metamorphosis. *J. Biol. Chem. 247*, 353–359.

GRISWOLD, M. D., FISCHER, M. S., AND COHEN, P. P. (1972). Temperature-dependent intracellular distribution of thyroxine in amphibian liver. *Proc. Nat. Acad. Sci. U.S.A. 69*, 1486–1489.

HAFFEN, K., CEDARD, L. AND SCHEIB, D. (1971). Biosynthesis of steroid hormones by the chick embryonic gonads. *In* "Hormones in Development" (M. Hamburgh and E. J. W. Barrington, eds.), pp. 705–718. Appleton-Century-Crofts, New York.

HAMBURGH, M. (1968). An analysis of the action of thyroid hormone on development based on *in vivo* and *in vitro* studies. *Gen. Comp. Endocrinol. 10*, 198–213.

HAMBURGH, M., MENDOZA, L. A., BURKART, J. F., AND WEIL, F. (1971). Thyroid-dependent processes in the developing nervous system. *In* "Hormones in Development" (M. Hamburgh and E. J. W. Barrington, eds.), pp. 403–415. Appleton-Century-Crofts, New York.

HARRIS, G. W., AND JACOBSON, D. (1952). Functional grafts of the anterior pituitary gland. *Proc. Roy. Soc. Lond. B 139*, 263–276.

HINTZ, R. L., CLEMONS, D. R., UNDERWOOD, L. E., AND VAN WYK, J. J. (1972). Competitive binding of somatomedin to the insulin receptors of adipocytes, chondrocytes, and liver membranes. *Proc. Nat. Acad. Sci. U.S.A. 69*, 2351–2353.

ILLIANO, G., AND CUATRECASAS, P. (1972). Modulation of adenylate cyclase activity in liver and fat cell membranes by insulin. *Science 175*, 906–908.

ILLIANO, G., TELL, G. P. E., SIEGEL, M. I., AND CUATRECASAS, P. (1973). Guanosine 3′,5′-cyclic monophosphate and the action of insulin and acetylcholine. *Proc. Nat. Acad. Sci. U.S.A. 70*, 2443–2447.

ISOBE, M., HASEGAWA, K., AND GOTO, T. (1973). Isolation of the diapause hormone from the silkworm *Bombyx mori. J. Insect Physiol. 19*, 1221–1239.

JENSEN, E. V., AND DE SOMBRE, E. R. (1972a). Estrogens and progestins. *In* "Biochemical Actions of Hormones" (G. Litwack, ed.), Vol. 2, pp. 215–255. Academic Press, New York.

JENSEN, E. V., AND DE SOMBRE, E. R. (1972b). Mechanism of action of the female sex hormones. *Ann. Rev. Biochem. 41*, 203–230.

JENSEN, E. V., AND JACOBSON, H. I. (1962). Basic guides to the mechanism of estrogen action. *Rec. Progr. Horm. Res. 18*, 387–414.

JENSEN, E. V., MOHLA, S., GORELL, T., TANAKA, S., AND DE SOMBRE, E. R. (1972). Estrophile to nucleophile in two easy steps. *J. Steroid Biochem. 3*, 445–458.

JOSSO, N. (1970). Effect of cyanoketone, an inhibitor of Δ_5-3β-hydroxysteroid dehydrogenase, on the reproductive tracts of male fetal rats in organ culture. *Biol. Reprod. 2*, 85–90.

JOST, A. (1970). Hormonal factors in the sex differentiation of the mammalian fetus. *Phil. Trans. Roy. Soc. Lond. B 259*, 119–130.

JUST, J. J. (1972). Protein-bound iodine and protein concentration in plasma and pericardial fluid of metamorphosing anuran tadpoles. *Physiol. Zool. 45*, 143–152.

KALTENBACH, J. C. (1968). Nature of hormone action in amphibian metamorphosis. *In* "Metamorphosis" (W. Etkin and L. I. Gilbert, eds.), pp. 399–441. Appleton-Century-Crofts, New York.

KARLSON, P. (1963). New concepts on the mode of action of hormones. *Perspectives Biol. Med. 6*, 203–214.

KATZENELLENBOGEN, B. S., AND GORSKI, J. (1972). Estrogen action *in vitro*. Induction of the synthesis of a specific uterine protein. *J. Biol. Chem. 247*, 1299–1305.

KNOBIL, E., AND HOTCHKISS, J. (1964). Growth hormone. *Ann. Rev. Physiol. 26*, 47–74.

KOLLROS, J. J. (1968). Endocrine influences in neural development. *In* "Growth of the Nervous System" (G. E. W. Wolstenholme and M. D. O'Connor, eds.), pp. 179–182. J. & A. Churchill, London.

KOSTYO, J. L. (1689). Rapid effects of growth hormone on amino acid transport and protein synthesis. *Ann N.Y. Acad. Sci. 148*, 389–407.

KOSTYO, J. L., AND RILLEMA, J. A. (1971). *In vitro* effects of growth hormone on the number and activity of ribosomes engaged in protein synthesis in the isolated rat diaphragm. *Endocrinology 88*, 1054–1062.

LEVEY, G. S., AND EPSTEIN, S. E. (1969). Myocardial adenyl cyclase–activation by thyroid hormones and evidence for 2 adenyl cyclase systems. *J. Clin. Invest. 48*, 1663–1669.

LI, C. H., EVANS, H. M., AND SIMPSON, M. E. (1945). Isolation and properties of the anterior hypophyseal growth hormone. *J. Biol. Chem. 159*, 353–366.

LUCKOCK, A. S., AND TIMIRAS, P. S. (1972). Changes in the subcellular localization of thyroxine during brain development. *Fed. Proc. 31*, 221, abstract.

LYONS, W. R., LI, C. H. AND JOHNSON, R. E. (1958). The hormonal control of mammary growth and lactation. *Rec. Progr. Horm. Res. 14*, 219–248.

MACDONALD, R. I., AND KORNER, A. (1971). Growth hormone stimulation of protein synthetic activity of membrane-bound ribosomes. *FEBS Letters 13*, 62–64.

MAJUMDER, G. C., AND TURKINGTON, R. W. (1971). Hormonal regulation of protein kinases and adenosine 3′, 5′-monophosphate binding protein in developing mammary gland. *J. Biol. Chem. 246*, 5545–5554.

MAJUMDER, G. C., AND TURKINGTON, R. W. (1972). Hormone-dependent phosphorylation of ribosomal and plasma-membrane proteins in mouse mammary gland *in vitro. J. Biol. Chem. 247*, 7207–7217.

MARGOULIES, M., AND GREENWOOD, F. C., eds. (1971). "Structure–Activity Relationship of Protein and Polypeptide Hormones." Excerpta Medica, Amsterdam.

MEANS, A. R., COMSTOCK, J. P., ROSENFELD, G. C., AND O'MALLEY, B. W. (1972). Ovalbumin messenger RNA of chick oviduct: partial characterization, estrogen dependence and translation *in vitro. Proc. Nat. Acad. Sci. U.S.A. 69*, 1146–1150.

MONEY, J., AND EHRHARDT, A. A. (1972). Gender dimorphic behavior and fetal sex hormones. *Rec. Progr. Horm. Res. 28*, 735–763.

MUELLER, G. C. (1971). Estrogen action: a study of the influence of steroid hormones on genetic expression. *In* "The Biochemistry of Steroid Hormone Action." Biochemical Society Symposium No. 32 (R. M. S. Smellie, ed.), pp. 1–29. Academic Press, New York.

MUELLER, G. C., VONDERHAAR, B., KIM, U. H., AND LE MAHIEU, M. (1972). Estrogen action: an inroad to cell biology. *Rec. Progr. Horm. Res. 28*, 1–50.

OKA, T., AND TOPPER, Y. J. (1971). Hormone-dependent accumulation of rough endoplasmic reticulum in mouse mammary epithelial cells *in vitro. J. Biol. Chem. 246*, 7701–7707.

OKA, T., AND TOPPER, Y. J. (1972). Is prolactin mitogenic for mammary epithelium? *Proc. Nat. Acad. Sci. U.S.A. 69*, 1693–1696.

O'MALLEY, B. W., MCGUIRE, W. L., KOHLER, P. O., AND KORENMAN, S. G. (1969). Studies on the mechanism of steroid hormone regulation of synthesis of specific proteins. *Rec. Progr. Horm. Res. 25*, 105–160.

O'MALLEY, B. W., SPELSBERG, T. C., SCHRADER, W. T., CHYTIL, F., AND STEGGLES, A. W. (1972). Mechanisms of interaction of a hormone-receptor complex with the genome of a eukaryotic target cell. *Nature (London) 235*, 141–144.

PFEIFFER, C. A. (1936). Sexual differences of the hypophyses and their determination by the gonads. *Amer. J. Anat. 58*, 195–225.

PRICE, D. (1970). *In vitro* studies on differentiation of the reproductive tract. *Phil. Trans. Roy. Soc. Lond. B 259*, 133–139.

PRICE, D., AND ORTIZ, E. (1965). The role of fetal androgen in sex differentiation in mammals. In "Organogenesis" (R. L. DeHaan and H. Ursprung, eds.), pp. 629–652. Holt, New York.

PRICE, D., AND PANNABECKER, R. (1959). Comparative responsiveness of homologous sex ducts and accessory glands of fetal rats in culture. *Arch. Anat. Microsc. Morphol. Exp. 48 bis*, 223–244.

PUCA, G. A., NOLA, E., SICA, V., AND BRESCIANI, F. (1971). Estrogen-binding proteins of calf uterus. Partial purification and preliminary characterization of two cytoplasmic proteins. *Biochemistry 10*, 3769–3780.

RAYNAUD-JAMMET, C., AND BAULIEU, E. E. (1969). Action de l'oestradiol *in vitro*: augmentation de la biosynthèse d'acide ribonucléique dans les noyaux utérins. *C. R. Acad. Sci. D268*, 3211–3214.

REICHLIN, S., MARTIN, J. B., MITNICK, M., BOSHANS, R. L., GRIMM, Y., BOLLINGER, J., GORDON, J., AND MALACARA, J. (1972).The hypothalamus in pituitary-thyroid regulation. *Rec. Progr. Horm. Res. 28*, 229–286.

RILLEMA, J. A., AND KOSTYO, J. L. (1971). Studies on the delayed action of growth hormone on the metabolism of the rat diapraghm. *Endocrinology 88*, 240–248.

ROBINSON, H. (1972). An electrophoretic and biochemical analysis of acid phosphatase in the tail of *Xenopus laevis* during development and metamorphosis. *J. Exp. Zool. 180*, 127–139.

ROBISON, G. A., BUTCHER, R. W., AND SUTHERLAND, E. W. (1971). "Cyclic AMP." Academic Press, New York.

SALAMON, D. F., BETTERID, S., AND KORNER, A. (1972). Early effects of growth-hormone on nucleolar and nucleoplasmic RNA synthesis and RNA-polymerase activity in normal rat liver. *Biochim. Biophys. Acta 272*, 382–395.

SALMON, W. D., JR., AND DAUGHADAY, W. H. (1957). A hormonally controlled serum factor which stimulates sulfate incorporation by cartilage *in vitro. J. Lab. Clin. Med. 49*, 825–836.

SCHALLY, A. V., ARIMURA, A., AND KASTIN, A. J. (1973). Hypothalamic regulatory hormones. *Science 179*, 341–350.

SCHOOLEY, R. A., GRIEDKIN, S., AND EVANS, E. S. (1966). Re-examination of the discrepancy between acidophil numbers and growth hormone concentration in the anterior pituitary gland following thyroidectomy. *Endocrinology 79*, 1053–1057.

SCHRADER, W. T., AND O'MALLEY, B. W. (1972). Progesterone-binding components of chick oviduct. IV. Characterization of purified subunits. *J. Biol. Chem. 247*, 51–59.

SCHRADER, W. T., TOFT, D. O., AND O'MALLEY, B. W. (1972). Progesterone-binding components of chick oviduct. VI. Interaction of purified progesterone-receptor components with nuclear constituents. *J. Biol. Chem. 247*, 1401–2407.

SCHWIND, J. L. (1933). Tissue specificity at the time of metamorphosis in frog larvae. *J. Exp. Zool. 66*, 1–14.

SHYAMALA, G. (1973). Specific cytoplasmic glucocorticoid hormone receptors in lactating mammary glands. *Fed. Proc. 32*, 453, abstract.

SIMPSON, M. E., EVANS, H. M., AND LI, C. H. (1949). The growth of hypophysectomized female rats following chronic treatment with pure pituitary growth hormone. I. General growth and organ changes. *Growth 13*, 151–170.

SNIPES, C. A. (1968). Effects of growth hormone and insulin on amino acid and protein metabolism. *Quart. Rev. Biol. 43*, 127–147.

SOKOLOFF, L., AND ROBERTS, P. A. (1972). Invalidity of alleged stimulations of protein syntheses by thyroxine *in vitro* in absence of mitochondria. *Fed. Proc. 31*, 485, abstract.

SOKOLOFF, L., ROBERTS, P. A., JANUSKA, M. M., AND KLINE, J. E. (1968). Mechanisms of stimulation of protein synthesis by thyroid hormone *in vivo*. *Proc. Nat. Acad. Sci. U.S.A.* 60, 652–659.

SPELSBERG, T. C., STEGGLES, A. W., CHYTIL, F., AND O'MALLEY, B. W. (1972). Progesterone-binding components of chick oviduct. V. Exchange of progesterone-binding capacity from target to nontarget tissue chromatins. *J. Biol. Chem.* 247, 1368–1374.

TANNER, J. M. (1972). Human growth hormone. *Nature (London)* 237, 433–439.

TATA, J. R. (1970a). Simultaneous acquisition of metamorphic response and hormone binding in *Xenopus* larvae. *Nature (London)* 227, 686–689.

TATA, J. R. (1970b). Regulation of protein synthesis by growth and developmental hormones. *In* "Biochemical Actions of Hormones" (G. Litwack, ed.), Vol. 1, pp. 89–133. Academic Press, New York.

TELL, G. P. E., CUATRECASAS, P., VAN WYK, J. J., AND HINTZ, R. L. (1973). Somatomedin: inhibition of adenylate cyclase activity in subcellular membranes of various tissues. *Science* 180, 312–315.

TOPPER, Y. D. (1970). Multiple hormone interactions in the development of mammary gland *in vitro*. *Rec. Progr. Horm. Res.* 26, 287–303.

TRUMAN, J. W., AND RIDDIFORD, L. M. (1970). Neuroendocrine control of ecdysis in silkmoths. *Science* 167, 1624–1626.

TURKINGTON, R. W. (1968). Hormone-dependent differentiation of mammary gland *in vitro*. *Curr. Top. Develop. Biol.* 3, 119–223.

TURKINGTON, R. W. (1972). Multiple hormonal interactions. The mammary gland. In "Biochemical Actions of Hormones" (G. Litwack, ed.), Vol. 2, pp. 55–80. Academic Press, New York.

TURNER, C. D., AND BAGNARA. J. T. (1971). "General Endocrinology." Saunders, Philadelphia.

UNSWORTH, B., AND COHEN, P. P. (1968). Effect of thyroxine treatment on the transfer of amino acids from aminoacyl transfer ribonucleic acid into protein by cell-free extracts from tadpole liver. *Biochemistry* 7, 2581–2588.

VAN DER WERFF TEN BOSCH, J. J., TUINEBREŸER, W. E., AND VREEBURG, J. T. M. (1971). The incomplete or delayed early-androgen syndrome. *In* "Hormones in Development" (M. Hamburgh and E. J. W. Barrington, eds.), pp. 669–675. Appleton-Century-Crofts, New York.

WEBER, R. (1967). Biochemistry of amphibian metamorphosis. *In* "The Biochemistry of Animal Development" (R. Weber, ed.), Vol. 2, 227–301. Academic Press, New York.

WEBER, R. (1969). Tissue involution and lysosomal enzymes during anuran metamorphosis. *In* "Lysosomes in Biology and Pathology" (J. T. Dingle and H. B. Fell, eds.), Vol. 2, pp. 437–461. North-Holland, Amsterdam.

WIGGLESWORTH, V. B. (1970). "Insect Hormones." Oliver and Boyd, Edinburgh.

WILLIAMS, C. M., AND KAFATOS, F. C. (1971). Theoretical aspects of the action of juvenile hormone. *Mitt. Schweiz. Entomol. Ges.* 44, 151–162.

WILLIAMS-ASHMAN, H. G., AND REDDI, A. H. (1971). Actions of vertebrates sex hormones. *Ann. Rev. Physiol.* 33, 31–82.

WITSCHI, E. (1967). Biochemistry of sex differentiation in vertebrate embryos. *In* "The Biochemistry of Animal Development" (R. Weber, ed.), Vol. 2, pp. 193–226. Academic Press, New York.

WIXOM, R. L., REDDY, J. K., AND COHEN, P. P. (1972). A concerted response of the enzymes of urea biosynthesis during thyroxine-induced metamorphosis of *Rana catesbeiana*. *J. Biol. Chem.* 247, 3684–3692.

WOOL, I. G., STIREWALT, W. S., KURIKARA, K., LAW, R. B., BAILEY, P., AND ØYER, D. (1968). Mode of action of insulin in the regulation of protein biosynthesis in muscle. *Rec. Progr. Horm. Res.* 24, 139–208.

WYATT, G. R. (1972). Insect Hormones. *In* "Biochemical Actions of Hormones" (G. Litwack, ed.), Vol. 2, pp. 385–490. Academic Press, New York.

YAMAMOTO, T. 1959). A further study on induction of functional sex reversal in genotypic males of the medaka, *Oryzias latipes*, and progenies of sex reversals. *Genetics* 44, 739–757.

CHAPTER SEVENTEEN

Developmental Genetics

DAVID T. SUZUKI

The relative importance of the nucleus versus the cytoplasm and environment to development and differentiation, long debated by geneticists and embryologists, has been unified by elucidation of nucleic acid and protein relationships. Genetic mutations and aberrations have been useful tools for probing the mechanisms of regulation of development and differentiation. This chapter focuses on specific examples of genetic approaches to development, rather than attempting an extensive review of the entire area. In particular, we are interested in the fundamental problem of developmental genetics today—which is to understand the mechanisms regulating gene function.

The most sophisticated models of genetic regulation come from studies on bacteria and their viruses. Jacob and Monod (1961a) derived their model of the operon by extensive analysis of the properties of mutations that affect the induction of enzymes concerned with the metabolism of lactose. The postulated role of regulator and operator regions on the control of structural genes has been extensively corroborated and extended by molecular biologists, and the operon remains the most definitive model for regulation in higher organisms.

Reproduction of bacteriophages presents another model system for the study of regulation. Using conditional lethal mutations that die under restrictive conditions but survive in a permissive environment, Epstein et al. (1963) studied morphogenesis of the T4 phage. They found that genes concerned with similar aspects of phage development (such as tail formation and DNA synthesis) were closely linked on the genetic map. Moreover, the positions of these functionally related genes were correlated with the temporal sequence of transcription of the genome after infection.

In chromosomes of eukaryotic, multicellular organisms, genes exhibit position effects; that is, the phenotypes of individuals heterozygous for two mutant genes differ when the two mutations are located on the same chromosome (cis) and when they are on different homologues (trans). This importance of chromosome location for genetic activity is most strikingly illustrated in altered activity of normal genes by virtue of their juxtaposition to different chromosomal regions in chromosome rearrangements.

Heterochromatin, a region of chromosomes cytologically defined by its intense staining properties, is genetically inactive. In some organisms, such as mealy bugs, parental source of chromosomes determines heterochromatization, and this imprint is then inherited somatically by all daughter cells (Brown and Nur, 1964). In the case of the X chromosome of the fly *Sciara*, a paternal imprint that determines the mitotic behavior of the centromere is imposed on a large heterochromatic element adjacent to the centromere (Crouse, 1960). In mammalian females, one of either of the X chromosomes is inactivated in all cells. While inactivation of the X appears to be random, once determined, all daughter cells inherit the same inactivated chromosome (Lyon, 1968).

Organisms that are mosaics for two or more genotypically distinct cells can be useful tools in the study of development (Nesbitt and Gartler, 1971). In *Drosophila*, a variety of methods permit the generation of mosaics experimentally: (1) the mutation, ca^{nd} enhances loss of chromosomes in the early mitotic divisions after fertilization (Sturtevant, 1929); (2) crossing-over during mitosis can generate "twin spots" of daughter cells homozygous for recessive markers (Stern, 1936); and (3) the ring X chromosome, $In(1)w^{vC}$, is somatically unstable and is often lost during early cleavage, thereby generating X/O:X/X gynandromorphs (Hinton, 1955). In all cases, loss or crossing-over early in development produces large spots of mosaic tissue, whereas progressively later occurrence results in correspondingly smaller patches. This allows the determination of the cell lineage of various structures as well as the autonomy of expression of various mutations.

In mammals, internal mosaicism can be studied by biochemical and cytological markers. Mintz (1962) perfected a method for fusing separate mouse embryos to form a single animal. Such tetraparental animals can be used to trace cell lineage of various structures. By fusing embryos of mouse strains having electrophoretically distinct isocitrate dehydrogenase, Mintz and Baker (1967) confirmed that multinucleated skeletal muscle cells arise by cell fusion rather than by mitotic division of nuclei within a single cell.

Mutations producing developmental abnormalities have been extensively studied in a variety of organisms (Goldschmidt, 1955; Hadorn, 1955). However, it is often difficult to determine the primary lesion produced by the genetic defect or its time of activity. The study of mutations whose phenotypic expression is affected by temperature has aided in such studies. Temperature-sensitive (ts) mutations have been most extensively exploited in *Drosophila* (Suzuki, 1970). Most ts lethal mutations act during development and the time of sensitivity to restrictive temperatures can be delineated by shifting cultures from restrictive to permissive temperatures and vice versa at successive stages in development. In studies of a single ts mutation affecting pigment production in several tissues, it was demonstrated that each tissue has a specific time of activation of the locus (Grigliatti and Suzuki, 1970).

Drosophila larvae carry clusters of cells called *imaginal discs*, which are predetermined to form adult structures. Hadorn (1965) found that serial transplants of discs in adults produced changes in the state of determination so that different adult structures would be produced by the disc cells. The programming of prospective disc fates can also be studied using *homeotic mutations*, which alter the determined state of a disc from one adult structure to another. A ts homeotic mutation, ss^{a40a}, produces a normal arista (the feathery distalmost component of the antennal complex) at 29°C but a leg at 17°C and has a ts period in the third larval instar (Grigliatti and Suzuki, 1971). These types of studies may provide an insight into the nature of the programming responsible for cell differentiation.

Temperature-sensitive lesions in the nervous system can be recovered by screening for the phenotype of ts paralysis. Thus the mutation $para^{ts}$ appears to affect nerve transmission in the cephalic and thoracic ganglia (Grigliatti et al., 1972). Several alleles of the gene *shi* have been recovered. The allele shi^{ts1} causes adult and larval paralysis as well as lethality at a number of stages and exhibits a variety of developmental anomalies in the adult (Poodry et al., 1973). A review of the developmental effects of several different ts mutations points to their utility in aiding study of the mechanism of regulation in higher organisms.

Genetic Analysis of Development

The intricate mechanisms regulating the organization and sequential regularity of the complex series of events observable in multicellular organisms during development and differentiation offer a staggering puzzle for solution by developmental biologists. One is immediately struck with such seemingly simple questions as the following: What controls the order of events during cell division? What are the signals initiating cell movement at gastrulation? What determines the divergent fates of muscle or nerve cells?

Geneticists long argued that the hereditary information of the nucleus controlled development, since any alteration in the genome such as a chromosome aberration or mutation could strikingly affect normal development (Stern, 1957). Embryologists, on the other hand, argued that the nuclear material remained constant from cell to cell and pointed to cytoplasmic elements as the most important factors (see review by Gurdon and Woodland, 1968). Moreover, environmental changes in oxygen concentration, temperature, or pH could alter or even arrest development. Geneticists countered that the egg cytoplasm is itself regulated by the maternal genotype.

This debate had the circularity of the which-came-first-the-chicken-or-the-egg conundrum. Happily, the unification of embryology and genetics into a common field of developmental biology has come about, largely as a result of the recognition of the molecular basis of genes and genetic function (see Watson, 1970). In eukaryotic organisms, DNA in the nucleus stores information in the form of the now-famous triplet code words based on a four-letter alphabet. The DNA sentences are duplicated in a molecule (messenger RNA) that copies the exact sequence of letters in DNA and is transmitted to the cytoplasm, where the transcribed message is translated by the synthetic apparatus into proteins (Watson, 1970).

Clearly, development and differentiation can proceed through a regulated sequence of reading of specific parts of the genetic blueprint in response to cytoplasmically transmitted cues.

The power of genetics to analyze development rests primarily on the ability of the experimenter to introduce a specific perturbation in the genetic blueprint with a view to finding out its net effect on the organism, and then proceeding backward to the primary cause. These genetic alterations are usually mutations or chromosomal aberrations. Thus, while molecular biologists studying hybridization of nucleic acids or the properties of repressor molecules, for example, are concerned with the properties of genetic material or its products, they do not necessarily utilize a genetic analysis. Moreover, mutations or chromosome aberrations are often used as nuclear or cell markers, just as radioactive isotopes function as molecular tags. Thus chromosome and nucleolar numbers have been used to verify the success of nuclear transplants (King and Briggs, 1956; Gurdon, 1963), and chromosome aberrations have been used to prove the clonal nature of regenerating tissues (Nowell and Cole, 1967). Again, such use of genetic markers does not fall under the topic of developmental genetics. This chapter will deal with the specific use of genetic analyses to probe developmental phenomena. With this restriction of my discussion, I shall present a few examples of genetic analyses in order to illustrate the utility of the genetic approach, the kinds of questions that may be posed, and the limitations of this work.

The demonstration that highly differentiated cells nevertheless possess the full genetic complement present at fertilization (Gurdon, 1963) has been corroborated by nucleic acid hybridization studies of somatic tissue. Thus differentiation must proceed by selective activation and inactivation of genes in a highly controlled sequential way. Indeed, studies of cytologically visible regions of genetic activity in giant polytene chromosomes of the Diptera verify the locus and tissue and temporal specificity of gene activity during development (Beermann, 1956, 1961; Pavan and Breuer, 1955). *The mechanism whereby genetic activity is regulated therefore constitutes the focal problem of developmental genetics.*

Models of Regulation Derived in Microorganisms

A classic model of genetic regulation was formulated by Jacob and Monod (1961a) in their analysis of the utilization of lactose by *Escherichia coli*. Wild-type *E. coli* cells grown on a medium supplied with the sugar glucose do not have significant levels of β-galactosidase, an enzyme required for the conversion of lactose to galactose. The existence of a gene for that enzyme is shown by the induction of β-galactosidase production when lactose is substituted for glucose (Fig. 1). That β-galactosidase production is a property of all cells rather than the selection and multiplication of mutant cells in the presence of lactose is proved by the same pattern of enzyme induction shown in cell cultures in which cell division is inhibited. Upon removal of galactose from the medium, β-galactosidase activity decreases. Thus the activity of a specific gene is regulated by the presence of the very substrate upon which its primary product acts.

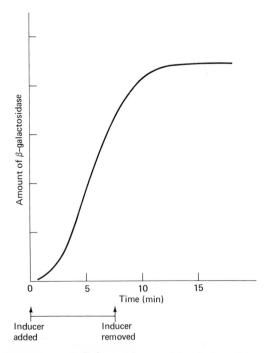

FIGURE 1. *β-Galactosidase activity induced in wild-type* E. coli.

The recovery of mutations that, in the presence of lactose, yield an enzymatically inactive form of β-galactosidase, which nevertheless cross-reacts with antiserum specific for the enzyme, permits the genetic mapping of the β-galactosidase cistron (called the *z* gene). In addition, a second enzyme that facilitates the transport and accumulation of lactose through the cell's membrane was detected by mutations. The enzyme, galactoside permease, concentrates lactose within the cell, and a mutation in the permease cistron nullifies this ability. Galactoside permease was found to be "coinduced" by lactose along with β-galactosidase, and genetic mapping of permeaseless mutants showed that the permease cistron (*y* gene) was adjacent to the *z* locus. Finally, a third cistron (*a* gene) determining the enzyme thiogalactoside transacetylase (which acetylates thiogalactosides) was detected, found to be induced by lactose, and mapped adjacent to *y*. The *z*, *y*, and *a* genes map genetically between the *ara* (arabinose) and *trp* (tryptophan) loci in the order *ara-z-y-a-trp*.

In addition to mutations in these three cistrons, two other types of mutations that affect regulation of those cistrons were recovered. One class, called *operator* (*o*) *mutants*, was found to map immediately adjacent to the *z* gene in the order *ara-o-z-y-a-trp*. Jacob and Monod (1961a) made use of the ability to construct *E. coli* cells partially diploid for certain genetic segments by the introduction of episomal elements, such as the fertility factor, that carry extra genetic pieces. Thus different combinations of operator mutations could be generated in partial diploids (Fig. 2). The operator mutations are of two types, those (o^o) that are never inducible in the presence of lactose (noninducible) and those (o^c) that produce β-galactosidase in the absence of inducer (constitutive). These mutations were found to be *cis-dominant;* that is, they affect only those cistrons to which they are linked. Thus the operator region controls the expression of all three cistrons physically adjacent to it.

The second class of mutations was found to map between *ara* and *o* and was designated the *regulator* or *i* gene; the genetic order is therefore *ara-i-o-z-y-a-trp*. Again, two types of regulator mutations were detected, noninducible and constitutive. In partial diploids these

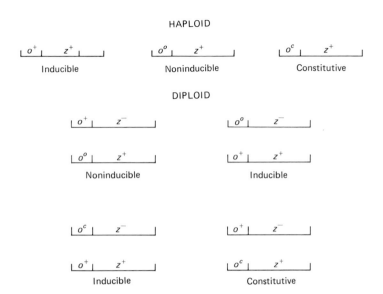

FIGURE 2. *Phenotypes of β-galactosidase production in different combinations of operator mutations. Only the z locus is indicated, as y and a behave similarly. See the text for an explanation of symbols.*

mutations exhibited a pattern of dominance different from the operator mutants (Fig. 3). Jacob and Monod (1961a) brilliantly suggested that the regulator locus specifies the production of a protein, the *repressor*, which acts via the cytoplasm, recognizes the operator region, and physically binds to it to prevent the transcription of the cistrons adjacent to the operator (Fig. 4). Finally, in addition to the operator recognition site, the repressor was postulated to have a site to which inducer could bind. The association of the repressor with inducer produces an "allosteric" change in the three-dimensional structure of the repressor, which prevents further binding with the operator (Monod et al., 1965). Thus, formation of the repressor–inducer complex frees the operator, thereby permitting transcription of its adjacent cistrons. As the enzymes concentrate and metabolize lactose, the inducer is depleted, so that the repressor molecules become free and reattach to the operator region, thereby shutting off further transcription. The regulator mutations, i^- and i^s, therefore produce defective repressors that fail to bind to the operator and to recognize inducers, respectively. The i^s repressor is dominant because once bound to the operator it cannot be dis-

lodged, whereas the repressor of i^- cannot bind and therefore an i^+ allele with a normal repressor will be dominant (Fig. 3). The unit of coinduction regulated by the i gene is called the *operon* and beautifully explains the control of enzyme production by the substrate upon which the enzymes act. In addition, the control of several cistrons by a single operator explains the coinduction of several enzymes involved in a related process. The model of the operon suggests that transcription proceeds in a polarized manner from the operator. The recovery of *polarity mutations*, which eliminate products from cistrons distal to the operator and mutational lesion, verified this prediction (Jacob and Monod, 1961b). Thus a polar mutation in z results in concomitant loss of y and a activity, whereas a polar mutant in a only eliminates the a product. Subsequent biochemical verification and extension of the initial operon model and predictions (Beckwith and Zipser, 1970) emphasize the power and elegance of the genetic studies by Jacob and Monod.

A number of different operons of varying size and different pathways have now been detected. In an extensive survey of the linkage relationships of genes involved in different bio-

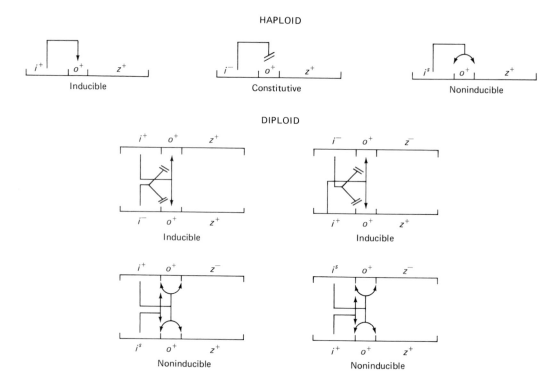

FIGURE 3. *Phenotypes of β-galactosidase production in different combinations of regulator mutations. Only the z locus is indicated, as y and a behave similarly. See the text for explanation of the symbols.*

synthetic pathways, Demerec (1964) found that the majority of cistrons involved in the same pathways were closely linked. This is suggestive of integrated controls of an operon type. Demerec and Demerec (1956) made the interesting observation that genes regulating enzymes involved in the tryptophan biosynthetic pathway were clustered and that the genetic order of these cistrons was the same as the sequence of reactions catalyzed by their respective enzymes. The histidine operon in *Salmonella* is comprised of eight cistrons, and the sequence of reactions catalyzed by their enzymes shows a very striking congruence with the genetic sequence (Ames and Hartman, 1963) (Fig. 5). These observations suggest a biological significance of this congruence that has yet to be understood.

The sequence of events following viral infection that lead to cell lysis and the release of mature viruses (Fig. 6) represents a microsystem of development and differentiation. Studies of the life cycle of the bacteriophage

T4, which infects *E. coli*, have been greatly aided by the existence of *conditional* lethal mutations.

"Conditional" lethal mutations are defined as those mutations whose lethal phenotype is dependent upon the conditions under which the organism is grown. Thus, under "restrictive" conditions, lethality results, whereas "permissive" conditions allow survival of the mutant. In microorganisms, the most extensively analyzed conditional lethal mutations have been (1) *auxotrophic*, which fail to grow on a minimal medium (restrictive) but which can survive when the medium is provided with a nutritional supplement (permissive) (Lederberg, 1950); (2) *nonsense*—a triplet codon that signals premature termination of reading of the genetic message in one genotype (restrictive) but that can be overcome by the presence of altered translational apparatus coded elsewhere in the genome (permissive) (Gorini and Beckwith, 1966); and (3) *temperature-sensitive*, which result from a single amino acid substi-

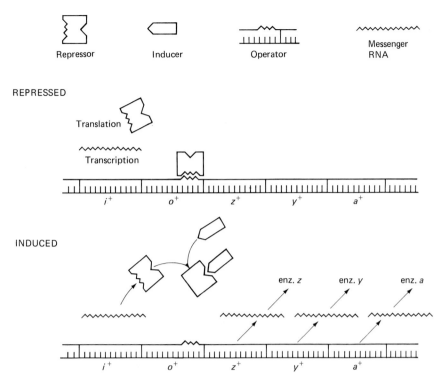

REPRESSED

Translation

Transcription

i^+ o^+ z^+ y^+ a^+

INDUCED

enz. z enz. y enz. a

i^+ o^+ z^+ y^+ a^+

Repressor   Inducer   Operator   Messenger RNA

FIGURE 4. *Model for regulation of the* lac *operon: enz.* z, *β-galactosidase, enz.* y, *permease; enz.* a, *transacetylase.*

tution in a protein (Wittmann and Wittmann-Liebold, 1966) that renders that protein inactive at one temperature (restrictive) but biologically active at a different temperature (permissive) (Jockusch, 1966a, b, c,).

A large number of conditional lethal mutations of the nonsense and temperature-sensitive (ts) classes were recovered in T4 phage (Edgar et al., 1964) and then grown under restrictive conditions (Epstein et al., 1963). Lysis of infected bacteria and electron microscope or biochemical examination of the debris permits a determination of phage-directed products present under restrictive conditions. For ex-

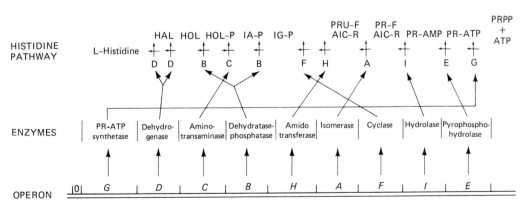

FIGURE 5. *Histidine operon and biosynthetic pathway regulated by its cistrons. For an explanation of the symbols, consult Ames and Hartman, 1963.*

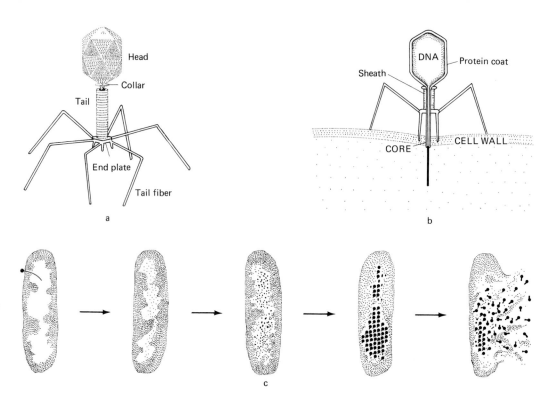

FIGURE 6. *Structure and life cycle of the T4 bacteriophage. (a) External morphology of the mature phage. (b) Section through infecting phage. (c) Cytological sequence of the phage life cycle from infection (left) to cell lysis (right). (From W. Wood and R. S. Edgar, Sci. Amer. 217, 5, 1967.)*

ample, only empty capsules of the viral head are recovered for mutations in genes 5 through 8 and 25 through 29. This implies that while head-forming ability is normal in each mutant genome, the genetic lesions prevent all other processes, such as DNA packaging and tail fiber formation, from occurring. Thus a lesion in a single cistron can have a pleiotropic effect on the expression of other loci. This is a striking illustration of the interdependence of gene function in normal morphogenesis.

In correlating the genetic positions of different cistrons with their mutant phenotypes under restrictive conditions, a striking pattern emerges (Fig. 7) (Epstein et al., 1963). Clearly, genes that exhibit similar phenotypes are not randomly scattered around the phage chromosome; rather, they are "clustered" together. This clustering pattern of morphogenetic cistrons is reminiscent of the clustering of cis-

trons involved in the same biosynthetic pathways (Demerec, 1964).

The time after phage infection during which restrictive temperatures prevent ts mutations from surviving can be determined by "shift" experiments. By infecting bacteria at the permissive temperature and then shifting different cultures to the restrictive at successive intervals after infection, mature phages can be recovered after a critical time interval has passed at the permissive temperature (Fig. 8) (Edgar and Epstein, 1965). The time at which shifts to restrictive temperature yield viable phages is specific for each mutation and suggests the time at which either the ts gene no longer functions or has functioned sufficiently to permit subsequent morphogenetic activity. The work on conditional lethal mutations in phages thus extends the concept of the clustering of genes related in their biosynthetic activities to

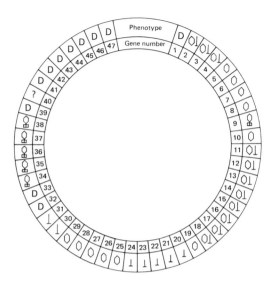

FIGURE 7. *Genetic map and corresponding mutant phenotype of each cistron of T4 phage.* D, *defective DNA synthesis;* ◖, *head capsids;* ⊥, *tail core and base plate;* ▭, *tail. (From R. S. Edgar and R. H. Epstein, Proc. 11th Int. Cong. Genet. 2, 10, 1965. Copyright © 1965, Pergamon Press Ltd.)*

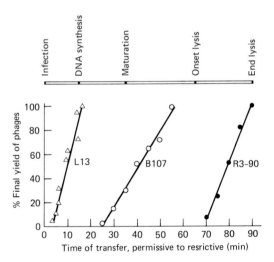

FIGURE 8. *Phage yields in cultures shifted from permissive to restrictive temperatures at different times after infection. L13, B107, and R3-90 are different mutations. (From R. S. Edgar and R. H. Epstein, Proc. 11th Int. Cong. Genet. 2, 12, 1965. Copyright © 1965, Pergamon Press Ltd.)*

the clustering of cistrons related in their morphogenetic activities and their temporal sequence of activity.

Gene Function in Relation to Chromosome Position in Eukaryotes

The obvious importance of chromosomal position of cistrons in operons in bacteria and the clustering of temporally and morphogenetically related genes in the T4 genome bring up the question of whether gene sequence in the chromosomes of eukaryotes also regulates their activity.

An illustration of the regulation of genetic activity by chromosomal position is illustrated by the so-called *stable* or *S-type* position effect in the fruit fly *Drosophila* (Lewis, 1950). Two mutations, Star (S) and asteroid (ast), are closely linked and located near the tip of chromosome 2. $\frac{S\ +}{+\ +}$ flies have eyes which exhibit a "rough" texture and are slightly smaller than wild type. The eyes of $\frac{+\ ast}{+\ ast}$ flies are small and rough. Eyes of $\frac{S\ +}{+\ ast}$ flies are very small and rough, whereas $\frac{S\ ast}{+\ +}$ flies are phenotypically identical with $\frac{S\ +}{+\ +}$ flies (Lewis, 1950). The difference in phenotype when S and ast are on the same chromosome (cis) and on different chromosomes (trans) shows that the chromosome is itself a unit of control in which the gene sequence in it plays a role in the regulation of genetic activity.

Cytologists recognized long ago that chromosomes were not uniformly staining organelles but, in fact, exhibited regions of intense staining (heteropycnosis) and of light staining (Heitz, 1933). Thus chromosome regions were classified cytologically as heterochromatin (darkly staining) and euchromatin which were assumed to be of functional significance. Generally, heterochromatin visible in mitotic chromosomes is localized around the centromeres, the nucleolus organizer, and the telomeres or tips. Heterochromatic bodies often remain visible in interphase when most of the chromosome arms are invisible.

The induction of chromosome aberrations such as inversions and translocations results in the displacement of genes from their normal chromosomal location to a different region of the genome. Often juxtaposition of a euchromatic locus to heterochromatin and vice versa results in an alteration of genetic activity of the locus even though the gene itself has not been altered by the rearrangement. A classic study of such an effect was carried out in *Drosophila* (Judd, 1955).

The white (*w*) locus resides in euchromatin near the tip of the X chromosome. The wild-type allele, *w+*, produces a bright red eye color and is dominant over the *w* allele, which causes a colorless white eye. Judd (1955) studied a translocation in which the tip of the X (including *w+*) was attached to the heterochromatin of chromosome 4 (Fig. 9). The eyes of females heterozygous for the rearrangement carrying *w+* and a normal X chromosome bearing *w* exhibit a "variegated" pattern of red and white tissue even though *w+* is normally dominant to *w*. The fact that the altered *location* rather than an alteration within the *w+* gene is responsible for the expression of the

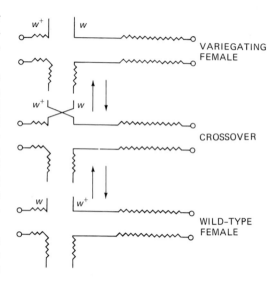

FIGURE 10. *Method of removal and insertion of* w+ *into a translocation.*

white phenotype was shown by reinsertion of the *w+* gene from the translocation into a normal X chromosome and the introduction of a *w+* gene from a normal X into the translocation by crossing-over. In the former case, the *w+* gene removed from the rearrangement is completely dominant and, in the latter, the *w+* inserted, variegates (Fig. 10). This demonstrates beautifully that the normal activity of the *w+* gene is altered by a change in chromosomal position.

In addition, when closely linked genes affecting the same tissue are simultaneously rearranged, a polarized pattern of variegation emerges. For example, in *Drosophila*, the mutations *w* and *rst* cause loss of color and roughening, respectively, and are closely linked at the tip of the X chromosome. The genes *w+* and *rst+* can be juxtaposed to heterochromatin in the order *w+*-*rst+*-heterochromatin. In females heterozygous for the rearrangement and a normal X carrying the recessive alleles *w* and *rst*, patches of red–rough and white–rough tissue, in addition to wild-type red–smooth tissue, can be found (Demerec and Slizynska, 1937) (Fig. 11). Patches of white–smooth tissue are never found. Such a "spreading effect" (reviewed by Lewis, 1950) reflects a polarized pattern of gene inactivation originating from heterochromatin and extending through *rst+* first

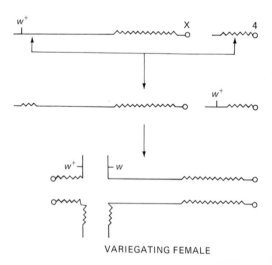

FIGURE 9. *The X-4 translocation, which variegates for* w. ——— , *euchromatin,* \/\/\ , *heterochromatin.*

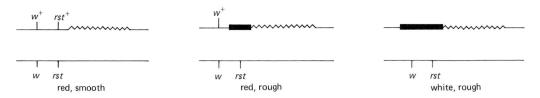

w^+ rst^+ — red, smooth
w^+ — red, rough
white, rough

w rst
w rst
w rst

FIGURE 11. *Nature of a spreading effect.* ——— , *euchromatin;* ⋀⋀ , *heterochromatin;* ▬ , *inactive euchromatin. See the text for detailed discussion and explanation of symbols.*

and then w^+. This observation must be an important clue to understanding the mechanism of gene control by chromosome position.

The eye phenotype of a female variegating for w is not a "salt-and-pepper" pattern of small patches of red and white tissue. Instead, the eye consists of relatively large sectors of red and white tissue. The patterns of red and white tissue were recorded carefully by Baker (1963) and were organized into complementary patterns. Remarkably, the sectors of variegated tissue correspond closely with the clonal origin of these areas. That is, cells in each sector indicated in Fig. 12 are clonally related in that they are all derived from the same original stem cell. The cell lineage was demonstrated genetically by Becker (1957) and will be discussed in a later section. Baker's observations suggest that at some point in development when a small number of cells exist in the tissue comprising the prospective adult eye, a deter-

minative event occurs in each cell that dictates whether the w^+ gene in the rearrangement will function at the time of pigment production. This event occurs when there are approximately eight prospective cells for the lower quadrant of the eye, and from that time on, all the dozens of daughter cells of each cell inherit a chromosome with the same functional state. The determination occurs during the larval stage, days before pigment is actually produced in pupae. Thus, after the decisive point, the on-or-off condition still to be realized is inherited somatically from cell to cell; yet in each generation of flies, the functional state is erased and again redetermined in each zygote.

Heterochromatin and Its Functional Significance

The pattern of somatic hereditability of functional activity and its erasure at each generation is reminiscent of the behavior of chromosomes in coccid mealy bugs (Brown and Nur, 1964; Brown, 1966). In coccid males, one entire set of chromosomes is heterochromatic and presumably nonfunctional (Fig. 13a). The factor which determines heterochromatization is the parental source of the chromosomes; that is, all paternally derived chromosomes in coccid males become heterochromatic. Proof of the paternal origin of the heterochromatized chromosomes was derived by Brown and Nelson-Rees (1961), who administered heavy doses of radiation to males and females and mated them to unirradiated animals. Male progeny of irradiated parents had breaks confined to the heterochromatic chromosomes when their

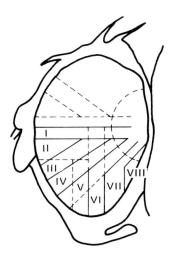

FIGURE 12. *Sectors of the lower half of a* Drosophila *eye which are clonally related.* (*From W. K. Baker,* Amer. Zool. 3, 60, 1963.)

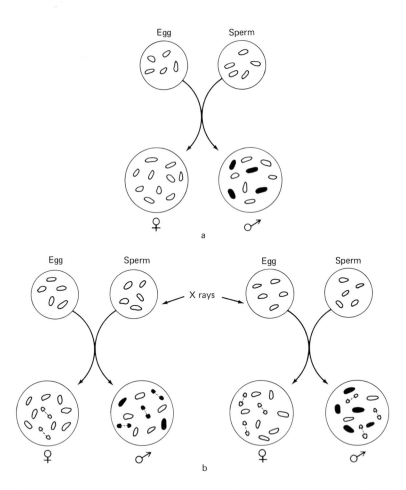

FIGURE 13. *Pattern of heterochromatization in coccids. (a) Unirradiated; (b) male or female parent irradiated.* ●, *heterochromatin;* ○, *euchromatin;* - - -, *chromosome break.*

father had been X-rayed, whereas euchromatic damage resulted after maternal treatment (Fig. 13b). Thus each paternally derived chromosome in a male zygote "remembers" its origin and passes that imprint on to each daughter cell. This paternally determined state acts also in germinal tissue, where the heterochromatic chromosomes assort together and eventually form nonfunctional meiotic products. Obviously then, the meiotic chromosomes that form functional sperm are euchromatic and maternally derived but will now be imprinted as paternally inherited chromosomes in the male offspring resulting from them. Here again we see the imprint of a *cellularly inherited* chromosome state (heterochromatization) that is changed at each generation.

The radiation experiments (Brown and Nelson-Rees, 1961) also indicate that the heterochromatic chromosomes of coccid males are genetically inert. Dominant lethal mutations are induced by irradiation as evidenced by extensive death of male and female zygotes after irradiation of their female parents. Irradiation of males, on the other hand, produced little lethality among their sons, but female progeny exhibited extensive death. Here clearly, the genetic inactivity of heterochromatic chromosomes prevents expression of any dominant lethality induced in them.

The genetic inertness of heterochromatized chromosomes has led to the proposal that those chromosomes or regions that are always heterochromatic in both chromosomes during division (constitutive heterochromatin) carry little or no genetic information (Muller and Painter, 1932). The Y chromosome of *Drosophila melanogaster* is always heterochromatic during cell division yet must carry genes necessary for fertility, since X/O males are sterile. By the recovery of radiation-induced deletions in the Y chromosome, Brosseau (1960) found seven different regions of the Y that were all necessary for male fertility. This still represents a limited number of loci in a chromosome that is cytologically almost as long as the X chromosome.

In a classic experiment, Crouse (1960) demonstrated a specific function of a heterochromatic element. Chromosome behavior in the fungus gnat, *Sciara coprophila*, is highly abnormal. One of the distinctive events in *Sciara* zygotes is the elimination of the paternally derived X chromosome at the sixth or seventh cleavage after fertilization. The paternally inherited X chromosome in somatic cells "remembers" its paternal origin and, at the proper developmental stage, migrates through the nucleus into the cytoplasm, where it disintegrates. Crouse (1960) asked what part of the chromosome was responsible for both the retention of the parental imprint and the control of chromosome elimination. The X chromosome of *Sciara* is acrocentric with a long arm and a short arm capped by a large heterochromatic knob. Crouse (1960) induced a number of X-autosome translocations with breakpoints at different places along the X chromosome (Fig. 14). She then mated males bearing each translocation and followed chromosome elimination in the zygotes produced. All progeny carrying a translocation with a break in the long arm (Fig. 14) showed elimination of the chromosome bearing the centromere of the X. This

would seem to suggest that elements very close to or at the centromere itself were responsible for X loss. However, one rearrangement, ORT_1, involved a break in the short arm of the X, thereby translocating the heterochromatic knob to an autosome. In this case, the X chromosome with its own centromere was *not* eliminated, and the autosomal centromere bearing the X heterochromatic knob *was* eliminated. The heterochromatic element adjacent to the centromere thus retains the information specifying its parental source and controls the movement of the chromosome at a specific time in development. Obviously, this heterochromatin contains information that does play a role in cell division.

One locus known to reside in heterochromatin is the nucleolus organizer (NO). In *Drosophila*, the NO region lies in the proximal heterochromatin of the X chromosome. The association of assembled ribosomes at this region prompted Ritossa and Spiegelman (1965) to ask whether the locus itself codes for ribosomal RNA. They utilized two X chromosome inversions (sc^4 and sc^8) that have distal breaks at very similar positions and proximal breaks on either side of NO (Fig. 15). In females heterozygous for both chromosomes, a single crossover between the inversions generates the recombinant chromosomes $sc^{4L}\ sc^{8R}$ and $sc^{8L}\ sc^{4R}$, which carry 0 and 2 NO regions, respectively (Fig. 15). By combining different chromosomes, Ritossa and Spiegelman (1965) could produce flies with different numbers of NO: 1 NO ($sc^{4L}\ sc^{8R}/X$), 2 NO (X/X), 3 NO ($sc^{8L}\ sc^{4R}/X$), and 4 NO ($sc^{8L}\ sc^{4R}/sc^{8L}\ sc^{4R}$). By measuring the amount of radioactive rRNA hybridizing with a known quantity of DNA extracted from flies with different NO dosages, they showed that ribosomal RNA hybridiza-

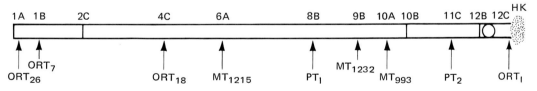

FIGURE 14. *X chromosome of* Sciara coprophila *and the positions of the translocation breakpoints.* HK, *heterochromatic knob;* ↑ *indicates position of an X break in one of the translocations; numbers above the chromosome are chromosome band positions. (From H. V. Crouse,* Genetics 45, *1432, 1960.)*

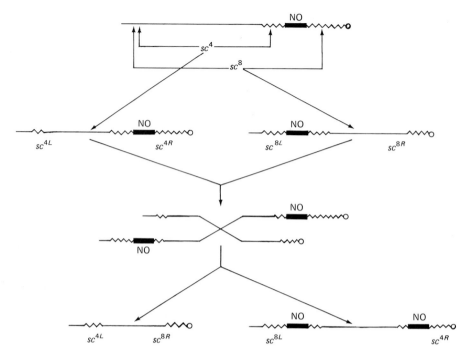

FIGURE 15. *Method of generating chromosomes carrying different numbers of nucleolus organizers. L and R refer to left and right parts of the inversions. See the text for further explanation of symbols.*

tion increases linearly with the number of copies of NO. Moreover, they calculated that 1 NO region corresponds to a duplicated sequence of 450 copies of each ribosomal DNA cistron. Thus regions of the genome exhibiting constitutive patterns of heterochromatization can be shown to contain genetically meaningful information that may be highly redundant.

In other cases, heterochromatic regions obviously carry genetic information and simply reflect a cytologically visible state of inactivity. Facultative heterochromatin is an induced state of heterochromatization in which both members of a pair of chromosomes do not show similar patterns. Paternally inherited chromosomes in male coccids are facultatively heterochromatized. Here chromosome condensation is not an intrinsic property of the genetic region but is imposed by other factors. In mammalian females, facultative heterochromatization of one of the two X chromosomes provides a graphic illustration of a phenomenon unique to diploid organisms with a heteromorphic chromosome mechanism of sex determina-

tion. X/X females and X/Y males obviously differ by a whole chromosome in genes carried on the X chromosome. Yet, in mice, for example, relatively normal X/X/Y males and X/O females can be obtained (for a fascinating review of the entire problem, read Stern, 1960). Since X/X/Y and X/Y individuals are males and X/X and X/O are females, it is clear, then, that some means must exist for regulating the level of activity of the X chromosomes such that the amount of genetic activity in cells with two Xs must be similar to the amount in cells with a single X. The similar extent of X chromosome activity in males and females was called "dosage compensation" (Muller, 1932).

The recovery of X-autosome translocations in mice provided a clue for the mechanism of dosage compensation in mammals. Females heterozygous for the translocation and a normal autosome carrying a recessive marker are often mottled with patches of mutant and wild-type tissue, a phenotype resembling position effects in *Drosophila*. Males, on the other hand, carrying the translocation and the recessive autosomal marker express the dominant, wild-type phenotype. This led Russell (1961) and Lyon (1961) to suggest independently, that during embryonic development of mammalian females, either of the two X chromosomes is *inactivated* by heterochromatization. A cytological basis for this hypothesis was the existence of a heterochromatic element (Barr body) in interphase nuclei only of females. The Russell–Lyon model suggests that translocation of autosomal genes to an X chromosome results in inactivation of the autosomal loci when the translocated X is inactivated, thereby allowing the expression of the recessive allele (Fig. 16). Inactivation of an X chromosome is thus a very effective method whereby the functional dosage of X-linked loci in females could be equated with the level of activity in males. Since males carrying a translocation do not variegate, it appears that compensation is affected by chromosome *number* and not sex per se. This has been verified by the demonstration that X/X/Y males variegate, whereas X/O females do not (Lyon, 1968).

Proof that one of the X chromosomes is inactivated in all mammalian female cells was provided by experiments with human cells (Davidson et al., 1963). Different alleles of the sex-linked locus specifying the enzyme glucose-6-phosphate dehydrogenase (G-6-PD) are known. In Negroes, two alleles are known

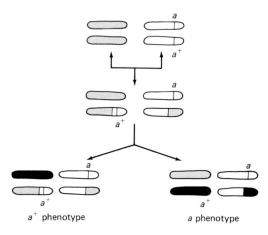

FIGURE 16. *Russell–Lyon explanation for variegation of mammalian females heterozygous for an X-autosome translocation.* ◯, X *chromosome;* ●, *heterochromatin;* ◯, *autosome.*

that produce G-6-PD with different electrophoretic mobility, thereby permitting their electrophoretic separation. In Caucasians, an allele exists with very low enzyme activity. Davidson et al. (1963) took samples of skin cells from women known to be heterozygous for different G-6-PD alleles. They then minced the sample and isolated single cells or small clusters of cells, each of which was cloned further in culture. Cells in each clone were then tested for the presence of the two separable types of G-6-PD in Negroes or for enzyme levels intermediate between the two allelic levels in Caucasians. In all clones tested, no "hybrid" patterns were recovered; that is, the phenotype of each clone reflected the genotype of one or the other X chromosome but never both simultaneously. For the X-linked G-6-PD locus, then, it is clear that in each cell of a female, only one of the two X chromosomes is functional. This has been corroborated for several other loci (Lyon, 1968). This experiment not only verifies the Lyon–Russell hypothesis, it points out two further developmentally interesting facts. One is that in skin cells, at least, once one of the X chromosomes is inactivated (either X may be inactivated) in a cell, then that state is inherited by all its daughter cells; otherwise the cloning test would not have been possible. This somatic cell heritability of the condition of a nonfunctional chromosome is reminiscent of the determination of variegation in *Drosophila* (Baker, 1963) and of paternal chromosomal heterochromatization in male coccids (Brown, 1964). The second point is that the determinative inactivation of the X in skin cells appears to occur late in development, since clones of both types were found within a single skin biopsy. This does not necessarily mean that both X chromosomes functioned in these cells until the late determinative event, only that *irreversible* inactivation occurs late.

In *Drosophila*, dosage compensation must occur by a mechanism different from mammals, since females heterozygous for sex-linked mutations do not normally variegate. That dosage compensation of X-linked genes does exist in *Drosophila* was strikingly demonstrated by Grell (1962). He studied enzyme activity of xanthine dehydrogenase, a protein polymer controlled by a sex-linked locus, maroon-like (*ma-l*), and an autosomal gene, rosy (*ry*). These two loci are recognized by mutant eye colors. He constructed duplications and deficiencies for the *ma-l* and *ry* loci so that both males and females carrying 1, 2, or 3 doses of *ma-l*+ or *ry*+ could be generated. The enzyme levels assayed in each type of fly were found to be as follows:

Gene dosage	Units of enzyme activity
1 *ma-l*+	0.144
2 *ma-l*+	0.147
3 *ma-l*+	0.152
1 *ry*+	0.084
2 *ry*+	0.172
3 *ry*+	0.202

These results illustrate that a sex-linked locus is dosage-compensated for its enzyme levels, whereas autosomal loci have no such regulation. The mechanism for this regulation of chromosomal activity is still an enigma (Stern, 1960) and may be one phenomenon for which models will have to be derived from diploid multicellular organisms.

Use of Genetic Mosaics to Study Development

Phenotypically, mammalian females are mosaics comprised of heterogeneous cell populations in which one or the other X chromosome is functional (Nesbitt and Gartler, 1971). Individual organisms composed of cells that are genetically different are called *genetic mosaics* and can be useful in a variety of developmental studies. Numerous examples of normally occurring genetic mosaicism exist in invertebrates where specific chromosomes or parts of chromosomes are eliminated in certain cells during development (see Wilson, 1925, and White, 1954, for reviews). In some mammals, blood cell chimeras or mosaics can result from exchange between twins through vascular anastomosis (Owen, 1945). Experimentally, inducible genetic mosaicism is a useful tool to study development and will be considered here.

Sturtevant (1929) generated mosaics in *Drosophila* utilizing the property of the mutation ca^{nd} whereby chromosome loss is induced in the early cleavages after fertilization. For example, loss of an X chromosome in an X/X zygote produced by a female homozygous for ca^{nd} yields a mosaic of X/X (♀) and X/O (♂) cells. If one of the X chromosomes is "marked" with a recessive mutation such as *y*, which produces a yellow body color, then loss of the other X would give a *y*+/*y* (wild-type ♀):*y*/O (yellow ♂) mosaic. Since simultaneous loss of an X chromosome in several cells of one embryo is rare, all yellow X/O tissue in an otherwise wild-type X/X individual presumably represents a clone derived from one original cell. Thus the cell lineage of the external surface of the fly can be deduced from a study of many such mosaics with varying sized mutant patches. It can be seen, then, that loss of a chromosome early in development will produce a relatively large proportion of X/O cells, whereas a loss later will be reflected in a correspondingly smaller patch of X/O tissue (Fig. 17).

Another useful method for inducing mosaics at later stages in development was provided by Stern (1936). He found that since homologous chromosomes pair during *mitosis* in *Drosophila*, occasional crossing-over could occur between them. For example, if two homologous chromosomes are marked with different recessive mutations, a mitotic crossover between the markers and the centromere produces two daughter cells that differ from each other *and* from the original cell in genotype (Fig. 18). Thus a "twin-spot" of mutant tissue surrounded by wild-type tissue is diagnostic of a mitotic crossover. Utilizing such a protocol, Becker (1957), by irradiating larvae, induced somatic crossovers between X chromosomes carrying different eye-color markers. By looking at the position and shape of twin spots of mutant tissue in a background of wild-type eye color, he was able to deduce the cell lineage of the eye surface (see Fig. 12). As noted earlier, the clonal sectors deduced for the eye correspond very closely with the patterns of eye variegation (Baker, 1963).

Besides providing a method for tracing cell lineage relationships between parts of an organism, the generation of mosaics also permits a test of autonomy of various mutations. Autonomy of genetic activity refers to the pheno-

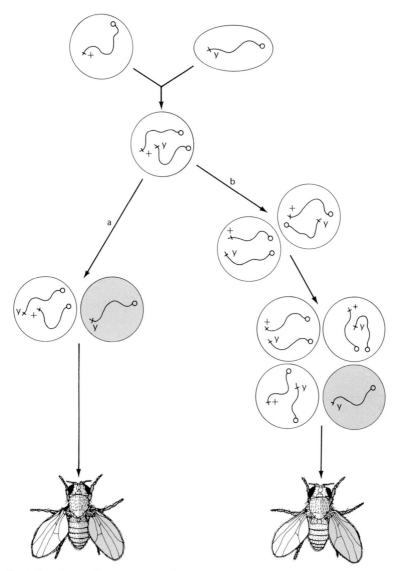

FIGURE 17. *Basis for formation of a genetic mosaic by chromosome loss. See the text for an explanation of symbols.*

typic expression of the genotype in a cell juxtaposed to cells of a different phenotype and genotype. For example, in a $y/+:y/O$ mosaic in *Drosophila*, the mutant phenotype yellow is expressed in X/O cells, even though wild-type cells are immediately adjacent. The mutation y is said, therefore, to be autonomous. Nonautonomy, on the other hand, would suggest that communication exists between cells such that factors regulated by the genotype of one cell could influence the phenotype of genotypically different cells. The sex-linked mutation v produces a bright scarlet mutant eye in *Drosophila*. A $v/+:v/O$ mosaic has near wild-type eye color in the v/O tissue, thereby proving that v is nonautonomous (Sturtevant, 1932). Subsequently, the v gene has been shown to specify the enzyme tryptophan pyrrolase, which converts tryptophan to formylkynurenine (Tatum and Haagen-Smit, 1941; Baglioni, 1960). A dif-

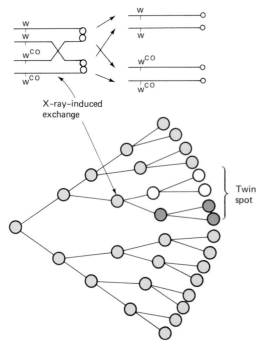

W
W
w^{co}
w^{co}

X–ray-induced
exchange

Twin
spot

FIGURE 18. *Generation of a somatic twin spot by mitotic crossing over.* ○, *heterozygous cell;* ◐, w^{co}/w^{co} *cell;* ○, *w/w cell. (From W. K. Baker,* Amer. Zool. 3, *61, 1963.)*

fusible end product, kynurenine, produced by v^+ cells can indeed be taken up by v cells, thereby producing a wild-type phenotype and corroborating Sturtevant's (1932) initial suggestion.

Stern (1954) analyzed mosaic patches in *Drosophila* in an attempt to demonstrate that nonautonomy of mutant tissue could be the consequence of disruption in a developmental "prepattern." Stern was struck by the fact that differentiated structures such as bristles always develop in a specific position in an adult fly, even though there do not appear to be significant features distinguishing the position of bristle-forming cells from their immediate neighbors. In other words, why is a specific bristle always formed at precisely the same place in a relatively homogeneous tissue? He suggested that there could be a genetically programmed overall prepattern or skeletal plan to which cells respond in accordance with their spatial distribution. For example, folding patterns of tissue underlying the cuticle could be the signal triggering differentiation of a cuticu-

lar cell into bristle production, either eliminating the bristle or changing its position—even when the overlying cuticular cells retain the normal bristle-forming capacity.

Stern's long search for such a mutation was rewarded by the discovery that the mutation ey^D on chromosome 4 alters the prepattern for sex comb development on the male foreleg (Stern and Tokunaga, 1967). Normal males carry a single sex comb comprised of 10 to 13 teeth, whereas ey^D-bearing males often have three or more times as many teeth arranged in several parallel rows. The gene ey^D was inserted into a third chromosome that also carried at its tip a duplication of y^+ translocated from the X chromosome. Stern and Tokunaga then X-rayed male larvae carrying y on the X chromosome and the third chromosome with y^+ and ey^D (Fig. 19).

A mitotic crossover induced in the region between ey^D and the centromere results in a daughter cell that no longer carries y^+ or ey^D and is detectable as a yellow patch on the exterior of the fly. Forty-seven mosaics with yellow sectors in the sex comb region were found. All 47 formed multiple sex combs characteristic of the ey^D phenotype, even though *the cells themselves did not carry the ey^D gene.* Here, clearly, is a case of nonautonomy where the surrounding $ey^D/+$ tissue induces an ey^D phenotype in $+/+$ cells. This provides evidence for Stern's interpretation of nonautonomy in terms of genetically determined prepatterns.

Nonautonomy has been interpreted in still other ways in terms of sequences of gene activity during development. Garcia-Bellido and Merriam (1971) induced mitotic crossing-over in flies ($y\ Hw/sta\ sn^3$) heterozygous for the dominant mutation Hw, which produces extra hairs on the wings of adults. Larvae were irradiated at different times prior to puparium formation. A mitotic crossover yields twin spots of $y\ Hw/y\ Hw$ and $sta\ sn^3/sta\ sn^3$ tissue. Extra chaetae were detected in $y\ Hw/y\ Hw$ cells but none in the $sta\ sn^3/sta\ sn^3$ cells in flies irradiated at any time before 8 hours prior to puparium formation. However, extra chaetae were present in $sta\ sn^3/sta\ sn^3$ tissue of flies irradiated 0 to 8 hours before puparium formation, even though the Hw gene responsible for the extra chaetae was absent. Eight hours before puparium formation, each cell in the wing disc has two further divisions to make before metamorphosis and differentiation. Garcia-Bellido and Merriam (1971) conclude, therefore, that the *Hw* gene is activated 8 hours before formation of a puparium, after which daughter cells have now been programmed to follow the subsequent sequence of events resulting in the *Hw* phenotype. In all these cases of nonautonomy (*v*, *ey*^{D}, *Hw*) detected in genetic mosaics, the model of a diffusible genetically regulated compound provides a simple way of interpreting the observations.

Whereas mitotic crossing-over permits the production of mosaics for autosomal as well as sex-linked genes in *Drosophila*, another useful technique exists for producing mosaics of sex-linked loci only. The ring X chromosome, *In(1)w*^{vC}, is somatically unstable and tends to be lost at a high frequency in early cleavage divisions (Hinton, 1955). *In(1)w*^{vC} carries wild-type alleles of all X loci. Thus, when an *In(1)w*^{vC}-bearing egg is fertilized by a sperm carrying an X chromosome marked by recessive mutations, loss of the ring permits expression of the recessive alleles in X/O tissue. The high frequency of loss of *In(1)w*^{vC} and the fact that it is an X chromosome aberration makes its usefulness much greater than *ca*nd. Utilizing *In(1)w*^{vC}, Hotta and Benzer (1970) determined that mutations affecting the pattern of electroretinograms are autonomous within the eye or its proximity. Similarly, Ikeda and Kaplan (1970) showed that a neural mutation affecting leg movement acted autonomously in leg tissue. Grigliatti et al. (1972) used *In(1)w*^{vC} to recover mosaics for *para*^{ts}, a mutation producing reversible, temperature-sensitive paralysis of adults. They found that the mutation is autonomous and that its site of action is anatomically localized in the area of the cephalic and thoracic ganglia. All three of these studies have utilized the somatic instability of *In(1)w*^{vC} to localize the sites of action of different mutations affecting the nervous system.

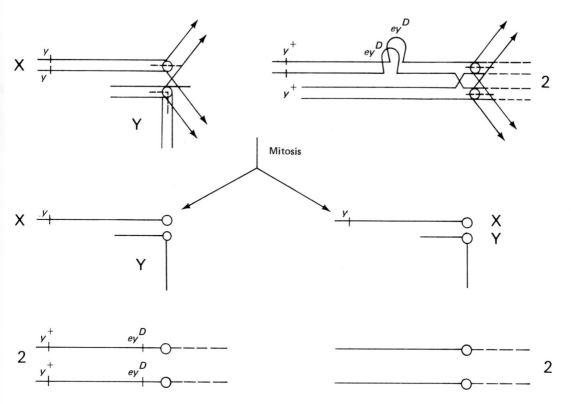

FIGURE 19. *Generation of mosaics for* ey^D *by mitotic crossing over. See the text for an explanation of symbols.*

All the *Drosophila* studies of mosaics just discussed rest on the assumption that phenotypically distinguishable mosaicism on the external surface of the flies reflects the genotype of tissue immediately underlying. There is some corroboration of this assumption, but methods are needed to recognize internal mosaicism. This is possible in mice using a powerful technique developed by Beatrice Mintz (1962). Mintz showed that the membrane surrounding mouse embryos in early cleavage could be mechanically removed and the cells of two separate embryos juxtaposed. Such fused embryos can form a single cohesive mass that, when introduced into a pseudopregnant female, can implant and develop into a single animal (Fig. 20). Such a tetraparental mouse can be a mosaic of two genotypes if the two embryos fused were genetically different.

While Mintz (1967) was able to trace the cell lineage of the surface of mice using mutations affecting hair color, she has been able to study internal structures as well using biochemically distinguishable genetic differences. In a very elegant study, Mintz and Baker (1967) were able to resolve the origin of multinucleate skeletal muscle. Two simple explanations can be advanced for the multinucleate nature of single muscle cells known to arise from mononucleate myoblasts (Fig. 21): Model 1, by successive nuclear division without cell division, and Model 2, by successive fusion of prospective muscle cells. Obviously, Model 1 predicts that the nuclei within each muscle cell having arisen by mitosis will be identical. On the other hand, Model 2 predicts that single muscle cells of a genetic mosaic could contain nuclei of different genotypes.

Mintz and Baker (1967) took two strains of mice differing in the properties of the enzyme isocitrate dehydrogenase (IDH). The two strains produce electrophoretically distinct forms of IDH. F_1 hybrids of the two strains produce the parental forms of IDH, and an ad-

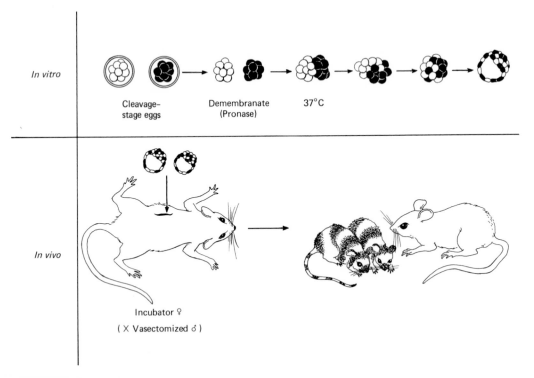

In vitro

Cleavage-
stage eggs

Demembranate
(Pronase)

37°C

In vivo

Incubator ♀
(× Vasectomized ♂)

FIGURE 20. *Method for embryonic fusion to form tetraparental mice.* ●, *cells for pigmented strain;* ○, *cells for albino strain. (From B. Mintz,* Proc. Nat. Acad. Sci. U.S.A. 58, 345, 1962.)

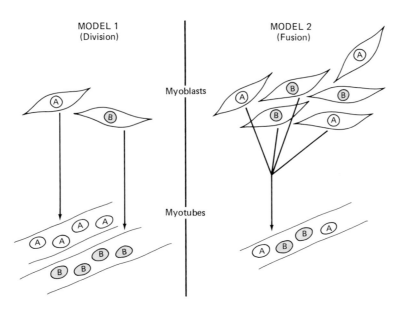

MODEL 1
(Division)

MODEL 2
(Fusion)

Myoblasts

Myotubes

FIGURE 21. *Models of the origin of multinucleate skeletal muscle cells in tetraparental mice. A and B represent genetically different nuclei.*

ditional hybrid form that can be seen to migrate between the two parental bands on electrophoresis. Embryos of the two strains were then fused to produce a tetraparental mouse. Analysis of mononucleate tissues such as liver of the tetraparental mouse revealed both parental forms of IDH, thereby verifying the mosaic nature of the animals. The absence of hybrid forms of IDH in the liver showed the autonomous behavior of the IDH genes. Skeletal muscle tissue, on the other hand, produced hybrid forms of IDH in addition to the parental types. This can only be explained by assuming the presence of both parental nuclei within a single cell, an observation that rules out Model 1 (Fig. 21). Thus, Mintz and Baker (1967) beautifully proved that skeletal muscle forms by cell fusion. The technique of embryonic fusion is being extensively exploited to study the developmental aspects of hermaphroditism (Tarkowski, 1964; Mintz, 1968), blood-producing systems (Mintz and Palm, 1969), neoplasms (Mintz and Slemmer, 1969), and numerous other problems. This beautiful technique promises to allow the exploitation of genetic mosaicism to answer a number of fundamental questions.

Use of Mutations to Analyze Developmental Phenomena

A discussion of developmental genetics of eukaryotes would be incomplete without considering the utility of specific mutations. The power of mutations as probes into regulatory phenomena has been amply demonstrated in microorganisms. The functional autonomy and the developmental stage and tissue sites of the activity of various mutations have been determined by the construction of genetic mosaics. Extensive studies of mutations exhibiting phenotypically visible developmental abnormality or lethality during development have been carried out (for reviews, see Goldschmidt, 1955, and Hadorn, 1955). However, such studies usually encounter considerable limitations imposed by the nature of the mutations. For example, it is most often extremely difficult to determine the primary biological defect caused by a developmental mutation. Most often, the phenotypically detectable abnormality is pleiotropically produced as a secondary consequence of the primary lesion. Moreover, the onset of a detectable deviation from normal development may occur long after the inferred time of the initial expression of a mutation. In addition, those mutations whose phenotypic effects are lethality during development prevent a determination of the duration over which such a locus might act and possible successive periods of activity. Finally, the lethal properties of many developmental mutations require maintenance of the defect in heterozygotes, a fact that can create considerable difficulty in experiments.

A method for enhancing the versatility and usefulness of mutations in eukaryotes is the selection for mutations whose mutant phenotype is conditionally expressed. The temporal regulation of mutant expression can then be determined by alternating permissive and restrictive conditions impinging on the developing organism. Temperature-sensitive (ts) mutations appear to be the most readily recoverable class of conditional defects since a defined medium or suppressor mutations are required for auxotrophs and nonsense mutations, respectively. Temperature-sensitive mutations have been recovered in fungi such as *Neurospora* (Horowitz, 1950) and yeast (Hartwell, 1967) as well as multicellular organisms such as the plant *Arabidopsis* (Langridge, 1965), *Drosophila* (Suzuki et al., 1967), the wasp *Habobracon* (Smith, 1968), and mammalian cells (Thompson et al., 1970). However, ts mutations have been most extensively exploited to study development in *Drosophila* (Suzuki, 1970) and the remaining discussion will deal with these studies.

In order to interpret experiments using ts mutations, the basis for temperature-sensitivity must be understood. At the present time, the assumed mechanism in multicellular organisms rests primarily on molecular studies done on ts mutations in phages and bacteria. In these organisms, most ts defects result from thermolability of proteins carrying a single amino acid substitution (Wittmann and Wittmann-Liebold, 1966; Jockusch, 1966a,b,c), although single base changes can produce ts transfer RNA molecules (Smith et al., 1970). In *Drosophila*, chemically-induced ts lethals do localize genetically as point mutations and map extensively throughout the genome (Suzuki, 1970). Temperature-sensitive activity of the enzyme tryptophan pyrrolase regulated by a ts allele of *v* in *Drosophila* has been detected (R. Camfield, unpublished). Thus ts mutations in higher organisms with a molecular basis similar to those in microorganisms could permit the use of temperature to affect the primary genetic lesion more precisely. Restrictive temperatures, then, are presumed to affect biological activity of mutant proteins or tRNAs, an effect that may be detectable long before any other visible abnormality in development.

One observation that may be studied by ts mutations is the discovery from nucleic acid hybridization studies that DNA sequences exist in multiple copies (Ritossa and Spiegelman, 1965; Laird and McCarthy, 1968). If such "redundant" material has a genetic function, recessive mutation in such regions could never be detected because they would be masked by duplicate wild-type sequences. Furthermore, if such redundant information is necessary for viability, then dominant alterations in these areas would be lethal. In fact, large segments of *Drosophila* chromosomes appear to be devoid of mutations (Muller and Painter, 1932) and have been found to contain the highly redundant DNA sequences (Botchan et al., 1971). A possible method permitting recovery of dominant mutations in such regions would be the detection of conditionally lethal dominants. While the frequency of induced dominant ts lethals (*DTS-L*) is low, they have been detected on the second and third chromosomes (Suzuki and Procunier, 1969; Suzuki et al., 1968). Although it remains to be shown whether any of the *DTS-L* are, in fact, located in the redundant areas of the chromosome, *DTS* lethality may be the only method to analyze such regions genetically.

An incidental use of *DTS-L* of considerable value to the geneticist is the selective elimination of flies of certain genotypes (Wright, 1970). For example, introduction by crossing-over of *DTS-L* into a second chromosome bearing the crossover suppressor, *Cy*, permits selective elimination of all *Cy*-bearing flies, so that only the required homozygotes survive. Thus, at restrictive temperatures, the cross,

$$\frac{Cy + DTS\text{-}L}{+ \quad a \quad +} \delta \times \frac{Cy + DTS\text{-}L}{+ \quad a \quad +} \female$$

yields only + *a* +/+ *a* + flies. Similarly, introduction of a DTS-L into X or Y translocations with autosomes can allow the production of unisexual offspring. An X-autosome translocation carrying a *DTS-L* can be maintained at permissive temperatures. When T(X;A) *DTS-L/Y;A* males (where T refers to the translocation, A to the autosomes) are crossed to normal X/X;A/A females at restrictive temperatures, all female zygotes are $\dfrac{T(X;A)DTS\text{-}L}{X;A}$ and die, whereas all male progeny are $\dfrac{X}{Y};\dfrac{A}{A}$ survivors. On the other hand, a Y-autosome translocation bearing a DTS-L yields only virgin females at restrictive temperatures. Thus a cross of

$$\frac{X;A}{T(Y;A)DTS\text{-}L} \, \male \times \frac{X}{X};\frac{A}{A} \, \female$$

results in death of the $\dfrac{X;A}{T(Y;A)DTS\text{-}L}$ males and survival of $\dfrac{X}{X};\dfrac{A}{A}$ females. In large experiments where mechanical separation of males and females and the collection of virgin females is time-consuming and tedious, such genetic tools are a significant aid.

Most ts lethal mutations recovered in *Drosophila* can be made homozygous at the permissive temperature and therefore maintained as a true breeding stock, thereby eliminating the complication of heterozygosity of parental organisms. When adult flies homozygous for ts lethals are transferred to restrictive temperatures, in most cases, they remain viable and fertile in spite of the fact that they are homozygous for a mutation that causes death at that temperature. Consequently, we assume that such ts mutations induce death at restrictive temperatures during development prior to eclosion of adults. After the adults hatch, the mutation is no longer operative. When eggs are collected and maintained at restrictive temperatures, the time of death—that is, the effective lethal phase (LP) (Hadorn, 1955) can be established. The LPs of different stocks are found to occur at many stages of development, although the majority die during embryogenesis in the egg or as pupae when metamorphosis into adults takes place (Suzuki, 1970). However, the LP indicates only the time of

expression of the lethal phenotype but does not delineate the actual developmental interval during which the restrictive temperature irreversibly commits the organism to death. That period is defined as the temperature-sensitive period (TSP) and can be established by "shift" experiments. Different cultures initiated at restrictive or permissive temperatures can be shifted to permissive or restrictive temperatures, respectively, at different successive times (Figs. 22a and b). The first shift from restrictive to permissive temperature that results in significant lethality defines the first part of the TSP (Fig. 21a). Similarly, the first shift from permissive to restrictive temperature that yields survivors defines the end of the TSP (Fig. 21b). The TSP so determined can be verified as a biologically real interval by "pulse" shifts (Figs. 21c and d). Thus, a culture maintained at permissive temperatures but pulsed with an exposure to restrictive conditions during the TSP results in death (Fig. 21c). The reciprocal pulse shift to permissive temperature during the TSP (Fig. 21d) allows survival. These experiments do not suggest the biological significance of a TSP, but the latter is assumed to be the time of thermal inactivation of the protein or tRNA gene products. Shift studies reveal that the TSP may coincide with the LP or precede the LP by hours to several days. Moreover, the TSP often precedes any cytologically detectable defects in development, thus coming closer to defining the time of gene activity.

In determining TSPs of various mutations, a variety of different patterns (such as single or multiple TSPs for individual mutations) has been detected (Tarasoff and Suzuki, 1970). However, without relating the TSP to a developmental phenotype other than death, little insight into the biological significance can be obtained. Therefore, mutations with effects on specific developmental systems permit more significant questions to be asked, and these will now be considered.

Numerous genes exist that are active in several different tissues. We can ask whether such a gene is activated in all tissues at one stage in development in response to a freely circulating signal or whether, in fact, tissue-specific stimuli control the locus at different stages. Chromatographically separable fluorescent pteridines and

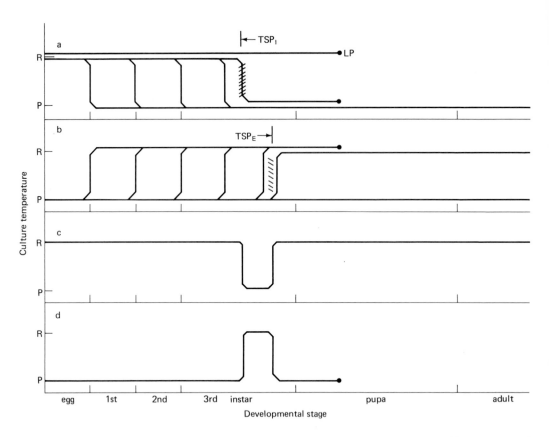

FIGURE 22. *Shift protocol for delineating the temperature-sensitive period (TSP) of a pupal lethal. (a) Restrictive to permissive; (b) permissive to restrictive; (c) pulse to permissive; (d) pulse to restrictive. LP, effective lethal phase; R, restrictive temperature; P, permissive temperature; TSP_I, initiation of the TSP; TSP_E, end of TSP.*

ommochromes are produced in the eyes, testes, and Malpighian tubules of *Drosophila* under the control of several genes (Hadorn and Mitchell, 1951). A ts lethal $(1(1)E6^{ts})$ was found to cause a ts loss of pigments in the three tissues as well as lethality. Shift studies and analysis of lethality and pigment production in each tissue revealed at least three temporally distinct TSPs for the mutation, thereby suggesting tissue-specific activation of the gene (Grigliatti and Suzuki, 1970).

Many organisms proceed through developmental stages as completely different animals. The genetic regulation of cell fates within such stages must require an intricate control system. Embryogenesis in *Drosophila* eggs, for example, produces a larval organism capable of sensing its environment, crawling, and eating.

However, the larva is predetermined to form a totally different adult organism equipped with eyes, legs, and wings. In anticipation of the adult, cells already programmed to differentiate into adult structures upon metamorphosis are packaged within the larvae as imaginal discs. Different discs can be identified on the basis of their size, shape, and anatomical location within the larva (Fig. 23). That disk cells are already determined in the larvae to differentiate into adult structures can be shown by transplantation of a specific disc into the abdomen of another larval host. Thus, for example, an eye disc transplant will differentiate into an eye within the abdomen of its host (see Hadorn, 1965).

The nature of the genetic programming that determines the fate of disc cells has been investigated with sophisticated transplantation experiments by Hadorn and his coworkers. Hadorn (1965) showed that fragments of discs injected into adult abdomens continue to proliferate but cannot differentiate into adult structures in the absence of the hormonal stimuli triggered during metamorphosis. Hence, a disc can be cultured within an adult, then removed and cut into pieces, and can again be reimplanted into adult hosts. Thus, the disc can be cloned extensively by serial transplantation. Fragments can be periodically implanted into larvae so that the cells differentiate and their programmed adult fate is determined. Surprisingly, after several serial transplants, some of the cloned material changes from one determined fate to produce a different adult structure. Hadorn (1965) called this phenomenon *transdetermination*. Interestingly, transdetermination of disc cells seems to be related to the number of serial transfers, and the changes in determination proceed in a defined direction (Fig. 24). This experimental observation promises a profound insight into the control of disc determination and the relationships between different types of discs.

That transdetermination represents an alteration in genetic regulation is shown by the existence of mutations (called *homeotic*) that change a disc from one fate to another. The kinds of disc changes in determination produced by homeotic mutations are similar to those brought about by transdetermination. Thus, for example, the mutation ey^{opt} changes some cells in the eye disc to form wing-like structures in the adult eye (Goldschmidt and Lederman-Klein, 1958). Homeotic mutations have been extensively studied by Goldschmidt et al. (1951). Interestingly, most of the homeotic loci known are genetically clustered on one chromosome within a region that is less than 2 percent of the genetic length of the genome, thus suggesting a chromosome segment concerned primarily with disc determination (Goldschmidt, 1955).

Most homeotic mutations have very low penetrance and expressivity, thereby rendering them difficult to work with. The homeotic mutation ss^{a40a}, however, is highly penetrant and expressive as well as ts (Grigliatti and Suzuki, 1971). At 29°C, the aristae of ss^{a40a} flies are wild type, whereas at 17°C, the aristae are

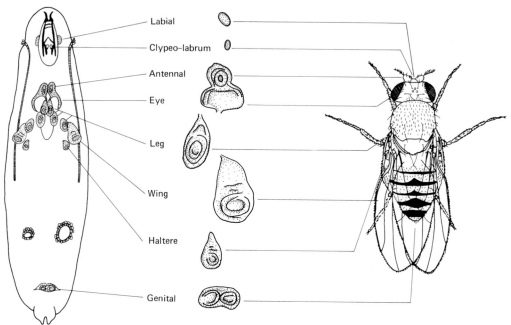

FIGURE 23. *Imaginal disks and their respective adult structures.* [*From J. W. Fristrom et al.* In *"Problems in Biology: RNA in Development,"* E. W. Hanly, ed., p. 382, Utah University Press, Salt Lake City, 1969.]

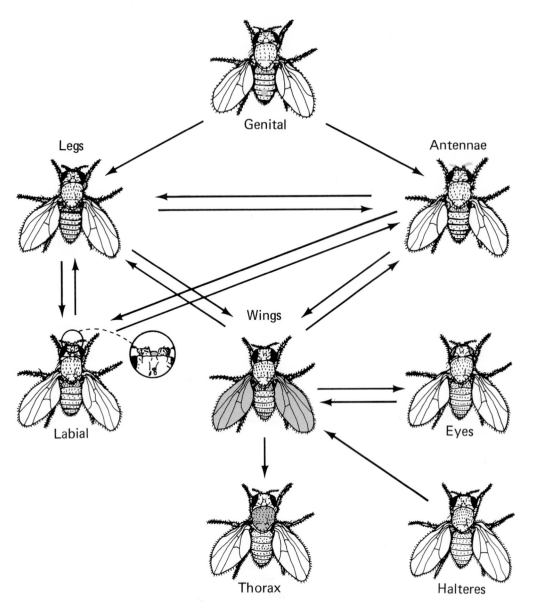

FIGURE 24. *Types of transdeterminations observed. The arrows indicate the direction of the disk changes, shading indicates organ changed. (From E. Hadorn, Sci. Amer. 219, 116, 1968.)*

transformed into distal segments of legs (Fig. 25). Shift studies of ss^{a40a} delineate a short TSP during the third larval instar and suggest that molecular studies of discs carried out at different temperatures during that interval may provide information on the nature of determination in imaginal discs.

Another useful property of mutations is a restriction of the activity in a specific group of cells at a specific time in development. In *Drosophila hydei*, the Y chromosome has been shown by nucleic acid hybridization, to produce RNA only in the testes of males and mainly in primary spermatocytes (Hennig, 1968). Thus an entire chromosome is restricted in its site and time of action. Temperature-

a

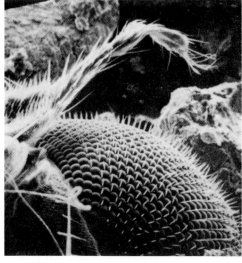

b

FIGURE 25. *Scanning electron micrographs of aristae of* ss[a40a] *at (a) 29°C and (b) 17°C.*

sensitive Y chromosome mutations that cause male sterility at restrictive temperatures have been detected in *D. melanogaster* (Ayles et al., 1973). Pulse shifts on such males at different stages of gonadal development show that the TSPs for sterility occur only in primary spermatocytes of meiotic cells of the testes. If techniques for separation of different cell types from testes (Lam et al., 1970) can be adapted for *Drosophila* testes, molecular studies at different temperatures of primary spermatocytes carrying ts Y chromosome steriles could be a model system to relate genetic activity with morphogenesis.

Occasionally, studies of developmental muta-

tions may reveal phenomena previously unknown. For example, in shift experiments on two different ts mutations, N^{60g11} (Foster and Suzuki, 1970) and shi^{ts1} (Poodry et al., 1973), a polarized pattern of eye facet organization in the compound eye was found. N^{60g11} causes disarray in the orderly arrangement of facets and shi^{ts1} creates a "scar" resulting from death of facet cells. The TSP for facet disruption is during the third larval instar for both mutations. Shifts of cultures to restrictive temperatures early in the TSP disrupt the facets along a vertical strip at the posterior rim of the eye. Later shifts reveal that the vertical wave of disruption proceeds across the eye toward the anterior rim. This pattern is completely unrelated to the cell lineage of the eye (Fig. 12) and is induced in larvae long before actual differentiation of the eye disc. The known effects of N alleles (Poulson, 1940) and the suggested role of shi^{ts} alleles (Poodry et al., 1973) on nervous tissue points to a possible polarized gradient of ennervation of the disc occurring during larval development that had not been anticipated at the start of the experiments.

Embryologists have long known that nervous tissue has inductive effects on other tissues during development. Mutations affecting both nervous as well as muscular tissues could provide insights for the developmental biologist. Wright (1968, 1970b) has recovered ts lethal mutations whose TSP occurs in a 4-hour interval during embryogenesis when myogenesis is known to occur. He has found several that exhibit defective muscle morphology at restrictive temperatures and promise to reveal new information about normal myogenesis. A different approach is to screen for flies that exhibit a ts reversible paralysis, a phenotype predicted for defects that are localized to muscles or nerves. This selected phenotype has yielded several new mutations including $para^{ts}$, a presumed central nervous system defect (Suzuki et al., 1971; Grigliatti et al., 1972), stn^{ts} (Grigliatti et al., 1973), and shi^{ts} (Poodry et al., 1973). Mutations in the *shi* locus appear to be especially interesting. The allele shi^{ts1} results in immediate paralysis of adults and larvae upon shifting to the restrictive temperature. Moreover, death occurs at several developmental stages after 3-hour pulse shifts to the

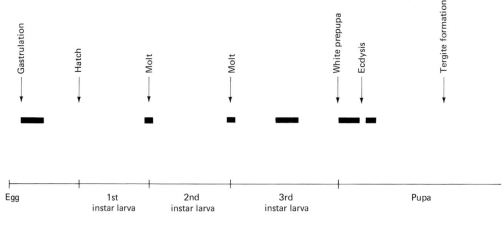

Developmental stage of *Drosophila*

FIGURE 26. *Temperature-sensitive periods for lethality of* shi[ts1].

restrictive temperature. Several TSPs for lethality have been delineated for *shi*[ts1] by pulse shifts (Fig. 26). In addition to lethality, *shi*[ts1] flies have several TSPs for morphologically visible defects in wings, legs, eyes, abdominal tergites, bristles, and hairs. The rapidity of the temperature effects on movement (less than 1 minute for paralysis or recovery) supports the supposition that preformed gene products are reversibly affected. Moreover, the rapidity with which an irreversible phenotype is induced by a pulse shift (3 hours for lethality, 4 to 6 hours for an anatomical defect) suggests that the defect noted is a consequence of the primary genetic lesion. The multiple phenotypic effects and the sensitivity to temperature at stages through development and in adults strongly indicate that the *shi* locus regulates a fundamental cellular process. Selection of mutations on the basis of paralysis clearly provides new defects that will greatly expand developmental studies.

In summary, it can be seen that mutations provide a refined method for altering normal development just as embryological surgery was a powerful tool in classical embryology. The added property of conditional expression of a mutant phenotype permits manipulation of expression of the mutation during development. Just as phage geneticists defined the temporal activity of ts lethal mutations, the developmental stage affected by ts defects in multicellular organisms can be determined. In com-

bination with cytological and biochemical methods of analysis, genetic studies of conditionally expressed mutations promise to provide profound insights into regulatory mechanisms of multicellular eukaryotic organisms.

References

AMES, B. N., AND HARTMAN, P. (1963). The histidine operon. *Cold Spring Harbor Symp. Quant. Biol. 28*, 344–356.

AYLES, G. B., SANDERS, T. G., KIEFER, B. I., AND SUZUKI, D. T. (1973). Temperature-sensitive mutations in *Drosophila melanogaster*. XI. Male sterile mutants of the Y chromosome. *Develop. Biol. 32*, 239–257.

BAGLIONI, C. (1960). The genetic control of tryptophan pyrrolase in *Drosophila melanogaster* and *D. virilis. Heredity 15*, 87–96.

BAKER, W. K. (1963). Genetic control of pigment differentiation in somatic cells. *Amer. Zool. 3*, 57–69.

BECKER, H. J. (1957). Über Rötgenmosaikflecken und Defektmutationen am Auge von *Drosophila* und die Entwicklungsphysiologie des Auges. *Z. Vererbungslehre, 88*, 333–373.

BECKWITH, J. R., AND ZIPSER, D. (1970). "The Lactose Operon." Cold Spring Harbor Laboratory, Cold Spring Harbor, New York.

BEERMANN, W. (1956). Nuclear differentiation and functional morphology of chromosomes. *Cold Spring Harbor Symp. Quant. Biol. 21*, 217–232.

BEERMANN, W. (1961). Ein Balbiani-Ring als Locus einer Speicheldrüsenmutation. *Chromosoma 12*, 1–25.

BOTCHAN, M., KRAM, R., SCHMID, C. W., AND HEARST, J. E. (1971). Isolation and chromosomal localization of highly repeated DNA sequences in *Drosophila melanogaster*. *Proc. Nat. Acad. Sci. U.S.A. 68*, 1125–1129.

BROSSEAU, G. (1960). Genetic analysis of the male fertility factors on the Y chromosome of *Drosophila melanogaster*. *Genetics 45*, 257–274.

BROWN, S. W. (1966). Heterochromatin. *Science 151*, 417–425.

BROWN, S. W., AND NELSON-REES, W. A. (1961). Radiation analysis of a lecanoid genetic system. *Genetics 46*, 983–1007.

BROWN, S. W., AND NUR, U. (1964). Heterochromatic chromosomes in the coccids. *Science, 145*, 130–136.

CROUSE, H. V. (1960). The controlling element in sex chromosome behavior in Sciara. *Genetics 45*, 1429–1443.

DAVIDSON, R. G., NITOWSKY, H. M., AND CHILDS, B. (1963). Demonstration of two populations of cells in the human female heterozygous for glucose-6-phosphate dehydrogenase variants. *Proc. Nat. Acad. Sci. U.S.A. 50*, 481–485.

DEMEREC, M. (1964). Clustering of functionally related genes in *Salmonella typhimurium*. *Proc. Nat. Acad. Sci. U.S.A. 51*, 1057–1060.

DEMEREC, M., AND DEMEREC, Z. E. (1956). Analysis of linkage relationships in Salmonella by transaction techniques. *Brookhaven Symp. Biol. 8*, 75–87.

DEMEREC, M., AND SLIZYNSKA, H. (1937). Mottled white 258–18 of *Drosophila melanogaster*. *Genetics 22*, 641–649.

EDGAR, R. S., DENHARDT, G. H., AND EPSTEIN, R. H. (1964). A comparative genetic study of conditional lethal mutations of bacteriophage T4D. *Genetics, 49*, 635–648.

EDGAR, R. S., AND EPSTEIN, R. H. (1965). Conditional lethal mutations in bacteriophage T4. *Proc. 11th Int. Congr. Genet. 2*, 1–16.

EPSTEIN, R. H., BOLLE, A., STEINBERG, C. M., KELLENBERGER, E., BOY DE LA TOUR, E., CHEVALLEY, R., EDGAR, R. S., SUSMAN, M., DENHARDT, G. H., AND LIELAUSIS, A. (1963). Physiological studies of conditional lethal mutants of bacteriophage T4. *Cold Spring Harbor Symp. Quant. Biol. 28*, 375–394.

FOSTER, G. G., AND SUZUKI, D. T. (1970). Temperature-sensitive mutations in *Drosphila melanogaster. IV*. A mutation affecting eye facet arrangement in a polarized manner. *Proc. Nat. Acad. Sci. U.S.A. 67*, 738–745.

GARCIA-BELLIDO, A., AND MERRIAM, J. R. (1971). Genetic analysis of cell heredity in imaginal discs of *Drosophila melanogaster. Proc. Nat. Acad. Sci. U.S.A. 68*, 2222–2226.

GOLDSCHMIDT, E., AND LEDERMAN-KLEIN, A. (1958). Reoccurrence of a forgotten homeotic mutant in *Drosophila. J. Hered. 49*, 262–266.

GOLDSCHMIDT, R. B. (1955). "Theoretical Genetics." University of California Press, Berkeley, California.

GOLDSCHMIDT, R. B., HANNAH, A., AND PITERNICK, L. (1951). The podoptera effect in *Drosophila melanogaster. Univ. Calif. Publ. Zool. 55*, 67–294.

GORINI, L., AND BECKWITH, J. R. (1966). Suppression. *Ann. Rev. Microbiol. 20*, 401–422.

GRELL, E. H. (1962). The dose effect of $ma\text{-}1^+$ and ry^+ on xanthine dehydrogenase activity in *Drosphila melanogaster. Z. Vererbungslehre, 93*, 371–377.

GRIGLIATTI, T., HALL, L., ROSENBLUTH, R., AND SUZUKI, D. T. (1973). Temperature-sensitive mutations in *Drosophila melanogaster. XIV*. Selection of immobile adults. *Mol. Gen. Genet., 120*, 107–114.

GRIGLIATTI, T., AND SUZUKI, D. T. (1970). Temperature-sensitive mutations in *Drosophila melanogaster. V*. A mutation affecting concentrations of pteridines. *Proc. Nat. Acad. Sci. U.S.A. 67*, 1101–1108.

GRIGLIATTI, T., AND SUZUKI, D. T. (1971). Temperature-sensitive mutations in *Drosophila melanogaster. VIII*. The homeotic mutant, ss^{a40a}. *Proc. Nat. Acad. Sci. U.S.A. 68*, 1307–1311.

GRIGLIATTI, T., SUZUKI, D. T., AND WILLIAMSON, R. (1972). Temperature-sensitive mutations in *Drosophila melanogaster. X*. Developmental analysis of the paralytic mutation, $para^{ts}$. *Develop. Biol. 28*, 352–371.

GURDON, J. B. (1963). Nuclear transplantation in amphibia and the importance of stable nuclear changes in promoting cellular differentiation. *Quart. Rev. Biol. 38*, 54–78.

GURDON, J. B., AND WOODLAND, H. R. (1968). The cytoplasmic control of nuclear activity in animal development. *Biol. Rev. 43*, 233–267.

HADORN, E. (1955). "Developmental Genetics and Lethal Factors." Wiley, New York.

HADORN, E. (1965). Problems of determination and transdetermination in genetic control of differentiation. *Brookhaven Symp. Biol. 18*, 148–161.

HADORN, E., AND MITCHELL, H. K. (1951). Properties of mutants of *Drosophila melanogaster* and changes during development as revealed by paper chromatography. *Proc. Nat. Acad. Sci. U.S.A. 37*, 650–665.

HARTWELL, L. H. (1967). Macromolecular synthesis in temperature-sensitive mutants of yeast. *J. Bacteriol. 93*, 1662–1670.

HEITZ, E. (1933). Cytologische Untersuchungen an Dipteren. III. Die somatische Heteropyknose bei *Drosophila melanogaster* und ihre genetische Bedeutung. *Z. Zellforsch. 20*, 237–287.

HENNIG, W. (1968). Ribonucleic acid synthesis of the Y chromosome of *Drosophila hydei. J. Mol. Biol. 38*, 227–239.

HINTON, C. W. (1955). The behavior of an unstable ring chromosome of *Drosophila melanogaster. Genetics 40*, 951–961.

HOROWITZ, N. H. (1950). Biochemical genetics of Neurospora. *Advan. Genet. 3*, 33–71.

HOTTA, Y., AND BENZER, S. (1970). Genetic dissection of the *Drosophila* nervous system by means of mosaics. *Proc. Nat. Acad. Sci. U.S.A. 67*, 1156–1163.

IKEDA, K., AND KAPLAN, W. D. (1970). Unilaterally patterned neural activity of gynandromorphs, mosaic for a neurological mutant of *Drosophila melanogaster. Proc. Nat. Acad. Sci. U.S.A. 67*, 1480–1487.

JACOB, F., AND MONOD, J. (1961a). Genetic regulatory mechanisms in the synthesis of proteins. *J. Mol. Biol. 3*, 318–356.

JACOB, F., AND MONOD, J. (1961b). On the regulation of gene activity. *Cold Spring Harbor Symp. Quant. Biol. 26*, 193–211.

JOCKUSCH, H. (1966a). Temperatursensitive Mutanten des Tabakmosaikvirus. I. *In vivo*-verhalten. *Z. Vererbungslehre 98*, 320–343.

JOCKUSCH, H. (1966b). Temperatursensitive Mutanten des Tabakmosaikvirus. II. *In vitro*-verhalten. *Z. Vererbungslehre 98*, 344–362.

JOCKUSCH, H. (1966c). Relations between temperature-sensitivity, amino acid replacements and quaternary structure of mutant proteins. *Biochem. Biophys. Res. Comm. 24*, 577–583.

JUDD, B. H. (1955). Direct proof of a variegated type position effect at the white locus in *Drosophila melanogaster. Genetics 40*, 739–744.

KING, T. J., AND BRIGGS, R. (1956). Serial transplantation of embryonic nuclei. *Cold Spring Harbor Symp. Quant. Biol. 21*, 271–290.

LAIRD, C. D., AND MCCARTHY, B. J. (1968). Nucleotide sequence homology within the genome of *Drosophila melanogaster. Genetics 60*, 323–334.

LAM, D. M. K., FURRER, R., AND BRUCE, W. R. (1970). The separation, physical characterization, and differentiation kinetics of spermatogonial cells of the mouse. *Proc. Nat. Acad. Sci. U.S.A. 65*, 192–199.

LANGRIDGE, J. (1965). Temperature-sensitive, vitamin-requiring mutants of *Arabidopsis thaliana. Australian J. Biol. Sci. 18*, 311–321.

LEDERBERG, J. (1950). Isolation and characterization of biochemical mutants of bacteria. *Methods Med. Res. 3*, 5–22.

LEWIS, E. B. (1950). The phenomenon of position effect. *Advan. Genet. 3*, 73–115.

LYON, M. J. (1961). Gene action in the X chromosome of the mouse *(Mus musculus L.) Nature (London) 190*, 372–373.

LYON, M. J. (1968). Chromosomal and subchromosomal inactivation. *Ann. Rev. Genet. 2*, 31–48.

MINTZ, B. (1962). Formation of genotypically mosaic mouse embryos. *Amer. Zool. 2*, 432, abstract.

MINTZ, B. (1967). Gene control of mammalian pigmentary differentiation. I. Clonal origin of melanocytes. *Proc. Nat. Acad. Sci. U.S.A. 58*, 344–351.

MINTZ, B. (1968). Hermaphroditism, sex chromosomal mosaicism and germ cell selection in allophenic mice. *J. Animal Sci. 27*, Suppl. 1, 51–60.

MINTZ, B., AND BAKER, W. W. (1967). Normal mammalian muscle differentiation and gene control of isocitrate dehydrogenase synthesis. *Proc. Nat. Acad. Sci. U.S.A. 58*, 592–598.

MINTZ, B., AND PALM, J. (1969). Gene control of hematopoiesis. I. Erythrocyte mosaicism and permanent immunological tolerance in allophenic mice. *J. Exp. Med. 129*, 1013–1027.

MINTZ, B., AND SLEMMER, G. (1969). Gene control of neoplasia. I. Genotypic mosaicism in normal and preneoplastic mammary glands of allophenic mice. *J. Nat. Cancer Inst. 43*, 87–95.

MONOD, J., CHANGEUX, J. P., AND JACOB, F. (1965). Allosteric proteins and cellular control systems. *J. Mol. Biol. 6*, 306–329.

MULLER, H. J. (1932). Further studies on the nature and causes of gene mutations. *Proc. 6th Int. Congr. Genet. 1*, 213–255.

MULLER, H. J., AND PAINTER, T. S. (1932). The differentiation of the sex chromosomes of *Drosophila* into genetically active and inert regions. *Z. Vererbungslehre 62*, 316–365.

NESBITT, M. N., AND GARTLER, S. M. (1971). The applications of genetic mosaicism to developmental problems. *Ann. Rev. Genet. 5*, 143–162.

NOWELL, P. C., AND COLE, L. V. (1967). Clonal repopulation in reticular tissues of X-irradiated mice: effect of dose and of limb-shielding. *J. Cell Physiol. 70*, 37–44.

OWEN, R. D. (1945). Immunogenetic consequences

of vascular anastomoses between bovine twins. *Science 102*, 400-401.

PAVAN, C., AND BREUER, M. E. (1955). Behavior of polytene chromosomes of *Rhynchosciara angelae* at different stages of larval development. *Chromosoma 7*, 371-386.

POODRY, C., HALL, L., AND SUZUKI, D. T. (1973). Developmental properties of *shibire*[ts1], a pleiotropic mutation producing larval and adult paralysis. *Develop. Biol. 32*, 373-386.

POULSON, D. F. (1940). The effects of certain X chromosome deficiencies on the embryonic development of *Drosophila melanogaster*. *J. Exp. Zool. 83*, 271-325.

RITOSSA, F. M., AND SPIEGELMAN, S. (1965). Localization of DNA complementary to ribosomal RNA in the nucleolus organizer region of *Drosophila melanogaster*. *Proc. Nat. Acad. Sci. U.S.A. 53*, 737-745.

RUSSELL, L. B. (1961). Genetics of mammalian sex chromosomes. *Science 133*, 1795-1805.

SMITH, J. D., BARNETT, L., BRENNER, S., AND RUSSEL, R. L. (1970). More mutant tyrosine transfer ribonucleic acids. *J. Mol. Biol. 54*, 1-14.

SMITH, R. H. (1968). Unstable temperature-sensitive mutations in Habobracon. *Genetics 60*, 227, abstract.

STERN, C. (1936). Somatic crossing over and segregation in *Drosophila melanogaster*. *Genetics 21*, 625-730.

STERN, C. (1954). Two or three bristles. *Amer. Sci. 42*, 213-247.

STERN, C. (1957). The scope of genetics. *Proc. Nat. Acad. Sci. U.S.A. 43*, 744-749.

STERN, C. (1960). Dosage compensation—development of a concept and new facts. *Canad. J. Genet. Cytol. 2*, 105-118.

STERN, C., AND TOKUNAGA, C. (1967). Nonautonomy in differentiation of pattern-determining genes in *Drosophila*. I. The sex-comb of eyeless-dominant. *Proc. Nat. Acad. Sci. 57*, 658-664

STURTEVANT, A. H. (1929). The claret mutant type of *Drosophila simulans*: a study of chromosome elimination and cell-lineage. *Z. Wiss. Zool. 135*, 323-356.

STURTEVANT, A. H. (1932). The use of mosaics in the study of the developmental effect of genes. *Proc. 6th Int. Congr. Genet. 1*, 304-307.

SUZUKI, D. T. (1970). Temperature-sensitive mutations in *Drosophila melanogaster*. *Science 170*, 695-706.

SUZUKI, D. T., GRIGLIATTI, T., AND WILLIAMSON, R. (1971). Temperature-sensitive mutations in *Drosophila melanogaster*. VII. A mutation *(para*[ts]*)*

causing reversible adult paralysis. *Proc. Nat. Acad. Sci. U.S.A. 68*, 890-893.

SUZUKI, D. T., PITERNICK, L. K., HAYASHI, S., TARASOFF, M., BAILLIE, D., AND ERASMUS, U. (1967). Temperature-sensitive mutations in *Drosophila melanogaster*. I. Relative frequencies among γ-ray and chemically induced sex-linked lethals and semilethals. *Proc. Nat. Acad. Sci. U.S.A. 58*, 907-912.

SUZUKI, D. T., AND PROCUNIER, D. (1969). Temperature-sensitive mutations in *Drosophila melanogaster*. III. Dominant lethals and semilethals on chromosome 2. *Proc. Nat. Acad. Sci. U.S.A. 62*, 369-376.

SUZUKI, D. T., PROCUNIER, D., CAMERON, J., AND HOLDEN, J. (1968). Dominant temperature-sensitive (DTS) lethal mutations in *Drosophila melanogaster*. *Proc. 12th Int. Congr. Genet. 1*, 144, abstract.

TARASOFF, M., AND SUZUKI, D. T. (1970). Temperature-sensitive mutations in *Drosophila melanogaster*. VI. Temperature effects on development of sex-linked recessive lethals. *Develop. Biol. 23*, 492-509.

TARKOWSKI, A. K. (1964). True hermaphroditism in chimaeric mice. *J. Embryol. Exp. Morphol. 12*, 735-757.

TATUM, E. L., AND HAAGEN-SMIT, A. J. (1941). Identification of *Drosophila v*[+] hormone of bacterial origin. *J. Biol. Chem. 140*, 575-580.

THOMPSON, L. H., MANKOVITZ, R., BAKER, R. M., TILL, J. E., SIMINOVITCH, L., AND WHITMORE, G. F. (1970). Isolation of temperature-sensitive mutants of L-cells. *Proc. Nat. Acad. Sci. U.S.A. 66*, 377-384.

WATSON, J. D. (1970). "Molecular Biology of the Gene," 2nd ed. W. A. Benjamin, Palo Alto.

WHITE, M. J. D. (1954). "Animal Cytology and Evolution." Cambridge University Press, New York.

WILSON, E. (1925). "The Cell in Development and Heredity." Macmillan, New York.

WITTMANN, H. J., AND WITTMANN-LIEBOLD, B. (1966). Protein chemical studies of two RNA viruses and their mutants. *Cold Spring Harbor Symp. Quant. Biol. 31*, 163-172.

WRIGHT, T. R. F. (1968). The phenogenetics of temperature-sensitive alleles of *lethal myospheroid* in *Drosophila*. *Proc. 12th Int. Congr. Genet. 1*, 141, abstract.

WRIGHT, T. R. F. (1970a). A short cut in making autosomes homozygous. *Drosophila Inform. Serv. 45*, 140.

WRIGHT, T. R. F. (1970b). The genetics of embryogenesis in *Drosophila*. *Advan. Genet. 15*, 261-395.

CHAPTER EIGHTEEN

Congenital Malformations

LAIRD JACKSON

Human development is a complex process requiring the orderly piecing together of a multitude of sequential biological steps that lead to the production of a mature human organism. This process has been a source of fascination and an object of study for man since early in scientific history. The first scientists and physicians were especially curious when there was a faulty organism produced—a congenital malformation. At first only objects of curiosity, these deviations were soon recognized to be "experiments of nature" that, through the study of their special abnormalities, might yield clues to normal pathways of development. At no time has this been more true than in the decade just past. Newer techniques and advances have made it possible to take a fresh look at the mechanism of production of many human congenital malformations. Although methods are now coming into use for powerful biochemical studies that may be performed on tissue-cultured representatives of the whole human, much of our knowledge has come largely from the study of abnormal people though the use of increasingly sophisticated clinical techniques.

The classic pattern in the development of this knowledge usually involves (1) the discovery of a new investigational technique, (2) the application of this technique to human malformations (3) the elucidation of a cause for one or more specific malformations through use of the technique, (4) more accurate diagnosis and prognosis of the malformation by clinical application of the new understanding, and (5) alteration of the usual clinical course or frequency of the malformation. Down's syndrome, or mongolism, is a relatively recent example of this sequence of events. First, J. H. Tjio and A. Levan (1956) demonstrated a more accurate technique for the study of human chromosomes. Then Lejeune et al. (1959) applied this to a study of mongoloid children and demonstrated the presence of an extra chromosome in these children. This made possible more accurate diagnosis and prognosis and soon led to the description of variant chromosomal mechanisms. The high-risk mother was defined, and further development of the chromosomal techniques led to prenatal diagnosis and prevention of Down's syndrome births (Valenti et al., 1968).

There are numerous other examples of such recent advances in the understanding of human congenital malformations on a developmental biological basis. These are important both for their correlative value with biological knowledge and for their clinical value in achieving better diagnosis and prognosis for the malformed child and the family. Some examples of human congenital malformations will be considered in the following pages in the light of these two areas of biological and clinical understanding. We have divided the discussion into four sections: (1) some general aspects of the phenotypic presentation of congenital malformations, (2) malformations caused by the influence of several genetic factors acting in concert, (3) malformations caused by a single genetic factor, and (4) malformations caused by an identifiable chromosome abnormality. A brief description of the clinical features of examples within each group is given, together with comments on the correlation of these features with the biological mechanism of their production and on the medical significance and approach to each of them.

Congenital Malformations: Clinical Presentation

Human malformation syndromes are very closely associated with the process of morphogenesis. In a series of excellent publications, D. W. Smith (1966, 1967, 1970) has developed the concept of two groups of malformation syndromes dependent on the alteration of morphogenesis involved. In the one group the malformation can usually be traced to a single major altered event in morphogenesis. This step can be fixed in time in the developmental sequence and may or may not be followed by the appearance of other secondary defects. Such defects are almost never traceable to a single genetic factor but are frequently the result of multiple abnormal factors. In contrast, the second group, composed of malformation syndromes in which no single developmental step appears to be primary, is frequently traceable to the influence of a single hereditary gene. In this group morphogenesis may or may not be involved at an observable level. Skeletal defects involving the gross absence of limbs, for example, may be determined by transmission of a single gene.

Children with neurological disorders, however, frequently begin extrauterine life as completely normal-appearing and functioning individuals only to manifest their defect later in life. Many of these have been traced to single genes, and on close investigation the effect of the gene may be seen even in the fetus, although this requires a sophisticated search. An example is Tay-Sachs disease, which is characterized by abnormal accumulation of lipid material in neuronal cells. This eventually destroys the neurons and produces a progressive syndrome of motor nervous system deterioration, mental retardation, seizures, and death (O'Brien et al., 1971). Although the child appears normal until 4 to 6 months of age, fetal studies have shown the genetically determined neuronal changes in early developmental life. Morphogenesis can be said to be altered on a microscopic and metabolic level.

The influence of altered morphogenesis in the production of congenital malformation syndromes is also important to the clinician. The morphologic characteristics of an individual are those immediately present to the clinicians examining eye; so his ability to recognize and diagnose a specific malformation syndrome depends on the fact that the same alteration in the developmental process usually produces the same recognizable end result. As these are malformation syndromes or collections of physical and other defects, diagnosis depends upon putting all the observable findings together. Reliance on a simple major recognizable problem can be confusing diagnostically. Just as a single altered step in morphogenesis may produce secondary effects and a distorted organism, so may an error in the primary step of diagnosis lead to the wrong prognosis for an individual or family and tragic mistakes in human lives.

To illustrate this point, let us consider a common defect in developmental retardation in

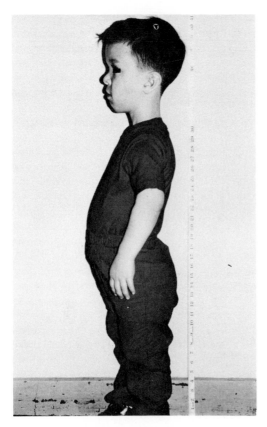

FIGURE 1. *Achondroplastic dwarfism. Note the relative shortening of trunk with exaggerated lumbar lordosis. Also, the head appears relatively large with prominent brow.*

growth. If growth retardation is moderately severe, it will produce individuals who are dwarfed or grossly below normal range for length or height. One of the most common types of dwarfism seen in humans is termed *rhizomelic* or *short-limbed dwarfism* (McKusick, 1966; Lamy, 1965). Achondroplasia is a specific and somewhat frequent dwarfing syndrome that produces this relatively greater shortening of the limbs as compared with the trunk. The syndrome is recognized by short limbs, by a tendency toward a sway-backed appearance and by a prominent head and brow (Fig. 1). X rays show that there are characteristic changes in multiple bones, especially in the hips, limbs, and spine. As this was one of the earliest recognized forms of dwarfism, many other short-limbed dwarfs have been errone-

ously diagnosed as achondroplastic dwarfs in the past (Lamy, 1965; Scott, 1972). Two such examples are illustrated in Figs. 2 and 3. The first (Fig. 2) is a child who is dwarfed and short-limbed but who has three other distinctive features. The feet are malformed or clubbed (and casted for correction), the hands have a malpositioned thumb, set low and abducted, and the external ears are cystic and dilated with hemorrhagic fluid. These are features of a dwarfing syndrome that progressively involves the spine in growth and results in a twisted trunk. The name *diastrophic* (bending) *dwarfing* has been given to this inherited malformation syndrome (Lamy and Maroteaux, 1970). The roentgenograms of a second dwarf resembling an achondroplast are pictured in Figure 3. Once again the child is dwarfed, the limbs are relatively more severely shortened than the trunk, and the head appears prominent. In addition, however, the chest is

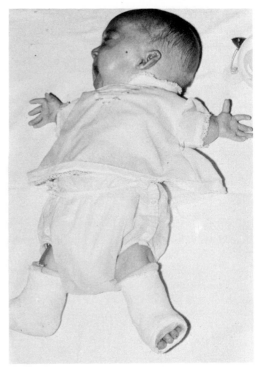

FIGURE 2. *Diastrophic dwarfism. Note malformed external ear with resolving cystic dilation, low-set and abducted or "hitch-hiker" thumb, and casted clubfeet.*

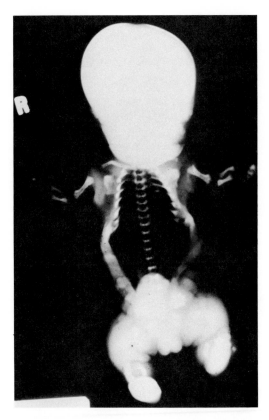

FIGURE 3. *Roentgenogram of thanatophoric dwarf. Note relatively prominent head, short and deformed limbs, and small narrow chest with thinned vertebral bodies.*

narrowed and the vertebral bodies are thinned. This child expired within 48 hours of birth and represents what is termed a *thanatophoric dwarf* (Maroteaux et al., 1967). In each of these cases the overall presenting feature of short-limbed dwarfism is the same; yet modifying features are different. Achondroplastic dwarfism is due to a single autosomal dominant gene, while diastrophic dwarfism is definitely an autosomal recessive disease, and thanatophoric dwarfism is probably also an autosomal recessive syndrome. Prognosis and genetic counseling are obviously markedly different for the three conditions.

Although the preceding examples show that reliance on one feature may confuse identification of the malformation syndrome, the analysis of the total presenting picture is usually quite characteristic of a distinct and identifi-able entity. Whether this is due to a single gene or a single primary event influenced by genes, the fact that the same abnormal developmental mechanism or pathway has been brought into play produces such a strong influence on the resulting individuals that they resemble each other strikingly. This has important diagnostic implications and can also lead to important investigational clues. An example of this in humans is the congenital malformation syndrome described initially by a Dutch physician, Cornelia de Lange (1933); the malformation syndrome bears her name. The fact that, in some families, siblings are affected suggests that the syndrome is caused by a single autosomal recessive gene (Opitz et al., 1965)—although the great majority of cases occur sporadically. In all instances one sees the presentation of the same striking facies, although the affected individuals may be genetically unrelated (Fig. 4). The syndrome is further characterized by marked retardation of growth and development, so that the child is small for his chronological age. Mental development is also markedly retarded. The feet and hands are small for the size of the body and frequently malformed. The eyes, eyebrows, and mouth have characteristic changes in which the brows grow across the forehead and bridge of the nose, the eyelids tend to look thin, the eyelashes are prominent and the mouth has a thin upper lip that is turned down at the corners. Comparison of the facies of three patients vividly demonstrates that similar changes in the developmental mechanism produce strikingly similar features of malformation.

Polygenetic Defects Resulting in Single Primary Defects

Many congenital malformation syndromes can be traced back in the embryological process and fixed in time with considerable accuracy (Smith, 1970). Although the malformation itself may be complex and involve several anatomical systems, a careful examination will show that one defect is primary and that the

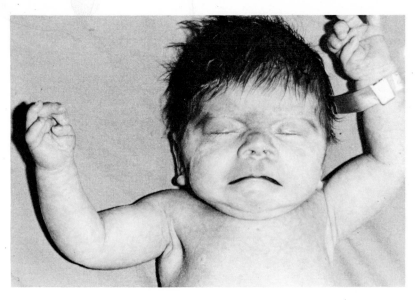

FIGURE 4. *Face of a child with the Cornelia-de-Lange syndrome. Note prominent features, such as confluent eyebrow and thin downturned upper lip.*

others are secondary to the primary defect. In this group of primary defects, relating the defect and the morphogenic step that has been disturbed should enable one to localize the time in development when the defect occurred. These single primary-defect syndromes may have environmental causes in some instances, but an important group are caused by genetic factors—usually multiple, as is shown by several lines of evidence (Carter 1969, Wynne-Davies, 1970). Let us examine some examples of such malformations.

DEFECTS IN FACIAL DEVELOPMENT

One relatively common malformation seen in several forms is the cleft defect of the upper jaw. This occurs as an isolated cleft lip, a cleft lip and palate, or a cleft defect in combination with other malformations. When the defect is isolated, it is found to be due to a relatively simple fusion failure in the upper-mouth-area tissues. As part of the morphogenesis of the facial region, the mouth forms from two lateral structures that then grow toward the midline and fuse at approximately the seventh to eighth week of fetal development (Smith, 1970). The

palate forms from the maxillary palatal shelves and the lip from the nasal processes. This fusion must take place while the skull and the remainder of the face are also growing; so timing is critical. Any factor that slows the growth and fusion process of the palate and lip will result in the production of a cleft defect. Despite continued subsequent growth of these processes toward the midline, the opportunity for fusion is missed if the latter does not occur by the eighth week, because thereafter growth only parallels the growth of the remainder of the face and head.

Although a cleft lip and palate may occur as isolated defects, they may also occur as secondary problems in a more severe malformation disorder. The development of the face and palate is dependent upon the prior normal development of the prechordal mesoderm, which should be completed at approximately $3\frac{1}{2}$ weeks of development. Gross failure of development of this portion of the body does occur under unknown influences and may cause secondary facial defects (De Myer, 1971). Two such clinical presentations are pictured in Figs. 5 and 6. Both involve severe maldevelopment of the forebrain. The clinical malformation presents with a midline cleft defect of the face, lip, and palate, a sloping fore-

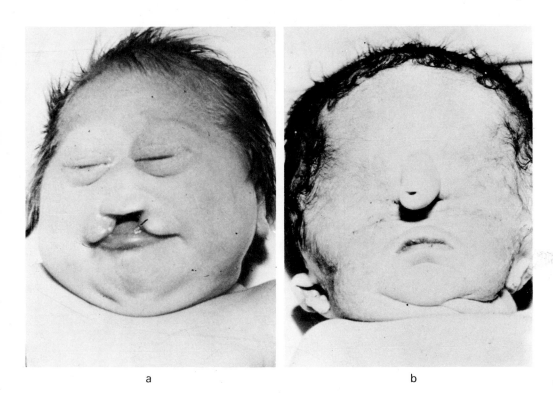

FIGURE 5. *Anterior facial defects associated with severe brain malformation, resulting from failure in development of prechordal mesoderm. (a) Holoproscencephaly with severe midline facial cleft and microphthalmia. (Courtesy of F. Moyer.) (b) Cyclopia with severe brain malformation.*

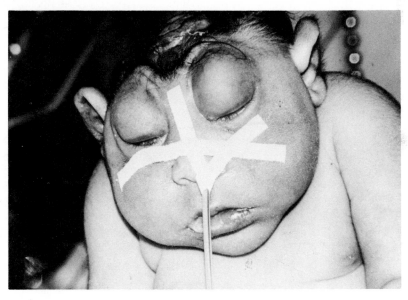

FIGURE 6. *Anencephaly resulting from fusion failure of cephalad neural tube. Note secondary eye and ear anomalies.*

head, and frequent maldevelopment of the eyes (microphthalmia). All this is secondary to abnormally poor growth and development of the forebrain and foremidline or prechordal mesoderm of the developing embryo. The resulting malformation varies widely in the presentation of facial defects from midline clefts to cyclopia (Fig. 5). There is a parallel variation in incomplete morphogenesis of the forebrain ranging up to complete alobar holoprosencephaly. The development of the anterior midline or prechordal mesoderm is a complex process apparently under multiple-gene control and subject to errors in developmental timing. As noted, the process should ordinarily be complete by $3\frac{1}{2}$ to 4 weeks of development and is critical to subsequent normal development of the face and forebrain. The variety of malformations resulting from the primary failure demonstrate the importance of this developmental step. They also emphasize the importance of the facial anomaly in predicting or suggesting severe cerebral involvement. Fortunately, these defects occur only uncommonly in newborn children.

NEURAL TUBE DEFECTS

A similar but separate problem occurs in the development of the posterior embryonic neural tube. This structure must develop from a plate of tissue that folds over and closes as a tube. Subsequently, this tissue develops into the central nervous system, the brain, and the spinal cord. The presentation of a congenital malformation syndrome that includes an open spine or "open" or undeveloped brain, therefore, points toward failure of this mechanism of morphogenesis. Anencephaly is the result of the failure of this mechanism in the cephalad portion of the neural tube. The face may again be affected secondarily with changes of the eyes, forehead, and ears (Fig. 6), but such manifestations are frequently a mechanical consequence of the position of the primary defect, which produces edema and distention, rather than true developmental anomalies. Similar problems in a caudad portion of the neural tube produce the child with meningomyelocele. As in the anencephaly defect, the overlying skeleton is involved, so that spina bifida is

present. The severity is variable—with the least affected child having only a closed skeletal defect. In more severe cases, there is an open lesion with involvement of the spinal cord function below the level of the defect (usually lumbosacral). Frequently, secondary anomalies at the level of the base of the skull produce obstructive hydrocephalus, complicating an already severe central nervous system problem. The development of the posterior embryonic neural tube should be complete by 28 days of development, at the same time that the closure of the anterior portion of the embryo takes place.

LIMB DEFECTS

Finally, abnormalities of limb development may also occur as a single primary developmental defect as well as in combination with other defects (Temtamy and McKusick, 1969). The limbs may be affected at the earliest stage of development through the failure of initiation or development of the limb bud and result in an amelia or phocomelia deformity (Fig. 7). The initiation of limb development begins at approximately 5 weeks, and critical growth continues through the sixth week. Insults in the later stages of development may be the cause of selective failure of development in the distal arm or hand. These malformations may be caused by environmental insult or by genetic abnormalities, or they may occur sporadically with no satisfactory explanation.

POLYGENIC RISK FACTORS

In dealing with the group of problems classified here as single major malformations, diagnosis and counseling are more difficult than with syndromal presentations of multiple abnormalities. As suggested earlier, any single entity may have several separate causes. Since a malformation represents the result of damage to the organism at a single point in time or morphogenesis, several types of insults could conceivably cause a similar if not identical end result. For example, many environmental factors are known to produce congenital malformations, and appropriate measures can and

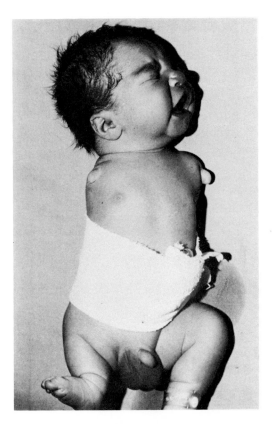

FIGURE 7. *Child with limb malformations, including amelia or absence of development of upper limbs combined with phocomelia of right lower limb. (Courtesy of F. Moyer.)*

in a range between 3 and 8 percent and is interpreted by most authorities as representing a polygenic effect. C. O. Carter (1965, 1969) has presented especially clear papers on this subject in recent years. The evidence against single-factor influence includes data on twins as well as on individual inheritance. An analysis of several common congenital malformations among identical twins fails to show the high degree of concordance expected in single-gene transmission. On the other hand, the concordance is much higher than for nonidentical twins, supporting the presence of genetic factors. Even subtle genetic differences, such as those between XX and XY genotypes, influence the risk for expression of common malformations. In at least two of these malformations there is a striking male–female difference in occurrence. Congenital hip dislocations occur approximately five times more frequently in females than in males, whereas the reverse is true for congenital pyloric stenosis.

In addition, there may be a variation between populations in the concentration of risk genes for one of these single malformations. An example of this is the unusual concentration of cases of meningomyelocele in South Western Wales. The incidence of this anomaly there, although still low, is twice that in the remainder of England and other areas of the world. Practically, of course, this can be used to alter the prognosis of recurrence risk in a couple originating from a population of unusual risk. The reasons for this type of gene concentration, although frequently understandable with single-gene diseases, are less clear in the polygenic problems. From a counseling standpoint, however, the only means of lowering one's risk is to find a mate from outside the population group and effectively dilute the pool of "risk" genes.

These data strongly support a polygenic determination of these malformations. There is probably a continuous distribution of some risk factor(s) for each malformation. Those individuals lying beyond a certain level (the threshold) on a normal distribution curve of this factor will exhibit the disorder. Accumulated experience shows that the distribution curve for relatives of persons with a given disorder will be shifted toward the threshold, indicating their increased incidence (and risk) of

should be taken by individuals and by society to reduce or eliminate exposure to such insults. The present discussion centers primarily on genetic causes, however, because of the implications for counseling and prevention of subsequent problems that hinge on proper recognition of the type and cause of the malformations.

What observations tell us that this group of single malformation defects has important genetic causes? The most obvious one is, of course, the analysis of cumulative data on their recurrence in families (Wynne-Davies, 1970). Observations of many family pedigrees for any of these malformations does not reveal a pattern of simple single-factor inheritance. Instead a low, but significant level of risk becomes apparent for those families in which such an anomaly has occurred. This risk falls generally the malformation. Falconer (1965) and Blyth and Carter (1969) have estimated the risks for many of these common malformations among relatives of affected persons. These data are extremely useful in counseling patients.

Single-Gene Defects

In a great many congenital malformation syndromes, each syndrome is composed of a variable number of defects, all traceable to the presence of a single Mendelian gene. Although there are many such entities, they are individually rare in their occurrence when compared to the more common polygenic abnormalities (McKusick, 1971). The single-gene etiology of these problems is evident primarily from analysis of the pedigree transmission of the abnormality. This follows the lines first described by Mendel (1865) and occurs in three primary modes of genetic transmission. These are the autosomal dominant and recessive, and the X-linked recessive patterns. Although there are others, these patterns account for the bulk of the clinically significant problems. Single-gene abnormalities range from simple to complex, but many present a clinical picture of multiple malformations or multiple defects that is relatively unique and recognizable. In contrast to the polygenic anomalies, the recognizable single-gene defects have a clear genetic prognosis. Genetic counseling can be given with mathematical assurance and frequently with specific biochemical identity of the gene carriers. A few examples of the primary types of Mendelian inheritance will illustrate the main features of single-gene diseases.

AUTOSOMAL DOMINANT INHERITANCE

As an example, we shall use a family with skeletal defects (Fig. 8) to illustrate the autosomal dominant pattern of transmission. *Autosomal* refers to the location of the gene on a non-sex chromosome or autosome. This is apparent from the occurrence of the disease in both sexes. *Dominant* refers to the transmission

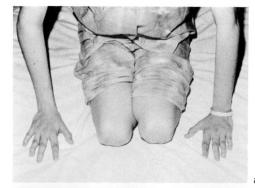

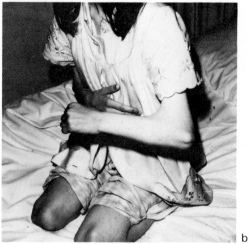

FIGURE 8. *Nail–Patella syndrome. (a) Showing absent patellae (kneecaps) and severe nail dysplasia but mild elbow webbing. (b) Also demonstrates absent patellae but milder nail dysplasia and moderately severe elbow webbing (note left arm).*

of the disease from an affected person to his or her offspring and to the fact that one mutant gene produces a clinical or phenotypic effect even in the presence of the normal gene. As there is a paternal and maternal gene for any phenotypic trait (whether at the clinical or molecular level), the affected person has a 50 percent risk of transmitting the mutant or abnormal gene to each offspring of each conception. The heterozygote is affected, and homozygotes do not usually occur. This example shows that the pedigree transmission is clearly that of a single gene. The effect of the gene does not appear to be limited to a single morphogenic developmental step, however, as several body areas are involved in the production of different phenotypic effects. This phenomenon is called *pleiotropism* and is commonly seen with autosomal dominant genes. For example, in the family that we are using as an example, the father and his three daughters present the very obvious symptoms of dysplastic finger- and toenails, absent to severely hypoplastic patellae, and webbing at the elbow. The latter is secondary to a partial subluxation of the head of the radius. This malformation is named the Nail–Patella syndrome after its two primary features (Lucas and Opitz, 1966). However, although these visible skeletal abnormalities are the most prominent, over one third of the families afflicted with this syndrome have an additional renal defect (Darlington and Hawkins, 1967). The family in question demonstrates this with every affected member having a progressive nephropathy as a part of the pleiotropic gene effect. From a clinical standpoint, this underscores the importance of thinking of all the manifestations of a gene when presented with one or two of its obvious effects. Searching for the complete signs of the syndrome may have practical significance in one's ability to help the patient medically. Discovery of the significant renal aspect of this disease in the three young girls in this family may allow preventive care that will alter the course of their disease and, consequently, their life expectancy.

Some other properties of Mendelian genes, particularly dominant genes, may also be seen in this pedigree. Many dominant syndromes are the result of freshly (newly) mutant genes. This is especially true where the phenotypic effect is severe and reduces the person's ability to reproduce. The propositus in our pedigree is such a mutant. His gene was not inherited, but it obviously was *heritable*. Had he had siblings, they would have had almost no chance of being affected, whereas he had a 50 percent risk of transmitting the gene to his children and was unfortunate enough to realize that possibility with each daughter.

Secondly, dominant genes vary in the severity of their effect on the subject. This is obviously partially due to the interaction of gene and environment. In this family the skeletal effects vary between individuals, and thus the *expressivity* of the gene is said to vary. One

daughter has a severe nail dystrophy and marked limitation of motion at the elbow, whereas another has no elbow limitation and mild nail dystrophy (Fig. 8). Similarly, one shows absence of the patella, but another has only hypoplasia of that bone. When the expression of a gene in the observable phenotype of an individual becomes mild enough, it can escape our detection. Our failure to observe this is usually due to our lack of proper tools to detect the effect of the gene—and we say that the gene is not *penetrant*. Penetrance might be illustrated by the renal disease in the pedigree that we are using as our example. By clinical observation and ordinary urinalysis only the propositus is affected in this family. None of the girls has clinical renal disease or pathologic signs in laboratory urinalysis, whereas the father required artificial dialysis at the time of ascertainment. If we were dependent on renal disease as the only manifestation of this gene and if we had reason because of their pedigree to believe that the girls possess the gene (Fig. 9), these individuals would be cases of nonpenetrance. When the renal tissues of these people are examined by electron microscopy, they show changes that indicate the presence of the gene, and thus it is no longer nonpenetrant. Observation of the direct mo-

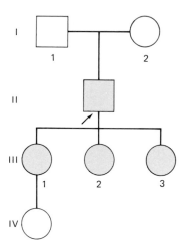

FIGURE 9. *Pedigree of S family with Nail–Patella syndrome. Propositus in generation II is apparently a new mutant. Daughters in generation III demonstrate obvious skeletal defects as pictured in Fig. 8 but require sophisticated techniques to reveal renal defect.*

lecular biological action of a gene would render many of our genetic terms obsolete.

AUTOSOMAL RECESSIVE INHERITANCE

Autosomal recessive genes are given this name because they do not produce a manifest clinical effect when present in a single dose—that is, when balanced by the normal allele for the gene in question. When the subject is a homozygote, and no normal balancing allele is present, the phenotypic effect is seen. "Autosomal," again, means that no sex chromosome is involved, so that both sexes should be equally affected. Therefore, in contrast to the dominant pedigree, neither parent of an affected child shows the effect of the gene, but both are carriers of the gene. Since each has one mutant and one normal allele, their chance of both transmitting their mutant gene to their child is 25 percent, and that represents the recurrence risk for autosomal recessive conditions with each child of a pair of heterozygous parents. Actually, many recessive genes control biochemical enzyme functions, and it is possible to detect the partial deficiency of the enzyme in the parents by appropriate biochemical carrier-detection tests. The term "recessive" is therefore artificial and applies to the observable phenotype.

Recessive genes, like dominant genes, may cause a wide variety of phenotypic effects. Although many are associated with rather precisely known biochemical defects, they may also produce gross anatomical malformations. One example is a condition in which congenital defects of limbs and blood elements are produced (Fig. 10). This condition is called the *Absent Radius–Thrombocytopenia syndrome*, a name that describes its cardinal manifestations (Hall and Levin, 1969). There may be other skeletal defects as well. The ulnae are almost always severely hypoplastic, as is shown, and the humeri are also frequently underdeveloped. The platelets are severely diminished, and the child shown required platelet transfusions for support for over 1 year. Many such children succumb during their first months of life. If they survive this period, the megakaryocytes begin to make platelets, and the children are able to meet all but critical situations with their own platelets. The limb malformation usually requires extensive orthopedic help.

Many times the pedigree of a family with a

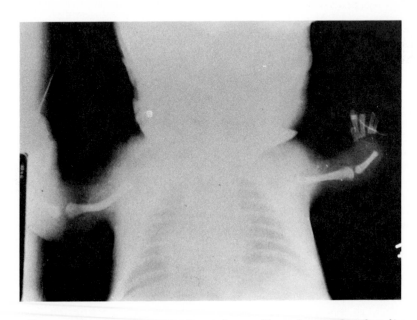

FIGURE 10. *Absent radius–thrombocytopenia syndrome. Note shortened and malformed upper limbs with X-ray evidence of absent radii and hypoplastic ulnae and humeri.*

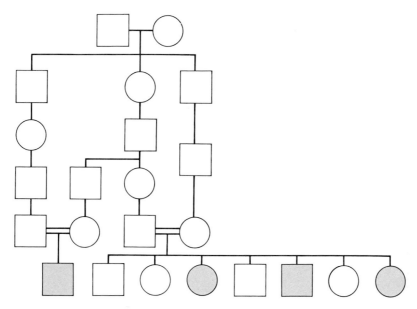

FIGURE 11. *Autosomal recessive inheritance. Partial pedigree showing four children with ataxia–telangiectasia in two related families.*

recessive genetic condition will show a consanguinous or blood-related marriage. Such unions are more likely to bring two identical rare genes together, because relatives share genes in common. The occurrence of recessive conditions is thus relatively more frequent in genetically isolated populations, because marriage occurs more frequently between relatives. Members of a genetic isolate owing to religious grounds, the Old Order Amish, have been examined by McKusick and his colleagues for rare recessive genes with considerable success (McKusick et al., 1964). One such gene is responsible for a rare neuro–immunological disorder, ataxia-telangiectasia (Boder and Sedgwick, 1958). The child develops signs of a cerebellar ataxia almost as soon as he is able to walk (Fig. 11). Later such children manifest small, dilated superficial blood vessels (telangiectasia) over the skin of the face and the sclera of the eye. The pedigree of the Amish family of the child shown in Fig. 11 demonstrates consanguinity. In this case the parents are not closely related, but each probably received the genes from one of the common ancestors. This condition also demonstrates something of the biochemical side of recessive genes, as the affected children all have a severe deficiency of

immune capacity associated with absence of immunoglobin A (IgA), as well as defective lymphocyte function (Peterson et al., 1964). Clinically, this makes them susceptible to repeated sinopulmonary infections and prone to develop lymphoid neoplasia. Either condition will usually lead to their death in early adolescence.

We mentioned earlier an example of a recessive genetic neurologic condition in which the child appears to be born with no defect at all. This condition, Tay-Sachs disease, is now known to be due to a specific enzymatic defect that renders the catabolic pathway for ganglioside or sphingolipids ineffective (O'Brien et al., 1971). This material is accumulated in the ganglion cells of the central nervous system and progressively destroys the brain, resulting in the child's death at age 3 to 5 years. The process begins in fetal life, but the infant appears well until 4 to 6 months of age when early neuronal degenerations may be seen in the retina. A deficiency of the enzyme hexosaminidase A in the disease is now well delineated, so that carrier detection and prenatal testing are possible—with the result that Tay-Sachs births can be prevented by selective therapeutic abortion (Kaback and Zeiger,

1972). Unfortunately, it is not yet possible to alter the course of the disease.

X-LINKED INHERITANCE

X-linked genes, like their autosomal counterparts, behave as dominant or recessive factors, although the majority of known X-linked genes are recessive. Since the gene is located on the X chromosome, the key difference in transmission is the lack of male-to-male transfer. A male may transfer the gene to his daughter, but he uses his Y chromosome to make a son and thus blocks transfer of his X-linked genes to male heirs. Recessive X-linked genes are carried by the female and expressed by their hemizygous male offspring, who have only the one X chromosome. This effectively unveils the recessive gene, as no normal allele is present to balance it. Since the carrier female has two X's, she has a 50 percent risk with each

girl of having a carrier daughter and a 50 percent risk with each boy of having an affected son. Several significant conditions follow this inheritance pattern, including Duchenne's muscular dystrophy, two forms of hemophilia, and color blindness. Since the female's X chromosome with its mutant gene is the sole determinant of her son's disease, we do not need a mating between heterozygotes to produce an "at risk" couple. Any sister of an affected male is "at risk" until proved otherwise. Carrier detection tests are needed to identify the carrier and to confidently eliminate the noncarrier from risk. Unfortunately, these are not available in most X-linked conditions. In Duchenne's muscular dystrophy, for example, one cannot reassure the sisters of affected boys that they may have sons without risking transmission of a hidden lethal gene.

A final principle of importance in genetic conditions and their diagnosis can be illustrated with an X-linked and an autosomal recessive

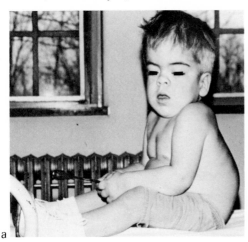

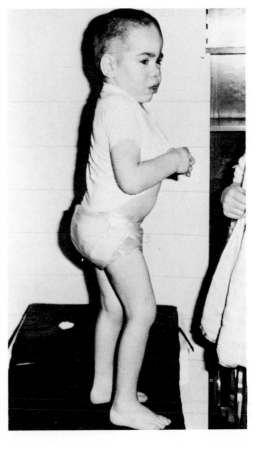

FIGURE 12. *Genetic heterogeneity. Two boys with different mucopolysaccharide storage diseases. Child in (a) has the autosomal recessive or Hurler's form; the boy in (b) has the X-linked or Hunter's form. Pedigree data show difference in inheritance, and the Hurler child also demonstrates a more severe and rapidly progressive form of disease. Both have hepatosplenomegaly and lumbar gibbus, but the Hurler child is much younger (3 years versus 5 years.) Hurler child also has more severe mental retardation and corneal clouding. Note similarity of facies.*

condition. The two children are shown (Fig. 12) who have nearly identical phenotypes but whose conditions are caused by separate and distinct genes. This phenomenon is called *genetic heterogeneity* (Childs and der Kaloustian, 1968). It is important to recognize this, because the differences are usually subtle, and yet these may be practically significant. Each of the two boys pictured has a form of mucopolysaccharide storage disease. One is the autosomal recessive Hurler syndrome, and the other is the X-linked recessive Hunter syndrome (McKusick, 1966). Both children excrete large amounts of the urinary mucopolysaccharides dermatan and heparan sulfate. Both have mental retardation, joint stiffness, bone destruction, hepatosplenomegaly, corneal clouding, and cardiopulmonary disease. The abnormality is apparently secondary to defective catabolism of the mucopolysaccharides, leading to their excessive intracellular accumulation (Fratatoni et al., 1968). Recently it has been discovered that normal plasma contains a factor that corrects the catabolic defect, and trials of plasma infusion therapy have shown some encouraging results (Neufeld, 1972).

Chromosome Abnormalities

A final group of malformations can be traced etiologically to the action of several genes, all of which are located on a single chromosome or piece of chromosome. If the chromosome segment is absent, duplicated, or even changed in position within the karyotype, this deviation may sufficiently alter the expression of its genes to cause severe congenital malformation. Chromosomal studies of malformed humans flourished in the early 1960s, when new techniques of tissue culture and lymphocyte blast transformation were being developed, and many chromosomally associated anomalies were delineated at this time (Carr, 1969). Today there is a second wave of enthusiastic work in progress, in which fluorescent (Caspersson et al., 1971) and hybridization-banding techniques (Hsu, 1972) are being applied to the identification of individual chromosomes. Although the phenotypic features of each of the chromosome anomaly syndromes are quite characteristic, thus far no specific biochemical gene abnormalities have been identified as a result of studies of such individuals. This is probably because the majority of these abnormalities are due to chromosomal (and thus gene) excess, and the activity of extra genes may not be detectable. This general absence of specific biochemical abnormalities may also mean that the principal effect of extra or abnormal chromosomes occurs in early fetal development. Whatever their primary effect, the mechanism of production of phenotypic malformations by chromosome abnormalities is poorly understood and thus constitutes a field for productive research.

The majority of chromosome malformation syndromes involve numerical abnormalities—usually extra, morphologically normal chromosomes. These chromosomes are present in excess because of maldistribution at either of the meiotic divisions of gametogenesis or because of a very early somatic mitotic division of the embryo. The mechanism usually invoked is "nondysjunction," or failure of the sister chromatids to separate, which leaves one resulting daughter cell deficient and one with an extra representative of that chromosome pair. This process could occur sequentially in meiosis or even into mitosis, so that several extra chromosomes of a given pair might be placed in the resulting zygote. The human embryo does not tolerate gross chromosome abnormalities very well. With the exception of the X chromosome, it does not usually withstand chromosomal loss, and even with the X, most monosomies are lost in early fetal death (Carr, 1971). Similarly, usually only one extra chromosome of any group is tolerated by the organism, again with the exception of the X. Of course, there have been isolated exceptions to both these general rules—which simply shows that rare extremes of biological variation occur in humans as well as elsewhere.

A few chromosome malformation syndromes occur as a result of morphological alteration of a chromosome. Two mechanisms appear to be important, and both involve chromosome breakage (Fig. 13). In the first, deletion, the broken chromosomal fragment is simply lost. As it is a piece without a centromere, it has no attachment at mitosis and presumably is lost at cell division. The second mechanism, translocation, involves the breakage of two chromosomes and an exchange of the fragments, resulting in morphologically altered chromosomes that also have genes altered in position. In deletion, the organism simply loses genes,

and this may give rise to phenotypic malformation effects. In translocation one of several effects may be produced as a result of segregation. The organism may be relatively normal, it may be deficient, or it may have an excess of genetic material as a result of the rearrange-

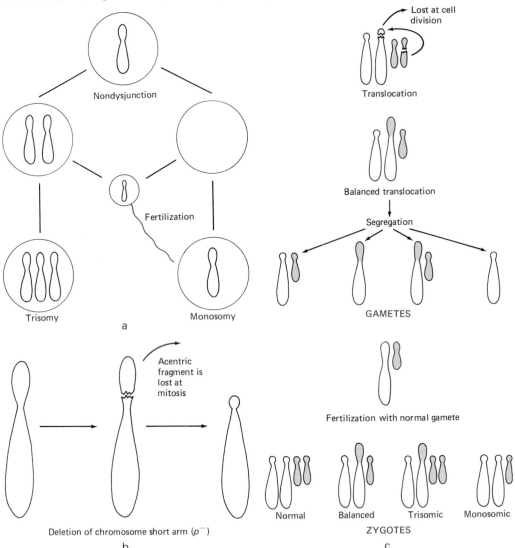

FIGURE 13. *Abnormal chromosome mechanisms. (a) Nondysjunction. Chromosome pair fails to separate (dysjunct) at meiotic reduction division, giving rise to gamete with missing chromosome (right) or with extra chromosome (left). Fertilization results in monosomy or trisomy. (b) Deletion. Chromosome may break and acentric fragment will be lost at division by failure of attachment to mitotic spindle. (c) Translocation. Breakage and rejoining of chromosome to form new chromosome arrangement. Balanced configuration retains majority of genes but abnormal pairing can arise at segregation. After fertilization one may have trisomy or monosomy for chromosome length (b) involved.*

ment and subsequent segregation. These events all take place at gametogenesis in one of the parents, of course. Such malformations are divided into sex chromosome abnormalities and non-sex or autosomal disorders.

SEX CHROMOSOME ANOMALIES

The sex chromosomes are the X and Y in the male and the two X's in the female. Developmentally, they are primarily responsible for the differentiation of the gonad from the genital ridge. In addition, the X carries many known genes, whereas the Y apparently does not. The Y chromosome contains male-determining genes that direct medullary proliferation of the gonad to form testes; absence of the Y, thus results in no testicular formation. Two normal X chromosomes are required for the differentiation of a normal ovary, and the absence of an X results in an undifferentiated fibrous streak. Finally, the presence of the Y is dominant to the X chromosome effect. Further sexual maturation is dependent upon gonadal products or hormones (Jost, 1953).

Klinefelter's Syndrome

Klinefelter, Reifenstein, and Albright described a specific form of male hypogonadism associated with gynecomastia in 1942. In 1956 the sex chromatin findings in such males (which were positive and identical to those of a normal female) were described from buccal smear cells. Subsequently, most of these males were discovered to have the XXY sex chromosome constitution (Jacobs and Strong, 1959). Thus variants may have more than one X and Y. This syndrome, the XXY Klinefelter syndrome, is probably the most common sex chromosome anomaly of men, occurring in approximately 1 out of 500 male births. Only the XYY syndrome has a comparable high incidence.

Clinically the features are variable (Rimoin and Schimke, 1971). The ductal and external genitalia are male, and the testes are small and firm. Postpubertally, they are firm and less than 2 cm in size; prepubertally, they are usually normal for age and are not helpful in the diagnosis. There is usually a lack of sperm formation. Gynecomastia occurs in only 25 to 40 percent of these males. Most patients have scant body and facial hair, decreased libido, or other signs of lack of androgen production. They are frequently taller than average with relatively long legs and a decreased upper segment (crown to pubis) to lower segment (pubis to floor) ratio. Intelligence is variable but tends toward the subnormal range. Severe mental retardation is not a feature of this disease, but victims of it do manifest delinquent or antisocial behavior.

Laboratory studies show a high total urinary gonadotropin concentration in postpubertal patients with increased LH and ICSH. 17-ketosteroids are usually normal, but plasma testosterone is mildly decreased. The testicular histology of the postpubertal stage shows hyalinization and atrophy of the seminiferous tubules and little or no evidence of spermatogenesis. Most cases show two maternal X chromosomes by linkage analysis (Race and Sanger, 1969), and this is compatible with the increased maternal age found in parents of Klinefelter patients. Diagnosis rests on chromosome studies or on testicular biopsy in chromatin-positive cases. The risk for recurrence of this nondysjunctional event is low.

Variations of the basic Klinefelter phenotype may occur in association with variant karyotypes. Thus 48 XXY and 49XXXXY cases are known and show many of the same phenotypic features of the 47XXY phenotype. There appears to be a correlation between the degree of mental retardation and the number of extra X chromosomes in the karyotype. The 49XXXXY child is therefore most severely retarded and also generally has obviously ambiguous external genitalia (Fig. 14). Skeletal abnormalities, such as radioulnar synostosis, are frequent and may lend a characteristic appearance to the phenotype.

Less clear are the effects of extra Y chromosomes in the male karyotype. Both 48XXYY and 47XYY males have been reported. Both are usually tall (over 71 in.) and tend to have aggressive, antisocial behavior patterns. These effects are extremely variable and warrant further study.

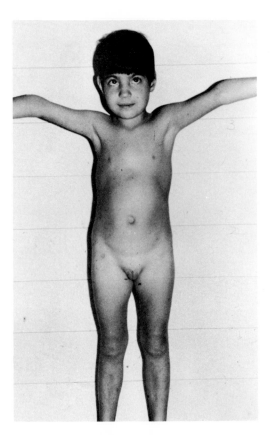

FIGURE 14. *XXXXY syndrome. Child with ambiguous genitalia, elbow deformities, and mental retardation. Child is retarded and has similar facies.*

ternal ears (80+ percent), cubitus valgus (exaggerated elbow carrying angle) (70+ percent), short fourth metacarpal or metatarsal or both (50 percent), renal anomalies (e.g., horseshoe kidney) (60+ percent) and cardiac defect (20 percent)—mostly coarctation of the aorta.

The lymphedema and loose neck-skin folds usually disappear rapidly following infancy. Linear growth is slow, and there is no adolescent growth spurt. At adolescence they will usually need cyclic replacement hormone therapy. Androgenic steroids appear to increase the growth rate and final height. Estrogens will reverse any masculinization but should be delayed as they may cause epiphyseal closure. This means a delay in the development of secondary sex characteristics but should mean the achievement of greater height. The patients and their families must be counseled that ovarian development in the patient will be incomplete—that she will be infertile and will thus have to adopt children if a family is desired. Other physical abnormalities, such as cardiac defects, should be handled as needed. The short stature is frequently the most severe psychological problem in adolescence.

The missing X is usually of paternal origin (Lindsten, 1963) and may well be due to zygotic chromosomal nondysjunction. There is no maternal age effect and family clusters do not occur, so the family may be counseled that the risk for recurrence is negligible.

Turner's Syndrome

Described by H. H. Turner in 1938, Turner's syndrome has come to be defined as one with short stature, sexual infantilism, streak gonads, and a missing, deleted, or abnormal X chromosome. The birth incidence is low, approximately 1 in 1500 female births—and many of these children are spontaneously aborted, as shown by chromosomal studies of abortuses. The somatic features, other than short stature, are variable. Birth weight is frequently low, demonstrating intrauterine growth retardation. Other somatic features include congenital lymphedema (80 percent), broad shield-like chest with widely spaced nipples (80+ percent), narrow maxilla (80 percent), anomalous ex-

AUTOSOMAL ABNORMALITIES

The human karyotype demonstrates several important autosomal trisomy syndromes as well as some caused by autosomal deficiency from either deletions or translocation deletions. The human does not appear to tolerate autosomal monosomy except in rare instances, nor does he tolerate more than one extra specific autosome. This is undoubtedly associated with the fact that the sex chromosomes contain a large amount of effectively inactive chromatin, whereas most autosomes do not. The exact reason why only a few autosomes are seen in triple number in live-born humans is not known. Size and heterochromatin (inactive genetically) content are probably involved, but specific gene content may also be responsible. Although many autosomal anomalies are known, three trisomy syndromes occur commonly and have consistent phenotypes; they will be described here. They are Down's syndrome, caused by trisomy of chromosome 21, and the trisomy-13 and trisomy-18 syndromes. In addition, two autosomal deletion syndromes demonstrate the relatively consistent differences conferred by very similar but chromosomally specific karyotypes.

Down's Syndrome

First described in 1866 by J. L. H. Down, Down's syndrome is a relatively common clinical malformation syndrome that occurs about once in every 660 newborns. The chromosome abnormality was first described by Lejeune in 1959. A full trisomy for chromosome 21 occurs in 94 percent of cases, trisomy-21/normal mosaicism in 2.4 percent, and translocations in 3.3 percent. In Down's syndrome cases that are born to mothers below the age of 30 the incidence of translocation cases is approximately 8.5 percent, with one third of those being inherited from a carrier parent. 14/21 and 21/22 translocations occur with nearly equal frequency. 21-isochromosomes are rare. Maternal age is a factor in trisomic cases: the risk of Down's syndrome births rises from 1 in 1500 at age 15 to 29; to 1 in 800 at 30 to 34 years; to 1 in 270 at 35 to 39 years; to 1 in 100 at 40 to 44 years; to 1 in 50 at 45 years and over. With a previous Down's syndrome birth the empirical risk for a second such birth is 1 to 2 percent regardless of age. Amniotic cell monitoring appears to be a wise course in any pregnant woman who has had a Down's syndrome child.

The diagnosis of Down's syndrome can be made in the newborn (Fig. 15), and the following ten signs are helpful—at least four of them being present in all of a series of Down children examined by Hall (1966): hypotonia (80 percent), poor startle reflex (85 percent), hyperflexibility of joints (80 percent), excess neck skin (80 percent), flat facial profile (90 percent), slanted palpebral fissures (80 percent), anomalous external ears (60 percent), dysplasia of pelvis (70 percent), dysplasia of

midphalanx of fifth finger (60 percent), and single palmar flexion crease (45 percent).

Hypotonia, open mouth with protruding tongue, and short stature are general signs. All cases are mentally retarded. The nose is small with a depressed nasal bridge. There are epicanthal folds with upward slanting palpebral fissures and speckling of the iris (Brushfield's spots). The ears are generally small with over-

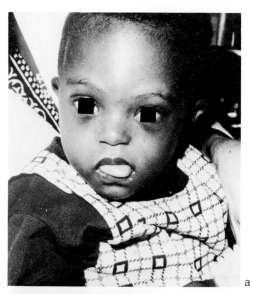

a

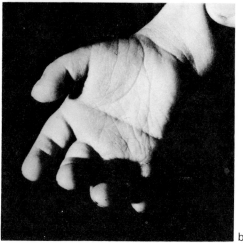

b

FIGURE 15. *Down's syndrome. (a) Facies showing depressed nasal bridge, slightly overrolled tips of ears, protruding tongue, upward lateral slant of eyes, and moderate epicanthus. (b) Hand showing single palmer flexion crease.*

folded upper helix. The hands have relatively short metacarpals and phalanges, hypoplasia of the midphalanx of the fifth digit, clinodactyly, and abnormal dermatoglyphics. There is a wide gap and a plantar furrow between the first and second toes. The pelvis shows X-ray changes, including a shallow acetabuar angle. About 40 percent have a congenital heart defect.

The muscle development improves with age, but neurological and intellectual development slows down and finally reaches a plateau (Penrose and Smith, 1966). Bone growth is also abnormally slow. Up to 3 years of age, many have an IQ above 50; the mean IQ for older patients, however, is 24. As infants they are happy, well-behaved, and enjoy mimicry, but as they grow older, behavioral problems may occur. In adolescence, sexual development is retarded: girls may become fertile, but males apparently do not. Although about one half of those with heart disease die in infancy, the remainder do enjoy a relatively normal life span. The incidence of acute leukemia, however, is much higher in a child suffering from Down's syndrome than it is in a normal child, although it is still an uncommon disease. In conclusion, we may list chronic rhinitis and dental disease as among the less serious problems to which such a child is subject.

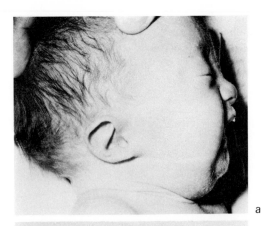

a

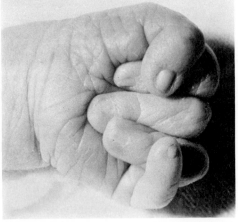

b

FIGURE 16. *Trisomy-18.* *(a)* *Lateral face showing hypoplastic mandible, small nose, and low-set simple ear.* *(b)* *Hand showing characteristic overlap at digits.*

Trisomy-18 Syndrome

First recognized as a chromosomally determined entity by Edwards (Edwards et al., 1960), trisomy-18 is the second most common autosomal malformation syndrome, with an incidence of 0.3 per 1000 newborns and a 3 to 1 preponderance of females to males. Multiple abnormalities are noted (Fig. 16); for example, each of the following symptoms are found in more than 50 percent of cases and is thus helpful in diagnosis:

General: feeble infantile activity; a history of pre- or postmaturity mental deficiency; low birth weight.
Craniofacial: low-set, malformed ears; short palpebral fissures; small mouth with high arched palate.
Hands and feet: clenched hand with the index finger overlapping the third finger and the fifth overlapping the fourth; low arch ridge pattern on the majority of fingers; hypoplasia of nails; short great toe that is frequently dorsiflexed.

Males are often cryptorchid. Many infants have congenital heart disease—especially ventricular septal defect or patent ductus arteriosus. Rocker-bottom feet are found in less than 50 percent of cases.

The child is very feeble and his survival is severely prejudiced, even with special care. In the first month 30 percent die, and only 10 percent survive the first year. There is no point, therefore, in expanding abnormal effort to keep those children alive.

This chromosome anomaly is almost always due to nondysjunction and increases in frequency with advancing maternal age. Many trisomy-18's die in fetal life and are the cause of spontaneous abortions (Carr, 1971). Risk for recurrence is probably much less than 1 percent.

Trisomy-13 Syndrome

Although the clinical features of the trisomy-13 syndrome were described as far back as 1657 (Warburg and Mikkelsen, 1963), the chromosomal etiology was not known until 1960 (Patau et al., 1960). It is a fairly rare event, occurring in 1 in 5000 births. The following characteristic defects appear in 50 percent or more of these infants:

CNS: maldevelopment of the forebrain with associated underdevelopment of the olfactory and optic nerves, giving a sloped forehead appearance; severe mental retardation associated with seizures and apneic spells in infancy.
Eyes: microphthalmia and/or colobomata of the iris (Fig. 17).
Ears: abnormal external ears that may or may not be low-set.
Face and Mouth: frequent bilateral clefts (Fig. 17).

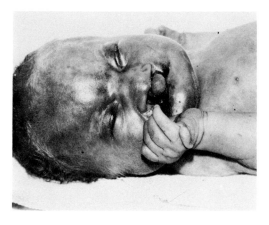

FIGURE 17. *Trisomy-13. Face of trisomy-13 child with severe bilateral facial cleft and microphthalmia. Hand shows characteristic polydactyly.*

Hands and Feet: distal axial triradius of the palm; simian crease; polydactyly (Fig. 17).
Genitalia: male with cryptorchidism and bifid scrotum; female with bicornate uterus.

There is frequently an increased incidence of nuclear projections on the circulating white cell nuclei. Fetal hemoglobins may persist into infant life.

These children are severely affected and 44 percent die in the first month. By the end of 6 months 69 percent die, and only 18 percent survive 1 year. Those that do are severely impaired, suggesting that no unusual means to prolong life should be used.

The chromosomal mechanism is usually nondysjunction, although translocations do occur. Usually these are *de novo* in the child and neither parent is a carrier. In such cases, the risk for recurrence is very low but rises slightly with increasing maternal age. The mechanism for malformation is probably an early single defect in the prechordal mesoderm, with the resulting malformations discussed previously.

B-Deletion Syndromes

The B chromosome deletion syndromes were both described in the mid-1960s. The 5-deletion (5p-) or *cri-du-chat* syndrome was described in 1963 by Lejeune and his colleagues. Children suffering from this disease have a relatively low birth weight; their growth, neuromuscular development, and mental development are all retarded, and they have a cat-like cry. The mewing cry is secondary to abnormal laryngeal development and disappears gradually with age. Most of the patients have microcephaly, a slight-to-prominent downward lateral slant to the eye, epicanthal folds, and hypertelorism (wide-set eyes) (Fig. 18). An unusually high percentage of cases have had translocation mechanisms involved in their production (Jackson and Barr, 1970).

The 4-deletion (4p-) syndrome was first described in a patient by Wolf and his associates in 1965 (Wolf et al., 1965). Subsequent autoradiographic studies identified the deleted chromosome as different from that found in the *cri-du-chat* syndrome and demonstrated the reason for the differences in clinical findings

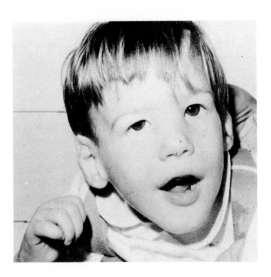

FIGURE 18. *Face of 5p- or* cri-du-chat *child. Eyes show epicanthal fold with downward lateral slant. Ears are low and nose shows depressed saddle.*

(German et al., 1964; Warburton et al., 1967). These patients all have a severely low birth weight (Fig. 19) and suffer from both growth and mental retardation as well. They frequently have seizures and infantile muscular

hypotonia. Epicanthal folds, cleft palate and lip (Fig. 19), and a peculiar mouth with down-turned corners are also frequently present, as well as hypospadias and a sacral dimple.

In conclusion, we shall summarize the common and differentiating features of B-deletion syndromes (Breg et al., 1970; Miller et al., 1970). Clinical features common to both B-deletion syndromes and found in most cases are as follows:

HYPOTONIA
PSYCHOMOTOR RETARDATION
MICROCEPHALY
HYPERTELORISM
DOWNWARD SLANT TO PALPEBRAL FISSURES
EPICANTHUS
BROAD BASE TO NOSE
NARROW EAR CANALS
LOW-SET EARS
MICROGNATHIA
HIGH ARCHED PALATE (50 PERCENT)
HEART DEFECT (50 PERCENT)
INGUINAL HERNIA (50 PERCENT)
FOOT DEFORMITIES
SIMIAN CREASE

The following defects are found commonly in 4p- but *not* in 5p-:

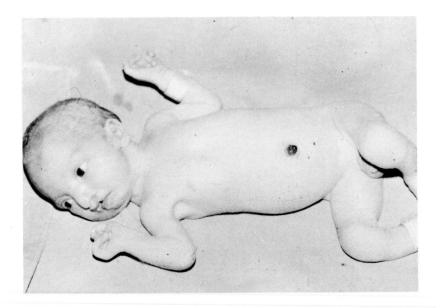

FIGURE 19. *4p- or* Wolf's *syndrome. Child shows thin, low-birth-weight appearance with severe midline facial cleft.*

PTOSIS

COLOBOMA

SCALP DEFECT

BEAKY NOSE

CARP-LIKE MOUTH

CLEFT PALATE

VERY SIMPLE EARS

PREAURICULAR DIMPLE OR SINUS

SACRAL DIMPLE OR SINUS

HYPOSPADIAS

SEIZURES

UNDERDEVELOPED DERMAL RIDGES

DELAYED OSSIFICATION OF PELVIS OR CARPALS

Finally, the two most obvious clinical features that distinguish the 4p- from the 5p- syndrome are (1) a cat-like cry in infancy—found in 5p- only, and (2) a difference in birth weight average—1866 g for 4p- but 2650 g for 5p-.

Summary

Malformation syndromes represent a broad spectrum of phenotypic variations that result from a diversity of etiologic mechanisms affecting human embryologic and biochemical development. The genesis of what finally appears at the termination of the gestational period, or even in adult life, as a well-defined phenotypic entity may come from a block in a significant developmental step, from an error that sensitizes the organism to what would otherwise be a normal environmental challenge, or from errors promoting an acceleration of degenerative process or causing a pathologic degenerative process. The same or similar end results may come from more than one of the generative mechanisms described. Thus skeletal defects of a similar nature can be caused by single-gene defects, by sporadic or multifactorial developmental processes, or they may occur as part of a chromosomal anomaly syndrome. A clear example of this heterogenic spectrum of causes is shown by the holoprosencephalies. As discussed in this chapter, these usually occur as the manifestation of a gross error in development of the forebrain with failure to cleave sagittally into cerebral hemispheres, transversely into the telencephalon and diencephalon, and horizontally into olfactory and optic bulbs. The degree of severity of this failure will dictate the severity of the associated facial defect. Although most of these occur sporadically and apparently as the result of sporadic errors in embryonic development, occasionally a familial aggregation suggests that they are the result of a single rare gene— at least in these instances. In addition, the same type of error is seen in chromosomal aberrations, including the number 13 and 18 chromosomes. Trisomy-13 presents with moderately severe holoprosencephaly, and, frequently, chromosome 18 short-arm deletions also demonstrate this effect. Whether this similar result is due to specific gene influence or a general retardation of early developmental steps is not known. In the former, the occurrence of extra gene dosage because of trisomy might lead to this defect, or the effect of a hemizygous abnormal gene uncovered by the deletion might also lead to this change. In the latter, the combined weight of the many gene duplications or deficiencies in the chromosomal anomaly could seriously but generally impair developmental timing. Although one can postulate a basic defect in the notochordal plate-neuroectodermal plate–oral plate interaction (Cohen et al., 1971), neither the normal nor the abnormal influences in this interaction are delineated. Studies of gene localization and chromosome mapping may help to further the understanding of these developmental mechanisms.

In summary, congenital malformations can be produced by a variety of mechanisms and etiologic agents. The extent of our understanding of these also varies widely. In some we can isolate the developmental step but not a discrete etiologic factor. In others the factor is known, but the developmental mechanism is not. From the standpoint of the clinician, the practical questions are how to correct, or predict and prevent, these congenital malformations. Between these islands of knowledge an ocean of understanding is waiting to be developed. Although much of the information needed will come increasingly from new and ever more sophisticated scientific techniques, continued stimulus can be provided by careful thought and analysis of the malformation syndromes as they present. Cooperative investigation by clinicians and research scientists should continue to improve our understanding of human congenital malformations and our ability to help them medically.

References

BLYTH, H., AND CARTER, C. (1969). "A Guide to Genetic Prognosis in Paediatrics." Spastics International Medical Publications, London.

BODER, E., AND SEDGWICK, R. P. (1958). Ataxia-telangiectasia. A familial syndrome of progressive cerebellar ataxia, oculocutaneous telangiectasia and frequent pulmonary infection. *J. Pediat. 21*, 526.

BREG, W. R., STEELE, M. W., MILLER, O. J., WARBURTON, D., MILLER, D. A., DeCAPOA, A., ALLERDICE, P. W., DAVIS, J., KLINGER, H. P., McGILVRAY, E., AND ALLEN, F. H. (1970). Partial deletion of the short arm of chromosome no. 5 (5p-): clinical studies in five unrelated patients. *J. Pediat. 77*, 282.

CARR, D. H. (1969). Chromosome abnormalities in clinical medicine. *Prog. Med. Genet. 6*, 1–61.

CARR, D. H. (1971). Chromosomes and abortion. *Advan. Hum. Genet. 2*, 201

CARTER, C. O. (1965). The inheritance of common congenital malformations. *Prog. Med. Genet. 4*, 59–84.

CARTER, C. O. (1969). Genetics of common disorders. *Brit. Med. Bull. 25*, 52.

CASPERSSON, T., LOMAKKA, G., AND ZECH, L. (1971). The 24 fluorescence patterns of the human metaphase chromosomes—distinguishing characters and variability. *Hereditas 67*, 1971.

CHILDS, B., AND DER KALOUSTIAN, V. M. (1968). Genetic heterogeneity. *New Eng. J. Med. 279*, 1205–1212, 1267–1274.

COHEN, M. M., JIRASEK, J. E., GUZMAN, R. T., GORLIN, R. J., AND PETERSON, M. Q. (1971). Holoprosencephaly and facial dysmorphia: nosology, etiology and pathogenesis. *Birth. Def. Orig. Art. Ser. 7*, 125–135.

DARLINGTON, D., AND HAWKINS, C. F. (1967). Nail-patella syndrome with iliac horns and hereditary nephropathy. Necropsy report and anatomical dissection. *J. Bone Joint Surg. 49-B*, 164.

DE LANGE, C. (1933). Sur un type nouveau de génération (typus Amstelodamensis). *Arch. Méd. Enfant. 36*, 713.

DE MYER, W. (1971). Classification of cerebral malformations. *Birth Def. Orig. Art. Ser. 7*, 78–93.

DOWN, J. L. H. (1866). Observations on an ethnic classification of idiots. *Clin. Lecture Rep. London Hosp. 3*, 259.

EDWARDS, J. H., HARNDEN, D. G., CAMERON, A. H., WOLFE, O. N., AND CROSS, Y. M. (1960). A new trisomic syndrome. *Lancet 1*, 787

FALCONER, D. S. (1965). The inheritance of liability to specific diseases, estimated from the incidence among relatives. *Ann. Hum. Genet. 29*, 51.

FRATATONI, J. C., HALL, C. W., AND NEUFELD, E. F. (1968). The defect in Hurler's and Hunter's syndromes: faulty degradation of mucopolysaccharide. *Proc. Nat. Acad. Sci. U.S.A. 60*, 699.

GERMAN, J. L., LeJEUNE, J., MacINTYRE, M. M., AND DEGROUCHY, J. (1964). Chromosomal autoradiography in the cri-du-chat syndrome. *Cytogenetics 3*, 347.

HALL, B. (1966). Mongolism in newborn infants. *Clin. Pediat. 5*, 4.

HALL, J. G., AND LEVIN J. (1969). Congenital amegakaryocytic thrombocytopenia with bilateral absence of the radius: radius platelet hypoplasia or RPH. *Birth Def. Orig. Art. Ser. 5*, 190–195.

HSU, T. C., ed. (1972). Section on chromosome banding techniques *Mamm. Chrom. Newsl. 13*, 21.

JACKSON, L., AND BARR, M. G. (1970). Case report: A45,XY,5—15—,t (5q15q) Cri-du-chat child. *J. Med. Genet. 7*, 161–163.

JACOBS, P. A., AND STRONG, J. A. (1959). A case of human intersexuality having a possible XXY sex determining mechanism. *Nat. 183*, 302–303.

JOST, A. (1953). Problems in fetal endocrinology: the gonadal and hypophyseal hormones. *Rec. Prog. Horm. Res. 8*, 379.

KABACK, M. M., AND ZEIGER, R. S. (1972). Heterozygote detection in Tay-Sachs disease: a prototype community screening program for the prevention of recessive genetic disorders. *In* "Sphingolipids, Sphingolipidoses and Allied Disorders" (B. W. Volk and S. M. Aronson, eds.), p. 613. Plenum, New York.

KLINEFELTER, H. F., JR., REIFENSTEIN, E. C., JR., AND ALBRIGHT, F. (1942). Syndrome characterized by gynecomastia and aspermatogenesis with a-Leydigism and increased excretion of follicle-stimulating hormone. *J. Clin. Endocrinal 2*, 615.

LAMY, M. E. (1965). Hereditary diseases of bone—an overview. *Birth Def. Orig. Art. Ser. 5*, 8–13.

LAMY, M., AND MAROTEAUX, P. (1970). Le nanisme diastrophique. *Presse Méd. 68*, 1977.

LeJEUNE, J., GAUTIER, M., AND TURPIN, R. (1959). Etude des chromosomes somatiques de neuf enfants mongoliens. *C. R. Acad. Sci. 248*, 1721–1722.

LeJeune, J., La Fourcade, J., Berger, R., Vialette, J., Boeswillard, M., Serlinge, P. J., and Turpin, R. (1963). Trois cas de délétion partielle du bras court du chromosome 6. *C. R. Acad. Sci.* 257, 3098.

Lindsten, J. (1963). "The Nature and Origin of X Chromosome Aberrations in Turner's Syndrome." Almquist and Wiskell, Stockholm.

Lucas, G. L., and Opitz, J. M. (1966). The nail-patella syndrome. Clinical and genetic aspects of 5 kindreds with 38 affected family members. *J. Pediat.* 68, 223.

Maroteaux, P., Lamy, M., and Robert, J. M.. (1967). Le nanisme thanatophore. *Presse Méd.* 75, 2519–2524.

McKusick, V. A. (1972). "Heritable Disorders of Connective Tissue." 4th ed. C. V. Mosby, Saint Louis, Missouri.

McKusick, V. A. (1971). "Mendelian Inheritance in Man." 3rd. ed. Johns Hopkins Press, Baltimore.

McKusick, V. A., Hostetler, J. A., and Egeland, J. A. (1964). Genetic studies of the Amish: background and potentialities. *Bull. Johns Hopkins Hosp.* 115, 203.

Mendel, G. (1865). Experiments in plant-hybridization. *Verhandl. Naturforsch. Ver. Brunn*, Abh. IV.

Miller, O. J., Breg, W. R. Warburton, D., Miller, D. A., DeCapoa, A., Allerdice, P. W., Davis, J., Klinger, H. P., McGilvray, E., and Allen, F. H. (1970). Partial deletion of the short arm of chromosome no. 4 (4p-): clinical studies in five unrelated patients. *J. Pediat.* 77, 292.

Neufeld, E. F. (1972). Mucopolysaccharidoses: the biochemical approach. *Hosp. Pract.*, Feb., 107.

O'Brien, J. S., Okada, S., Ho, M. W., Fillerup, D. L., Venth, L., and Adams, K. (1971). Ganglioside storage diseases. *Fed. Proc. 30*, 956.

Opitz, J. M., Segal, A. T., Lehrke, R. L., and Nadler, H. L. (1965). The etiology of the Brachmann-de-Lange syndrome. *Birth Def. Rep. Ser.*, p. 22.

Patau, K., Smith, D. W., Therman, E., Inhorn, S. L., and Wagner, H. P. (1960). Multiple congenital anomaly caused by extra autosome. *Lancet i*, 990.

Penrose, L., and Smith, G. I. (1966). "Down's Anomaly." Little, Brown, Boston.

Peterson, R. D. A., Kelly, W. D., and Good, R. A. (1964). Ataxia-telangiectasia. Its association with a defective thymus, immunological-deficiency disease and malignancy. *Lancet i*, 1189.

Race, R. P., and Sanger, R. (1969). Xg and sex-chromosome abnormalities. *Brit. Med. Bull. 25*, 99–103.

Rimoin, D. L., and Schimke, R. M. (1971). "Genetic Disorders of the Endocrine Glands." C. V. Mosby, Saint Louis, Missouri.

Scott, C. I. (1972). The genetics of short stature. *Prog. Med. Genet. 8*, 243.

Smith, D. W. (1966). Dysmorphology (Teratology). *J. Pediat. 69*, 1150–1169.

Smith, D. W. (1967). Compendium on shortness of stature. *J. Pediat. 70*, 463–519.

Smith, D. W. (1970). Recognizable Patterns of Human Malformation. Saunders, Philadelphia.

Temtamy, S., and McKusick, V. A. (1969). Synopsis of hand malformations with particular emphasis on genetic factors. *Birth Def. Orig. Art. Ser. 5*, 125–184.

Tjio, J. H., and Levan, A. (1956). The chromosome number of man. *Hereditas 42*, 1–6.

Turner, H. H. (1938). A syndrome of infertilism, congenital webbed neck, decubitus valgus. *Endocrinology 23*, 566.

Valenti, C., Schutta, E. J., and Kehaty, T. (1968). Prenatal diagnosis of Down's syndrome. *Lancet ii*, 220.

Warburg, M., and Mikkelsen, M. (1963). A case of 13–15 trisomy or Bartholin-Patau's syndrome. *Acta Ophthalmol. 41*, 321.

Warburton, D., Miller, D. A., Miller, O. J., Breg, W. R., DeCapoa, A., and Shaw, M. W. (1967). Distinction between chromosome 4 and chromosome 5 by replication pattern and length of long and short arms. *Amer. J. Hum. Genet. 19*, 399.

Wolf, U., Reinwein, H., Porsch, R., Schröter, R., and Baitsch, A. (1965). Defizienz an den kurzen Armen eines Chromosomes Nr. 4. *Humangenetik 1*, 397.

Wynne-Davies, R. (1970). The genetics of some common congenital malformations. In "Modern Trends in Human Genetics" (A. E. H. Emery ed.), Vol. 1, pp. 316–338. Appleton-Century-Crofts, New York.

CHAPTER NINETEEN

Cellular Basis of Regeneration

ELIZABETH D. HAY

Regeneration is the process by which the body grows back parts that are lost either by injury or in the course of normal wear and tear. In mammals, regrowth of missing parts after injury is not as dramatic as in some lower vertebrates and invertebrates. We cannot grow back an amputated limb as can the lowly salamander, or a missing head as does the tiny flatworm. Yet in a lifetime we do repair hundreds of skin wounds and small injuries to our internal parts, and we can probably replace as much as half of an injured or diseased liver with new liver cells. Moreover, we undoubtedly surpass many vertebrates and invertebrates in our daily physiological regeneration—that is, in the daily replacement of cells that are lost in the ordinary course of their duties. For example, 300×10^9 red and white blood cells per 70-kg man are lost and replaced every day defending the body against invaders, and the entire epithelium of the intestine turns over every 2 to 3 days in order to ensure an intact and functional absorptive surface (Hay, 1966). Both physiological and reparative regeneration have been

the subjects of an enormous amount of research, and the reader who is interested in studying the literature in depth will find available a large number of review articles and bibliographies, some of the most recent and useful of which are as follows: Thornton (1959, 1968); Vorontsova and Liosner (1960); Singer (1960, 1968); Schotté, (1961), Berrill (1961); Rudnick (1962); Lender (1962); Tardent (1963); Herlant-Meewis (1964); Goss (1964, 1969); Rose (1964, 1970); Kiortsis and Trampush (1965); Hay (1966, 1970); Bucher (1967); Schmidt (1968); Ursprung (1968); Dunphy and Van Winkle (1969); Goss and Hay (1970); Mauro et al. (1970); Faber (1971); Carlson (1972a,b); Polezhaev (1972); and Manasek (1972). The purpose of this chapter is to introduce the student to one aspect of the subject, the cellular basis of regeneration. We shall not attempt an exhaustive review of the literature because our aim is simply to interest the reader in the subject and to acquaint him with problems to be solved. In keeping with the title and spirit of this volume, we shall emphasize what is current and rely on the reviews listed above to document the past.

The principal problem and theme of our discussion of the cellular basis of regeneration is the question of the source of the regeneration cell. The weight of the evidence stemming from work on vertebrate regeneration is against the older idea that reserve cells are set aside in the embryo for the purpose of reparative regeneration. However, as we shall see, the idea of the reserve cell has been revived recently by students of vertebrate muscle regeneration and has never been extinguished in the case of planarian regeneration.

Definitions of Cytodifferentiation

In evaluating the extent to which differentiated cells, as compared to reserve cells, participate in regeneration, it is important to define exactly what is meant by these two extremes. To most minds, a reserve cell is morphologically and physiologically "undifferentiated," a simplified embryonic cell type lacking specialized molecules, functionless until called on to regrow a new part. It is said to be *totipotent*, that is, capable of differentiating into any cell type in the body or, at the very least, *pluripotent*—capable of giving rise to more than one cell type.

On the other hand, a differentiated cell possesses specialized structures, molecules, and functions that enable it to be distinguished from cells that have taken some other pathway of specialization. A cartilage cell differs from a muscle cell by structural modifications, by the specialized molecules it produces, and by the role it performs in the body. These identifying characteristics are often spoken of as "overt" signs of differentiation because embryologists have observed, in culture, seemingly undifferentiated cells (by morphological and biochemical criteria) that behave true to their original form when allowed to redifferentiate (Grobstein, 1959). The fact that such cells seem to know who they are even in the absence of overt products of differentiation has been attributed to a heritable *epigenetic* state of differentiation of the nucleus that is remarkably stable; that the *epigenotype* is a reversible genetic regulation is assumed because differentiated intestinal nuclei transplanted to oocyte cytoplasm have been shown to support development (see Ursprung, 1968). The concept referred to by the term "epigenotype" has in the past had such names as "determination" and "covert differentiation" (Grobstein, 1959). More recently, seemingly undifferentiated endoderm "determined" to become pancreas when supplied with an appropriate "inductor" has been shown, in fact, already to contain amylase and other specialized molecules typical of fully differentiated pancreatic acinar cells. Such cells, which possess minute quantities of differentiated products detectable only by the most sensitive biochemical methods, are said to be "protodifferentiated" (Rutter et al., 1968). For all practical purposes, a protodifferentiated cell corresponds to a covertly differentiated cell, and the epigenetic state of differentiation might also be protodifferentiation.

Whether or not differentiated cells, be they overt or covert in their specialization, and the regeneration cells to which they give rise, are totipotent or pluripotent under conditions other than transplantation of their nuclei to an ovum is an unsolved question that has been discussed in a number of recent reviews (Hay, 1966, 1970; Thornton, 1968; Ursprung, 1968). Until better experimental evidence is forthcoming, it no longer is useful, however, to think of differentiation as an irreversible condition, even though we know that somatic cell differentiation is very stable. That is, we should not make the reversibility issue part of the definition of cytodifferentiation but keep it in a separate category as a question having more to do with the cell's capacity for metaplasia. This is difficult to do because our concepts so overlap that we almost invariably think of differentiation as a restriction of developmental potency, even though we know that cells like the ovum that are truly totipotent actually are differentiated, highly specialized cells.

The quesion of developmental potency has clouded the usage of the term "dedifferentiation." On the one hand, there is no doubt at all that fully differentiated cells can give rise in culture and in regeneration to morphologically and biochemically unspecialized cells that are perhaps best referred to as "relatively un-differentiated cells" until we use microtechniques to look for specialized molecules. The term "dedifferentiation" will be used here to refer to such losses, however great or little, of differentiated products. In the literature, on the other hand, one constantly sees this term in quotation marks when it is used in this context; the implication is that the true meaning of the term "dedifferentiation" should be "the regaining of supposedly lost developmental potencies by the cells." Since no one has proved that developmental potency is lost during differentiation and we know that somatic nuclei can be totipotent when transplanted to ova, it seems pointless to insist that this "potency" definition be fulfilled before releasing the term "dedifferentiation" from quotation marks. The reader should understand, in any case, that we are not referring here either to "loss" or to "gain" in developmental potency when we use the terms "differentiation" and "dedifferentiation" and that we use these terms only in a relative sense.

In order that we may proceed to discuss the phenomenon of regeneration, we must define still another cell category, the *progenitor cell*. The progenitor or "stem" cell is a cell that divides to give rise to a progeny or "stem" of new cells. In the older embryo and in renewing adult tissues, the progenitor cell can be said to be "partially differentiated," for want of a better term. It contains specialized products that are not as abundant as in the final cell of the line. The important thing about the progenitor cell is that it is actively growing and giving rise to progeny; its state of differentiation is irrelevant to the definition and can only be implied by such awkward modifying expressions as "partially differentiated" and "relatively undifferentiated."

Physiological and Reparative Regeneration in Renewing Tissues

The first generalization about the cellular basis of regeneration that we would like to establish is that the normal progenitor cell in the vertebrate is a partially differentiated cell that divides in regular cycles under the careful control of humoral or local "feedback" factors. The blood and lymphoid organs of mammals provide a model system for discussion of this question.

There are three principal renewing cell lines in the mammalian bone marrow: the erythropoietic or red blood cell line, which is under the control of circulating factors (hormones) such as erythropoietin (Goss, 1964; Marks and Kovach, 1966): the granulocytic or myeloid series; and the lymphoid series. It was once believed that the reticular cell, a rather undifferentiated-appearing cell associated with the supporting reticular fibers and sinusoids of the marrow, was the progenitor or "stem" cell for all these lines. The reticular cell has now been shown not to be dividing under normal (physiological) conditions. It is the young blood cells of the erythropoietic series (proerythroblasts, erythroblasts), granulocytic series (promyelocytes, myelocytes), and lymphoid series (large lymphocytes) that incorporate tritiated thymidine and are the progenitor cells. Autoradiographic studies show that these partially differentiated progenitor cells, not the seemingly undifferentiated reticular cells, normally perpetuate the marrow population (Cronkite and others, see Hay, 1966). Red and white blood cells that have almost completed their maturation (normoblasts, metamyelocytes) and small lymphocytes do not synthesize DNA, nor do the fully differentiated erythrocytes and granulocytes and the lymphocytes that are released into the blood stream. The withdrawal from the mitotic cycle by erythroblasts that have begun to synthesize overt hemoglobin occurs 1 to 2 cell divisions later; the regularity of withdrawal of such cells has given rise to the misleading impression that progeny with a phenotype different from that of the parent cell can arise only after a terminal or "quantal" mitosis (Holtzer, in Ursprung, 1968). In fact, environmental and humoral factors control and can even decrease the number of divisions a cell such as a myoblast undergoes after beginning to differentiate overtly (O'Neill and Stockdale, 1972; Chapter 9, this volume).

The old idea of a multipotential stem cell for the renewal of lymphocytes, granulocytes, and erythrocytes is incompatible not only with autoradiographic studies of normal marrow but also with recent work demonstrating that the large lymphocyte once thought to represent an omnipresent circulating "stem" cell is, in fact, a thoroughly committed cell in the adult. Perhaps the best example of the postulated epigenetic state of differentiation, the large circulating lymphocyte, has but one known fate, and that is to respond to foreign antigen either by transforming into plasma cells or by becoming a sensitized tissue lymphocyte (Humphrey and White, 1970). The result of a remarkable succession of embryonic home organs (yolk sac, thymus, spleen, bone marrow), the lymphocyte does not transform into myeloblasts or erythroblasts (Weiss, 1972). The question of whether or not a "potential" stem cell common to the erythroid and myeloid lines exists that is not normally proliferating and so not revealed by the autoradiographic studies is unresolved. Marrow cells containing radiation-induced marker chromosomes injected into unmarked anemic mice colonize the spleen with clones that contain both erythrocytes and granulocytes. It is believed that each clone comes from one cell, because in mitotic cells where the chromosome marker can be identified, the same one is present in 90 percent of the cells; the likelihood of two cells containing the same marker populating the same clone seems remote (Wu et al., 1967). The experiment may, like the nuclear transplant studies, show that under unusual conditions either a differentiated cell or an undifferentiated somatic cell (if one exists) can transform into another cell or more than one cell type. The ongoing rule in the normal adult organism, however, seems to be that for every differentiated, renewing cell population there is a specific precursor cell that is already partially differentiated.

The cells of the adult body can be divided into three main groups, depending on the degree to which they normally are growing or turning over.

1. In the renewing (labile) group, there is always a well-defined progenitor cell cycle and the end-product cell is terminally differentiated and often dispensable, as in the case of the blood cells discussed above. The surface epithelia comprise the other class of renewing cells in the adult mammal. The progenitor cell in stratified epithelia is a basal cell characterized by such differentiated products as tonofilaments and desmosomes; as the progeny move out to the surface they become flattened and more keratinized. The turnover time of the mammalian epidermis is 1 to 2 weeks. Local factors are more important than humoral factors in controlling its growth (Goss, 1964). The progenitor cell in the gastrointestinal tract is a partially differentiated cell in the necks of the glands (stomach) or crypts (intestine). The intestinal lining turns over every 2 to 3 days. Reparative regeneration of renewing tissues probably involves the same progenitor cells. The reader may be interested to know that, after alcohol or aspirin doses sufficient to cause the stomach surface epithelium to slough, it takes 2 to 3 days for cells to migrate from the protected progenitor site in the gland necks to cover the exposed connective tissue surface (Hingson and Ito, 1971).

2. The stable (expanding) group of cells, as classified by Leblond (Goss, 1964), shows some incorporation of tritiated thymidine in young adults. These are the connective tissue and epithelial secretory cells (fibroblasts, chondroblasts, osteoblasts; epithelia of the liver, kidney, pancreas, and other glands). The end product cell rarely dies and the cell population can increase in number (hyperplasia) under stress.

3. The static (permanent) group of cells shows no cell division in the normal adult. These are the nerve cells, which never divide after a certain point in their embryonic history, and the muscle cells. They are capable of hypertrophy after stress (increased production of cytoplasm), but they never (nerve) or rarely (muscle) exhibit hyperplasia (proliferation that increases cell numbers in response to physiological conditions).

Having established that regeneration in a renewing tissue is generally accomplished, not by undifferentiated reserve cells, but by partially differentiated and very specific progenitor cells, let us inquire as to the origin of regeneration cells in stable and static tissues that are subjected to injury or stress.

Hyperplasia and Reparative Regeneration in Stable Tissues

The generalization that this section will support is that regeneration cells in stable tissues, which lack progenitor cells, always arise directly from differentiated cells rather than from reserve cells. Some degree of dedifferentiation, as defined earlier in this chapter, is often but not always involved.

The liver is one of the best-studied examples of this class of tissues. After excision of three quarters of the rat liver, 30 percent of the remaining liver cells have begun to synthesize DNA at 20 hours, and at 26 hours the mitotic index reaches the peak of 3.6 percent (Bucher, 1967). There is no massive dedifferentiation. Such dedifferentiation as occurs is limited to the first day and seems to consist merely of vesiculation of the endoplasmic reticulum accompanied by an increase in free ribosomes, RNA, and protein synthesis. Three days after injury, only 3 percent of the cells are still synthesizing DNA. Proliferation is under the control of humoral factors that restrict the total liver mass that can be created (Bucher, 1967). Many of the other epithelial glands are probably under similar hormonal control. For example, one kidney will show hyperplasia if the other is removed (Goss, 1964).

All the connective tissues, bone, cartilage, and fibrous connective tissue are capable of reparative regeneration after injury and hyperplasia in response to local or humoral stimulants. In a bone fracture, few if any of the injured osteocytes are seemingly released from the encompassing rigid matrix. The main source of regenerating cells is the supporting bone sheath. These cells proliferate and give rise to a cartilaginous callus that is subsequently replaced by new bone. Dedifferentiation of cartilage has been documented by direct observation in at least one vertebrate (Hay, in Rudnick, 1962). In skin wounds, fibrocytes proliferate and transform into fibroblasts with abundant granular endoplasmic reticulum (Ross, in Dunphy and Van Winkle, 1969). They do not arise, as was once believed, from totipotent blood cells (Fischman and Hay, 1962; Ross, 1968).

Hypertrophy and Regeneration in Static Tissues

The static tissues show no capacity at all for cell proliferation (neurons) and must either dedifferentiate drastically or call on a hypothetical reserve cell in order to regenerate (skeletal muscle). We shall discuss the latter point in some detail in this section. It should be made abundantly clear at the beginning that smooth and cardiac muscle cells are mononucleated and do retain the capacity to synthesize DNA and divide after their differentiation has begun (Hay, 1966; Weinstein and Hay, 1970). In skeletal muscle, the mononucleated myoblasts stop making DNA in order to fuse into a syncytium that then differentiates into a muscle fiber (Fig. 1). Other syncytia, such as osteoclasts, also exhibit no DNA synthesis (Fischman and Hay, 1962). In the static tissue—nerve—true syncytia are not formed as was once believed, but the mononucleated cells link up via synapses into an enormously complex "functional syncytium." Both skeletal muscle fibers and neurons retain the ability to make extra cytoplasm. Nerves are constantly renewing their terminal ends and exercised muscle is capable of enormous hypertrophy. Most injured nerve fibers eventually regenerate if the cell lives.

The current controversy as to whether or not fully formed skeletal muscle retains a reserve cell for use in reparative regeneration began in 1961 when Mauro illustrated a small, inactive-looking mononucleated cell with a dense nucleus under the basement membrane of three frog muscle fibers. He called the mononucleated cell, a *satellite cell* (Mauro et al., 1970). Since then, the term has been badly misused to refer both to myoblasts (Fig. 1) and to regeneration cells (Reznick, in Mauro et al., 1970). Reznick is of the opinion that the 'satellite' cells that appear in the vicinity of an injured mammalian muscle fiber arise *de novo* by the formation of new cell membranes in the juxtanuclear cytoplasm. A small segment of cytoplasm, together with a nucleus, then leaves the fiber as a mononucleated cell (Fig. 2). Cell turnover studies support his hypothesis, but a statistical survey of the mouse muscle that Rez-

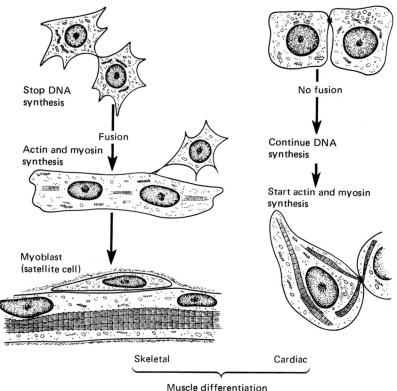

Stop DNA
synthesis

Fusion

Actin and myosin
synthesis

Myoblast
(satellite cell)

No fusion

Continue DNA
synthesis

Start actin and myosin
synthesis

Skeletal   Cardiac

Muscle differentiation

FIGURE I. *The term "satellite cell" has had various usages. It has been used, as shown here, to refer to mononucleated muscle progenitor cells (myoblasts) in embryos. It is necessary for growing skeletal muscle to possess such mononucleated myoblasts because the nuclei, once they fuse into a syncytial fiber, can no longer synthesize DNA. In cardiac muscle, mononucleated differentiated cells do divide, and so undifferentiated-appearing mononucleated cells (satellite cells) do not occur. The term "satellite cell" has been used not only to refer to skeletal muscle myoblasts but also to refer to regeneration cells (Fig. 2). Still a third usage is embodied in the concept that nongrowing adult muscle retains mononucleated reserve cells to use for regeneration. The latter concept is unlikely to be true, at least in newts (Table 1).*

nick used needs to be done before injury, in order to rule out absolutely the participation of preexisting satellite cells.

Most, if not all, of the reports of satellite cells in nonregenerating muscle refer to embryonic or young adult muscle that is still growing and can be expected to contain mononucleated myoblasts (Ishikawa, Venable, in Mauro et al., 1970; Flood, Carlson and Rogers, in Carlson, 1972a). Such mononucleated cells have been shown to label with tritiated thymidine and divide; their progeny may fuse with the fiber or divide again. If the regeneration of such a muscle were studied, it would be difficult to

rule out the mononucleated myoblasts, or satellite cells, as the source of the regeneration cells. A careful study at the electron microscope level remains to be done of regeneration in an old mammal, which might lack the satellite cells that correspond to myoblasts. Most of the statistical studies of satellite cells, moreover, have been done at the light microscope level, where it is difficult to distinguish satellite cells from pericytes; the latter lie close to, but outside, the basement membrane.

Unlike the rodent most often used in the studies referred to above, the salamander quickly reaches its adult form and does not

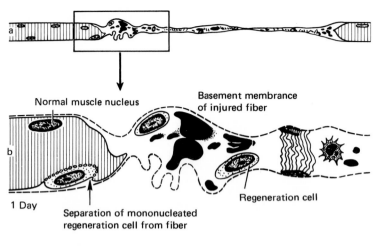

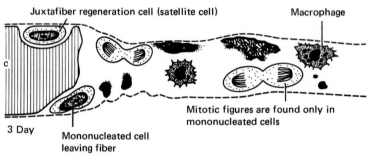

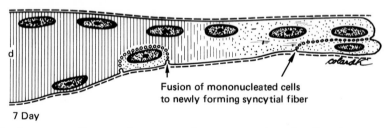

FIGURE 2. *During mammalian muscle regeneration, mononucleated cells arise from the muscle fiber. Small vesicles appear which then seem to fuse to separate the regeneration cell from the myofiber, as depicted here. These mononucleated cells then proliferate and later fuse, as in embryogenesis (Fig. 1), to produce new muscle fibers. Although Reznick called the regeneration cells satellite cells, he has published electron micrographs to support the view that they arise* de novo *from the fiber during regeneration. (Modified from E. D. Hay,* New Eng. J. Med. *284, 1033, 1971.)*

continue to grow in size for so many years. Like the mammalian muscle, its myofibers give rise to mononucleated cells during regeneration (Hay, 1966). A preliminary electron microscope study by Hay and Doyle (1973) revealed a few curious pericytes seemingly enclosed in the basement membrane of the muscle fiber but separated from the muscle plasmalemma by basement membrane material. In 200 nuclei within muscle fibers (that is, beneath the basement membrane), one qualified as satellite cell in the sense that it lay under the muscle basement membrane immediately next to the muscle plasmalemma (Table 1). Examination of serial sections of the area revealed overt evidence of muscle injury and myofibril regeneration. The cell itself resembled a myoblast in that its nucleus was more euchromatic and its muscle-type polysomes more abundant than in Muir's satellite cell. True satellite cells in Muir's sense were not seen (resting cells of no apparent function). The electron microscope reveals stages in the formation of mononucleated cells from muscle during regeneration of the amputated newt limb that are similar to those observed in regenerating mammalian muscle (Figures 2, 6, 7, 8). A far greater number of mononucleated regeneration cells arise from the cut muscle fibers in the amputated limb than can be accounted for by preexisting satellite cells (for further discussion, see the footnote to Table I).

The only good example in a vertebrate of a reserve cell, that is, an undifferentiated-appearing cell set aside with no apparent function other than to replenish an injured organ, is the testis, which normally has a clearly defined dividing progenitor cell, yet can call on a relatively inactive type of spermatogonium in case of dire need (Dym and Clermont, 1970). The germ cell is a very specific kind of cell, however. It is the only cell capable of undergoing the kind of differentiation that leads to the full expression of "totipotency." Both spermatogonia and oogonia (which never proliferate in the adult vertebrate) are set aside prior to gastrulation (as early as the first cleavage in some animals).

For the somatic cell type, it seems fair to conclude that the weight of the evidence is against the idea of the retention, in the adult, of a totipotent reserve cell for regeneration. Only in the case of skeletal muscle can the theory be seriously entertained, and here the evidence is that mononucleated cells can occur within the muscle fiber basement membrane in embryos, young adults, and diseased and regenerating animals. These cells are not thought of as totipotent but as muscle cells that have not yet fully differentiated. The question of reten-

TABLE 1

CLASSIFICATION OF CELLS ASSOCIATED WITH MYOFIBERS IN ADULT NEWTS[a]

| | Unamputated Limb | | | | Regenerating Limb | | |
Animal	Muscle	Mono-nucleated	Total	Age	Muscle	Mono-nucleated	Total
1	33	0	33	11 days	54	16	70
2	55	0	55	14 days	24	21	45
3	23	0	23	17 days	50	9	59
4	16	1	17	17 days	21	7	28
5	10	0	10				
6	96	0	96				
	233	1 (0.5%)	234		149	53 (26%)	202

[a]The term "satellite cell" does not appear in this table because its meaning has been confused by various usages. The "mononucleated cell" tabulated here is a cell lacking myofibrils that is located under the basement membrane and separated from the myofiber plasmalemma by a narrow (100 to 200 Å) space. A "muscle" nucleus lies within the myofiber cytoplasm. The identification was made with the electron microscope, and comparable areas were studied in amputated and unamputated limbs (Hay and Doyle, 1973). The single mononucleated cell identified in the nonamputated adult was identical in appearance to the mononucleated cells in regenerating muscle; it probably arose as the result of injury to the myofiber because other evidence of regeneration was found in the vicinity. Even if this mononucleated cell in the nonamputated adult were counted as a reserve cell, one cell could not give rise to 50 by day 11 since there is essentially no cell division among nuclei associated with myofibers (Chalkley, in Hay, 1966). The mononucleated cells in the regenerating muscle fibers must have arisen from differentiated muscle nuclei.

tion of such cells in healthy adult muscle has not been solved for mammals. Truly adult mammalian muscle known to be lacking in disease and with no history of injury has yet to be surveyed carefully with the electron microscope. The young rat, mouse, and bat muscles that have been studied so extensively are growing tissues containing a progenitor cell, the mononucleated myoblast, which is dividing and in no sense is a reserve cell. As we have seen here, there is good evidence to believe that during regeneration in the adult newt, the myofiber itself gives rise to the mononucleated cells that repair the fiber.

Regeneration of Amphibian Limb

The amphibian possesses, among vertebrates at least, truly remarkable regenerative powers. One group can regenerate a lens from the dorsal iris, the only clearcut case of transformation from one differentiated state (pigment cell) to a completely different one (lens fiber). The tissue, however, remains epithelial so that in a sense the transformation is not that much more dramatic than the modulation of fibroblast to chondroblast or osteoblast. We shall examine here only limb regeneration, and for other examples of organ regeneration in amphibians, refer the reader to the review articles we listed at the beginning of this chapter. In particular, we want to see how the principles of regeneration cell origin derived above for tissue regeneration apply to regeneration of a whole organ system. Does the whole limb follow the same rules as do the individual tissues, even though it forms a much larger mass of regeneration cells?

We can begin with the epidermis of the amputated limb. We classified the epidermis of amphibian and terrestial vertebrates as a continuously renewing cell system with a clearly defined progenitor cell cycle. Within a few hours after the amputation of the limb of the adult newt *Notophthalmus viridescens*, a wound epithelium that is not synthesizing DNA has covered the cut surface, just as in the case of mammalian wounds. The epithelial cells pile up and become phagocytic (Singer

and Salpeter, in Hay, 1966), just as in mammalian wounds (Ross, in Dunphy and Van Winkle, 1969). The source of the piled-up epithelium or apical cap is the proximal limb epidermis (Figs. 3 and 4), which increases the number of partially differentiated progenitor cells that are in the S phase of the progenitor cycle. Cells continue to move from proximal epithelium to the wound apical cap throughout the period of early regeneration (Hay, 1966). Thus, in its origin and apparent function in cleaning up debris, the amphibian wound epidermis is not unlike that in the mammal and it may even secrete collagenase as well (Grillo, in Dunphy and Van Winkle, 1969). The wound epithelium may also direct the accumulation of underlying mesenchymatous cells (Thornton, 1968). The other epithelium in the limb, the endothelium, also grows in from previously existing epithelial cells.

The stable cell group is represented in the amphibian limb by fibrocytes, osteocytes, and chondrocytes. As in mammalian wounds, collagenase is released, which seemingly helps to remove the preexisting matrix from around these cells. As the matrix disappears, the released cells take on the characteristics of growing mesenchymatous cells (Fig. 5). Interestingly, hyaluronic acid is produced in greater than normal amounts during the period of formation of regeneration cells and may provide an environment conducive to the growth and migration of the regeneration cells derived from the connective tissues and muscle of the limb stump (Toole and Gross, 1971). The most impressive aspect of the dedifferentiation of the connective tissue of the stump of adult newts is the replacement of the old "differentiated" matrix by a more "embryonic" matrix, which is probably produced by the mesenchymatous blastema cells; such cells are rich in secretory organelles (Fig. 5).

The morphological changes in the injured limb muscle are similar to, but more extensive than, those in injured mammalian muscles. The process of replacement of the syncytial muscle fibers by mononucleated regeneration cells proceeds proximally from the amputation site a considerable distance into the limb stump (Figs. 3 and 4). New cell membranes can be observed to form across the dedifferentiating muscle fibers (Fig. 6). The dedifferentiation of

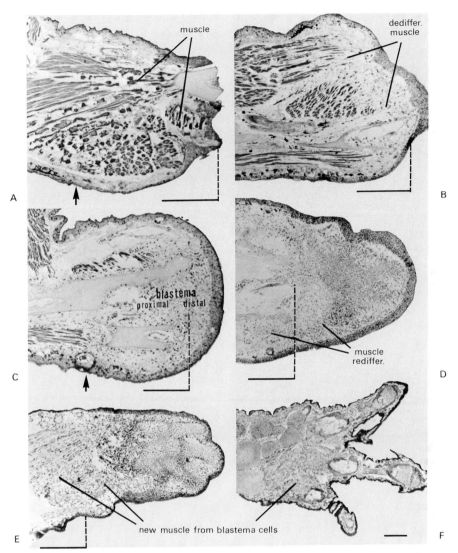

FIGURE 3. *Series of light micrographs depicting muscle dedifferentiation and redifferentiation in the limb of the newt* (Notophthalamus viridescens) *following amputation through the radius and ulna. The dotted line indicates the approximate level of amputation. The black bar (0.5 mm long) calls attention to the same region of the limb stump in parts A to E. The plane of the section in A shows to good advantage the large amount of muscle present in the forearm at the at the time of amputation. The process of muscle dedifferentiation will extend proximally to the level indiacted by the arrow (A). By 18 days (B), the distal 0.5 mm of the stump contains blastema cells derived from dedifferentiating (dediffer.) muscle. By 21 days (C), the level of dedifferentiation extends to the arrow (compare C and A). The blastema derives from various formed tissues of the old stump (proximal blastema, C). The proliferating regeneration cells form an outgrowth, the distal blastema (C). In the most proximal part of the blastema (arrow, C), dedifferentiation of muscle is incomplete and these fibers redifferentiate by reorganization of their internal structure. In the blastema proper, cartilage differentiates first, then muscle (muscle rediffer., D). The sections at E and F show the large amount of new muscle that forms from blastema cells during the later periods of limb redifferentiation.*

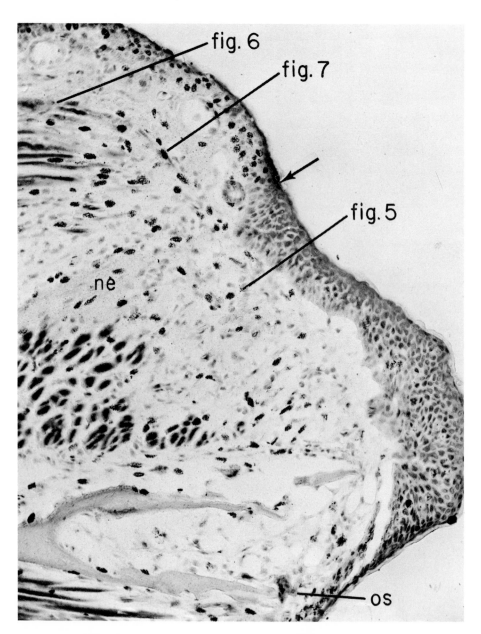

FIGURE 4. *Autoradiograph of a section of the regenerating limb shown at B in Fig. 3. Tritiated thymidine was administered to the newt intraperitoneally shortly before fixation. The blastema cell nuclei are radioactive, indicating that the cells are preparing to divide (area labeled "fig. 5"). Mononucleated cells released from muscle are also synthesizing DNA (areas labeled "fig. 6" and "fig. 7"). The apical epithelium is not significantly radioactive, but epidermis proximal to the arrow is actively growing. An osteoclast (os) can be seen resorbing the bone. × 150. (From E. D. Hay and D. A. Fischman, Develop. Biol. 3, 26, 1961.)*

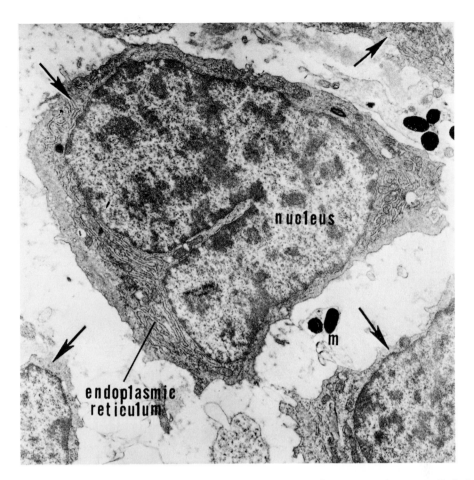

FIGURE 5. *Electron micrograph of blastema cells (arrows) in the regenerating newt limb from the area labeled "fig. 5" in Fig. 4. Regeneration cells in adult newt blastemas have considerable endoplasmic reticulum and are probably producing extracellular matrix at the same time that they are dividing. Part of a melanocyte process (m) can be seen here. × 10,000.*

skeletal muscle consists of at least two different processes in urodele limbs. On the one hand, myofibrils are partially lost from fibers that shrink in size (Fig. 6) but do not break up entirely (Fig. 3B). On the other hand, mononucleated cells that synthesize DNA (Fig. 7) arise from the muscle and come to resemble the fibroblast-like mesenchymatous cells derived from the connective tissues. They acquire the endoplasmic reticulum typical of newt regeneration cells, often before even leaving the muscle fiber (Fig. 8). The evidence that these cells arise *de novo* rather than from preexisting satellite (reserve) cells has been considered in a previous section of this chapter.

The remaining limb tissue, the nerve, follows the same rules as does its mammalian counterpart. Axons from the cut ends of the nerves grow into the wound area (Singer, 1960, 1968). The bud of embryonic-appearing tissue thus formed is called a *blastema*, and the regeneration cells derived from muscle and connective tissue are called *blastema cells*. Blastema cells resemble the mesenchymal cells from which the connective and muscular tissues were derived. They have large vesicular nuclei and prominent nucleoli, reflecting active RNA, DNA, and protein synthesis, a change which in the adult at least (Dresden, 1969; Lebowitz and Singer, 1970) requires the trophic presence of nerves and an appropriate hormone balance. In the adult newt, the mesenchymatous blastema cells (upper right, Fig. 9) grow slower and have more endoplasmic reticulum than those of the larva (upper left, Fig. 9).

The cellular origin of the amphibian blastema, then, is essentially the same as that which would be expected in a mammal. With the exception of cells from mesenchymal tissues, which come to look remarkably alike whether derived from muscle or connective tissue, and which can apparently substitute for each other (Carlson, 1972a), no dramatic cell transformations are observed, and there is no reason to believe that the cells are any more magically totipotent than those of the mammal (for a more complete discussion of this point, see Hay, 1966, 1970; Thornton, 1968; Ursprung,

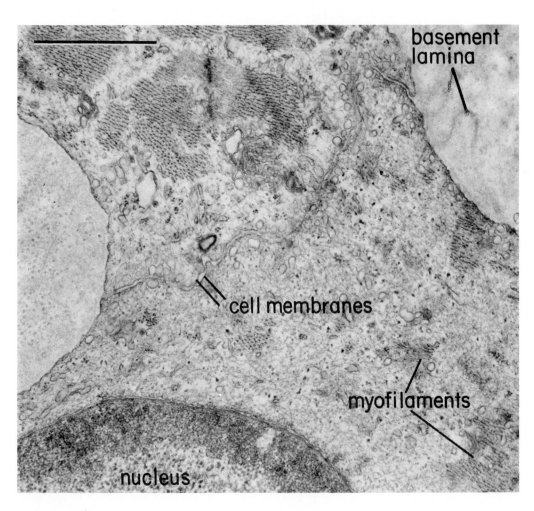

FIGURE 6. *Electron micrograph of part of a dedifferentiating muscle fiber from a regenerating newt limb fixed 17 days after amputation. In areas where the muscle fibers are dissociating, the basement lamina appears highly irregular in outline, as shown at the left and right in this figure. Basement laminas in areas of the proximal blastema where the muscle completely dissociates (Fig. 3, D) are resorbed. The micrograph shows to good advantage two newly formed cell membranes which separate the muscle unit illustrated here into separate fragments. In the cytoplasm of the dedifferentiating fiber, myofibrils are breaking up into myofilaments. The bar equals 1 μm.*

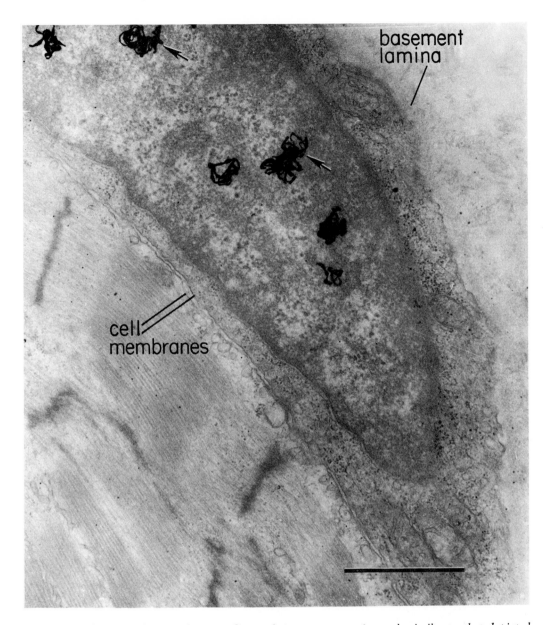

FIGURE 7. *Electron micrograph–autoradiograph from an area of muscle similar to that depicted in the area labeled "fig. 7" in Fig. 4. The newt was injected with tritiated thymidine 1 hour before fixation. Silver grains (arrows) over nuclei indicate DNA synthesis. Where labeled nuclei appear within the myofibers (Fig. 4), they are seen by electron microscopy to be in mononucleated cells demarcated from the fibrillar part of the fiber by cell membranes, as shown here. These labeled cells in muscle are located under the basement lamina and therefore clearly are derived from the muscle. Many will become part of the blastema. The bar equals 1 μm.*

1968; Goss and Hay, 1970). Why then does the urodele amphibian regenerate a limb when even his close relative the frog does not? A point of some interest is that the urodele cells, which regenerate so remarkably well, have very large nuclei and enormous amounts of

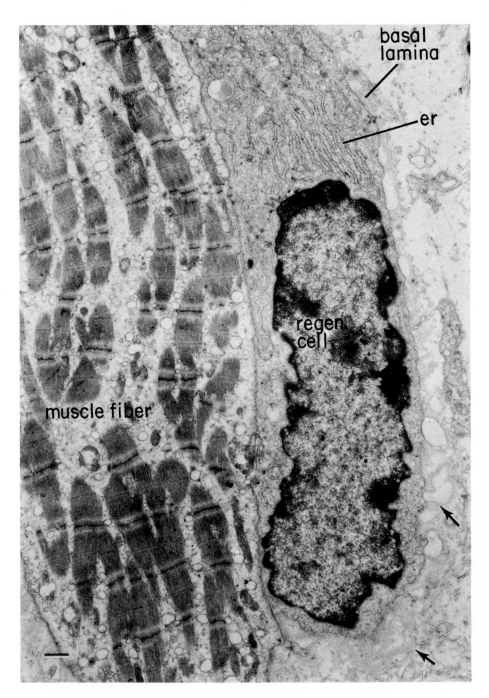

FIGURE 8. *Electron micrograph of a mononucleated cell under the basal lamina (arrows) of a dedifferentiating muscle fiber in an 18-day regenerating newt limb. Were this an autoradiograph, the regeneration cell (regen. cell) would probably be labeled with tritiated thymidine. The picture shows that the endoplasmic reticulum (er) characteristic of blastema cells has appeared in this mononucleated cell prior to its complete separation from the muscle fiber of origin. The bar equals 1 μm.*

redundant DNA, as compared to the cells of fishes, frogs and mammals (Rees and Jones, 1972). There are, however, a number of differences in tissue organization that are likely to be more important than this cytological difference. The number of nerves per unit limb area is critical and is far greater in salamanders than in frogs and mice (Singer, 1968). The connective tissue is relatively less conspicuous in the fast regenerating aqueous urodeles than in the slower terrestial forms. Clearly, the terrestial animals have perfected a process of scar

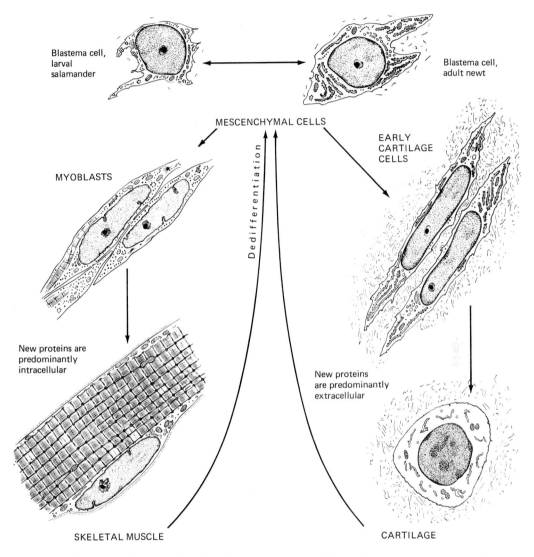

FIGURE 9. *Summary diagram showing the cytology of differentiating muscle and cartilage in the regenerating salamander limb. Following amputation of the limb, formed muscle and cartilage dedifferentiate to give rise to mesenchymatous blastema cells which proliferate and redifferentiate. Blastema cells in the adult newt (upper right) have considerable endoplasmic reticulum and probably are making hyaluronate. In the larval salamander, blastema cells have little endoplasmic reticulum (upper left). It is not known whether or not cells derived from muscle fibers can redifferentiate into cartilage, or vice versa. (From E. D. Hay, "Regeneration," Ronald Press, New York, 1962, p. 177.)*

formation that is to their advantage and, to paraphrase Goss (1969), not something they are likely to want to get rid of in terms of the value to survival on land. Other reactions of advantage to the survival of the warm-blooded mammal are those of the immune and inflammatory systems. Inflammation is disastrous, however, to the newly regenerated minced muscles of the rat leg, even when the minced muscle is an autotransplant and has already regenerated well (Carlson, 1972b). The dramatic regeneration of opossum limbs reported recently by Mizell occurs long before the connective tissues are fully formed or the immune sytsem operative; even then, nerves must be added to booster the growth potential of the embryonic cells (Mizell, in Goss and Hay, 1970). Attempts to induce limb regeneration in adult terrestial vertebrates have produced partial outgrowths unlikely to be of much benefit to the animal (Polezhaev, 1946, 1972; Singer, 1960; Schotté, 1961; Rose, 1964, 1970; Becker and Spadaro, 1972). It seems unrealistic to think that the study of amphibian regeneration is going to make it feasible to replace human limbs with useful regenerates in the near future.*

What the regenerating organ systems in amphibians do provide us with is a remarkable

*The data of Becker and Spadaro (1972), for example, do not support the conclusion that electrical stimulation induces either limb or tissue regeneration. The fact that their evidence was accepted as an advance toward discovering alternatives to prosthetic implantation in man (Becker, 1972) reflects the regrettable assumption in medical circles (Hay, 1971) that mammalian tissues such as muscle cannot regenerate. Figure 5 (Becker and Spadaro, 1972) shows a regenerating rat muscle fiber "stimulated" electrically, but no control is provided; Figure 4 does not show blastema formation or adequate controls. Normal rat muscle without electrical stimulation shows more evidence of regeneration after injury than that illustrated in Figures 4 and 5, as does human muscle (Mauro et al., 1970). Minced rat muscles can dissociate into mononucleated cells that proliferate and redifferentiate into muscles aligned with the limb axis (Carlson, 1972b). The capacity of mammalian bone and other connective tissues to regenerate is also well established (Dunphy and Van Winkle, 1969). What the illustrations by Becker and Spadaro (1972) seem to show is the normal (limited) regenerative recovery of mammalian tissues.

opportunity to study the control of cell proliferation and especially, of cell redifferentiation. The mesenchymatous cells released from the cartilage, muscle, and fibrous connective tissues of the inner limb tissues begin to redifferentiate as soon as an outgrowth can be detected on the end of the amputated limb (Fig. 3). Following cues that in large part derive from the overlying skin and a poorly understood longitudinal axis of polarity (Rose, 1964, 1970), the cells proceed to recreate the same general pattern of limb differentiation that occurred (without nerves or hormones) in the embryo. It has been suggested that even cancer cells can be persuaded to become benign in the differentiating environment of the limb (Rose, 1970 and others; but see Ruben et al., 1966). It is to be hoped that future investigators will be more interested in this phase of limb regeneration than past investigators, who have studied primarily the dedifferentiative phase.

Regeneration in an Invertebrate

Our chapter is already too long to begin a serious consideration here of the various regenerative phenomena among the invertebrates. Some of these are reviewed by Berrill (1961), Lender (1962), Tardent (1963), Herlant-Meewis (1964), Hay (1966), and Goss (1968). It would be incomplete, however, to conclude a chapter that attempts to discredit the reserve cell hypothesis of regeneration without some mention of the planarian. The "neoblasts" credited to the flatworm can be considered as the historical prototype for the reserve cell hypothesis. As originally described by Wolff and Lender with the light microscope, the neoblast is said to be a large, intensely basophilic cell with an "embryonic"-looking nucleus characterized by a prominent nucleolus (Wolff, in Rudnick, 1962; Lender, 1962). We now know that all cells making ribosomes have a prominent nucleolus, which produces the ribosomal RNA, and a euchromatic nucleus; these features are just as typical of nongrowing gland cells as of growing embryonic cells. The light micrographs of "neoblasts" published by

Wolff, Lender, and their associates reveal intense basophilia of the type characteristic of the granular endoplasmic reticulum of plasma cells and gland cells. Such photographs often show a clear, nonbasophilic juxtanuclear area that we believe is the Golgi apparatus (Fig. 10). Electron microscopy reveals secretory granules in such cells (X, Figs. 10 and 11). We conclude that many of the "neoblasts" described by light microscopists are, in fact, differentiated gland cells (Hay, 1968).

The electron microscope has revealed, however, that a remarkably undifferentiated appearing cell is associated with these basophilic gland cells in the parenchyma of the adult worm. These cells, smaller than the average gland cells counted as "neoblasts" by light microscopists, have free ribosomes, mitochondria, and very large chromatoid bodies in their cytoplasm, and an unusual clumped nuclear chromatin pattern reminiscent of an early prophase figure (beta cell, Figs. 10 and 11). The chromatoid bodies (arrows, Figs. 10 and 11) are so called because they have a density reminiscent of nuclear chromatin. Whether they have the same chemical makeup as the chromatoid bodies or polar granules of differentiating gametes (Fawcett et al., 1970) is as yet unknown. We have arbitrarily called these undifferentiated cells containing chromatoid bodies *beta cells* (category B in our classification) so as not to give them a name that would imply that we knew their function.

While we noted in our earlier work the presence of these relatively undifferentiated cells, we discounted their numbers (Hay, 1966, 1968) until we undertook a statistical survey of the planarian cell population at the electron microscope level (in collaboration with S. J. Coward). Puzzled as to their function and unconvinced of the utility of laying away so many cells in reserve for an occasional head amputation, we first established that beta cells are not being used for asexual reproduction. They are as common, in fact, in sexual animals as in asexual animals. Recalling the enormous numbers of parenchymal gland cells that produce the surface slime and disagreeable flavor of this defenseless creature, we reasoned that the beta cell might be the progenitor for physiological turnover of the gland cell. The clumps of "neoblasts" illustrated by light microscopists (inset, Fig. 10) appear in the electron microscope to be nests ("clones") of basophilic gland cells and smaller beta cells that can become gland cells under appropriate stimuli (Coward and Hay, 1972).

It seems likely that during regeneration in turbellaria, both the progenitor cells and the more differentiated cells participate. During the formation of the regeneration blastema following amputation of the head, dedifferentiation of muscle (Figs. 12 and 13) and other formed types can be observed. The dedifferentiating muscle cells acquire chromatoid bodies. Beta cells and gland cells can also be observed to enter the blastema. The same rules seem to be followed as in vertebrata, since the most marked dedifferentiation is observed in tissues like muscle that normally have no progenitor cells. Regeneration cells of the flatworm blastema (Fig. 14) are similar to blastema cells of the amputated amphibian limb, except for the presence of small chromatoid bodies (Hay, 1968; Sauzin, 1968; Pedersen, 1972; Coward and Hay, 1972).

What are we to say of the developmental potency of the regeneration cells? Just because the regeneration blastema of the amputated flatworm has the capacity to reform a head, we cannot conclude that its component cells are totipotent (Lender, 1962; Pederson, 1972), any more than we could say that the cells of a regeneration blastema in the amphibian are totipotent just because they look alike and together regrow a limb (Hay, 1968; Thornton, 1968; Steen, in Mauro et al., 1970; Namenwirth, 1968). Each cell might still retain the obligation to redifferentiate into the same kind of cell type from which it originated. That is, it might carry over the so-called epigenetic state of differentiation we mentioned in the introduction to this chapter. Even the undifferentiated appearing cells of insect imaginal discs breed remarkably true in culture, rarely transforming (transdetermination) into a cell type other than the one originally determined *in vivo* (see Ursprung, 1968).

What should become increasingly apparent, however, is that there is no need to postulate the existence of a "totipotent" cell to explain regeneration in either flatworms or amphibians. Even in theories of embryogenesis, it is no longer prudent or necessary to postulate the

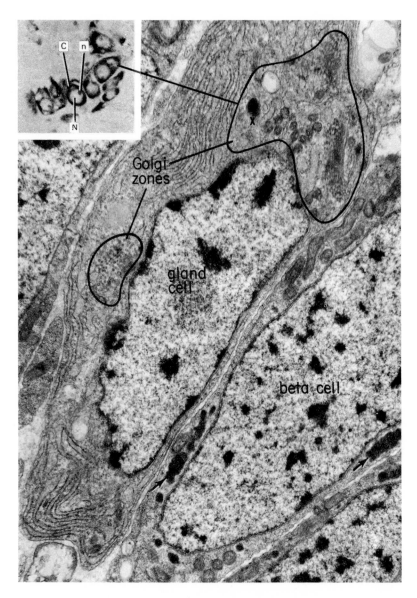

FIGURE 10. *The inset is a light micrograph from Lender (1962) showing "neoblasts" in the parenchyma of a flatworm* (Dugesia lugubris), *stained with methyl green-pyronine. The worm was not regenerating. The electron micrograph shows part of a similar nest of basophilic cells at high magnification. The gland cell would be highly basophilic as viewed in the light microscope after appropriate staining. The two Golgi zones encircled would appear as clear areas in a light micrograph (inset), because they contain no ribonucleic acid. A secretory granule, X, identifies the gland cell as a mucous cell. The relatively undifferentiated cell labeled "beta cell" has large chromatoid bodies (arrows) next to its nucleus and the nucleus has a finely clumped chromatin pattern. The beta cell seems to be a progenitor cell for the gland cells. A nest of basophilic cells, as in the inset, would contain gland cells (with clear Golgi zones) and beta cells. Inset, × 375 (From T. Lender, Advan. Morphogen. 2, 305, 1962). Main section, × 15,000.*

existence of an undifferentiated cell, in any sense of the definition. For example, the cells of the primitive streak in the chick embryo are already making actin and myosin (Orkin, et al., 1973) and also are clearly not totipotent as judged by their immediate fate. It is obvious that the cells of the regenerating newt limb or flatworm head do possess a remarkable plasticity as to their eventual reorganization into tissues and organs, for they do reform a missing part to which they did not previously belong. The same kind of plasticity on a tissue level exists in most early embryos. It is this level of organization to which we should now

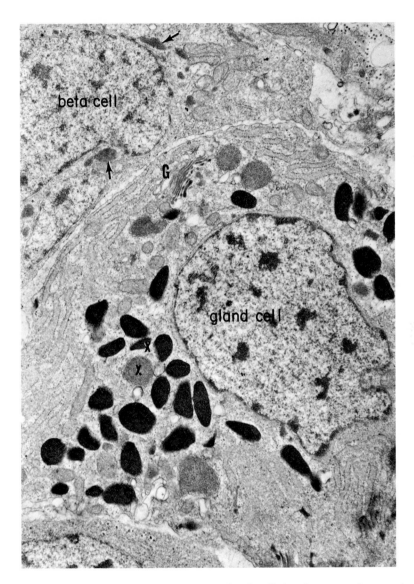

FIGURE 11. *Electron micrograph of a mucous gland cell in the parenchyma of a flatworm (Dugesia tigrina). Mucous granules (X) vary in density depending upon the degree of compaction. The abundant granular endoplasmic reticulum would make this cell very basophilic as viewed in the light microscope with an appropriate stain. A Golgi zone is labeled "G." The adjacent beta cell has characteristic chromatoid bodies (arrows). × 18,500.*

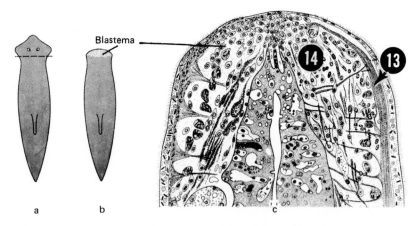

Blastema

a   b   c

FIGURE 12. *Level of amputation of the flatworm (dotted line, A) and formation of the regeneration blastema 24 hours later (B). A section through the blastema as seen in the light microscope appears at C. The number 13 shows an area of dedifferentiating muscle, and 14 indicates the blastema cells. [Based on Lang, from E. D. Hay, "The Stability of the Differentiated State" (H. Ursprung, ed.), Springer-Verlag, New York, 1968.]*

devote our attention. Embryogenesis may well involve a series of controlled metaplasias, rather than a graded transition from "undifferentiated" to "differentiated" cells. The significance of dedifferentiation of static tissues during regeneration and the relatively simplified appearance of many embryonic cells probably have to do with the metabolic requirements for cell proliferation. It well may be (Swann, p. 110, in Hay, 1966) that the so-called undifferentiated state often exhibited by actively growing cells is, in fact, a state of differentiation, a specialized condition for synthesis and turnover of the proteins, nucleic acids, and organelles needed for cell division.

Summary

We have in this chapter reviewed the current status of the problem of the origin of regeneration cells in vertebrates, in physiological as well as in reparative regeneration. We conclude that *renewing tissues*, such as blood and epithelia, depend on a partially differentiated progenitor cell for physiological renewal and for most, if not all, reparative regeneration. The epithelium of the amputated salamander limb also originates from such partially dif-

ferentiated progenitor cells. The *stable cells* of the body, such as the gland and connective tissue cells, continue to proliferate slowly in growing young mammals. They can proliferate faster to effect reparative regeneration with only a minimal amount of dedifferentiation. In the regenerating salamander limb, the extracellular matrix regresses more dramatically than in mammals, to release fibrocytes, chondrocytes, and osteocytes that proliferate to form a blastema. The changes that occur in the stable cells and their extracellular products during dedifferentiation are best described as preliminary to active cell proliferation. The *static cell population* in the vertebrate consists of two principal cell types: the neuron, which never divides, but does regenerate its axon; and the syncytial skeletal muscle fiber, which gives rise to mononucleated proliferating cells during reparative regeneration. We believe that the regeneration cells arise directly from differentiated muscle fibers by a process that seems to be the reverse of the original fusion of mononucleated cells to produce the syncytial fiber. During fusion, cell membranes disappear. During regeneration, cell membranes reappear to separate the proliferating cell from the rest of the muscle fiber. This issue is still controversial, as we point out here. The fundamental process, however, is likely to be the same in amphibians and in mammals. We cannot explain, on a cellular basis, the remarkable difference in the capacity of limbs of salamanders and mammals to regenerate. Higher levels of organization must be involved. Finally, we allude briefly to one of our mysterious invertebrate relatives, the planarian, and suggest that the undifferentiated appearing "reserve" cells described by electron microscopists are progenitor cells for gland renewal. They are certainly not the only cells that participate in regeneration because dedifferentiation does take place in the formed tissues of the injured flatworm.

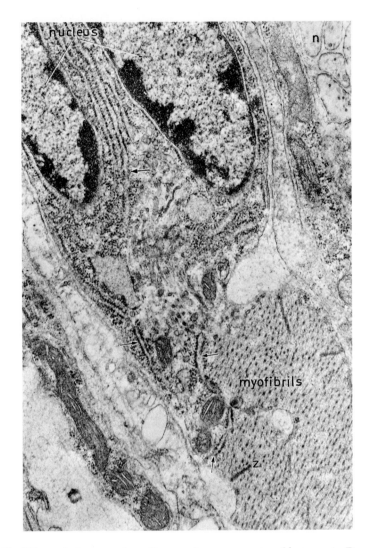

FIGURE 13. *Dedifferentiating muscle cell near the regeneration blastema in* Dugesia tigrina, *24 hours after amputation of the head. This cell has a bilobed nucleus and considerable endoplasmic reticulum (arrows). The cell body would be basophilic as viewed in the light microscope and counted as a "neoblast," yet it is clearly part of a muscle cell. The myofibrillar part demarcated by a cytoplasmic constriction will probbly be pinched off and left behind when the cell body becomes a regeneration cell. A Z band is labeled. Several nerve fibers (n) appear at the upper right. [From E. D. Hay, in "The Stability of the Differentiated State" (H. Ursprung, ed.), p. 95, Springer-Verlag, New York, 1968.] × 26,000.*

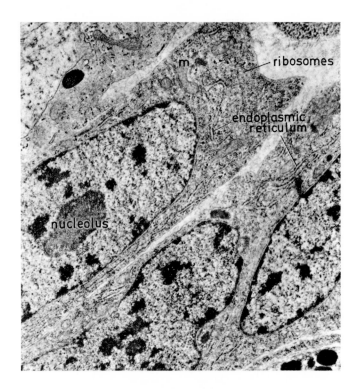

FIGURE 14. *Electron micrograph of regeneration cells in the blastema of* Dugesia tigrina *1 day after amputation of the head. The regeneration cells vary in the amount of endoplasmic reticulum they contain. Cytoplasmic chromatoid bodies are present but are smaller than in typical beta cells (Fig. 13). Cells seemingly in transition from muscle, gland, and beta cell morphology are present. Although gland progenitor cells (beta cells) probably participate in reparative as well as physiological regeneration, they may only supply the cell types which they were producing prior to amputation. ✕ 15,000. [From E. D. Hay, in "The Stability of the Differentiated State" (H. Ursprung, ed.), p. 96, Springer-Verlag, New York, 1968.]*

References

BECKER, R. O. (1972). Augmentation of regenerative healing in man. A possible alternative to prosthetic implantation. *Clin. Orthop. 83*, 255–262.

BECKER, R. O., AND SPADARO, J. A. (1972). Electrical stimulation of partial limb regeneration in mammals. *Bull. N.Y. Acad. Med. 48*, 627–641.

BERRILL, N. J. (1961). "Growth, Development, and Pattern." W. H. Freeman, San Francisco.

BUCHER, N. R. L. (1967). Experimental aspects of hepatic regeneration. *New Eng. J. Med. 277*, 686–696.

CARLSON, B. M. (1972a). Muscle morphogenesis in axolotl limb regenerates after removal of stump musculature. *Develop. Biol. 28*, 487–497.

CARLSON, B. M. (1972b). "The Regeneration of Minced Muscles." Karger, New York.

COWARD, S. J., AND HAY, E. D. (1972). Fine structure of cells involved in physiological and reconstitutive regeneration in planarians. *Anat. Rec. 172*, 296.

DRESDEN, M. H. (1969). Denervation effects on newt regeneration: DNA, RNA, and protein synthesis. *Develop. Biol. 19*, 311–320.

DUNPHY, J. E., AND VAN WINKLE, H. W., eds. (1969). "Repair and Regeneration: The Scientific Basis for Surgical Practice." McGraw-Hill, New York.

DYM, M., AND CLERMONT, Y. (1970). Role of spermatogonia in the repair of the seminiferous epithelium following x-irradiation of the rat testis. *Amer. J. Anat. 128*, 265–282.

FABER, J. (1971). Vertebrate limb ontogeny and limb regeneration: morphogenetic parallels. *Advan. Morphogen. 9*, 127–148.

FAWCETT, D. W., EDDY, E. M., AND PHILLIPS, D. M. (1970). Observations on the fine structure and relationships of the chromatoid body in mammalian spermatogenesis. *Biol. Reprod. 2*, 129–153.

FISCHMAN, D. A., AND HAY, E. D. (1962). The origin of osteoclasts from mononuclear leucocytes in regenerating newt limbs. *Anat. Rec. 143*, 329–338.

GOSS, R. J. (1964). "Adaptive Growth." Academic Press, New York.

GOSS, R. J. (1969). "Principles of Regeneration." Academic Press, New York.

GOSS, R. J., AND HAY, E. D., eds. (1970). Metazoan regeneration. *Amer. Zool. 10*, 90–186.

GROBSTEIN, C. (1959). Differentiation of vertebrate cells. *In* "The Cell" (J. Bracket and A. E. Mirsky, eds.), Vol. 1, pp. 437–496. Academic Press, New York.

HAY, E. D. (1966). "Regeneration." Holt, New York.

HAY, E. D. (1968). Dedifferentiation and metaplasia in vertebrate and invertebrate regeneration. *In* "The Stability of the Differentiated State" (H. Ursprung, ed.), pp. 85–108. Springer-Verlag, Berlin.

HAY, E. D. (1970). Reversibility of the differentiated state. *In* "Third Congress on Congenital Malformations," pp. 91–105. Excerpta Medica, Amsterdam.

HAY, E. D. (1971). Skeletal-muscle regeneration. *New Eng. J. Med. 284*, 1033–1034; *285*, 120.

HAY, E. D., AND DOYLE, C. M. (1973). Absence of reserve cells (satellite cells) in nonregenerating muscle of mature newt limbs. *Anat. Rec. 175*, 339–340.

HERLANT-MEEWIS, H. (1964). Regeneration in annelids. *Advan. Morphogen. 4*, 155–216.

HINGSON, D. J., AND ITO, S. (1971). Effect of aspirin and related compounds on the fine structure of mouse gastric mucosa. *Gastroenterology 61*, 156–177.

HUMPHREY, J. H., AND WHITE, R. G. (1970). "Immunology for Students of Medicine." Blackwell Scientific, Oxford, England.

KIORTSIS, V., AND TRAMPUSCH, H. A. L., eds. (1965). "Regeneration in Animals and Related Problems." North-Holland, Amsterdam.

LEBOWITZ, P., AND SINGER, M. (1970). Neurotrophic control of protein synthesis in the regenerating limb of the newt *Triturus. Nature (London) 225*, 824–827.

LENDER, T. (1962). Factors in morphogenesis of regenerating fresh-water planaria. *Advan. Morphogen. 2*, 305–331.

MANASEK, F., ed. (1972). Current literature: regeneration, cell division. *Develop. Biol. 28*, 455–456.

MARKS, P. A., AND KOVACH, J. S. (1966). Development of mammalian erythroid cells. *Curr. Top. Develop. Biol. 1*, 213–252.

MAURO, A., SHAFIG, S. A., AND MILHORAT, A. T., eds. (1970). "Regeneration of Striated Muscle, and Myogenesis." Excerpta Medica, Amsterdam.

NAMENWIRTH, M. R. (1968). Is cell differentiation inherited? *Am. Zool. 8*, 785.

O'NEILL, M. C., AND STOCKDALE, F. E. (1972). Differentiation without cell division in cultured skeletal muscle. *Develop. Biol. 29*, 410–418.

ORKIN, R. W., POLLARD, T. D., AND HAY, E. D. (1973). SDS gel analysis of muscle proteins in embryonic cells. *Develop. Biol. 35*, 388–394.

PEDERSEN, K. J. (1972). Studies on the regeneration blastemas of the planarian *Dugesia tigrina. Wilhelm Roux Arch. Entwicklungmech. Organismem 169*, 134–169.

POLEZHAEV, L. V. (1946). The loss and restoration of regenerative capacity in the limbs of tailless amphibia. *Biol. Rev. 21*, 141–147.

POLEZHAEV, L. V. (1972). "Organ Regeneration in Animals." Charles C. Thomas, Springfield, Illinois.

REES, H., AND JONES, R. N. (1972). The origin of the wide species variation in nuclear DNA content. *Int. Rev. Cytol. 32*, 53–92.

ROSE, S. M. (1964). Regeneration. *In* "Physiology of the Amphibia" (J. Moore, ed.), pp. 545–622. Academic Press, New York.

ROSE, S. M. (1970). "Regeneration: Key to Understanding Normal and Abnormal Growth and Development." Appleton-Century-Crofts, New York.

ROSS, R. (1968). The fibroblast and wound repair. *Biol. Rev. 43*, 51–96.

RUBEN, L. N., BALLS, M., AND STEVENS, J. (1966). Cancer and super-regeneration in *Triturus viridescens* limbs. *Experientia 22*, 260–261.

RUDNICK, D., ed. (1962). "Regeneration." Ronald Press, New York.

RUTTER, W. J., CLARK, W. R., KEMP, J. D., BRADSHAW, W. S., SAUNDERS, T. G., AND BALL, W. D. (1968). Multiphasic regulation in cytodifferentiation. *In* "Epithelial-Mesenchymal Interactions" (R. Fleischmajer and R. E. Billingham, eds.), pp. 114–131. Williams & Wilkins, Baltimore.

SAUZIN, M. J. (1968). Présence d'émissions dans les cellules différenciées et en différenciation de la planaire adulte *Dugesia gonocephala. C. R. Acad. Sci. 267*, 1146–1148.

SCHMIDT, A. J. (1968). "Cellular Biology of Vertebrate Regeneration and Repair." University of Chicago Press, Chicago.

SCHOTTÉ, O. E. (1961). Systemic factors in initiation of regenerative processes in limbs of larval and adult salamanders. *In* "Molecular and Cellular Synthesis" (D. Rudnick, ed.). Ronald Press, New York.

SINGER, M. (1960). Nervous mechanisms in the regeneration of body parts in vertebrates. *In* "Developing Cell Systems and Their Control" (D. Rudnick, ed.), pp. 115–133. Ronald Press, New York.

SINGER, M. (1968). Some quantitative aspects concerning the trophic role of the nerve cell. *In* "Systems Theory and Biology" (M. D. Mesarovic, ed.), pp. 233–245. Springer-Verlag, New York.

TARDENT, P. (1963). Regeneration in the hydrozon. *Biol. Rev. 38*, 293–333.

THORNTON, C. S., ed. (1959). "Regeneration in Vertebrates." University of Chicago Press, Chicago.

THORNTON, C. S. (1968). Amphibian limb regeneration. *Advan. Morphogen. 7*, 205–249.

TOOLE, B. P., AND GROSS, J. (1971). The extracellular matrix of the regenerating newt limb. *Develop. Biol. 25*, 57–77.

URSPRUNG, H., ed. (1968). "The Stability of the Differentiated State." Springer-Verlag, New York.

VORONTSOVA, M. A., AND LIOSNER, L. D. (1960). "Asexual Propagation and Regeneration." Pergamon Press, Elmsford, New York.

WEINSTEIN, R. B., AND HAY, E. D. (1970). DNA synthesis and mitosis in differentiated cardiac muscle cells of chick embryo. *J. Cell Biol. 47*, 310–316.

WEISS, L. (1972). "The Cells and Tissues of the Immune System. Structure, Functions, Interactions." Prentice-Hall Inc., Englewood Cliffs, New Jersey.

WU, A. M., TILL, J. E., SIMMINOVITCH, L., AND McCULLOCH, E. A. (1967). A cytological study of the capacity for differentiation of normal hemopoietic colony-forming cells. *J. Cell Physiol. 69*, 177–184.

CHAPTER TWENTY

Aging

VINCENT J. CRISTOFALO

Aging, or senescence, is the final stage in the life cycle of the organism, and for most organisms it represents a relatively long developmental period. In general, however, little is known about the process of aging. As a sub-science within biology, gerontology has matured very little during the last 25 years or so, while, at the same time, such great strides were being made in other areas such as bacteriology, virology, and genetics. Moreover, during this same period of scientific growth in biology, much effort has been expended in attempts to understand the basis for many of the age-associated diseases such as cancer and the cardiovascular diseases. It seems evident, however, that a complete understanding of the biology of these and other age-associated diseases would require an understanding of the fundamental changes that occur in the cells and tissues of the organism and that result in the increased susceptibility of the individual to develop these diseases.

Despite the lack of a well-defined, reproducible body of experimental data about the aging process, the literature does contain a large number of reports derived from wide-ranging experimental approaches. These include investigations at every level of biological organization and based on a variety of modulating factors. For example, radiation biology alone has generated a sizable literature all its own. Yet the question of whether radiation accelerates aging by acceleration of the natural aging process has not been resolved.

It would be impossible in the space allotted to adequately review the literature of the field of biological gerontology. What I have attempted to do in the following pages, however, is to present an overview of some of the more important research in the field as background and as an introduction to our own work and to that of others interested in the control of cell proliferation during aging. For more extensive treatments of the other areas not presented here in detail, the reader is referred to the excellent reviews and monographs of Lansing (1952), Strehler (1962), Comfort (1964), Walford (1969), and Kohn (1971).

The Aging Organism: Observations and Correlations

CHARACTERISTICS OF BIOLOGICAL AGING

No really satisfactory definition has been formulated for biological aging; however, one can list certain characteristics that describe and delineate the aging process.

As a working scheme we can consider biological aging as a process that (1) occurs in all members of a population, thus distinguishing this process from some diseases and various accidental causes of death; (2) is progressive and eventually harmful to the individual; (3) is irreversible under usual conditions; and (4) is characterized by a decline in the ability of the organism to respond to environmental change (i.e., loss of homeostasis).

The scope of these characteristics can be extended by analyzing biological aging mathematically. Figure 1a shows the survival curve for a population in which the proportion of individuals dying is constant with the passage of time. This curve is characteristic of populations in which the causes of death are independent of age—that is, occur randomly irrespective of the age of the individual. Figure 1b depicts a curve in which the frequency of death increases with increasing age. One would expect a curve of this type in populations in which there is an increasing probability of dying or an increasing vulnerability of the individual as a function of time; that is, individuals die in old age from causes that would not have killed them in their youth. This latter survival curve is characteristic of biological senescence. Actually, in most populations in nature, survival curves lie somewhere between these two extremes.

Figure 2 shows survival curves for populations ranging from British India in the 1920s to populations in the United States and New Zealand in the 1930s. Analysis of the curves suggests that the improved medical care and sanitation of the more highly developed countries allowed more people to live to the upper ages. Note, however, that the upper limit of 90 to 100 years is the same for all those populations. This upper limit appears to have re-mained constant throughout recorded history. Thus the increased life-span really reflects an increased life expectancy at birth; more people live to the upper ages. At 65 years, life expectancy is not really strikingly different anywhere in the world.

This apparent upper limit on life-span (maximum life-span) varies widely among different species, even closely related species. Within a

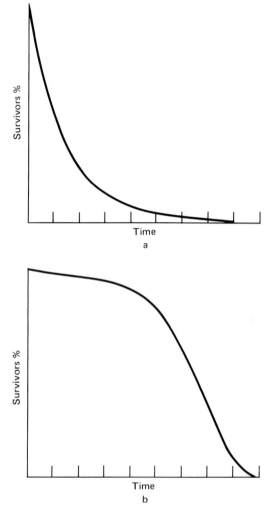

FIGURE 1. *(a) Survival curve at a constant rate of mortality (50 percent per unit of time). (b) Survival curve of a population that exhibits senescence. (Adapted from "Ageing: The Biology of Senescence" by Alex Comfort. Copyright © 1956, 1964, by Alex Comfort. Reprinted by permission of Holt, Rinehart and Winston, Inc.)*

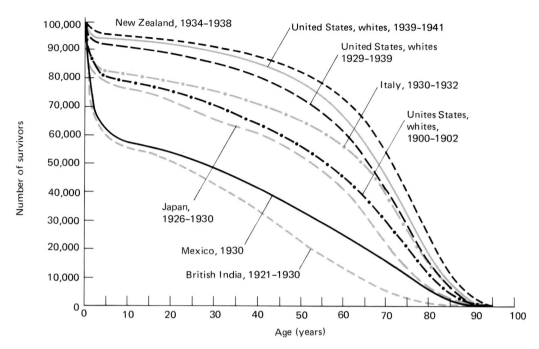

FIGURE 2. *Survival curves for various human populations. (Adapted from "Ageing: The Biology of Senescence" by Alex Comfort. Copyright © 1956, 1964, by Alex Comfort. Reprinted by permission of Holt, Rinehart and Winston, Inc.)*

species, however, it appears to be highly consistent, which suggests that underlying the various causes of death, such as disease and accidents, there is a genetically based "program" involving biological clocks that determine the overall maximum life-span of that species.

LIFE-SPAN OF ANIMALS

It appears that, in general, all multicellular animals age. (Unicellular or acellular species, under appropriate environmental conditions, also show senescence.)

A number of attempts have been made to extract correlations between life-span and other physiological characteristics of a species to gain insight into possible common mechanisms of aging. Although, in general, these have yielded disappointing results, there are some interesting correlations. For mammals, for example, there appears to be a relationship between body weight and maximum life-span. Brain weight seems to show an even better correlation with

life-span than body weight. However, the significance and meaning of these correlations is not clear. Correlations recognized for mammals do not seem to hold true necessarily for other classes of vertebrates. Moreover, it is difficult to derive from these relationships reasonable hypotheses of the possible mechanisms that could account for the data.

Components of the Aging Animal

I mentioned above that one of the most salient features of biological aging is the loss of capacity for adaptive response—that is, a loss of homeostasis. This appears to occur across a broad spectrum of organization within the animal organism and, as such, must be very complicated in origin and mechanism of development. Thus if we are to understand the overall process, we should begin by dissecting and analyzing the parts involved.

For the purposes of our discussion, then, we can arbitrarily consider the organism to be made up of three kinds of components: (1) the extracellular material such as collagen and elastin, (2) fixed postmitotic cells, which are best represented by nerve and muscle, and (3) the continually proliferating elements such as skin, the lining of the gut, and the blood-forming elements.

Each of these components certainly ages. In fact, age-related deficiencies in functional capacity due to changes or failures in all three components are well documented in the literature (e.g., see Kohn, 1971). It is likely, then, that each component ages at characteristic rates and mechanisms and that each resulting aging process contributes to the functional failure of the organism in its own unique way. In this section we shall consider, in general terms, some of the characteristics and properties of aging in each of these components and their effects on the integrity of the organism.

EXTRACELLULAR COMPONENTS

Among the extracellular components we include collagen, elastin, and the mucopolysaccharides; however, we will limit our discussion to the role of collagen in the process of aging.

Collagen is a triple-helical, fibrous protein that comprises 20 to 30 percent of the entire body protein. It is a unique protein in that one third of its amino acid residues consist of proline and hydroxyproline, while another 25 to 30 percent are glycine. It is well known that as time passes, aggregation and cross-linking of the collagen molecule occur, resulting in changes in the properties of the molecule that are significant to the organism. The changes include an increased rigidity of the collagen and a decrease in its thermal contracility. It is clear, of course, that some increased rigidity of mature collagen that results from cross-linking is essential to the integrity of the animal. However, increasing rigidity of mature collagen eventually reaches a point where it interferes with normal or optimal function. Muscle, for example, contracts in a matrix of collagen. As increased rigidity develops, it would interfere with contraction and thus impair muscular

function. In addition, deposition and aggregation of collagen could interfere with the diffusion processes between cells and capillaries and thus impair the passage of vital nutrients and essential humoral factors to and from the cells.

A different aspect of the relationship between collagen cross-linking and aging involves the study of the physiochemical details of the cross-linking process, using collagen as a model system for the cross-linking of macromolecules in general. It is easy, for example, to visualize how cross-linking in DNA could interfere with normal genetic transcription. Understanding the chemical details of the mechanism of cross-linking in collagen and the factors regulating this process might be of value, then, in elucidating the role that macromolecular cross-linking plays in aging in general.

Sinex (1968) has reviewed some aspects of the evidence relating collagen cross-linking as a basic factor in the mechanism of aging. In general, the evidence suggests that although maturation of collagen is important to the survival of the organism, it is probably not a basic controlling factor. Detailed discussions of this material can be found in Kohn (1971) and in Sinex (1968).

FIXED POSTMITOTIC CELLS

Fixed postmitotic cells are the nonproliferating elements in the mature animal and would include a large number of cell types. We shall consider only muscle and nerve cells in our discussion. More detailed information is contained in the article by Buetow (1971) and in the monographs by Kohn (1971), Strehler (1962), and Comfort (1964).

If we accept the notion that aging changes result from an accumulation of errors in the genome, then nondividing cells would seem to be ideal candidates for the age-dependent changes that are critical to the survival of the organism. In postmitotic cells there is no reshuffling or renewal of the genetic material; nor is there pressure for selection of the cells without errors. Thus errors in the genome can accumulate until the level of faulty information is so high as to interfere with the survival of the cells.

In considering the aging of nonrenewing cell

types, we may view the deficiencies in the aging individual as resulting from (1) losses in the functional capacity of individual cells in the animal, and/or (2) a reduction in the number of cells in a given organ or tissue so that the functional capacity of the entire element is compromised. Both these aspects of aging have been investigated, although neither as thoroughly as might be desired.

For muscle, there appears to be no evidence of a pattern of cell loss (Buetow, 1971). On the other hand, loss of functional capacity in muscle during aging is well documented and is presumably based at the level of the muscle cell.

Very little is known about the functional capacity of neurons during aging. Various psychological tests have indicated a decline in learning ability. In addition, there is evidence of a decline in the conduction velocity of certain nerves, although it has not been ascertained whether this is a primary or secondary effect (Kohn, 1971).

The numbers of nerve cells surviving as a function of age have been studied at various sites and in several mammalian species. In many cases the number of nerve cells appears to decrease during aging, but in other cases this is not so. In general, there appears to be a wide variability in cell number during aging, and loss of neurons does not seem to be universally characteristic of the aging process.

One other aspect of aging in nonproliferating cells involves the so-called age pigment. The occurrence of pigmented inclusion bodies (age pigment or lipofuscin) in several nondividing cell types has been recognized for many years and has been studied extensively by a number of workers (e.g., Strehler et al., 1959). There appears to be a linear relationship between the intracellular volume occupied by age pigment and the age of human myocardium. Similar pigment accumulation has been noted in the myocardium of the dog. It is of particular interest that the accumulation of lipofuscin in the dog proceeds at a faster rate than in the human and that the difference in rate is proportional to the difference in life-span. Thus, age pigment accumulation is not simply related to time but also to the metabolism and aging rate of the organism.

The origin of lipofuscin is unknown. Chemically, the granules are heterogeneous and contain many types of lipids. They also contain protein, and a portion of this protein exhibits the activity characteristic of the hydrolytic enzymes found in lysosomes such as acid phosphatase and the cathepsins (Strehler, 1962).

Unfortunately, no specific decline in the functional capacity of nerve or muscle can be related to the accumulation of lipofuscin, and the role of this substance is not understood. Current thinking favors the view that the lipofuscin deposits are the products of lipid peroxidation for which no satisfactory mechanism of removal is available in the cell. The relationship of this material and its postulated origin to aging remains to be elucidated.

PROLIFERATING CELLS

The third component of aging systems is that group of cells that maintain the ability to proliferate throughout the life-span of the animal. This category encompasses cells of the skin, the lining of the gut, and the blood-forming elements, including the immune system. Also included here are certain groups of cells that maintain the capacity to divide throughout the life-span of the animal but do so only when called upon and under special circumstances—for example, regenerating liver, fibroblasts during wound healing, and salivary gland cells when appropriately stimulated.

It is reasonably well documented for some populations of proliferating cells that, after an initial period of rapid cell division, there is a decline in the proliferative capacity as the animal ages (LeBlond, 1964; Buetow, 1971). It is significant here that age-associated changes *in vivo* are not limited to fixed postmitotic cells. Buetow (1971) has recently tabulated and reviewed the literature on age-associated changes in cellular proliferation rates. In general, there is a decline in mitotic activity in a wide variety of tissues of humans and various rodent species. More recently, Lesher and Sacher (1968), Thrasher (1971), and Cameron (1972a,b) have confirmed these reports in detailed studies with mouse tissues.

Another group of studies that bears on this point involves the serial transplantation of normal somatic tissues to new, young inbred hosts each time the current recipient approaches old age. For example, the work of Daniel et al. (1968), in which mouse mammary glands were transplanted into gland-free mammary fat pads of young, isogenic female mice, shows that the growth rate of a normal gland declines with time. Thus the normal mammary gland has a limited ability to proliferate *in vivo* even under these most favorable growth conditions. Similar findings have been published for transplanted skin (Krohn, 1962, 1966) and for bone marrow cells (Siminovitch et al., 1964; Cudkowicz et al., 1964). Thus, in general, normal cells serially transplanted to inbred hosts seem to show a decline in proliferative capacity and probably cannot survive indefinitely.

In addition, age-associated declines in proliferative capacity have been documented for isoproterenol-stimulated salivary gland (Adelman et al., 1972) and for elements of the mouse immune system (Price and Makinodan, 1972a,b). Thus a decline in proliferative capacity represents one more kind of gradual, functional failure that occurs in the aging animal.

The relationship of this loss of proliferative capacity to the aging of the organism is not yet clear, however. It is evident that animals do not age and die because they "run out" of cells of various kinds. Several transplant studies have shown that the transplant itself has a longer life-span than that of the organism from which it was derived (Krohn, 1966). In fact, it is impossible to relate the time course of this loss of proliferative capacity *in vivo* to the total life-span of the animal. In addition, it is not really clear whether a reduced rate of proliferation would ordinarily result in any survival-compromising shortage of cells. It is possible, however, that failures in the regulation of proliferative capacity could be accompanied by failures in other cellular functions, which would then compromise the ability of the animal to survive in a rapidly changing environment.

The following section will describe studies carried out on the regulation of cell proliferation in two different model systems that show a decline in proliferative capacity during aging —one *in vivo* and one in cell culture. At present, there is no reason to believe, or to expect, that failures in the regulation of cell proliferation are either more or less important than age-related changes in the other two components we have discussed. The more detailed presentation of this aspect of aging simply reflects the interest and experience of the author.

Studies on Proliferative Capacity as a Function of Age

AGING OF CELLS SERIALLY PROPAGATED *IN VIVO*

Different models have been used to examine in detail whether the loss of proliferative capacity is truly intrinsic to cells, or if it is wholly dependent on the various environmental factors present in old animals. If environmental influences that cause aging and death could be controlled, would cells nevertheless lose their proliferative capacity and cease normal function?

Typically in the *in vivo* studies, selected somatic tissues have been serially transplanted into new, young inbred host animals each time the current recipient approaches old age. Thus the host is always young. As mentioned above, experiments of this kind have been carried out with skin cells (Krohn, 1962, 1966) and hematopoietic cells (Siminovitch et al., 1964; Cudkowicz et al., 1964), both of which clearly showed a decline in proliferative capacity with cell age.

The mammary transplant system as described by DeOme et al. (1959) has been found to be particularly useful for these aging experiments and has been employed by Daniel and his coworkers (Daniel, 1972) to study the proliferative capacity of mammary epithelial cells as a function of age.

In these experiments, female mice at 3 weeks of age are anesthetized; short incisions are made in the ventral skin and the inguinal fat pads are exposed. After cauterization of the appropriate blood vessels and the nipple area (the point of origin of ductal growth) and removal of the small piece of growing mammary parenchyma, the now gland-free fat pad can receive transplants of other mammary gland tissue. Thus the grafted tissue is placed in its proper environment and with adequate space for proliferation. Primary transplants will grow and fill the fat pad within about 8 to 12 weeks. Daniel and his coworkers (Daniel et al., 1968, 1971; Young et al., 1971) have used this system to address four key questions about the characteristics of mammary cell aging.

The primary question was whether, in fact, the growth rate of these transplanted mammary epithelial cells declined during serial subcultivation. For these experiments the extent of mammary outgrowth was measured at each transplant generation and the percentage of the fat pad occupied by parenchyma was recorded. Figure 3 shows that there was a gradual but continual decline in the percentage of the fat pad filled after a fixed time interval. In no case could indefinite growth be achieved. Thus, under these *in vivo* conditions, in which the normal environment of the cell was maintained, the growth rate was limited.

The second question was concerned with the effects of both the host and tissue age on the growth rate and life span of the transplanted mammary cell; that is, does mammary gland transplanted from old animals have a reduced prolifertive capacity compared with that of young animals? What is the effect of host age on proliferative capacity?

The growth potential of transplants from 26-month-old and 3-week-old animals were compared by transplanting the respective tissues on contralateral sides of the same young host. The results showed that, surprisingly, there was essentially no difference in the proliferative capacity of young and old transplants in young hosts. One possible explanation for this result is that the cells do not "recognize" the passage of calendar time but rather simply count cell divisions.

To investigate this question (Daniel and Young, 1971; Daniel et al., 1971; Daniel, 1972), advantage was taken of the fact that, once a fat pad is filled with cells, no more proliferation takes place, although the tissue remains metabolically active. Two transplant lines were started from the same donor; one was transplanted every 3 months so that it remained in a proliferative state almost continuously. The other line was transplanted at 12-month intervals so that for most of the period it was

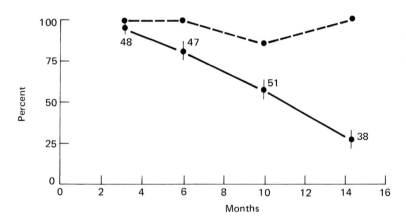

FIGURE 3. *Serial transplantation of normal mouse mammary epithelium.* ●—● , *Mean percent of available fat occupied by transplants;* ●–● *percent successful transplants. Each point denotes a transplant generation, and small numbers beside points refer to the number of successful transplants in each generation. Vertical lines indicate 95 percent confidence intervals. The transplant line was initiated at time 0 and the first passage occurred at 3 months. Note the lack of correlation beween growth and percentage of successful transplants; the latter appears to be a function of various technical difficulties experienced during transplantation: opacity of the fat, imperfect staining with trypan blue, or difficulties experienced in removing transplants from very small outgrowths. (After Daniel et al., 1968. Reprinted from C. W. Daniel, Advan. Gerontol. Res. 4, 167–198, 1972, with permission of Academic Press, Inc., New York.)*

metabolically active but proliferation was inhibited. The results showed that it was the number of division events that the cells had experienced that determined their remaining future proliferative capacity.

A second procedure was concerned with clarifying whether the shortened life-span of the continually proliferating series was the result of the trauma of more frequent transplantations or was truly the result of the increased number of divisions. The approach to this question was based on the fact that in a filled fat pad the cells at the periphery of the pad have undergone more divisions than the cells at the center of the pad. Thus, transplant lines were initiated from a donor and then two serial transplants were made, one from the center of the pad and the other from the periphery. This pattern of serial transplantation was continued for each subsequent passage. The results showed that cells that had undergone more divisions were aging faster. This result supports the interpretation that it is the cumulative number of cell divisions that determine the future proliferative capacity of the cell.

In summary, then, mammary cells transplanted *in vivo* have a limited proliferative capacity. The mechanism underlying this limited proliferative capacity depends on the number of doublings that the population has experienced and not the passage of chronological time.

PROLIFERATIVE CAPACITY OF ANIMAL CELLS *IN VITRO*

It is now reasonably well established that populations of normal diploid human fibroblasts can proliferate in culture for only limited periods of time. Typically, after explanation, there is a period of rapid proliferation during which the cultures can be subcultivated relatively often, followed by a period of declining proliferative capacity during which the cells become granular, debris accumulates, and ultimately the culture is lost. The reader is referred to the work of Swim and Parker (1957), Hayflick and Moorhead (1961), and others in establishing the generality of this phenomenon.

Figure 4 shows a typical set of data obtained

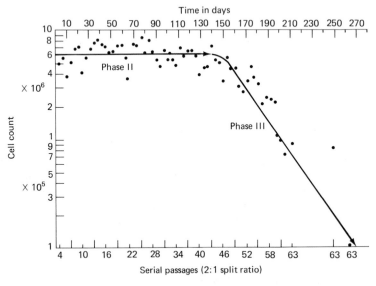

FIGURE 4. *Cell counts determined at each passage of human diploid cell, strain WI-44. This figure results in a curve suggestive of multiple-hit or multiple-target inactivation phenomena as an explanation for the mechanism of the occurrence of phase III. The initial plateau during phase II, with no apparent loss of biological function as measured by constant doubling time, is followed by phase III, where doubling time increases exponentially. (From L. Hayflick, Exp. Cell Res. 37, 614–636, 1965. Reprinted with permission of Academic Press, Inc., New York.)*

for human fetal lung cells. After the period of rapid growth (about 150 days under these conditions), there is a declining proliferative rate and the culture is eventually lost. A similar limitation on the growth of chick cells has been noted by several authors (e.g., Haff and Swim, 1956; Hay and Strehler, 1967). During the last 10 years (in a large number of laboratories throughout the world), this observation has been noted for both chick and human cells, as well as cells from other mammalian species. The doubling capacities of various populations of cells in culture are reproducible within relatively narrow limits. For example, about 50 population doublings occur with human embryonic fibroblasts, while embryonic chick fibroblasts have a doubling potential of about 25 passages. Thus the notion that isolated animal cells are capable of unlimited proliferation in culture, as proposed by Carrel and his co-workers (Ebeling, 1913, 1921), does not appear to be the case.

Initially, this inability of cell cultures to proliferate indefinitely was ascribed to various technical difficulties, such as inadequate nutrition, pH variation, toxic metabolic products, and microcontaminants. However, in an extensive series of experiments, Hayflick and Moorhead (1961) showed that clonal degeneration was unrelated, at least in any simple, direct way, to any of these factors. When mixtures of young and old populations (male and female), distinguishable by karyotypic markers (Barr bodies), were grown in the same pool of medium, the older population was lost after it had undergone a total of approximately 50 population doublings, while the younger population continued to proliferate until the 50 or so expected doublings had been completed. Such results would seem to rule out any direct effect of the composition of the medium, or the presence of contaminating microorganisms or toxic end products of metabolism.

Loss of proliferative capacity cannot be related to depletion of some essential, nonreplicating metabolite, since the initial presence of 2^{50} molecules of even the lightest element, hydrogen, would have a mass in excess of that of a single cell.

Hayflick and Moorhead (1961; Hayflick, 1965) concluded from these data that the limited life-span phenomenon must be pro-grammed, and they interpreted their observations as a cellular expression of senescence.

The suggestion that aging changes are reflected in various properties of tissue cultures is not new. For example, it has long been known that age-associated changes that occur in plasma can inhibit cell growth in vitro (Carrel and Ebeling, 1921). In addition, the time elapsing prior to cell migration from explanted tissue fragments increases with increasing age (Carrel and Burrows, 1910a,b, 1911). These are both examples of the expression, in vitro, of aging in vivo. The more recent studies of Hayflick and Moorhead (1961) and Hayflick (1965) focus attention on the occurrence of senescence in vitro.

To one interested in the study of senescence, the important questions are: (1) Does aging in vitro bear any relationship to aging of the whole organism? And (2) Do studies carried out in cell culture have any relevance to aging in the whole animal? In approaching these questions, the reader should bear in mind that it is very difficult within the framework of our present state of knowledge to pinpoint what relationship, if any, exists between aging in vivo and aging in vitro.

We are not suggesting that animals die because they exhaust the supply of, for example, lung fibroblasts, bone marrow, or skin cells. In fact, as in the in vivo studies of Daniel and others described above, it is impossible to directly relate the time course of this loss of proliferative capacity in vitro to the life-span of the animal. On the other hand, it seems reasonable to expect that the study of the regulation of cell proliferation in vitro would lend important insight into the mechanism of regulation of cell proliferation in vivo and its ramifications for aging in the intact animal.

Perhaps one of the strongest sources of support for the relationship between in vivo and in vitro aging derives from the now well-documented studies (Soukupovà and Holečkovà, 1964; Soukupovà et al., 1970) of the increase in the latent period of tissue explants. The latent period is defined as the time required for cells to migrate from an explanted tissue that has been placed in an appropriate culture medium. They demonstrated that for a number of rat and chicken tissues the length of the latent period is directly proportional to the age of the donor; this direct proportionality extends not only during the growth period but also into the old age of the animals. This, then, is an *in vitro* expression of *in situ* aging.

Another and perhaps more striking line of evidence for the relationship of a limited lifespan *in vitro* to aging *in situ* springs from a variety of works that suggest a cumulative effect of *in situ* plus *in vitro* aging by showing a relationship between the age of the cell donor and the proliferative capacity of the cells derived from that donor. For example, Hayflick (1965) found a dramatic difference between cell lines derived from human embryo and human adult lung.

More recently, Goldstein and his coworkers (1969) found that an inverse correlation existed between the age of the donor and the number of population doublings that a series of skin cultures could achieve.

Martin et al. (1970) carried out an extensive study in which over 100 mass cultures of fibroblast-like human diploid cells from a variety of donors varying in age from newborn to 90 years were used. They found a significant negative regression of growth potential as a function of the age of the donor.

Goldstein et al. (1969) and Martin et al. (1970) have shown that cells from patients with progeria or with Werner's syndrome (both diseases associated with premature aging) have a reduced proliferative capacity, as compared with cells from normal (control) donors of the same ages. In addition, cells derived from diabetic individuals have a reduced ability to grow and survive in culture, as manifested by a reduced plating efficiency (Goldstein et al., 1969).

Correlative metabolic findings show that cell cultures derived from adult lung have lysosomal enzyme characteristics after only 12 to 14 passages similar to those of fetal cells after 35 to 40 passages (Cristofalo et al., 1967) and that adult skin cells show cell-cycle characteristics after only a few passages, similar to those of degenerating fetal cell cultures (Macieira-Coelho and Pontén, 1969; Macieira-Coelho, 1970).

Despite these well-documented findings, a number of questions still remain unresolved. If the limited life-span *in vitro* is truly relevant to aging, then it should be characteristic of cultures from all vertebrate species. This spontaneous degeneration of cells in culture has been carefully documented for human (Hayflick and Moorhead, 1961; Hayflick, 1965) and avian cell cultures (Hay and Strehler, 1967) and for cultures derived from the marsupial *Potorous tridactylis* (Simons, 1970). In general, rodent cell cultures also show a decline in proliferative capacity. However, a number of reports involving a variety of rodent cell cultures (Petursson et al., 1964; Krooth et al., 1964; Yaffe, 1968; Valenti and Friedman, 1968), as well as cultures of marine fish cells (Regan et al., 1968), indicate that these cells maintain their proliferative capacity for what seem to be indefinite periods without any major alteration in karyotype. In addition, certain cultures derived from human lymphoid elements seem to retain the normal diploid karyotype and have an indefinite life-span (Moore and McLimans, 1968; Levy et al., 1968). However, none of these appear to meet the full criteria of being karyologically normal (Hayflick, 1972).

A second problem involves the interpretation of data on the life-span of cell cultures. Figure 4 shows a typical set of results in which the yield/culture vessel is plotted as a function of the number of 1:2 subcultivations (1 subcultivation = 1 population doubling). The progressive decline in cell yield eventually leads to loss of the culture. The results of this experiment are highly reproducible but tell us nothing about the mechanism by which the process occurs—that is, whether this decline in proliferative capacity is due to a uniform increase in the generation time of the cells or to an increasing heterogeneity in which some cells can divide normally while others are incapable of division. For example, Merz and Ross (1969) have shown that for very low density populations the percentage of cells that undergo division within a fixed time period decreased as a function of the number of divisions. This increase in the fraction of nondividers in the population is clearly an age-related phenomenon *in vitro*.

Macieira-Coelho et al. (1966a,b; Macieira-Coelho and Pontén, 1969) have shown an increased heterogeneity in the length of the division cycle in late-passage cells, due principally to an increased length of the G_1 and G_2 periods and to the fact that the fraction of cells included in DNA synthesis decreased as a function of culture age. It is interesting in this regard that Holečková and her coworkers (Soukupová et al., 1970), in studying the latent period, have shown that explants of tissues from young donors appears to contain a large quantity of cells capable of migration and division, while the tissues of older donors behave like mosaics of tissue regions with fewer, or no cells capable of division.

In more recent studies, we (Cristofalo and Sharf, 1973) have examined this problem further. Using autoradiography we determined the percentage of cells incorporating ^{3}H-labeled thymidine into their nuclei. The results showed that the percentage of cells with labeled nuclei declined as a function of the cumulative population doublings. If the ability to incorporate [3]HTdR can be equated with ability to carry out cell division, then these data confirm the interpretations above (Macieira-Coelho et al., 1966a,b; Merz and Ross, 1969) and establish that the decline in proliferative capacity of these human cells is due, in part, to a declining fraction of the population able to carry out cell division.

This pattern of gradual continuous decline in labeling index is in contrast to that recently reported for chick cells. Close to 100 percent of chick cells entered DNA synthesis throughout almost the entire life-span of the culture (Lima and Macieira-Coelho, 1972). Possibly some of the key steps involved in the in vitro aging of these two species may be different.

There was an exponential increase in the cells unable to incorporate label as a function of time (Fig. 5). This relationship suggests to us that labeling indices might represent a method for accurate determination of the biological age of these populations.

In addition, if the [3]HTdR labeling procedure just described can be shown similarly accurate and age-correlated for adult-derived skin cultures, then this would enable a rapid method for evaluating their proliferative state. If a correlation is found, it may represent a method for evaluating one aspect of the biological age of an individual and the relationship, if any, between biological and chronological age.

As in the in vivo transplantation experiments discussed above, a related question is whether the rate of decline in proliferative capacity in vitro really depends on the number of cell doublings, as reported by Hayflick (1965), or simply on the passage of so-called "metabolic time" that the cell has been in culture. Hay et al. (1968) attempted to test these alternatives experimentally by using an agar-overlay technique that maintained a chick cell monolayer attached to the glass surface but unable to divide. After various periods in this nondividing state, the agar was removed, substrains were established, and subcultivation was resumed. There was a progressive loss of proliferative capacity directly related to the length of maintenance time under agar. Death of the culture occurred at essentially the same calendar time in all substrains, irrespective of the number of divisions. Thus total life-span ap-

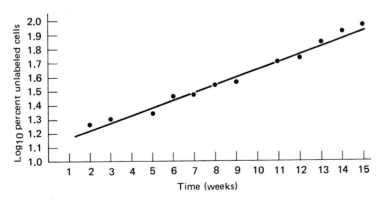

FIGURE 5. *Percentage of unlabeled cells as a function of chronological time. Curve generated by method of least squares. (From V. J. Cristofalo and B. B. Sharf, Exp. Cell Res. 76, 419–427, 1973, with permission of Academic Press, Inc., New York.)*

peared to be a function of the total time in culture, rather than of population doublings.

More recently, additional evidence for this conclusion came from the work of McHale et al. (1971). Using the human diploid cell line WI-38, they have shown that reduction of the serum in the medium from 10 to 0.1 percent (v/v) maintained the cells in a viable stationary phase for up to 5 weeks. Upon subcultivation, these cells underwent fewer population doublings than the continuously subcultivated controls; all the cultures reached the degenerative stage at approximately the same calendar time, irrespective of the number of cell doublings.

An alternative explanation for the findings of Hay et al. (1968) and McHale et al. (1971)

would be that the various environmental factors involved in maintenance culture induced a loss in proliferative capacity. We can include among these factors the effect of the agar on cell division and the effect of nutritional deprivation in the medium supplemented with 0.1 percent serum. There is the additional possibility of cell loss during the maintenance period.

More recently, Dell'Orco et al. (1972) have repeated these studies using human fibroblast-like cells maintained in medium supplemented with 0.5 percent serum. The cultures were monitored continually for cell loss, protein loss, and cell division. In 0.5 percent serum, some cell loss did occur but less than 1 per-

TABLE 1

SOME METABOLIC PROPERTIES OF DIPLOID CELLS
STUDIED DURING AGING[a]

Parameter	Variation with cell age[b]
Glycolysis	o
Glycolytic enzymes	o; —
Pentose phosphate shunt	—
Permeability to glucose	o
Glycogen content	+
Mucopolysaccaharide synthesis	—
Respiration	o
Respiratory enzymes	o
Lipid content	+
Lipid synthesis	+
Protein content	+; o
Permeability to amino acids	o
Transaminases	o; —
Glutamic dehydrogenase	o
Nucleohistone content	o
Collagen synthesis	—
DNA content	o; —
RNA content	+
Nucleic acid synthesis	—
RNA turnover	+
Lysosomes and lysosomal enzymes	+
Alkaline phosphatase	o

[a]From V. J. Cristofalo, *Advan. Gerontol. Res. 4*, 45–79, 1972. Reprinted with permission of Academic Press, Inc., New York.
[b](+), increase with age; (−), decrease with age; (o), no change.

cent of the cells synthesized DNA under these maintenance conditions. When the cells were returned to proliferating conditions, all cultures resumed proliferation and completed the same number of population doublings as control cultures that were continually subcultured. These results were confirmed using the cell labeling technique described above (Cristofalo and Sharf, 1973). The data of Dell'Orco et al. (1972), then, are similar to those reported by Daniel (1972) for mammary transplants, which were described above. The cells appear to have a counting mechanism for division events rather than a calendar mechanism for monitoring accumulated metabolic time. The differences between the results of McHale et al. (1971) and Dell'Orco et al. (1972) may depend on the differences in the serum concentration used. Possibly division potential is lost only when the environmental conditions fall below a certain minimal level required to effect a truly quiescent, but nondegenerating state.

A number of studies have been carried out in attempts to relate significant qualitative or quantitative changes in cell function with changes in proliferative capacity. Most of the investigations have employed chick or human cells studied at various stages in their overall life cycle. The data obtained were then related to the number of doublings the population had undergone. The results of these studies are summarized in Table 1 and have been reviewed by Cristofalo (1972).

In the following paragraphs we shall consider only one aspect of the studies that have been done—that is, changes in the lysosomes and lysosomal enzymes during aging.

The lysosomes appear to be involved in a variety of degenerative processes, including, for example, autolysis during inflammatory reactions, as well as programmed involutional changes, such as the resorption of the tadpole's tail. Allison and Paton (1965) have reported that the lysosomes of WI-38 cells contain a DNA depolymerase that can cause chromosomal breaks, and since it has two active sites, it can destroy both strands of DNA with a single hit.

We (Cristofalo et al., 1967; Cristofalo, 1970) have studied the activity of two enzymes, usually considered "marker" enzymes for lysosomes, during the aging of diploid human cells.

For these studies, enzyme activities of WI-38 cells were measured at passage levels ranging between 15 and 45. The data obtained were arbitrarily divided into young, intermediate, and senescent groups. The senescent group included cultures with a decreased proliferation rate and cultures that no longer grew to confluency. For purposes of comparison two other cultures were included: (1) human fetal lung fibroblasts (A-11-L) of undetermined but presumably diploid karyotype (these were measured before reaching passage 10 and thus were "younger" than the earliest WI-38 cultures measured), and (2) a culture of diploid fibroblasts derived from human adult lung (WI-1006) obtained from a 58-year-old donor (Hayflick, 1965). (These adult lung fibroblasts have a doubling potential of only about 20 generations, and in terms of their life-span were equivalent to the oldest cultures of WI-38.) They were studied at passages 13 through 15.

Figure 6 shows that cells at passage levels above 35 had a significantly higher activity than cells of less than 25 passages (p < 0.001). The intermediate-age cells showed activity in-

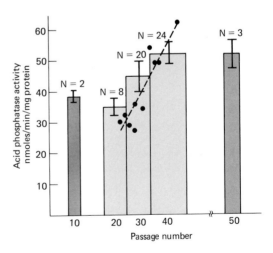

FIGURE 6. *Acid phosphatase activity during the aging of human diploid cells. Mean enzyme activity is expressed by the heights of the bars and the standard error of the mean by the brackets. N is the number of determinations. (From V. J. Cristofalo, N. Parris, and D. Kritchevsky, J. Cell Physiol. 69, 263–272, 1967. Reprinted with permission from The Wistar Institute Press, Philadelphia, Pennsylvania.)*

termediate between the young and old cells and were not significantly different from either. Cultures of human fetal lung fibroblasts studied between passages 6 and 10 had activities similar to the young WI-38 cultures, while the lung fibroblasts of adult origin showed acid phosphatase activity similar to that of the oldest WI-38 cells.

Superimposed on the bars in Fig. 6 and representing the mean acid phosphatase activities are the activities found for a single series of continuous mother–daughter subcultivations between passages 22 and 42. These data also showed a rising level of acid phosphatase activity with an increasing number of subcultivations, indicating that the increase in the mean value of the activity was not an artifact of the method of designating cell population age. Similar data were obtained from β-glucuronidase activity.

More recently, morphological evidence for the biochemical findings of Cristofalo et al. (1967) has been presented by Brock and Hay (1971) for chick cells and by Robbins et al. (1970), Lipetz and Cristofalo (1972), and Brandes et al. (1972) for human cells. Of particular interest is the work of Robbins et al., who examined the ultrastructural changes developing in a number of human fibroblast strains from embryonic and infant skin during serial propagation and compared these with human skin biopsies from four donors of different ages. Their findings indicate that the serial propagation of cells that ultimately become senescent in culture is accompanied by major changes in the lysosomes, including a progressive increase in both number and size, and profound "degenerative" changes. Fresh skin biopsies from aged subjects did not show these changes upon initial growth in culture; however, this is to be expected, since, presumably, cells that grow from any explant are "young" cells, not "aged" cells; that is, only cells capable of proliferation ("young" cells) would survive explanation and outgrowth.

The finding that lysosomes and lysosomal enzyme activity increased during senescent degeneration of diploid cells in culture represents one of the best described biochemical markers, thus far, for following cellular senescence and also suggests approaches to elucidating the mechanism of senescent degeneration.

Cristofalo (1970) has shown that the increased activity of acid phosphatase and β-glucuronidase in older cells is attributable, in part, to an increase in lysosomal enzyme mobility. In attempts to stabilize the lysosomal membrane of aged cells, we have treated cultures with hydrocortisone (5 μg/ml) and have compared the subcellular distribution of these enzymes as a function of serial passages. The results showed that the activity of these enzymes as assayed in crude homogenates increased with aging whether hydrocortisone was present or not. In addition, the subcellular distribution of acid phosphatase and β-glucu-

TABLE 2

INCREASE IN CELL DOUBLINGS WITH CONTINUOUS EXPOSURE TO HYDROCORTISONE (MOTHER–DAUGHTER SUBCULTIVATIONS)

| Cell Type | Passage at start of experiment | Cumulative cell doublings from start of experiment | | Percent increase with hydro-cortisone |
		Control	Hydrocortisone	
WI-38$_j$	19	42.0	56.0	33
WI-38$_h$	18	37.0	49.0	32
WI-38$_p$	18	35.0	48.0	37
WI-26	22	46.0	60.0	30
WI-38$_x$[a]	21	18.0	26.0	44
WI-38$_y$[a]	21	17.0	25.0	48

[a]Subcultivated once a week; others subcultured at confluency.

ronidase was not significantly affected by the hydrocortisone treatment. The presence of hydrocortisone, however, produced a striking prolongation of the life-span of the cells.

Macieira-Coelho (1966) had previously reported a similar extension of the life span of human cell cultures using the steroid hormone cortisone.

We questioned whether the hydrocortisone effect was reproducible and whether it was one of improved maintenance of the cultures or a real increase in proliferative capacity.

In other experiments in which subcultivations were made both at confluency, and at fixed time intervals, the hydrocortisone-treated cultures always averaged at least a 30 to 50 percent increase in total population doublings (Table 2). We have also shown that the effectiveness of hydrocortisone in extending the life-span of the culture was directly proportional to the duration of hydrocortisone exposure, and that no rescue effect could be shown by treatment of already senescent cultures with the hormone.

We have ruled out the possibility that the hydrocortisone effect is simply an artifact of improved maintenance during subcultivation or increased adherance to the glass surface. The increase in cell doublings could result then either from an increased rate of cell division, and/or an increase in the percentage of cells capable of dividing. Using standard techniques of autoradiography, we have compared the effect of hydrocortisone on the fraction of cells able to incorporate ^{3}H-labeled thymidine into their nuclei under a standard set of conditions. For both cell lines derived from a human female (WI-38) and a human male (WI-26) fetus, the results showed that the hydrocortisone-treated cells undergo more doublings. Autoradiographic studies have shown that for populations of different ages there is always a higher percentage of cells incorporating [3]HTdR in the presence of hydrocortisone. It appears that hydrocortisone maintains cells in the proliferating pool for longer periods of time. Interestingly, it is in the older cultures that the magnitude of the differences in labeling index and cell yield are most striking. On the other hand, although the hydrocortisone effect is much more striking in older cultures, in fact, the responsiveness of the cell to the

hormone is reduced with age. This is reminiscent of the findings with the hormonal stimulation of DNA synthesis in mammary cells, cultivated *in vivo*, and with the isoproterenol stimulation of cell division in salivary gland. There is an age-related loss in the responsiveness of the population to the signals initiating division.

One of the most intriguing aspects of this hydrocortisone study thus far has been the knowledge that this hormone typically inhibits rather than stimulates cell division. The antimitotic effect of hydrocortisone and other glucocorticoids on cell cultures is well documented (Ruhman and Berliner, 1965). It is difficult to relate our results to these other data in the literature. In our studies we have shown that the mitogenic effect is reproducible, specific for human diploid cells (i.e., transformed and tumor-derived cell cultures show mitotic inhibition), and specific with respect to the chemical structure of the hormone (i.e., all steroids do not elicit the mitogenic reactions in human diploid cells).

It is known that hormones may work differently on different tissues. Of interest here, however, is the possible mechanism of action of this hormone in this system.

Proposals for the control of gene expression based on the possible interaction of regulatory proteins with hydrocortisone are supported by reports that interaction of hydrocortisone with rat liver chromatin has been observed as has an associated increase in the activity of RNA polymerase (Beato et al., 1968).

Much of the basis for the suggested relationship between histones and gene expression derives from the demonstration that histones inhibit transcription by inhibiting RNA polymerase synthesis (Allfrey et al., 1963; Huang and Bonner, 1962; Spelsberg et al., 1969). Chemical modification of histones (in cell-free systems) has been shown to reduce their effectiveness as inhibitors, thus allowing a possible reversible mechanism for the control of gene expression. In this regard, Ryan and Cristofalo (1972) have observed a reduction in histone acetylation in old human cells *in vitro*. There is evidence in the literature suggesting that hydrocortisone may interact specifically with the arginine-rich fraction of the histone (Sluyser, 1969).

Alternatively, considerable evidence has accumulated that suggests that the nonhistone, acidic nuclear proteins are involved in the regulation of gene expression (Shelton and Allfrey, 1970; Baserga et al., 1971). Indeed, the acidic nuclear proteins show considerable tissue heterogeneity as well as tissue specificity—both characteristics that would be expected of proteins involved in regulation of gene expression (Hnilica, 1967; DeLange et al., 1969). Moreover, Lavrinenko et al. (1971) have reported a preferential binding of labeled hydrocortisone to nonhistone proteins, and Shelton and Allfrey (1971) have also shown specific binding of hydrocortisone to the nonhistone chromosomal proteins. Thus both these questions will require resolution.

Concluding Remarks

In this chapter we have described some of the characteristics of aging and the nature of aging research. We have discussed some current research based on three arbitrarily designated components of the organism: the extracellular material—that is, collagen; the "postmitotic" cells; and cells that are able to proliferate throughout the life-span of the organism.

Collagen cross-linking appears to eventually reach a stage where it can interfere with several vital functions—for example, muscular contraction and diffusion processes.

Decrease in the number of nondividing cells as well as loss of the functional capacity of nondividing cells seem well documented, at least for a few tissues. The potential effects of these losses to the declining capabilities of the organism are obvious.

It seems clear that there is a loss of proliferative capacity during aging. However, the meaning of this loss of the proliferative capacity in terms of the organism's survival remains obscure.

A number of laboratories are studying cellular proliferative capacity and its regulation as a function of age. We have considered two model systems that are currently being used in these investigations: the mammary cells, serially transplanted in vivo, and human fibroblast-like cells, serially transplanted in vitro. A decline in proliferative capacity has been well documented for both these systems.

Steroid hormones have been used as modulators of proliferative capacity in many systems. In cultures of diploid human cells in vitro, hydrocortisone stimulates DNA synthesis and cell division.

The relationship between cell proliferation and the life-span of the individual has not yet been determined. On the other hand, any experiments that lead to a better understanding of how cell proliferation is regulated as a function of time or of the age of the animal will be highly significant to an understanding of the overall developmental process.

References

ADELMAN, R. C., STEIN, G., ROTH, G. S., AND ENGLANDER, D. (1972). Age-dependent regulation of mammalian DNA synthesis and cell proliferation in vivo. Mech. Ageing Develop. 1, 49-59.

ALLFREY, V. G., LITTAU, V. G., AND MIRSKY, A. E. (1963). On the role of histones in regulating ribonucleic acid synthesis in the cell nucleus. Proc. Nat. Acad. Sci. U.S.A. 49, 414-421.

ALLISON, A. C., AND PATON, G. R. (1965). Chromosome damage in human diploid cells following activation of lysosomal enzymes. Nature (London) 207, 1170-1173.

BASERGA, R., ROVERA, G., AND FARBER, J. (1971). Control of cellular proliferation in human diploid fibroblasts. In Vitro 7, 80-87.

BEATO, M., HOMOKI, J., LUKACS, I., AND SEKETIS, C. E. (1968). Increased template activity for RNA synthesis of rat liver nuclei incubated with cortisol in vitro. Hoppe-Seyler Z. Physiol. Chem. 349, 1099-1104.

BRANDES, D., MURPHY, D. G., ANTON, E., AND BARNARD, S. (1972). Ultrastructural and cytochemical changes in cultured human lung cells. J. Ultrastruc. Res. 39, 465-483.

BROCK, M. A., AND HAY, R. J. (1971). Comparative ultrastructure of chick fibroblasts in vitro at early and late stages during their growth span. J. Ultrastruc. Res. 36, 291-311.

BUETOW, D. E. (1971). Cellular content and cellular proliferation changes in the tissues and organs of the aging mammal. In "Cellular and Molecular Renewal in the Mammalian Body" (I. L. Cameron and J. O. Thrasher, eds.), pp. 87-104. Academic Press, New York.

CAMERON, I. L. (1972a). Minimum number of cell doublings in an epithelial cell population during the life span of the mouse. *J. Gerontol.* 27, 157–161.

CAMERON, I. L. (1972b). Cell proliferation and renewal in aging mice. *J. Gerontol.* 27, 162–172.

CARREL, A., AND BURROWS, M. T. (1910a). Cultivation of adult tissues and organs outside of the body. *J. A.M.A.* 55, 1379–1381.

CARREL, A., AND BURROWS, M. T. (1910b). Cultivation of sarcoma outside of the body. *J. A.M.A.* 55, 1554.

CARREL, A., AND BURROWS, M. T. (1911). On the physicochemical regulation of the growth of tissues. *J. Exp. Med.* 13, 562–570.

CARREL, A., AND EBELING, A. H. (1921). Age and multiplication of fibroblasts. *J. Exp. Med.* 34, 599–623.

COMFORT, A. (1964). "Ageing: The Biology of Senescence." Holt, Rinehart and Winston, New York.

CRISTOFALO, V. J. (1970). Metabolic aspects of aging in diploid human cells. *In* "Aging in Cell and Tissue Culture" (E. Holečkovà and V. J. Cristofalo, eds.), pp. 83–119. Plenum Press, New York.

CRISTOFALO, V. J. (1972). Animal cell cultures as a model system for the study of aging. *Advan. Gerontol. Res.* 4, 45–79.

CRISTOFALO, V. J., PARRIS, N., AND KRITCHEVSKY, D. (1967). Enzyme activity during the growth and aging of human cells *in vitro*. *J. Cell Physiol.* 69, 263–272.

CRISTOFALO, V. J., AND SHARF, B. B. (1973). Cellular senescence and DNA synthesis: thymidine incorporation as a measure of population age in human diploid cells. *Exp. Cell Res.* 76, 419–427.

CUDKOWICZ, G., UPTON, A. C., SHEARER, G. M., AND HUGHES, W. L. (1964). Lymphocyte content and proliferative capacity of serially transplanted mouse bone marrow. *Nature (London)* 201, 165–167.

DANIEL, C. W. (1972). Aging of cells during serial propagation *in vivo*. *Advan. Gerontol. Res.* 4, 167–198.

DANIEL, C. W., DEOME, K. B., YOUNG, J. T., BLAIR, P. B., AND FAULKIN, L. J. (1968). The *in vivo* life span of normal and preneoplastic mouse mammary glands: a serial transplantation study. *Proc. Nat. Acad. Sci. U.S.A.* 61, 53–60.

DANIEL, C. W., AND YOUNG, L. J. T. (1971). Influence of cell division on an aging process. Life span of mouse mammary gland during serial propagation *in vivo*. *Exp. Cell Res.* 65, 27–32.

DANIEL, C .W., YOUNG, L. J. T. MEDINA, D., AND DEOME, K. B. (1971). The influence of mammogenic hormones on serially transplanted mouse mammary gland. *Exp. Gerontol.* 6, 95–100.

DELANGE, R. J., FAMBROUGH, D. M., SMITH, E. L., AND BONNER, J. (1969). Calf and pea histone IV. III. Complete amino acid sequence of pea seedling histone IV. Comparison with the homologous calf thymus histone. *J. Biol. Chem.* 244, 5669–5679.

DELL'ORCO, K. T., MERTINS, J. G., AND KRUSE, P. F., JR. (1972). Doubling potential, calendar time, and senescence of human diploid cells in culture. *In Vitro* 7, 273, abstract.

DEOME, K. B., FAULKIN, L. J., JR., BERN, H. A., AND BLAIR, P. B. (1959). Development of mammary tumors from hyperplastic alveolar nodules transplanted into gland-free mammary fat pads of female C3H mice. *Cancer Res.* 19, 515–521.

EBELING, A. H. (1913). The permanent life of connective tissue outside of the organism. *J. Exp. Med.* 17, 273–285.

EBELING, A. H. (1921). Measurement of the growth of tissues *in vitro*. *J. Exp. Med.* 34, 231–243.

GOLDSTEIN, S., LITTLEFIELD, J. W., AND SOELDNER, J. S. (1969). Diabetes mellitus and aging: diminished plating efficiency of cultured human fibroblasts. *Proc. Nat. Acad. Sci. U.S.A.* 64, 155–160.

HAFF, R. F., AND SWIM, H. E. (1956). Serial propagation of three strains of rabbit fibroblasts: their susceptibility to infection with vaccinia virus. *Proc. Soc. Exp. Biol. Med.* 93, 200–204.

HAY, R. J., MENZIES, R. A., MORGAN, H. P., AND STREHLER, B. L. (1968). The division potential of cells in continuous growth as compared to cells subcultivated in stationary phase. *Exp. Gerontol.* 3, 35–44.

HAY, R. J., AND STREHLER, B. L. (1967). The limited growth span of cell strains isolated from the chick embryo. *Exp. Gerontol.* 2, 123–135.

HAYFLICK, L. (1965). The limited *in vitro* lifetime of human diploid cell strains. *Exp. Cell Res.* 37, 614–636.

HAYFLICK, L. (1972). International symposium "Cell Aging and Cell Differentiation." *In* "Aging and Development" (H. Bredt and J. W. Rohen, eds.), pp. 1–15. Mainz Academy of Science and Literature, Mainz, Germany.

HAYFLICK, L., AND MOORHEAD, P. S. (1961). The serial cultivation of human diploid cell strains. *Exp. Cell Res.* 25, 585–621.

HNILICA, L. S. (1967). Proteins of the cell nucleus. *Progr. Nucleic Acid Res. Mol. Biol.* 7, 25–106.

HUANG, R. C., AND BONNER, J. (1962). Histone, a

suppressor of chromosomal RNA synthesis. *Proc. Nat. Acad. Sci. U.S.A. 48,* 1216-1222.

KOHN, R. R. (1971). "Principles of Mammalian Aging." Foundation of Developmental Biology Series (C. L. Markert, ed.). Prentice-Hall, Englewood Cliffs, New Jersey.

KROHN, P. L. (1962). Review lectures on senescence. II. Heterochronic transplantation in the study of aging. *Proc. Roy. Soc. Med. 157,* 128-147.

KROHN, P. L. (1966). Transplantation and aging. *In* "Topics of the Biology of Aging" (P. L. Krohn, ed.), p. 125. Wiley-Interscience, New York.

KROOTH, R. S., SHOW, M. W., AND CAMPBELL, B. K. (1964). A persistent strain of diploid fibroblasts. *J. Nat. Cancer Inst. 32,* 1031-1040.

LANSING, A. I. (1952). General physiology. *In* "Cowdry's Problems of Aging" (A. I. Lansing, ed.), pp. 3-22. Williams & Wilkins, Baltimore.

LAVRINENKO, I. A., MOROZOVA, T. M., AND USCHKOVA, L. F. (1971). Binding of 1,2-³H-hydrocortisone by chromatin and proteins and histones *in vitro. Mol. Biol. (Russian) 5,* 17-21.

LEBLOND, C. P. (1964). Classification of cell populations on the basis of their proliferative behavior. *Nat. Cancer Inst. Monogr. 14,* 119-145.

LESHER, S., AND SACHER, G. A. (1968). Effects of age on cell proliferation in mouse duodenal crypts. *Exp. Gerontol. 3,* 211-217.

LEVY, J. A., VIROLAINEN, M., AND DEFENDI, V. (1968). Human lymphoblastoid lines from lymph node and spleen. *Cancer 22,* 517-524.

LIMA, L., AND MACIEIRA-COELHO, A. (1972). Parameters of aging in chicken embryo fibroblasts cultivated *in vitro. Exp. Cell Res. 70,* 279-284.

LIPETZ, J., AND CRISTOFALO, V. J. (1972). Ultrastructural changes accompanying the aging of human diploid cells in culture. *J. Ultrastruc. Res. 39,* 43-56.

MACIEIRA-COELHO, A. (1966). Action of cortisone on human fibroblasts *in vitro. Experientia 22,* 390-391.

MACIEIRA-COELHO, A. (1970). The decreased growth potential *in vitro* of human fibroblasts of adult origin. *In* "Aging in Cell and Tissue Culture" (E. Holečková and V. J. Cristofalo, eds.), pp. 12-132. Plenum Press, New York.

MACIEIRA-COELHO, A., AND PONTÉN, J. (1969). Analogy in growth between late passage human embryonic and early passage human adult fibroblasts. *J. Cell Biol. 43,* 374-377.

MACIEIRA-COELHO, A., PONTÉN, J., AND PHILIPSON, L. (1966a). The division cycle and RNA synthesis in diploid human cells at different passage levels *in vitro. Exp. Cell Res. 42,* 673-684.

MACIEIRA-COELHO, A., PONTÉN, J., AND PHILIPSON, L. (1966b). Inhibition of the division cycle in confluent cultures of human fibroblasts *in vitro. Exp. Cell Res. 43,* 20-29.

MARTIN, G. M., SPRAGUE, C. A., AND EPSTEIN, C. J. (1970). Replicative life span of cultivated human cells, effect of donor's age, tissue and genotype. *Lab. Invest. 23,* 86-92.

McHALE, J. S., MOUTON, M. L., AND McHALE, J. (1971). Limited culture life-span of human diploid cells as a function of metabolic time instead of division potential. *Exp. Gerontol. 6,* 89-93.

MERZ, G. S., AND ROSS, J. D. (1969). Viability of human diploid cells as a function of *in vitro* age. *J. Cell Physiol. 74,* 219-221.

MOORE, G. E., AND McLIMANS, W. F. (1968). The life-span of the cultured normal cell: concepts derived from studies of human lymphoblasts. *J. Theoret. Biol. 20,* 217-226.

PETURSSON, G., COUGHLIN, J. I., AND MEYLAN, C. (1964). Long-term cultivation of diploid rat cells. *Exp. Cell Res. 33,* 60-67.

PRICE, G. B., AND MAKINODAN, T. (1972a). Immunologic deficiencies in senescence. I. Characterization of intrinsic deficiencies. *J. Immunol. 108,* 403-412.

PRICE, G. B., AND MAKINODAN, T. (1972b). Immunologic deficiencies in senescence. II. Characterization of extrinsic deficiencies. *J. Immunol. 108,* 413-417.

REGAN, J. D., SIGEL, M. M., LEE, W. H. LLAMAS, K. A., AND BEASLEY, A. R. (1968). Chromosomal alterations of marine fish cells *in vitro. Canad. J. Genet. Cytol. 10,* 448-453.

ROBBINS, E., LEVINE, E. M., AND EAGLE, H. (1970). Morphologic changes accompanying senescence of cultured human diploid cells. *J. Exp. Med. 131,* 1211-1222.

RUHMAN, A. G., AND BERLINER, D. L. (1965). Effects of steroids on growth of mouse fibroblasts *in vitro. Endocrinology 76,* 916-927.

RYAN, J. M., AND CRISTOFALO, V. J. (1972). Histone acetylation during aging of human cells in culture. *Biochem. Biophys. Res. Commun. 48,* 735-742.

SHELTON, K. R., AND ALLFREY, V. G. (1970). Selective synthesis of a nuclear acidic protein in liver cells stimulated by cortisol. *Nature (London) 228,* 132-134.

SIMINOVITCH, L., TILL, J. E., AND McCULLOCH, E. A. (1964). Decline in colony-forming ability of marrow cells subjected to serial transplantation into irradiated mice. *J. Cell. Comp. Physiol. 64,* 23–31.

SIMONS, J. W. I. M. (1970). A theoretical and experimental approach to the relationship between cell variability and aging *in vitro. In* "Aging in Cell and Tissue Culture" (E. Holečková and V. J. Cristofalo, eds.), pp. 25–39. Plenum Press, New York.

SINEX, F. M. (1968). The role of collagen in aging. *In* "Treatise on Collagen" (B. S. Gould, ed.), pp. 409–448. Academic Press, New York.

SLUYSER, M. (1969). Interaction of steroid hormones with histones *in vitro. Biochim. Biophys. Acta 182,* 235–244.

SOUKUPOVÀ, M., AND HOLEČKOVÀ, E. (1964). The latent period of explanted organs or newborn, adult and senile rats. *Exp. Cell Res. 33,* 361–367.

SOUKUPOVÀ, M., HOLEČKOVÀ, E., AND HNEVKOVSKY, P. (1970). Changes of the latent period of explanted tissues during ontogenesis. *In* "Aging in Cell and Tissue Culture" (E. Holečková and V. J. Cristofalo, eds.), pp. 41–56. Plenum Press, New York.

SPELSBERG, T. C., TANKERSLEY, S., AND HNILICA, L. S. (1969). Inhibition of RNA polymerase activity by arginine-rich histones. *Experientia 25,* 129–130.

STREHLER, B. L. (1962). "Time, Cells and Aging." Academic Press, New York.

STREHLER, B. L., MARK, D., MILDVAN, A. S., AND GEE, M. (1959). Rate and magnitude of age pigment accumulation in the human myocardium. *J. Gerontol. 14,* 430–439.

SWIM, H. E., AND PARKER, R. F. (1957). Culture characteristics of human fibroblasts propagated serially. *Amer. J. Hygiene 66,* 235–243.

THRASHER, J. O. (1971). Age and the cell cycle of the mouse esophageal epithelium. *Exp. Gerontol. 6,* 19–24.

WALFORD, R. L. (1969). "The Immunologic Theory of Aging." Williams & Wilkins, Baltimore, 248 pp.

YOUNG, L. J. T., MEDINA, D., DEOME, K. B., AND DANIEL, C. W. (1971). The influence of host and tissue age in life span and growth rate of serially transplanted mouse mammary gland. *Exp. Gerontol. 6,* 49–56.

Acknowledgments

LONGO AND ANDERSON: *Gametogenesis*

We are grateful to Drs. James N. Dumont, Erwin Huebner, David M. Phillips, and Herbert E. Potswald for discussions of various aspects associated with this work and for their thoughtful and constructive criticism. The assistance of Mrs. Ellen Looney and Mr. Edward Rutkowski is gratefully acknowledged. Portions of this article were supported by grants HD-05846, HD-04924, HD-36162, and HD-06822 from the National Institutes of Child Health and Human Development.

RAPPAPORT: *Cleavage*

The author's original investigations described in this chapter were supported by grants from the National Science Foundation. I thank Dr. T. E. Schroeder for generously providing the electron micrographs used in Figure 11.

NEMER: *Molecular Mechanisms of Cellular Differentiation*

This work was supported by U.S.P.H.S. grants HD-04367, CA-06927, and RR-05539 from the National Institutes of Health, and by an appropriation from the Commonwealth of Pennsylvania.

NEMER: *Molecular Basis of Embryogenesis*

This work was supported by U.S.P.H.S. grants HD-04367, CA-06927, and RR-05539 from the National Institutes of Health, and by an appropriation from the Commonwealth of Pennsylvania.

WHITTAKER: *Aspects of Differentiation and Determination in Pigment Cells*

Research of the author described in this paper was supported by grants from the National Institutes of Health (EY-00776 and RR-05540) and the National Science Foundation (GB-15960).

KONIGSBERG AND BUCKLEY: *Regulation of the Cell Cycle and Myogenesis by Cell-Medium Interaction*

The original research reported in this chapter was supported by National Science Foundation Grant GB-5963X (to I. R. K.) and a National Institute of Health Traineeship (5 TOI HD 00029) to P. A. B.

LASH: *Tissue Interactions and Related Subjects*

Research of the author described in this chapter was supported by grants from the National Institutes of Health (HD-00380) and the National Science Foundation (GB-3674 and GB-6748).

SPOONER: *Morphogenesis of Vertebrate Organs*

I thank J. F. Ash, M. Coughlin, K. M. Yamada, and N. K. Wessells for collaboration in various studies described here. The research was supported by U.S.P.H.S. grant GM-19289. Some of the work reviewed here was carried out at Stanford University in the laboratory of Dr. Norman K. Wessells, where I was an NIH Postdoctoral Fellow. Further thanks are due to S. R. Hilfer and N. K. Wessells for critical comment on the text of this chapter.

WILDE: *Time Flow in Differentiation and Morphogenesis*

The author wishes to acknowledge with heartfelt thanks and pleasure the critical evaluation of, and suggestions toward improvement of, the manuscript by the following: Drs. R. Piddington, G. Levenson, R. J. Schwartz, and E. Macarak. However, all statements and interpretations are the responsibility of the author. Grateful thanks are due to Ms. L. L. Wallis for careful preparation of the manuscript.

Recent research reported here from the author's laboratory were supported by grants from NIH and NSF both to him and to The Mount Desert Island Biological Laboratory. Current research from the author's laboratory is supported by a research grant from The National Foundation and the Pennsylvania Plan of the University of Pennsylvania.

BILLINGHAM: *The Phenomenon of Immunological Tolerance and Its Possible Role in Development*

The expenses entailed in producing this article and in performing some of the work cited were defrayed in part by U.S.P.H.S. Grant AI-07001.

HAY: *Cellular Basis of Regeneration*

The original research by the author reported in this paper was supported by a National Institutes of Health Grant, HD-00143.

CRISTOFALO: *Aging*

This work was supported in part by United States Public Health Service Grants HD-02721 and HD-06323.

Indices

Simpson, M. E., 326, 346, 347
Sinex, F. M., 432, 447
Singer, M., 404, 412, 415, 419, 420, 427, 428
Singer, R., 113, 118
Sirlin, J. L., 25, 43
Slater, D. W., 112, 118, 120, 125, 127, 254, 260
Slater, I., 112, 118, 125, 127, 254, 260
Slavaja, I. L., 231, 239
Slemmer, G., 369, 378
Slizynska, H., 358, 377
Sluyser, M., 443, 447
Smiles, J., 61, 72
Smiles, A. E., 114, 118
Smith, D., 340, 345
Smith, D. W., 381, 383, 384, 399, 403
Smith, E. L., 444, 445
Smith, G. I., 398, 403
Smith, J. A., 191, 193
Smith, J. D., 109, 115, 370, 379
Smith, J. M., 281, 291
Smith, K. D., 102, 118
Smith, P. A., 11, 13, 45
Smith, R. H., 370, 379
Smithies, O., 268, 271
Snelgrove, L. E., 57, 74
Snell, G. D., 282, 291
Snipes, C. A., 327, 347
Snyder, B. W., 331, 345
Soderwell, A. L., 59
Soeldner, J. S., 438, 445
Sokoloff, L., 333, 347
Solomon, J. B., 265, 271, 275, 291
Sorenson, G. D., 154, 161
Soukupovà, M., 437, 439, 447
Soupart, P., 57, 74
Sox, H. C., 191, 192
Spadaro, J. A., 420, 426
Spargo, B., 282, 291
Speake, R. N., 220, 236
Spelsberg, T. C., 247, 259, 341, 342, 347, 348, 443, 447
Spemann, H., 198, 208, 212, 242, 246, 260
Spiegelman, S., 104, 117, 120, 127, 158, 160, 361, 370, 379
Spielman, A., 14, 15, 43
Spirin, A. S., 110, 111, 118, 127, 254, 260
Spohr, G., 110, 112, 118
Spooner, B. S., 201, 214, 215, 216, 220, 222, 224, 226, 227, 229, 230, 231, 233, 234, 235, 236, 239, 240
Sporn, M. B., 115, 117
Sprague, C. A., 438, 446
Sprent, J., 262, 263, 270
Springer, A., 234, 238
Spudich, J. A., 235, 240
Srinivasan, P. R., 110, 117
Srivastava, P. N., 61, 74
Stambaugh, R., 61, 67, 73, 74
Staples, D. H., 114, 116
Stavy, L., 121, 126, 127

Steele, M. W., 400, 402
Stefanini, M., 62, 74
Steggles, A. W., 247, 259, 341, 342, 347, 348
Stein, G., 434, 444
Steinberg, C. M., 349, 355, 356, 377
Steinberg, M., 142, 143, 148, 202, 212
Steinert, G., 14, 46
Steitz, J. A., 114, 118
Stephenson, J. R., 157, 161
Steptoe, P. C., 56, 73
Stern, C., 350, 351, 363, 364, 366, 379
Stern, H., 9, 46
Sterzel, J., 264, 271
Stevens, A. R., 21, 46
Stevens, J., 420, 427
Stewart, J. A., 153, 161
Stirewalt, W. S., 333, 348
Stites, D. P., 263, 265, 271
Stockdale, F. E., 181, 182, 187, 191, 192, 193, 203, 205, 211, 212, 215, 240, 406, 427
Stocum, D. L., 256, 260
Stohlman, F., Jr., 158, 161
Stone, W. H., 277, 291
Storb, R., 283, 290
Strauss, F., 52, 75
Strehler, B. L., 181, 182, 193, 429, 432, 437, 438, 439, 440, 445, 447
Streilein J. W., 281, 285, 287, 291, 292
Strohman, R. C., 181, 193
Stroman, J., 234, 239
Strong, J. A., 395, 402
Stuart, F. P., 282, 291
Stumpf, W. E., 248, 259
Sturtevant, A. H., 350, 364, 365, 366, 379
Subramanian A. R., 113, 118
Sud, B. N., 36, 46
Sugiyama, M., 80, 96
Summerbell, D., 174, 177
Susman, M., 349, 355, 356, 377
Sussman, M., 114, 118
Sutherland, E. W., 325, 344, 347
Suzuki, D. T., 350, 367, 370, 371, 372, 373, 375, 376, 377, 379
Suzuki, T., 248, 259
Suzuki, Y., 303, 320
Swann, M. M., 83, 84, 90, 91, 97, 98
Swanson, E. W., 277, 291
Swartzendruber, D., 265, 270
Swift, H., 154, 160
Swim, H. E., 436, 437, 445, 447
Szent-Györgyi, A., 233, 240
Szollosi, D., 26, 46, 88, 98
Szollosi, D. G., 63, 75

T

Taber, R., 158, 160
Taderera, J. V., 216, 219, 240
Takaku, F., 158, 159
Takeda, K., 235, 238

Tamm, C., 220, 239
Tanaka, S., 341, 346
Tankersley, S., 443, 447
Tanner, J. M., 327, 328, 348
Tapiero, H., 102, 116
Tarasoff, M., 370, 371, 379
Tardent, P., 404, 420, 428
Tarkowski, A. K., 200, 212, 369, 379
Tartof, K. D., 104, 118
Tata, J. R., 330, 331, 332, 348
Tattrie, B., 158, 159
Tatum, E. L., 365, 379
Taylor, E. L., 220, 234, 240
Taylor, E. W., 220, 237
Taylor, J. H., 34, 46
Tchen, T. T., 171, 177
Teitelman, G., 254, 260
Telfer, W. H., 11, 14, 15, 43, 46, 47
Tell, G. P. E., 326, 327, 328, 345, 348
Temple, G. F., 158, 162
Temtamy, S., 386, 403
Terada, M., 152, 153, 155, 158, 160, 161, 162
Terman, S. A., 122, 127
Therman, E., 399, 403
Thiersch, J. B., 51, 74
Thomas, A. V., 285, 290
Thomas, C. A., Jr., 104, 118
Thomas, E. D., 283, 290
Thomas, N. S. T., 93, 97
Thomas, W., 125
Thompson, E. B., 110, 112, 118, 317, 320
Thompson, L. H., 370, 379
Thompson, D. W., 255, 260
Thorbecke, G. J., 265, 270
Thornton, C. S., 404, 405, 412, 416, 421, 428
Thrasher, J. O., 433, 447
Threlfall, G., 103, 117
Tiedemann, H., 244, 254, 260
Tiegler, D., 158, 159
Tigelaar, R., 263, 269
Till, J. E., 157, 160, 162, 370, 379, 407, 428, 434, 446
Till, T. E., 191, 193
Tillack, T. W., 145, 148
Tilney, L. G., 28, 47, 220, 237, 240
Timiras, P. S., 329, 346
Tjio, J. H., 380, 403
Toft, D. O., 340, 342, 345, 347
Tokunga, C., 366, 379
Toivonen, S., 199, 208, 212, 244, 246, 259
Tolis, H., 67
Tomkins, G. M., 110, 112, 118, 317, 320
Toole, B. P., 412, 428
Tootle, M., 179, 182, 192
Topper, Y. D., 203, 211, 212, 334, 335, 347, 348
Townes, P. L., 141, 148
Trampusch, H. A. L., 404, 427

Yunis, J. J., 102, 104, 118
Yuyama, S., 78, 98

Z

Zakarian, S., 281, 285, 292
Zamboni, I., 154, 162

Zamboni, L., 26, 47, 62, 63, 74, 75
Zanjani, E. D., 150, 157, 158, 160
Zech, L., 393, 402
Zeiger, R. S., 391, 402
Zetterquist, P., 69, 72

Ziegler, H. E., 83, 98
Zimmerman, A. M., 92, 98
Zipser, D., 353, 376
Zwaan, J., 220, 221, 239
Zwilling, E., 207, 309, 211, 212

Subject Index

A

absent radius-thrombocytopenia
    syndrome, 390
accessory cell functions, 11–15
achondroplastic dwarfism, 382
acid phosphatase
    of aging cells, 441
    isozyme changes, metamorphosis,
    333
acromegaly, 326
acrosome
    enzymes, 61
    formation, 37
acrosome reaction, 61-62
actinomycin D
    DNA-directed RNA synthesis,
    inhibition of, 300
    enzyme synthesis, inhibition of,
    307–315
    hormone-stimulated growth, ef-
    fect on, 339
    preformed mRNA, evidence for,
    120, 122, 156
    superinduction of tyrosine am-
    inotransferase, 112, 113
    ribosomal RNA synthesis, inhibi-
    tion of, 25
    teleost development, inhibition
    of, 251–255
    tyrosinase synthesis, inhibition
    of, 167
    uterine estrogen response, inhib-
    its, 339
activator RNA, 107
actomyosin
    at cell surface, 201
    in cleaving sea urchin eggs, 93
    in microfilamentous band, 93
    in platelets, 233
    in primitive streak cells, 423
    in slime moulds, 233
adenohypophysis hormones, 322
adenosine deoxyribonucleotide, 9
adenyl cyclase
    growth hormone effects on level,
    327

insulin inhibition, 326
theophylline inhibition of, 327
thyroxine inhibits myocardial,
326
adrenalectomy, 313
adrenocorticotrophic hormone
    (ACTH)
    activity, 325
    release-stimulating factors, 322
Ag-B: histocompatibility gene
    locus, 275–276
age pigment, 433
aging
    cells in vitro, 436–444
    characteristics of, 430
    extracellular materials, 432
    general problem, 429–444
    life span of animals, 431
    tissues in vivo, 434–436
agouti hair patterns, 169, 170
allophenic. See tetraparental and
    chimerism
allosteric change, 353
anencephaly, 385
animal pole, 11, 28, 131
annulate lamellae, 26
antibody molecule, gene coding,
    268
antibody theories, 268–269
antilymphocytic serum, 281
ants marching, 251
asters, 78, 94
ataxia-telangiectasia, 391
ATPases: activated receptors, 245
autoimmunity, 272, 287
autotolerance in frog development,
    288
auxotrophic mutations, 354
avidin synthesis, progesterone in-
    duced, 341
axon initiation and elongation, 226
axonemal complex, 41

B

bacteriophage
    chromosome, gene clustering,
    356–357
    mRNA, 113
    reproduction, 349, 354, 356, 357
    tail assembly, 243
Balfour's law, 79
Barr body, 363, 437

basophilic erythroblast, 155
Bidder's organ, 337
bivalent, 8
blastema, 256, 415
blastema cells, 415
blastocoel, 128
blastoderm, 135
blastomere separation studies, 200,
    246
blastula, 128
blastula formation, 129
blood cells, daily loss, 404
blood islands, 151
blood proteins, in oogenesis, 15
bone marrow cells
    development, 267
    limited proliferation in vivo, 434
bone regeneration, 408, 420
bottle cells
    formation, 136
    nature of, 132
    transformations, 137
brain cells, numbers during aging,
    433
bristle production in Drosophila,
    366
bromodeoxyuridine effects, 167
bursa of Fabricius, 262, 267

C

calcitonin, 323
cancer cells in regeneration envi-
    ronment, 420
capacitation, 41, 56–57
carbamyl phosphate synthetase, in
    metamorphosis, 333
cardiac cells, differentiation in
    clonal culture, 215
casein synthesis in mammary tissue,
    203–204, 334
cathepsin, 333
cell adhesiveness, 142
cell autonomy, 169
cell contact behavior, 138
cell culture studies, aging, 436–444
cell cycle
    density effects, 187–191
    DNA synthesis, 181–182, 186–191
    G_1 and G_2 periods in aging cells,
    439
    mathematical analysis, 189
    regulation, 179–192

insect metamorphosis, steroid hormones, 305
instructive antibody theory, 268
insulin
blood sugar regulation, 323
receptors, 311
RNA stimulation in epithelial cells, 335
intercellular adhesion, models, 145
intercellular bridges, in spermiogenesis, 12, 31–33
intercellular information exchange, 244
interkinesis, 9
intracellular organelles in morphogenesis, 219–220
invagination, 131
iris tissue dedifferentiation, 171–172
isocitrate dehydrogenase in muscle, 268
I-somes, 111
isozymes, 302

J

juvenile hormone, 343

K

kidney
hyperplasia, 408
tubulogenesis, induction of, 208
kinoplastic droplet, 57
Klinefelter's syndrome, 395

L

δ-lactalbumin, in mammary development, 334
lactose operon, 108, 296, 352–355
lactose synthetase, in mammary development, 334
lampbrush chromosomes, 10, 21–24, 107, 120
"last man on the right" problem, 250
lens
antigens in iris tissue, 172
regeneration, 171, 172, 173, 199, 413
vesicle formation, 221
leptotene, 8
lethal mutations
conditional, 354
temperature-sensitive, 371
life expectancy at birth, 430
life-span of animals, 431
limb defects, 386
lipofuscin, 433
lipoidal hormones, 343
lithium chloride, pigment cell stimulation, 172
liver
dedifferentiation of tissue, 203
morphogenesis, 216, 266
regeneration, 408

lung development
branching of bronchial epithelium, 214, 216, 218
collagen fibrils, role of, 218
mesodermal requirement, 217
microfilament contributions, 236
in organ culture, 217, 236
luteinizing hormone (LH), 322
luteinizing hormone-releasing hormone (LHRH), 338
lymphoid cells
limited proliferation, 434
immunoglobin production, 191
ontogeny of, 266–269, 406
T and B types, 262–264
lysosomal enzymes
in aging cells, 438, 441–42
in tail regression, 333
lysozyme synthesis, estrogen induced, 341

M

macrophage
immunological reactions, 286
red cell destruction, 404
maintenance of differentiation. See stability
mammary gland
aging of serial transplants, 434–436
development in vitro, 203–204
hormone effects on development, 203, 333–336
maturation divisions, 6–10
megakaryocytic cell line, 157
meiosis
in germ cells, 6–10
stages of, 7
melanin polymer, 165
melanoblast, 173
melanocyte
dendritic process development, 242
differentiation, environmental influence, 169
neural crest origin, 164
melanocyte-stimulating hormone (MSH), 322
melanoprotein, 166, 244
melanosome
development, 166, 168
Golgi apparatus, role in formation, 166
proteins, 165
structure, 165, 242
melanotic expression, 165–168
melanotic system, 164–166
meroblastic cleavage, 133
meroistic ovariole, 11, 14
meromyosin, 93, 233
mesoblast, 135
mesodermal requirement in organogenesis, 217–219

metabolic activity, postfertilization, 63
metamorphosis,
amphibian, 329–333
ascidian, 220, 234
insect, 305, 343
metaplasia, 171, 246, 424
1-methyladenine, 55
microfilamentous band, 88
microfilaments
actin-like properties, 233
cell shape changes, 137, 220
contractile function, 137, 232–235
heavy meromyosin binding, 233
nature of, 88, 220
micropyle, 49, 60
microtubule
around nucleus, 35
cell elongation, 137
cell shape changes, 220
protein, 41
in sperm flagellum, 49
milk proteins
autoantigenic, 289
synthesis, hormonal effects, 334
mitochondria
heterogeneity, 19
in oogenesis, 19
of sperm, 39
mitosis
in germ cells, 5–6
stages of, 6
mitotic activity, decline in aging animals, 433
mitotic apparatus
cleavage in absence of, 81
mode of action, 94
size differences, 78
stimulus of cleavage, 84
mitotic cycle. See cell cycle
mitotic spindle protein, 120
modulation, 171, 199, 246, 297
morphogenesis
abnormal, 380–402
cytodifferentiation correlation, 213–216
extracellular materials role, 201, 216–219
general problems of, 213–236, 241–258
microtubule and microfilament roles, 219–236
model of, 251
multicellular, 221–225
of organs, 235–236
of single cells, 225–232
theoretical considerations, 241–258
time flow in, 241–258
tissue interactions in, 207–210, 216–219
morphogenetic cell movements, 128
morphogenetic movements, motive forces, 219–220